SYMPOSIA OF THE
SOCIETY FOR EXPERIMENTAL BIOLOGY

NUMBER XXXXII

PROCEEDINGS OF A MEETING
HELD AT THE UNIVERSITY OF ESSEX, ENGLAND
8–10 SEPTEMBER 1987

SYMPOSIA OF THE SOCIETY FOR EXPERIMENTAL BIOLOGY

I Nucleic Acid
II Growth, Differentiation and Morphogenesis
III Selective Toxicity and Antibiotics
IV Physiological Mechanisms in Animal Behaviour
V Fixation of Carbon Dioxide
VI Structural Aspects of Cell Physiology
VII Evolution
VIII Active Transport and Secretion
IX Fibrous Proteins and their Biological Significance
X Mitochondria and other Cytoplasmic Inclusions
XI Biological Action of Growth Substances
XII The Biological Replication of Macromolecules
XIII Utilization of Nitrogen and its Compounds by Plants
XIV Models and Analogues in Biology
XV Mechanisms in Biological Competition
XVI Biological Receptor Mechanisms
XVII Cell Differentiation
XVIII Homeostasis and Feedback Mechanisms
XIX The State and Movement of Water in Living Organisms
XX Nervous and Hormonal Mechanisms of Integration
XXI Aspects of the Biology of Ageing
XXII Aspects of Cell Motility
XXIII Dormancy and Survival
XXIV Control of Organelle Development
XXV Control Mechanisms of Growth and Differentiation
XXVI The Effects of Pressure on Organisms
XXVII Rate Control of Biological Processes
XXVIII Transport at the Cellular Level
XXIX Symbiosis
XXX Calcium in Biological Systems
XXXI Integration of Activity in the Higher Plant
XXXII Cell–Cell Recognition
XXXIII Secretory Mechanisms
XXXIV The Mechanical Properties of Biological Materials
XXXV Prokaryotic and Eukaryotic Flagella
XXXVI The Biology of Photoreceptors
XXXVII Neural Origin of Rhythmic Movements
XXXVIII Controlling Events in Meiosis
XXXIX Physiological Adaptations of Marine Animals
XXXX Plasticity in Plants
XXXXI Temperature and Animal Cells

SYMPOSIA OF THE
SOCIETY FOR EXPERIMENTAL BIOLOGY

NUMBER XXXXII

PLANTS AND TEMPERATURE

EDITED BY

S. P. LONG AND F. I. WOODWARD

Published for the Society for Experimental Biology
by The Company of Biologists Limited,
Department of Zoology, University of Cambridge,
Downing Street, Cambridge CB2 3EJ

Typeset, Printed and Published by The Company of Biologists Limited,
Department of Zoology, University of Cambridge,
Downing Street, Cambridge CB2 3EJ

ISBN 0 948601 20 5

SOCIETY FOR EXPERIMENTAL BIOLOGY SYMPOSIA

SEB Symposia form a long-standing series of volumes first published in 1947. The series is annual and each publication is a collection of authoritative articles on an aspect of modern experimental Biology. The contributors are all invited and speak on specific topics within the chosen field. Meetings are held annually, in September, over a period of two or three days.

The aims of the Symposium series are to stimulate discussion and communication between scientists of all nationalities, to foster the development of, and research on, modern aspects of plant, animal, and cell biology.

Cover pictures are reproduced by kind permission of Prof. Felix Franks of Pafra Limited, 150 Science Park, Cambridge CB4 4GG.

CONTENTS

CONTENTS

PREFACE

The relationships of plant distributions, crop productivity and physiological processes with temperature have been the subject of many studies over the last three decades. More recently, there has been an increased understanding of temperature effects at the cell and molecular level paralleled by the development of new areas of study, such as temperature shock proteins and low temperature dependent photoinhibition. As yet, however, there have been limited attempts to draw together studies at these different levels of organization. This volume arises from a Symposium of the Society for Experimental Biology held at the University of Essex from 7–9 September 1987 which aimed to bring together information on temperature effects in plants at three major levels of organization: the population, the whole plant and the cell, this grouping providing the structure for this volume. The topic also forms a logical progression from the subject of the 1986 Symposium of the Society 'Temperature and Animal Cells'.

The meeting attracted an international group of over 200 participants from more than 30 countries. Formal presentations were complemented by 60 poster contributions, extended discussion sessions and a trade exhibition. Speakers were invited from disciplines ranging from molecular biology and membrane physiology to crop physiology and plant ecology. They reviewed current knowledge of the effects of temperature in their own discipline and highlighted current advances, outstanding questions and future directions of research in their own disciplines. The extended discussion and poster sessions at the Symposium encouraged considerable cross-disciplinary debate on the possible significance of cellular phenomena at the whole plant level and the likely mechanisms involved in population responses to temperature.

The Symposium volume begins with chapters considering the physical laws governing the temperature of plant tissues and the global patterns of plant distribution in relation to temperature, and interactions between temperature and other environmental variables in determining the ecological patterns of distribution. A general review of the effects of temperature on plant productivity is followed by specific chapters on effects of interactions between temperature and other climatic variables on seed germination, carbon acquisition and partitioning, respiration, cell elongation and the cell cycle. Much understanding of the effects of temperature at the subcellular level has stemmed from developments in the knowledge of effects on plant membranes. The temperature sensitivity of membranes, therefore, provided the logical starting point for considerations of subcellular effects and a basis for further chapters on ion relations and the effects of low and high temperatures on photosynthesis. Photosynthesis is not only fundamental to plants, but also provides an example of a process combining metabolic and membrane effects. In these chapters, particular emphasis is placed on the effects of temperature towards the extremes of the normal ranges experienced by a species in its natural environments, including effects of high temperature, chilling and freezing. Developments in understanding of how protein synthesis is modified by temperature are discussed in a further chapter providing an insight into how these 'shock-proteins' may relate to acclimation. The final two chapters aim to relate subcellular and whole plant aspects of temperature effects. The first shows by the example of photosynthesis how mathematical modelling provides a framework for exploring the quantitative implications of biochemical effects to integrated processes at the whole plant level. The final contribution draws together the threads of the Symposium to look at the common ground, discrepancies and key questions raised by the preceding chapters, particularly with respect to understanding of the molecular basis of temperature effects. The publication of the Symposium proceedings aims to provide, in a single volume, a basis for cross-disciplinary interest and research into the effects of temperature on plants. It is timely because the recent advances in understanding of effects at the cellular level are now sufficiently advanced to provide exciting new opportunities for understanding effects at the whole plant and population levels.

While the meeting was organized through the S.E.B., the British Ecological Society also provided substantial financial contributions to bring speakers from overseas and assisted with advertising the meeting. Additional support was provided by the Environmental Physiology Group of the S.E.B., who initiated the proposal for the Symposium. We thank Dr Lynton Incoll of the University of Leeds and convenor of the Environmental Physiology Group, whose help, advice, interest and encouragement through all stages of the planning and organization proved invaluable. We are equally indebted to Dr Richard Jurd of the University of Essex for his unfailing efficiency as local organizer.

S. P. Long
F. I. Woodward
July 1988

PRINCIPLES UNDERLYING THE PREDICTION OF TEMPERATURE IN PLANTS, WITH SPECIAL REFERENCE TO DESERT SUCCULENTS

PARK S. NOBEL

Department of Biology and Laboratory of Biomedical and Environmental Sciences, University of California, Los Angeles, California, USA

Summary

The thermal motion of molecules responsible for the property we know as temperature influences essentially every aspect of plant biology. Emphasis in this review is on principles of wide biological applicability, with specific examples being chosen from research on agaves and cacti.

Plant temperatures can be predicted using energy budgets incorporating shortwave and longwave radiation, heat conduction and convection, latent heat, and heat storage. Energy budgets are used here to show the effect of plant size and shortwave absorptance on tissue temperature. The importance of air temperature or transcription rate for tissue temperature is calculated quantitatively. The influences of surface appendages, such as apical pubescence and spines, on minimal temperatures near the apical meristem of cacti are predicted, such minimal temperatures influencing the geographical distribution of various species. The Boltzmann energy distribution and Arrhenius plots are presented and used to analyse thermal responses in terms of energy barriers and activation energies. The many ways that temperature can influence CO_2 diffusion into a leaf are also considered. Tolerances to low and high temperatures, such as the ability of agaves and cacti to tolerate extremely high tissue temperatures of 70°C, are discussed from both cellular and ecological perspectives. Although the influences of temperature on plants are multitudinous, many can be predicted, or at least analysed, based on well-established physical principles.

Introduction

Temperature results from the motion of molecules and, moreover, is a manifestation of that motion. Absence of molecular motion is theoretically possible in a system and would occur at 0 K (i.e. the temperature of absolute zero). The addition of energy, such as by the absorption of radiation or by contact with another system whose molecules are in motion, causes molecules in the first system to move and thereby to collide with other molecules. These collisions cause a random, chaotic exchange of energy among the molecules, whose mean kinetic energy affects the state and density of the matter.

For biological considerations, the key molecule from a thermal point of view is water, whose characteristics dictate the temperature range for life as we know it.

To heat 1 kg of ice from 0 K to 273 K (0°Celsius) takes about 300 kJ of energy (Eisenberg & Kauzmann, 1969). If we now add the heat of fusion to the ice, 330 kJ kg^{-1}, a transition occurs to the liquid state where the molecules can relatively freely slide past each other. Movement of other molecules in such a liquid can then take place by the random thermal motion known as diffusion, allowing biochemical reactions to proceed. An additional 250 kJ kg^{-1} added to liquid water at 0°C raises the temperature to 60°C, resulting in such vigorous molecular motion that the three-dimensional structure of protein molecules changes and specific reactions can no longer be catalysed in an orderly fashion, except for certain thermophilic blue-green bacteria (Berry & Björkman, 1980). Thus higher plants operate within a fairly narrow range of kinetic energy per unit mass, a range that is finely tuned to the position of the Earth's orbit in our solar system.

Energy can enter or leave a system in various ways. The first law of thermodynamics states that energy is not created or destroyed but simply changed from one form to another. Thus, plant temperatures can be predicted by means of a bookkeeping device known as an energy budget. Energy budgets allow us to quantify the effect of environmental variables and plant parameters on tissue temperature. After indicating how temperature can be predicted, we will discuss methods for analysing the effect of temperature on specific plant processes. Indeed, sometimes the temperature response can help us decide between alternative choices for mechanisms underlying a process. We will briefly explore how extremely low or extremely high temperatures can damage plants. Most of the examples will be from the monocot genus *Agave* and the dicot family Cactaceae, massive plants that can survive in extremely hot environments, generally without the benefit of day-time transpirational cooling. We will show that such unrelated taxa can have similar responses to temperature and yet closely related plants can differ greatly in their responses to temperature.

Components of energy budgets

Temperatures, which influence nearly every process in biology, can be both predicted and measured. Accurate prediction entails a detailed knowledge of how plants exchange energy with the environment and thus is a worthwhile goal, but biology is predominantly an empirical science and so measurements are more convincing to most people. Yet organization of the component energy fluxes into an energy budget can help interpret the influence of a morphological feature such as apical pubescence, spines, or stem diameter of cacti on tissue temperature and hence on plant distribution. Moreover, such models can allow us to do 'experiments' that are virtually impossible in nature. For example, we can decrease or increase the length of the apical pubescence, change spine length or spine frequency, or vary stem diameter while keeping other morphological parameters constant; the ensuing simulations indicate what happens to plant temperature. We can likewise predict what happens to tissue temperature as we change windspeed or stomatal aperture and, hence, transpiration rate. Combining the insights of modelling with the information generated by empirical testing will lead to greater understanding of plant responses to their thermal environment. In this section we will describe the six specific terms that can enter into an energy budget (Raschke, 1960; Monteith, 1973; Gates, 1980; Nobel, 1983).

Shortwave radiation

The energy necessary to raise temperatures at the Earth's surface far above absolute zero is mainly provided directly or indirectly by the sun. Sunlight has 98 % of its energy in wavelengths less than 4000 nm. Such shortwave radiation incident on a surface, S, is termed the shortwave irradiation or irradiance, of which a fraction *a* (the shortwave absorptance) is absorbed:

$$\text{Shortwave irradiation absorbed} = a\,(S^{\text{direct}} + S^{\text{diffuse}}) \qquad \text{(Eqn 1)}$$

where we recognize that the sunlight can directly strike the surface, S^{direct}, or it can be scattered by molecules and particles in the air or be scattered from surrounding surfaces before striking the surface of interest (S^{diffuse}). The units of S can be $J\,m^{-2}\,s^{-1}$ ($W\,m^{-2}$). All equations given here will be presented on a total surface area basis so that we can consider three-dimensional objects such as the massive stems of cacti; for flat leaves, the total surface area is twice that of one side, and gas exchange data are usually presented as total amount per unit area of one side only.

Longwave radiation

All entities above absolute zero emit radiation, the amount and wavelengths emitted varying greatly with surface temperature. For temperatures of plants and their environment, such energy emission generally occurs at long wavelengths. For instance, about 99 % of the radiation from a plant because of the thermal motion of its molecules is emitted at wavelengths longer than 4000 nm. Such longwave radiation, which is also termed infrared or thermal radiation, can be predicted by the Stefan–Boltzmann law:

$$\text{Longwave radiation emitted} = \varepsilon\sigma\, T_K^{\text{surf}})^4 \qquad \text{(Eqn 2)}$$

where ε is the emissivity of the surface, a dimensionless parameter with a maximum value of unity for a perfect or 'blackbody' radiator; σ is the Stefan–Boltzmann constant ($5{\cdot}67\times10^{-8}\,W\,m^{-2}\,K^{-4}$); and T_K^{surf} is the surface temperature in Kelvin units. For agaves and cacti, ε is generally 0·96 to 1·00 (Idso *et al.* 1969; Arp & Phinney, 1980), so their emitted longwave radiation is near the maximum possible, as is also true for surfaces of other plants.

Because the emission of electromagnetic radiation involves the same sort of considerations of atomic and molecular energy levels as does its absorption, ε is numerically nearly identical to a_{IR}, the fraction of longwave radiation absorbed. In any case, we can represent the net longwave exchange as follows:

$$\text{Net longwave radiation exchange} = a_{IR}\,\sigma\,(T_K^{\text{surr}})^4 - \varepsilon\sigma\,(T_K^{\text{surf}})^4 \quad \text{(Eqn 3)}$$

where T_K^{surr} is the effective temperature of the surroundings. The first term on the right-hand side of Eqn 3 is the longwave energy input, which on clear nights can be a consequence of an effective sky temperature that is much lower than the minimum air temperature measured in a standard weather bureau enclosure.

Heat conduction

Whenever a temperature difference exists between two adjacent regions, heat is conducted from the warmer to the cooler region:

$$\text{Heat conduction} = K\,\Delta T/\Delta x \qquad \text{(Eqn 4)}$$

where K is the thermal conductivity coefficient ($W\,m^{-1}\,°C^{-1}$) and ΔT is the temperature drop across the distance Δx. For applications of energy budgets to leaves of most plants, heat conduction across the leaf or along its petiole is ignored, because such fluxes are generally small or are transitory. On the other hand, heat fluxes can be substantial within the massive leaves of agaves or the massive stems of cacti. Heat is also conducted across the relatively still layers of air adjacent to the various surfaces of a shoot, regions known as the air boundary layers.

Heat convection

Heat conducted across an air boundary layer is moved away from the shoot surface in small packets by the adjoining, flowing, generally turbulent air, a process referred to as heat convection. Replacing Δx in Eqn 4 by the boundary layer thickness, δ^{bl}, heat convection for a particular surface can be represented as follows:

$$\text{Heat convection} = K^{air}(T^{surf} - T^{air})/\delta^{bl} \qquad \text{(Eqn 5}a\text{)}$$

$$= h_c(T^{surf} - T^{air}) \qquad \text{(Eqn 5}b\text{)}$$

where K^{air} is the thermal conductivity coefficient of air ($0{\cdot}0257\,W\,m^{-1}\,°C^{-1}$ for dry air at 20°C) and h_c is a heat convection coefficient ($W\,m^{-2}\,°C^{-1}$).

Relations have been developed empirically to estimate the thickness of air boundary layers for various wind speeds and various geometrical shapes (Nobel, 1974, 1975, 1983):

$$\text{Flat plate:}\quad \delta^{bl}_{(mm)} = 4{\cdot}0\sqrt{\frac{l_{(m)}}{v_{(m\,s^{-1})}}} \qquad \text{(Eqn 6}a\text{)}$$

$$\text{Cylinder:}\quad \delta^{bl}_{(mm)} = 5{\cdot}8\sqrt{\frac{d_{(m)}}{v_{(m\,s^{-1})}}} \qquad \text{(Eqn 6}b\text{)}$$

$$\text{Sphere:}\quad \delta^{bl}_{(mm)} = 2{\cdot}8\sqrt{\frac{d_{(m)}}{v_{(m\,s^{-1})}}} + \frac{0{\cdot}25}{v_{(m\,s^{-1})}} \qquad \text{(Eqn 6}c\text{)}$$

where $\delta^{bl}_{(mm)}$ is the average boundary layer thickness in mm, $l_{(m)}$ is the average distance across a flat plate in the wind direction in m, $d_{(m)}$ represents the diameter of the cylinder or sphere in m, and $v_{(m\,s^{-1})}$ is the ambient wind speed in $m\,s^{-1}$, as measured just outside the air boundary layer. Eqn 6 indicates that the air boundary layers are thinner and hence heat convection per unit area (Eqn 5) is greater for smaller objects and at higher wind speeds.

Latent heat

Energy is required to evaporate water. This energy is generally referred to as the latent heat for this change of state from a liquid to a gas, so transpiration represents a mode for heat loss. We can represent the water vapour flux density, J_{wv} ($mg\,m^{-2}\,s^{-1}$ or $mmol\,m^{-2}\,s^{-1}$), comprising transpiration as follows:

$$J_{wv} = g_{wv}\,\Delta c_{wv} = g'_{wv}\,\Delta N_{wv} \qquad \text{(Eqn 7)}$$

where g_{wv} and g'_{wv} are water vapour conductances, which generally reflect the degree of stomatal opening, Δc_{wv} is the drop in concentration of water vapour from the sites of evaporation to the ambient air outside the boundary layer, and ΔN_{wv} is the drop in mole fraction of water vapour over this pathway. Because the heat of vapourization H_{vap} is the energy required to vapourize a unit amount of water at the local shoot temperature, we can represent the energy flux density by latent heat as follows:

$$\text{Latent heat} = H_{vap}\,J_{wv} = H_{vap}\,g'_{wv}\,\Delta N_{wv} \qquad \text{(Eqn 8)}$$

At 20°C, H_{vap} is 44·2 $kJ\,mol^{-1}$ or 2450 $kJ\,kg^{-1}$.

Heat storage

For the thin leaves of most plants with their low mass per unit surface area, very little energy is stored as leaf temperature changes. However, for massive plant parts, such as the leaves of agaves, the stems of cacti, and the trunks of trees, considerable amounts of energy can be stored as their temperatures change. For a region of volume V undergoing a temperature change ΔT in time Δt, we can represent the rate of heat storage as follows:

$$\text{Rate of heat storage} = C_p\,V\,\Delta T/\Delta t \qquad \text{(Eqn 9)}$$

where C_p is the volumetric heat capacity ($J\,m^{-3}\,°C^{-1}$).

Energy can also be stored or released by metabolic processes such as photosynthesis and respiration. A typical net rate of CO_2 uptake by a well-illuminated leaf is about 10 $\mu mol\,m^{-2}\,s^{-1}$. Because approximately 480 kJ of energy are stored by changes in bond energy per mole of CO_2 fixed into carbohydrates, photosynthesis then represents $(10\times10^{-6}\,mol\,m^{-2}\,s^{-1})(480\times10^{3}\,J\,mol^{-1})$, or 4·8 $W\,m^{-2}$. Because such a leaf might be absorbing 500 $W\,m^{-2}$ shortwave irradiation, photosynthesis can generally be neglected as a form of heat storage. In most cases respiration is even less important than photosynthesis on an energy basis, so it too can usually be ignored in energy budget calculations. However, metabolic heat production must be reckoned with for the inflorescences of many members of the *Arum* family. For instance, the 2–9 g spadix of *Symplocarpus foetidus* (eastern skunk cabbage) has an exceptionally high heat production of about 0·10 W g^{-1}, which is about the same rate as an active mammal of the same size (Knutson, 1974). In this case, metabolism makes a substantial contribution to the energy budget, leading to tissue temperatures that are 15°C to 30°C above ambient air temperatures.

Examples of energy budget predictions

The basic philosophy of an energy budget is quite simple: energy in − energy out = energy stored. Calculation difficulties occur because the various terms have different dependencies on temperature. Absorption of shortwave irradiation is independent of temperature, emission of longwave radiation depends on surface temperature (in Kelvin units) raised to the fourth power, heat conduction depends on a temperature gradient, heat convection depends on the difference between T^{surf} and the temperature of the ambient air, latent heat loss depends approximately exponentially on temperature (because of the temperature dependency of the water vapour content for the nearly saturated air in the tissue), and heat storage depends on the rate of change of temperature with time. The advent of relatively inexpensive, high speed computers greatly facilitates energy budget calculations under such non-linear conditions, especially when heat conduction and heat storage cannot be ignored.

Environmental conditions and hence tissue temperature can vary considerably at different locations on a shoot. Compared with a shaded leaf at the bottom of the canopy, a sunlit leaf at the top of a tree canopy receives more shortwave radiation but generally less longwave radiation and is exposed to higher wind speeds. If we also incorporate morphological differences, such as the two- to sixfold greater area that can occur for shade leaves compared with sun leaves on the same plant of some species, the boundary layer would be much thicker for the shade leaves because of both their greater size and the lower ambient wind speed (Eqn 6). Yet heat convection (Eqn 5) can still be greater for sun leaves, because they are often much further from air temperature than shade leaves. For massive tree trunks (Derby & Gates, 1966) or cactus stems (Lewis & Nobel, 1977), temperature can vary greatly from one location to another on the same organ. Such variation can be dealt with by dividing up the structure into hundreds of smaller pieces and calculating an energy budget for each of these pieces, which are termed nodes, a procedure that has been successfully applied to various species of cacti (Lewis & Nobel, 1977; Nobel, 1978).

Size

Because of boundary layer effects, size can have a major influence on plant temperature. The small, terete, ephemeral leaves on many species of *Opuntia* are often only about 2 mm in diameter. For a moderate wind speed of 0·5 m s^{-1}, Eqn *6b* indicates that such cylinders would have a boundary layer thickness of 0·37 mm. Cacti in the primitive genus *Pereskia* have much larger leaves that are often 3 cm across. Using Eqn 6*a*, we find that their boundary layer thickness would be 0·98 mm at the same wind speed, a typical value for the leaves of many plants. Let us next consider an approximately spherical barrel cactus 30 cm in diameter, as can occur for *Eriosyce ceratistes* or *Ferocactus acanthodes*; δ^{bl} for this case is 2·67 mm (Eqn 6*c*). Because of their thin boundary layers, we thus expect the small, terete leaves to have high heat convection (Eqn 5*a*), which causes their temperatures to be tightly coupled to that of the air (generally within 1 °C). The larger leaves of *Pereskia* lead to thicker boundary layers and hence a longer conduction distance in the relatively still air layer. Indeed, *Pereskia columbiana* can be up to 5 °C above day-time air temperatures (Schnetter, 1971). Parts of the stem of the much more massive barrel cacti are often 15 °C or more above air temperature (Herzog, 1938; Mooney, Weisser & Gulmon, 1977; Smith, 1978; Ehleringer *et al.* 1980). In the later case we have to consider not only the much larger boundary layer thickness but also heat storage and conduction within the stem, so the energy budget analysis is considerably more complicated for succulent stems than for thin leaves of cacti.

Even using a comprehensive energy budget, all the subtle effects of size on plant temperature (Fig. 1) are difficult to interpret, but certain general features can be recognized. At vanishingly small stem diameters, the air boundary layers become

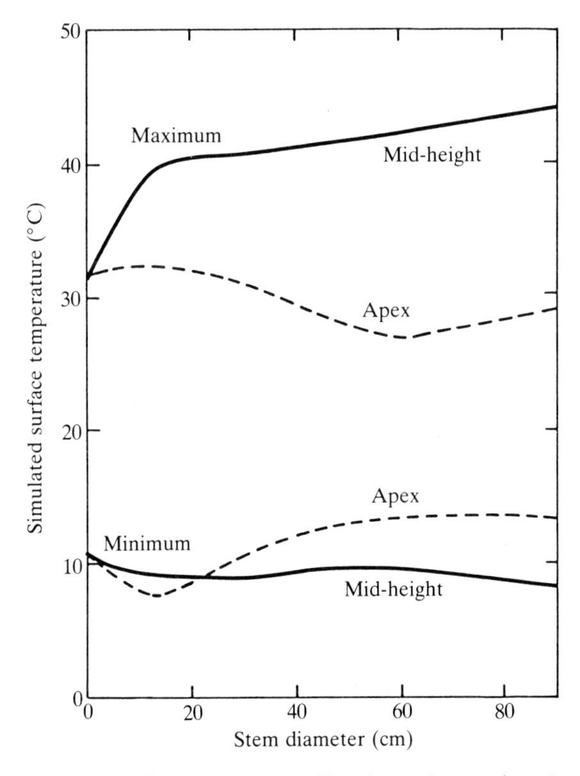

Fig. 1. Influence of stem diameter on predicted maximum (continuous line) and minimum (broken line) stem surface temperatures over a 24-h period at the apex and at mid-height on *Carnegiea gigantea*. Adapted from Nobel (1978).

extraordinarily thin, the heat convection coefficient becomes great, and surface temperatures converge on air temperatures. Increases in the stem diameter of *Carnegiea gigantea* (saguaro) cause its maximum temperatures at mid-height on the sides to increase over 10°C (stem surface temperatures and the temperatures of the underlying chlorenchyma or apical meristem usually differ by less than 0·6°C for this species). As diameter increases up to about 20 cm, maximal surface temperatures increase as the heat convection coefficient decreases. Above this diameter, heat conduction into the shoot becomes proportionally more important, and the increase in maximal temperature up to stem diameters of 40 cm is slight (Fig. 1). At larger stem diameters, shortwave irradiation cannot be as readily dissipated and thus more of it becomes stored in the larger plants, causing a rise in maximal temperature (Nobel, 1978). Yet maximal temperatures at the stem apex, where the meristem is located and hence a region critical for growth, tend to decrease as stem diameter increases (Fig. 1). As stem diameter increases, the minimum temperatures first decrease and then increase at the apex but tend to stay fairly constant at mid-height. Energy budgets, where the stem is divided into hundreds of nodes and which allow simulations for such obviously complex situations, generally predict average surface temperatures that are within 1°C of those measured in the field (Lewis & Nobel, 1977; Nobel, 1978).

Apical pubescence and spine shading

The depth of apical pubescence and the degree of spine shading affect tissue temperatures near the apical meristem of cacti by influencing air flow, heat conduction, heat convection, shortwave absorptance, and longwave radiation (Hadley, 1972). For instance, spines decrease the heat convection coefficient (Eqn 5*b*) by about 20 % for *Ferocactus acanthodes* compared with the spineless case (Lewis & Nobel, 1977). Spines also intercept radiation and thus decrease the shortwave radiation reaching the stem; in addition, they provide a longwave radiator at approximately air temperature, because the small diameter of the spines causes them to be tightly coupled to air temperature. The apical pubescence, which ranges up to 15 mm in thickness for barrel and columnar cacti, acts like an insulating layer that traps air. Both spines and pubescence moderate daily extremes in stem temperature, with the moderation of the minimum temperature for *Carnegiea gigantea* (Fig. 2) leading to protection of the apical meristem from freezing damage and hence extending its distributional boundaries into colder regions. We will next consider such matters in a little more detail.

Under a particular set of environmental conditions determined in the field, such as minimum air temperature at 2 m above the ground of 10·5°C, simulations using Eqns 1–9 (Nobel, 1980*a*, 1988) indicate that minimum apical stem temperatures at night are about 5°C in the absence of spines and pubescence but 12°C when the apical pubescence is 10 mm thick and the spines lead to 100 % shading of the apex (Fig. 2). Toward the northern part of the distribution of *C. gigantea* in Arizona, minimum air temperatures at a particular elevation decrease 1°C for approxi-

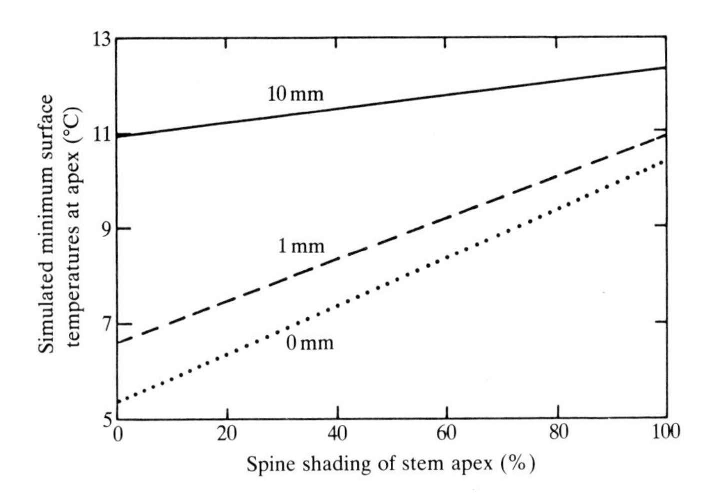

Fig. 2. Effect of apical spine shading and apical pubescence depth (indicated next to the curves) on predicted minimum apical stem surface temperatures for *Carnegiea gigantea*. Adapted from Nobel (1980*a*).

mately every 100 km moved northward. Also, minimum temperatures decrease about 0·6 °C for each 100-m increase in elevation (the dry adiabatic lapse rate is 0·98 °C per 100 m). In northern Arizona (at 34°50′ N, 113°39′ W), apical pubescence averages 10 mm in thickness and apical spine coverage averages 41 % for *C. gigantea*. Such apical coverages raise the simulated minimum apical temperature by 6 °C compared with their absence (Fig. 2). Although the depth of apical pubescence is fairly constant throughout its geographical distribution, the spine shading increases northward for *C. gigantea*. For instance, the percentage shading of the apex by spines increases from 9 % at 30°25′ N to 21 % at 31°58′ N to 35 % at 33°31′ N to 41 % at the northern Arizona site (Nobel, 1980*a*). This increase in spine shading can lead to a 0·5 °C increase in minimum apical temperatures (Fig. 2), which translates into a 50-km northward extension of the distributional boundary of *C. gigantea*, whose northern limit has long been recognized as being caused by low temperature (Shreve, 1911). The diameter of *C. gigantea* also increases from south to north (Niering, Whittaker & Lowe, 1963). For the four latitudes just considered, the mean diameter increases northward from 33 to 36 to 41 to 44 cm (Nobel, 1980*a*). Such a diameter increase raises the minimum apical temperature by about 0·8 °C (Fig. 1), which could allow for an 80-km extension to the north. Thus, the observed variation in stem diameter and in apical spine coverage could extend the distribution of this species by over 100 km northward.

So far we have considered only mature plants, although the seedling stage can be more vulnerable with respect to freezing damage. Again, insights into the interplay of morphological and environmental factors have been obtained using energy budgets. Seedlings of various agaves and cacti tend to become established under the canopy of other perennials known as nurse plants. Nurse plants, such as the small tree *Cercidium microphyllum* (palo verde), increase the winter survival of small *C. gigantea* (Steenbergh & Lowe, 1976) and so have been proposed to extend the distribution of this species northward (Steenbergh & Lowe, 1977). The thin branches of a nurse plant are tightly coupled to air temperature and thereby increase the incident longwave radiation compared with that from the sky, which can often have a very low effective temperature. For seedlings of *C. gigantea* up to 0·5-m tall at the northern part of its range, nurse plants can raise the effective temperature of the surroundings (T^{surr} in Eqn 3) by 16°C (Nobel, 1980*b*). Near 0°C, nurse plants would thus increase the longwave radiation received by the cactus seedlings by about 75 W m^{-2}; this can raise the apical temperatures of a small seedling by nearly 2°C, which can prevent freezing damage during a particular low-temperature episode that would kill the cactus seedlings in exposed locations.

What happens for other species of cacti? Actually, no other columnar cactus extends as far north as *C. gigantea*. The two other columnar cacti that occur in the southwestern United States, *Lophocereus schottii* (senita) and *Stenocereus thurberi* (organ pipe cactus), have only about 0·1 mm of apical pubescence (Nobel, 1980*a*), which would lead to much lower apical temperatures under a particular set of environmental conditions (Fig. 2) and hence greater vulnerability to freezing damage than for *C. gigantea*. One of the four species of barrel cacti that occur in the southwestern United States, *Ferocactus viridescens*, has only about 2 mm of apical pubescence and 20 % apical spine shading, and it is found in moderate coastal environments (Nobel, 1980*c*). The other three *Ferocactus* species occur in colder, inland environments and all have at least 7 mm of apical pubescence. The northernmost distributional limits of these three increase from *F. covillei* to *F. wislizenii* to *F. acanthodes*, as does the spine shading of the apex (15 %, 50 %, and 93 %, respectively, at the northernmost limit of each; Nobel, 1980*c*). Also, an increase in apical spine coverage occurs northward for *F. acanthodes* similar to that for *C. gigantea*.

Populations of the barrel cactus *Eriosyce ceratistes* and the columnar cactus *Trichocereus chilensis* have distinct upper elevational limits over wide differences in latitude in central Chile (Nobel, 1980*c*). Energy budget calculations help quantitatively account for the effect of the observed variations in diameter, apical pubescence, and apical spine shading over the geographical distribution on minimum apical temperatures. Using measured temperature lapse rates with elevation and measured changes in minimum temperature as a function of latitude at a particular elevation, minimum plant temperatures can be predicted at the upper elevational limits for a series of sites (Nobel, 1988). On slopes of the same aspect over the 500-km transect examined, the predicted minimum stem temperature at the upper elevational limit is the same within 0·2°C for *E. ceratistes* at all four sites and within 0·3°C for *T. chilensis* at all five sites (Nobel, 1980*c*). For these two species, whose upper elevational limit is determined by freezing damage, we can thus surmise that discrimination of temperatures of 0·3°C is possible at the population level.

Shortwave absorptance

The shortwave radiation absorbed increases in proportion to increases in *a* (Eqn 1), leading to higher maximum plant temperatures. Simulations show that for each 0·1 increase in *a* from 0·30 to 0·75, the maximum temperature of leaves of *Agave deserti* increases about 2 °C for a partially cloudy winter day at 34° N (Woodhouse, Williams & Nobel, 1983). The increase would be nearly twice as great on a clear summer day with its greater shortwave irradiation. A study of 14 agave species found that the two with the lowest *a* (0·49 for *A. deserti* and 0·50 for *A. lechuguilla*) occur in open desert habitats and those with the highest *a*s (0·65 for *A. multifilifera* and *A. rhodacantha*) tend to occur in cooler regions where much shading by other vegetation occurs (Nobel & Smith, 1983).

Shortwave absorptances can also differ substantially for cacti. For three sympatric dwarf cacti from the Chihuahuan Desert, *a* is 0·34 for *Epithelantha bokei*, 0·49 for *Mammillaria lasiacantha*, and 0·67 for *Ariocarpus fissuratus* (Nobel *et al.* 1986). The lower *a* for the former two species is a consequence of their highly reflective spines, especially for *E. bokei*. Under the same environmental conditions, the predicted maximal stem temperatures are 2°C higher for *M. lasiacantha* and 5 °C higher for *A. fissuratus* than for *E. bokei*, mainly because of the differences in *a* among the species. The high temperatures tolerated, a topic we will discuss later, are about 2°C higher for *M. lasiacantha* and 6°C higher for *A. fissuratus* than for *E. bokei* (Nobel *et al.* 1986), indicating an interesting tradeoff between surface morphology and tissue tolerances allowing these species to occupy the same habitat.

Environmental factors

Now that we have highlighted some of the morphological features that affect plant temperature, let us turn to environmental factors. Effects of longwave radiation have been indicated when discussing nurse plants, changes in shortwave radiation have analogous effects to changes in shortwave absorptance, and decreases in wind speed have similar effects on tissue temperature as do increases in plant size. Therefore, we will next consider air temperature.

Simulations show that holding all other environmental variables constant and changing air temperature leads to nearly the same change in plant temperature. Such sensitivity analyses show that a 5·0°C decrease in air temperature decreases the average shoot surface temperature by 4·7°C for *Agave deserti* (Woodhouse *et al.* 1983), 4·0°C for *Carnegiea gigantea*, and 4·4°C for *Mammillaria dioica* (Nobel, 1978). If the soil temperature and the temperature of the surroundings were also decreased, as would occur under field conditions when the air temperature decreases, then the decrease in plant temperature would be nearly 5·0°C in all three cases. It is thus with good reason that air temperatures have received the most attention over the years when discussing thermal responses of plants. Nevertheless, the equations presented allow calculations of exactly how much plant temperatures would differ from ambient air temperatures, and indeed we have emphasized such differences when discussing effects of plant size.

Transpiration – sample calculation

Shortwave radiation always represents a net energy input, as does net longwave radiation (Eqn 3) when T_K^{surr} is greater than T_K^{surf} and heat convection (Eqn 5) when T^{surf} is less than T^{air}. However, the latent heat term always represents a mode for energy loss when transpiration is occurring (we will avoid situations of frost or dew formation, which cause the heat of fusion to be released to a leaf and hence represent an energy input; Nobel, 1983). A representative transpiration rate for a well-illuminated leaf of a C_3 or C_4 plant is 2 to 5 mmol m^{-2} s^{-1} (Nobel, 1983), which equals 36 to 90 mg m^{-2} s^{-1} (in keeping with convention, transpiration rates represent the total rate of leaf water loss from both sides expressed per unit area of one side). Using the value of H_{vap} appropriate for 20 °C, such a J_{wv} corresponds to 90 to 220 W m^{-2}. Of course, as the stomata close and hence the water vapour conductance decreases, the latent heat loss decreases proportionally.

To be specific, let us reconsider the 3-cm wide *Pereskia* leaf and assume that the air temperature is 20 °C, the effective temperature of the surroundings is 10 °C, the shortwave radiation is 600 W m^{-2} (about 60 % of S when the sun is directly overhead on a clear day), a is 0·50, and ε is 0·98. The shortwave irradiation absorbed is then 300 W m^{-2} and the longwave irradiation absorbed is much greater, 714 W m^{-2} (357 W m^{-2} for each side of the leaf). To dissipate this energy input in the absence of transpiration, the leaf temperature must be above air temperature, leading ot a convective heat loss, and the higher leaf temperature leads to more emitted longwave radiation; in the present case, the leaf temperature is 23 °C, or 3 °C above air temperature. Let us next recalculate the leaf temperature for the case where transpiration leads to a dissipation of 150 W m^{-2}, a rather typical value when the stomata are open. In this case, energy budget calculations indicate that the leaf temperature is 19 °C, which is 1 °C below air temperature. Thus, not only is transpiration a cooling process, but in the present case it changes the leaf from above to below air temperature.

Analysis of temperature influences

Now that we have indicated how temperature can be predicted, we turn to a consideration of the influences of temperature on important processes in plants and how such influences can be analysed. Again, our main concern is with principles of wide biological applicability. We will begin by considering how the kinetic energy of molecules varies with temperature. This leads naturally to a consideration of energy barriers, which underlie many processes in biology, from diffusion to catalysis of biochemical reactions. To illustrate the many effects of temperature, we will use net CO_2 uptake, a crucial plant process.

Boltzmann energy distribution and Q_{10}

To determine what fraction of the molecules has the requisite energy for a particular process, we need an expression describing the distribution of energy among the molecules as a function of temperature. We recognize that very few molecules possess extremely high kinetic energies, indicating that the number with a particular energy [n(e)] must drop off rapidly with increasing energy. The specific kinetic energies possessed by molecules at equilibrium in an aqueous solution are described by the Boltzmann energy distribution (Bull, 1964; Nobel, 1983):

$$n(E) = n_{total}\, e^{-E/kT_K} \quad \text{molecule basis} \qquad \text{(Eqn 10}a\text{)}$$

$$n(E) = n_{total}\, e^{-E/RT_K} \quad \text{mole basis} \qquad \text{(Eqn 10}b\text{)}$$

where n_{total} in Eqn 10*a* is the total number of molecules, k is the Boltzmann constant ($1{\cdot}381 \times 10^{-23}$ J molecule^{-1} K^{-1}), and T_K is the temperature in Kelvin units. For most plant applications, it is more convenient to consider energy per mole instead of per molecule. Multiplying the Boltzmann constant k by Avogadro's number N yields the gas constant R (8·31 J mol^{-1} K^{-1}). For Eqn 10*b*, n(E) is the number of moles with a particular energy E out of the total number of moles (n_{total}).

Many processes, including diffusion and biochemical reactions, can be portrayed as requiring a certain minimum energy (E_{min}) in order to proceed. For molecules diffusing across a membrane, which can be represented as a one-dimensional process, the number of moles with a kinetic energy of E or greater resulting from velocities in some particular direction is proportional to $\sqrt{T}\, e^{-E/RT}$ (Davson & Danielli, 1952). This application of the Boltzmann energy distribution can indicate how the rate of diffusion is influenced by a 10°C increase in temperature. For this, we need to introduce a temperature coefficient known as the Q_{10}:

$$Q_{10} = \text{rate of process at } T_K + 10°\text{C}/\text{rate of process at } T_K \qquad \text{(Eqn 11)}$$

For the case of diffusion in one dimension, Q_{10} is as follows (Nobel, 1983):

$$Q_{10} = \sqrt{\frac{T_K + 10}{T_K}}\, e^{\frac{10\, E_{min}}{RT_K(T_K+10)}} \qquad \text{(Eqn 12)}$$

A Q_{10} near unity is characteristic of processes with no energy barrier to surmount. For diffusion of potassium ions in water at 20°C, E_{min} is about 17 kJ mol^{-1}, which leads to a Q_{10} of 1·3. Thus, K^+ diffuses more readily within a cell as the temperature increases. For the passive efflux of K^+ from many cells, E_{min} is 63 kJ mol^{-1} near 20°C (Stein, 1967), which leads to a Q_{10} of 2·5. Thus, even a passive process can have a high Q_{10} if there is a substantial energy barrier to surmount.

Activation energy and Arrhenius plots

Energy barriers are related to the concept of activation energy, which is the minimum amount of energy necessary for some reaction to take place (Lehninger, 1982; Nobel, 1983). If we represent the activation energy per mole by A, then the

Arrhenius equation indicates that the rate constant for a chemical reaction varies with temperature as follows:

$$\text{rate constant} = \text{B e}^{-\text{A}/\text{RT}_\text{K}} \qquad \text{(Eqn 13)}$$

where B is generally considered to be independent of temperature, or at least nearly so. For a first-order reaction, the rate of change of a reactant or a product is directly proportional to the rate constant.

A plot of the logarithm of the rate constant, or of the rate for a first-order reaction, versus $1/T_K$ is commonly known as an Arrhenius plot. Using Eqn 13, we obtain

$$\ln(\text{rate constant}) = \ln \text{B} - \frac{\text{A}}{\text{RT}_\text{K}} \qquad \text{(Eqn 14)}$$

and so the slope of an Arrhenius plot is $-A/R$. Many reactions of importance in biochemistry have activation energies of 50 to 100 kJ mol^{-1}, which lead to Q_{10}s of 2 to 4.

Besides allowing a determination of activation energies, Arrhenius plots have been used to demonstrate the effects of phase transitions in the membranes of chilling-sensitive plants (Lyons, Graham & Raison, 1979). Chilling-sensitive plants such as corn, cotton, cucumber, rice, soybean, and tomato are often severely injured by exposure to temperatures near 10°C, which is far above freezing. Changes in the spatial organization of membrane lipids can occur at such transition temperatures, leading to changes in the activation energies of enzymes that are detected as changes in slopes (and even discontinuities in the curves) on Arrhenius plots.

Net CO_2 uptake

We will first examine the overall influence of temperature on the instantaneous rate of net CO_2 uptake for two species of agave, such CO_2 uptake occurring at night for these CAM plants. We will then discuss some of the ways that temperature can influence such patterns.

The optimal temperature for net CO_2 uptake for plants grown at day/night air temperatures of 10°C/10°C is 12°C for *Agave americana* (Fig. 3A) and 15°C for *A. deserti* (Fig. 3B); such low optimal temperatures are characteristic of CAM plants (Kluge & Ting, 1978). When the growth temperature is raised to 30°C/30°C, the optimal temperature shifts upward by 7°C for *A. americana* and 3°C for *A. deserti*. Such temperature acclimation is commonly observed for other agaves and for cacti as well as for C_3 and C_4 species (Berry & Björkman, 1980; Nobel & Hartsock, 1981). Shifting *A. americana* to the higher growth temperature causes the maximum rate of net CO_2 uptake at the optimal temperature to increase, whereas a similar shifting of *A. deserti* causes it to decrease (Fig. 3). In this regard, the former species is native to warmer, subtropical regions and the latter is native to cooler regions in the northwestern Sonoran Desert (Gentry, 1982).

Specific effects of temperature on many aspects of net CO_2 uptake can be identified. First of all, CO_2 diffusion in the gas phase is based on random thermal

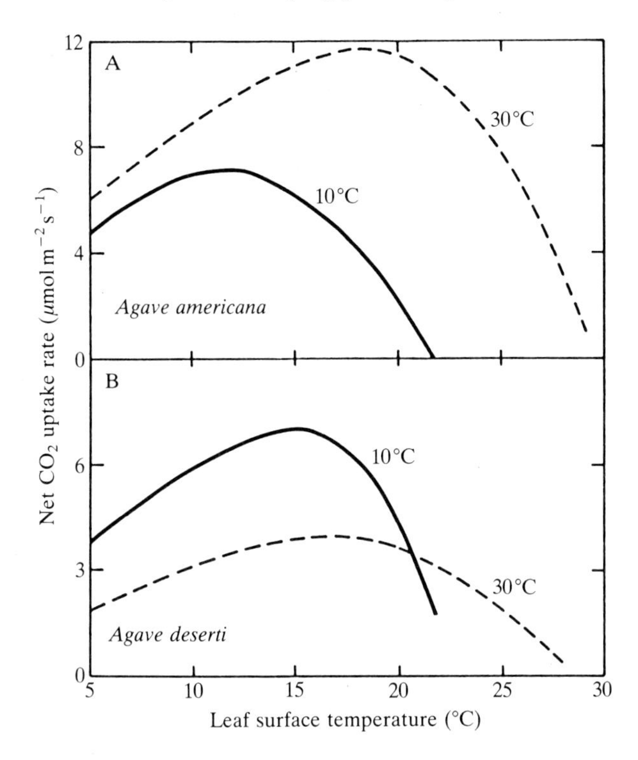

Fig. 3. Temperature dependence of nocturnal CO_2 uptake for two species of agave maintained for 4 weeks at day/night air temperatures of 10°C/10°C or 30°C/30°C. Adapted from Nobel & Hartsock (1981).

motion of the molecules. For gases in air, the diffusion coefficients depend approximately on $T_K^{1.8}$ (Fuller, Schettler & Giddings, 1966; Nobel, 1983). Although the mole fraction of CO_2 in air is essentially independent of temperature, the partitioning of CO_2 between air and water depends on temperature. For instance, the partition coefficient $c_{CO_2}^{water}/c_{CO_2}^{air}$ is 1·71 at 0°C, 0·88 at 20°C, and 0·53 at 40°C (Nobel, 1983). The solubility of CO_2 as a gas in water decreases as temperature rises; but the solubility of bicarbonate, which can be readily formed from CO_2, especially if carbonic anhydrase is present ($CO_2 + H_2O \rightleftharpoons HCO_3^- + H^+$), increases with temperature. We have already discussed how to analyse the diffusion of CO_2 across a membrane or in aqueous solutions once the respective energy barriers are known. We must also interpret why individual enzymes such as carbonic anhydrase or groups of enzymes such as those involved in net CO_2 fixation exhibit an optimal temperature. The rising portion below the optimal temperature presumably is a reflection of the Boltzmann energy distribution, for which temperature increases cause more molecules to have enough kinetic energy to surmount the energy barrier. As temperature exceeds the optimal temperature, the increasing thermal motion near the active site may disrupt the enzyme–substrate binding necessary for catalysis by inducing slight distortions in the three-dimensional structure of the enzyme. This opens the further question concerning the changes in enzyme structure that accompany acclimation.

The net rate of CO_2 uptake decreases to half of the maximal value at an average of 12 °C below and 7 °C above the optimal temperature for the two agave species at the two growth temperatures (Fig. 3). To help explain such an asymmetry, we first examine the stomatal response to temperature. For these two agave species, the water vapour conductance (Eqn 7) steadily decreases as the temperature rises, the decrease averaging 20 % for the 12 °C below the optimal temperatures for net CO_2 uptake and 35 % for the 7 °C above the optimal temperatures (P. S. Nobel, unpublished observations). Such a monotonic response of water vapour conductance to temperature helps account for the faster drop in net CO_2 uptake above compared with below the optimal temperature. Net CO_2 uptake represents the algebraic sum of gross CO_2 fixation and CO_2 release by respiration; because respiration for *A. deserti* has a Q_{10} of about 2 from 5 °C to 30 °C (Nobel, 1984*a*), the CO_2-releasing process is favoured at higher temperatures, which also causes net CO_2 uptake to drop faster above compared with below the optimal temperature. The fact that a peak is observed in the temperature response curves (Fig. 3) is ascribed to the biochemical reactions of CO_2 fixation.

As just indicated for *A. americana* and *A. deserti* and as commonly occurs for cacti (Nobel, 1988), the water vapour conductance decreases at an increasing rate as temperature rises. This can offset the approximately exponential increase in water vapour content in the tissue as the temperature rises, so that transpiration becomes fairly independent of temperature for these succulents. As would be expected for a process occurring at night, the acclimation of the optimal temperature for nocturnal CO_2 uptake in response to changing growth temperatures is in response to the night-time temperature, at least for *Coryphantha vivipara* (Nobel & Hartsock, 1981). A complicating factor that can affect shoot gas exchange by many plants is root temperature, although this has so far not been demonstrated for agaves and cacti.

For CAM plants, the total CO_2 uptake during the whole night is generally more important than the instantaneous rate at a particular time during the night. For instance, higher temperatures in the field at dusk can delay stomatal opening until later in the night when the temperatures are lower and hence the transpiration rate is lower; but the total CO_2 uptake during the whole night can be nearly the same as for lower temperatures at dusk, in which case partial stomatal closure generally occurs during the latter part of the night. Also, the high temperatures that can occur in the field in the early afternoon generally have little effect on nocturnal CO_2 uptake by CAM plants. For *Agave deserti* in the northwestern Sonoran Desert, net CO_2 uptake over 24-h periods is reduced 33 % for the average day/night temperatures of the hottest month, 6 % for the coldest month, and 12 % for all 12 months compared with net CO_2 uptake under optimal day/night temperatures (Nobel, 1984*a*). Similarly, net CO_2 uptake over 24-h periods averaged for each of the 12 mean monthly day/night air temperatures in the field is reduced only 10 % to 18 % for four other species of agave and two species of cacti

(Nobel, 1988). We can thus reach the important conclusion that for various agaves and cacti, both native and cultivated, monthly field temperature is not a major limiting factor for net CO_2 uptake over 24-h periods.

Temperature tolerances

As our final topic we will consider plant responses to extreme temperatures, both low and high, which can have even more importance to a particular species than intermediate temperatures, because extreme temperatures affect survival. We have already indicated that low temperatures influence the distribution of cacti, and we have shown how to predict the effects of apical pubescence, spines, and stem diameter on minimum tissue temperatures. To complete such an analysis, we need to know the tissue sensitivity to low temperatures. For this we require criteria by which to assess plant damage and death. Some plants of *Carnegiea gigantea* exposed to a low-temperature episode in January 1937 did not die until 3 years later (Steenbergh & Lowe, 1976, 1977). However, waiting around for a few years is not practical for laboratory experimentation on temperature tolerances. Uptake of the vital stain neutral red [3-amino-7-dimethyl-amino-2-methylphenazine (HCl)] by cells has proved quite useful for early determination of whether a particular temperature treatment will eventually lead to cell death and local necrosis (Gurr, 1965). Neutral red accumulates in the central vacuoles of living cells only, so microscopic examination of tissue slices can demonstrate the location of red colouring and thus reveal whether the cells have survived a certain treatment.

Low temperature tolerance

We can obtain considerable insight into freezing tolerances and the mechanism of low temperature damage by examining a cooling curve and the accompanying change in the ability of cells to take up neutral red (Fig. 4). When the stem of *Coryphantha vivipara* is gradually cooled below 0 °C, the chlorenchyma temperature steadily decreases to −6 °C without any ice formation, which indicates considerable supercooling because the cellular constituents of this cactus would freeze at about −1 °C (Nobel, 1981). Even though the air temperature is still being lowered, the tissue temperature of *C. vivipara* next increases (Fig. 4A), indicating that heat is being released within its stem. Such temperature increases represent the release of the heat of fusion as ice forms, in this case extracellularly (Uphof, 1916). The rise in temperature observed for *C. vivipara* represents the heat released as approximately 10 % of the stem water freezes (Nobel, 1981). Water next diffuses out of the chlorenchyma cells and becomes incorporated into the extracellular ice crystals. Yet all of these processes are nonlethal, as the fraction of chlorenchyma cells taking up neutral red is unaffected (Fig. 4B).

What happens as the stem is cooled further? Once the chlorenchyma temperature is lowered below −8 °C, the percentage of cells taking up neutral red decreases substantially (Fig. 4). This percentage decreases to 50 % at −14 °C and

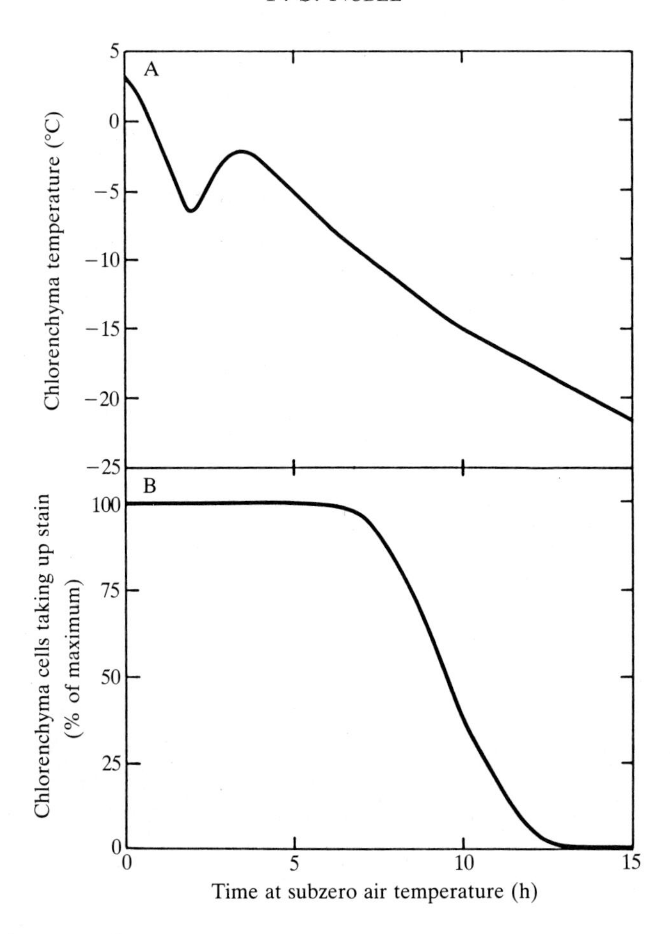

Fig. 4. Influences of temperature on low temperature responses of *Coryphantha vivipara*: (A) Cooling curve, showing the time course of chlorenchyma temperature as the air temperature is progressively lowered through the night; and (B) Influence of subzero temperatures on uptake of neutral red by chlorenchyma cells in tissue isolated at various times along the cooling curve in (A). Adapted from Nobel (1981).

0 % at −19°C. The protoplasts of *C. vivipara* appear basically normal during the initial decrease in chlorenchyma temperature, the subsequent rise, and then the fall to −5°C. Yet at −10°C the protoplasts appear shrunken, and they are severely shrunken at −15°C (Nobel, 1981). The decreased protoplast volume is a consequence of water diffusion out of the cells and its crystallization onto the extracellular ice crystals. The eventual death of the cells is similar to that caused by desiccation during drought. Moreover, at temperatures where the fraction of cells taking up stain was reduced to zero (−20°C), no net CO_2 uptake was observed over the next 2 months and all the plants eventually died. At −15°C, where some cells exhibited uptake of neutral red (Fig. 4), net CO_2 uptake occurred after 2 months for about half of the plants and these plants survived (Nobel, 1981).

Coryphantha vivipara ranges from Mexico to Canada and exhibits substantial cold-hardening, meaning that it becomes more tolerant of sub-zero temperatures after exposure to such low temperatures. In particular, its low temperature for death decreases 1·7°C as the growth temperature is reduced by 10°C (Nobel, 1981). The halftime for cold hardening in response to a sudden decrease in growth temperatures is 3 to 4 days for various agaves and cacti (Nobel, 1982; Nobel & Smith, 1983). Taking both low-temperature tolerances and their acclimation into consideration, *Carnegiea gigantea* can tolerate about the same low temperatures as *Stenocereus thurberi*, and *Ferocactus acanthodes* about the same as *F. wislizenii* (Nobel, 1982), so the more northerly distribution for the former in each pair is, presumably, the result of the differences in morphology that we have already discussed. Cold tolerance actually varies considerably among cacti; species like *C. vivipara*, *Opuntia polyacantha*, and *Pediocactus simpsonii* are able to tolerate −30°C when properly acclimated, whereas *Ferocactus viridescens* and *Opuntia ramosissima* are killed by −10°C. Indeed, breeding of various platyopuntias for increased cold tolerance is an important aspect of the expansion of their agronomic use for cattle fodder.

High-temperature tolerance

For a long time agaves and cacti have been known to be able to tolerate high temperatures [see Nobel (1988) for a review], but only recently have we recognized that they possess the greatest high-temperature tolerance among higher vascular plants so far examined. Most vascular plants succumb to high-temperature damage at tissue temperatures of 50°C to 55°C (Larcher, 1980; Kappen, 1981). Protein denaturation was originally thought to be the cause of high-temperature damage, but more recently attention has been focused on the disruption of membrane function (Steponkus, 1981).

For plants maintained at high day/night air temperatures of 50°C/40°C, the 1-h treatment temperature leading to halving of the fraction of chlorenchyma cells taking up neutral red averages 62°C for 14 species of agaves (Nobel & Smith, 1983) and 64°C for 18 species of cacti (Smith, Didden-Zopfy & Nobel, 1984; Nobel *et al.* 1986). Plants generally die at above 3°C above the temperature leading to such inhibition of stain uptake. Both taxa also exhibit considerable heat hardening, the temperature for 50% inhibition of the fraction of chlorenchyma cells taking up neutral red increasing 4·0°C for agaves and 5·3°C for cacti as the growth temperature is raised by 10°C. When properly acclimated to high temperature, nearly half of the species tested could survive 1 h at 70°C, *Opuntia ficus-indica* could survive 73°C, and *Ariocarpus fissuratus* could survive 74°C (Nobel *et al.* 1986; Nobel, 1988). Such temperatures are lethal to all other groups of higher vascular plants reported so far.

Many C_3 and C_4 plants tolerate higher temperatures as their osmotic pressures increase (Levitt, 1980). However, as osmotic pressure increases from 0·5 to 1·5 MPa for various agaves and cacti, the high temperature tolerated decreases about 2°C (Nobel, 1984*b*, 1988). Also, length of the photoperiod does not appear to have a significant effect on the high-temperature tolerance of agaves and cacti.

Membrane lipids of *Agave deserti*, *Carnegiea gigantea*, and *Ferocactus acanthodes* are more saturated (fewer double bonds per carbon atom) than those of C_3 plants (Kee & Nobel, 1985). Because greater saturation leads to less membrane fluidity, the membranes of these three desert succulents should be better adapted to high temperatures than are membranes of C_3 plants. Yet shifts in gross fatty acid composition are not consistent with the observed heat hardening; for *A. deserti*, saturation decreases at higher temperatures, no change occurs for *C. gigantea*, and saturation increases slightly for *F. acanthodes* (Kee & Nobel, 1985). These three species synthesize various new proteins (including 25 to 27 kD ones, typical of many in membranes) over a period of days during the acclimation to higher growth temperatures (Kee & Nobel, 1986), but whether such proteins are involved in the acclimation is not known. Indeed, much remains to be learned about the role of heat-shock and other proteins synthesized in response to temperature stress with respect to how they might increase plant tolerances to high temperatures.

The photosynthetic apparatus can also be damaged by high temperature. In this regard, fluorescence increases indicative of damage occur at an average temperature of 55 °C for *Agave americana* and 9 species of cacti compared with 47 °C for the other 35 species considered (Downton, Berry & Seemann, 1984). For *Agave deserti* and *Opuntia ficus-indica*, electron transport associated with Photosystem II is more sensitive to disruption by high temperature than is that associated with Photosystem I and both processes can acclimate to higher growth temperature (Chetti & Nobel, 1987). At day/night temperatures of 45 °C/35 °C, 19 % of Photosystem I activity remains after 1 h at 70 °C.

Although the molecular mechanisms for the extraordinary tolerance of high temperatures by agaves and cacti are not understood, certain ecological consequences are clear. High-temperature damage is difficult to observe in the field for both taxa, even for species native to hot deserts, and their massiveness coupled with the general absence of daytime stomatal opening by these CAM plants leads to tissue temperatures far above air temperatures. Even though they tolerate extremely high temperatures, such tolerance may not be sufficient to withstand the temperatures near the soil surface, which often exceed 70 °C in desert regions (Nobel *et al.* 1986). In fact, the plants are most vulnerable during the seedling stage. Seedlings of agaves and cacti can be a few degrees Celsius more sensitive to high temperatures than adult plants (Nobel, 1984*b*); because of their small size, the temperature of seedlings is tightly coupled to that of the soil surface. Indeed, the shade provided by nurse plants allows seedlings of many desert succulents to survive this vulnerable phase (Steenbergh & Lowe, 1977; Nobel, 1984*b*, 1988). The high temperatures near the soil surface also affect the roots, which for *Agave deserti* and *Ferocactus acanthodes* are a few degrees Celsius more sensitive to high temperature than are the shoots (Jordan & Nobel, 1984). Although the roots of both species are shallow with a mean depth of only 10 cm, they are absent from the upper 2 cm of the soil (except directly under the plants) as a consequence of the extremely high temperatures that can occur in the upper part of the soil. We can thus conclude that even with their tolerance of extremely high tissue temperatures, agaves and cacti must make certain concessions to the high temperatures that can occur in their native habitats.

In conclusion, it should be noted that temperatures influence nearly every aspect of the biology of plants. Temperatures can be accurately predicted for plants of widely differing morphology and under various environmental conditions. Also, the physical principles underlying the analysis of plant thermal responses are well known. Nevertheless, the mechanisms of action of temperature are often unknown. The future should bring greater insight into the mechanisms of various physiological processes, such as the stomatal response to temperature, high temperature tolerances, optimal temperatures for enzyme action, and the molecular basis for acclimation to changing temperatures.

Financial support from the Ecological Research Division of the US Department of Energy is gratefully acknowledged.

References

ARP, G. K. & PHINNEY, D. E. (1980). Ecological variations in thermal infrared emissivity of vegetation. *Environ. exp. Bot.* **20**, 135–148.
BERRY, J. A. & BJÖRKMAN, O. (1980). Photosynthetic response and adaptation to temperature in higher plants. *A. Rev. Pl. Physiol.* **31**, 491–543.
BULL, H. B. (1964). *An Introduction to Physical Biochemistry*, 433pp. Philadelphia: F. A. Davis.
CHETTI, M. & NOBEL, P. S. (1987). High-temperature sensitivity and its acclimation for photosynthetic electron transport reactions of desert succulents. *Pl. Physiol.* **84**, 1063–1067.
DAVSON, H. & DANIELLI, J. F. (1952). *The Permeability of Natural Membranes*, 2nd edn, 365pp. Cambridge: Cambridge University Press.
DERBY, R. & GATES, D. M. (1966). The temperature of tree trunks, calculated and observed. *Am. J. Bot.* **53**, 580–587.
DOWNTON, W. J. S., BERRY, J. A. & SEEMANN, J. R. (1984). Tolerance of photosynthesis to high temperature in desert plants. *Pl. Physiol.* **74**, 786–790.
EHLERINGER, J., MOONEY, H. A., GULMON, S. L. & RUNDEL, P. (1980). Orientation and its consequences for *Copiapoa* (Cactaceae) in the Atacama Desert. *Oecologia* **46**, 63–67.
EISENBERG, D. & KAUZMANN, W. (1969). *The Structure and Properties of Water*, 296pp. New York and Oxford: Oxford University Press.
FULLER, E. N., SCHETTLER, P. D. & GIDDINGS, J. C. (1966). A new method for prediction of binary gas-phase diffusion coefficients. *Ind. Engng. Chem.* **58**, 19–27.
GATES, D. M. (1980). *Biophysical Ecology*, 611pp. New York, Heidelberg, Berlin: Springer-Verlag.
GENTRY, H. S. (1982). *Agaves of Continental North America*, 670pp. Tucson: University of Arizona Press.
GURR, E. (1965). *The Rational Uses of Dyes in Biology and General Staining Methods*, 422pp. Baltimore: Williams & Wilkins.
HADLEY, N. F. (1972). Desert species and adaptation. *Am. Scientist* **60**, 338–347.
HERZOG, F. (1938). Formgestalt und Wärmehaushalt bei Sukkulenten. *Jahrbucher für wissenschafliche Botanik* **87**, 211–243.
IDSO, S. B., JACKSON, R. D., EHRLER, W. L. & MITCHELL, S. T. (1969). A method for determination of infrared emittance of leaves. *Ecology* **5**, 899–902.
JORDAN, P. W. & NOBEL, P. S. (1984). Thermal and water relations of roots of desert succulents. *Ann. Bot.* **54**, 705–717.
KAPPEN, L. (1981). Ecological significance of resistance to high temperature. In *Physiological Plant Ecology I Responses to the Physical Environment. Encyclopedia of Plant Physiology,*

New Series, Volume 12A (ed. O. L. Lange, P. S. Nobel, C. B. Osmond & H. Ziegler), pp. 439–474. Berlin, Heidelberg, New York: Springer-Verlag.
Kee, S. C. & Nobel, P. S. (1985). Fatty acid composition of chlorenchyma membrane fractions from three desert succulents grown at moderate and high temperatures. *Biochim. Biophys. Acta* **820**, 100–106.
Kee, S. C. & Nobel, P. S. (1986). Concomitant changes in high temperature tolerance and heat-shock proteins in desert succulents. *Pl. Physiol.* **80**, 596–598.
Kluge, M. & Ting, I. P. (1978). *Crassulacean Acid Metabolism. Analysis of an Ecological Adaptation*, 209pp. Berlin, Heidelberg, New York: Springer-Verlag.
Knutson, R. M. (1974). Heat production and temperature regulation in eastern skunk cabbage. *Science* **186**, 746–747.
Larcher, W. (1980). *Physiological Plant Ecology*, 2nd edn, 252pp. Berlin, Heidelberg, New York: Springer-Verlag.
Lehninger, A. L. (1982). *Principles of Biochemistry*, 1011pp. New York: Worth.
Levitt, J. (1980). *Responses of Plants to Environmental Stresses, 2nd edn, Volume 1, Chilling, Freezing, and High Temperature Stresses*, 497pp. New York: Academic Press.
Lewis, D. A. & Nobel, P. S. (1977). Thermal energy exchange model and water loss of a barrel cactus, *Ferocactus acanthodes. Pl. Physiol.* **60**, 609–616.
Lyons, J. M., Graham, D. & Raison, J. K. (eds.) (1979). *Low Temperature Stress in Crop Plants: The Role of the Membrane*, 565pp. New York: Academic Press.
Monteith, J. L. (1973). *Principles of Environmental Physics*, 241pp. New York: American Elsevier.
Mooney, H. A., Weisser, P. J. & Gulmon, S. L. (1977). Environmental adaptations of the Atacaman Desert cactus *Copiapoa haseltoniana. Flora* **166**, 117–124.
Niering, W. A., Whittaker, R. H. & Lowe, C. H. (1963). The saguaro: a population in relation to environment. *Science* **142**, 15–23.
Nobel, P. S. (1974). Boundary layers of air adjacent to cylinders. Estimation of effective thickness and measurements on plant material. *Pl. Physiol.* **54**, 177–181.
Nobel, P. S. (1975). Effective thickness and resistance of the air boundary layer adjacent to spherical plant parts. *J. exp. Bot.* **26**, 120–130.
Nobel, P. S. (1978). Surface temperatures of cacti – Influences of environmental and morphological factors. *Ecology* **59**, 986–996.
Nobel, P. S. (1980*a*). Morphology, surface temperatures, and northern limits of columnar cacti in the Sonoran Desert. *Ecology* **61**, 1–7.
Nobel, P. S. (1980*b*). Morphology, nurse plants, and minimum apical temperatures for young *Carnegiea gigantea. Bot. Gaz.* **141**, 188–191.
Nobel, P. S. (1980*c*). Influences of minimum stem temperatures on ranges of cacti in southwestern United States and central Chile. *Oecologia* **47**, 10–15.
Nobel, P. S. (1981). Influence of freezing temperatures on a cactus, *Coryphantha vivipara. Oecologia* **48**, 194–198.
Nobel, P. S. (1982). Low-temperature tolerance and cold hardening of cacti. *Ecology* **63**, 1650–1656.
Nobel, P. S. (1983). *Biophysical Plant Physiology and Ecology*, 608pp. New York: W. H. Freeman.
Nobel, P. S. (1984*a*). Productivity of *Agave deserti*: measurement by dry weight and monthly prediction using physiological responses to environmental parameters. *Oecologia* **64**, 1–7.
Nobel, P. S. (1984*b*). Extreme temperatures and the thermal tolerances for seedlings of desert succulents. *Oecologia* **62**, 310–317.
Nobel, P. S. (1988). *Environmental Biology of Agaves and Cacti*, 270pp. New York: Cambridge University Press.
Nobel, P. S., Geller, G. N., Kee, S. C. & Zimmerman, A. D. (1986). Temperatures and thermal tolerances for cacti exposed to high temperatures near the soil surface. *Pl. Cell Environ.* **9**, 279–287.
Nobel, P. S. & Hartsock, T. L. (1981). Shifts in the optimal temperature for nocturnal CO_2 uptake caused by changes in growth temperature for cacti and agaves. *Physiologia Pl.* **53**, 523–527.

Nobel, P. S. & Smith, S. D. (1983). High and low temperature tolerances and their relationships to distribution of agaves. *Pl. Cell Environ.* **6**, 711–719.

Raschke, K. (1960). Heat transfer between the plant and the environment. *A. Rev. Pl. Physiol.* **11**, 111–126.

Schnetter, R. (1971). Untersuchungen zum Wärme- und Wasserhaushalt ausgewähtler Pflanzenarten des Trockengebietes von Santa Maria (Kolumbien). *Beitr. Biol. Pfl.* **47**, 155–213.

Shreve, F. (1911). The influence of low temperature on the distribution of the giant cactus. *Pl. World* **14**, 136–146.

Smith, S. D., Didden-Zopfy, B. & Nobel, P. S. (1984). High-temperature responses of North American cacti. *Ecology* **65**, 643–651.

Smith, W. K. (1978). Temperatures of desert plants: another perspective on the adaptability of leaf size. *Science* **201**, 614–616.

Steenbergh, W. F. & Lowe, C. H. (1976). Ecology of the saguaro. I. The role of freezing weather in a warm-desert plant population. In *Research in the Parks*, National Park Service Symposium Series No. 1, pp. 49–92. Washington, DC: US Government Printing Office.

Steenbergh, W. F. & Lowe, C. H. (1977). Ecology of the saguaro. II. Reproduction, germination, establishment, growth, and survival of the young plant. National Park Service Scientific Monograph Series, No. 8. 242pp. Washington, DC: US Government Printing Office.

Stein, W. D. (1967). *The Movement of Molecules across Cell Membranes*, 369pp. New York: Academic Press.

Steponkus, P. L. (1981). Responses to extreme temperatures. Cellular and subcellular bases. In *Physiological Plant Ecology I Responses to the Physical Environment. Encyclopedia of Plant Physiology, New Series, Volume 12A* (ed. O. L. Lange, P. S. Nobel, C. B. Osmond & H. Ziegler), pp. 371–402. Berlin, Heidelberg, New York: Springer-Verlag.

Uphof, J. C. Th. (1916). *Cold-Resistance in Spineless Cacti*, Bulletin 79, 144pp. Tucson: University of Arizona Agricultural Experiment Station.

Woodhouse, R. M., Williams, J. G. & Nobel, P. S. (1983). Simulation of plant temperature and water loss by the desert succulent, *Agave deserti. Oecologia* **57**, 291–297.

PLANT LIFE IN COLD CLIMATES

Ch. KÖRNER and W. LARCHER
Institut für Botanik, Sternwartestrasse 15, A-6020 Innsbruck, Austria

Summary

Structural and functional features of plants from cold regions such as high mountain and tundra environments are characterized. Cold climates are not necessarily cold for plants at all times and influences of plant-growth form on canopy climate are substantial. Extreme low temperatures can cause temporal cessation of metabolic processes or partial tissue losses, but rarely represent an existential problem for plants native to cold regions. Low temperatures induce drastic changes in plant physiognomy and leaf anatomy, but dry matter allocation to the different plant compartments does not show a uniform trend. This suggests that generalizations of optimization and adaptation theories are not appropriate in this respect and that the functional importance of carbohydrate budgets is commonly over-estimated. Inconsistent with widespread beliefs, photosynthetic capacity in plants from cold regions is not essentially different from that in temperate regions, when comparable life forms are considered. Prevailing leaf temperatures exert minor limitations to seasonal photosynthetic carbon gain. Low temperatures come into play primarily in two ways: indirectly, *via* the length of the growing season and perhaps mineral nutrient availability, and directly, through influences on the growth process *per se*. Inherited slow or temporally restricted growth resembles an evolutionary response to long-term expectations of low resource availability. Within these genetic constraints the mitotic rate seems to be an essential point of action where low temperature determines the growth of an individual under respective local climates. Evidence is provided to support this view of an overruling significance of developmental processes in plant performance under cold conditions.

Introduction

The global availability of thermal energy for life processes varies latitudinally and altitudinally, and does not necessarily correspond to the abundance of other essential resources. Thus, low temperatures can co-occur with virtually all possible magnitudes of shortwave radiation, moisture and nutrient availability. 'Cold climates' encompass such different situations as the peak of a humid or arid tropical mountain, the windswept foggy ranges of a sub-antarctic island and cold desert areas beyond the polar circles. Hence it becomes difficult to separate low-temperature effects from simultaneous influences of other environmental factors.

Moreover, 'cold' is a very subjective term, interpreted quite inconsistently. Here we restrict our scope to those polar and high mountain areas that are too cold for natural tree growth. Some climatic data for such areas at different latitudes are compiled in Table 1. In some instances we shall also refer to trees at treeline. We shall further restrict our considerations mainly to vascular plants and to those areas where low temperature effects are not overruled by severe water stress, as in some of the South-American (e.g. Ruthsatz, 1978) or Asiatic mountain ranges (Breckle, 1973; Agakhanyantz & Lopatin, 1978).

The literature about mechanisms of plant life in cold climates is extensive and reference to some core papers should suffice here. The first attempts to characterize plant life in cold regions are the classical surveys by Schröter (1905) and Turesson (1925). Early reviews that included physiological aspects of arctic and alpine plant function are those by Pisek (1960), Scott & Billings (1964), Billings & Mooney (1968), Bliss (1971) followed by the volumes edited by Ives & Barry (1974), H. Franz (1979) and Bliss *et al.* (1981). Reviews of arctic tundra vegetation were provided by Tikhomirov (1963), Wielgolaski (1975), Alexandrova & Mateeva (1979), Sonesson (1980), Tieszen *et al.* (1981) and Kallio (1984), about plant life on mountains of the temperate zone by Larcher (1983), Nakhutsrishvili & Gamtsemlidze (1984) and Shibata (1985), on afro-alpine plants by Hedberg (1964), high-altitude plants of the Himalaya by Mani (1978) and alpine timberlines by Wardle (1974) and Tranquillini (1979). Physiological processes under low temperatures have been reviewed by Larcher & Bauer (1981) and Öquist & Martin (1986). A general text about plant survival at low temperature stress was published by Sakai & Larcher (1987).

Earlier attempts to characterize plant responses to cold climates have focussed on either selected cold regions (e.g. arctic tundra or temperate alpine areas) or specific physiological properties (e.g. metabolic processes or nutrient relations). Here we attempt to integrate some of these views and supply new and recent insights. Yet, the heterogeneity of cold environments which we have already mentioned and the multitude of plant functions forces us to restrict our survey to some key questions which, however, will be treated in a comparative manner as far as possible.

Table 1. *Influence of plant life form on leaf temperature (measurements on clear days in midsummer in the Australian Snowy Mountains by Körner & Cochrane, 1983)*

	Mean (maximum) midday leaf temperature (°C)		
	Low altitude (940 m)	High altitude (2040 m)	Difference (1100 m)
Air temperature (°C)	33·7	20·0	−13·7
Broad leaved sclerophyllous tree (2–3 m above ground)	38·7 (40·7)	20·0 (21·7)	−18·7 (−19·0)
Nanophyll dwarf shrub (20–50 cm above ground)	35·9 (37·3)	30·0 (32·5)	−5·9 (−5·2)
Small grass tussock (1–2 cm above ground)	46·2 (48·8)	39·2 (44·0)	−7·0 (−4·8)
Surface of dark humic soil	75·5 (81·0)	73·1 (81·9)	−2·4 (+0·9)

Six aspects of plant life in cold climates will be addressed in the following sections: (1) plant temperatures, (2) plant temperature resistance, (3) structural and (4) nutritional characteristics as well as (5) metabolic and (6) growth processes.

Are cold environments cold for plants?

The answer to this question is not self-evident. The common ranking of temperature conditions often reflects human experience and meteorological data collected in meteorological screens, both not relevant for the temperature conditions in the active plant tissue. Numerous investigations have shown that normal meteorological data deviate substantially from those collected in the vicinity of leaves and flowers in a low dense plant cover and from roots (Fig. 1).

Besides ambient air temperature, the most important factors that affect plant temperatures are the radiation input and the extent of aerodynamic coupling between the plant layer and the free atmosphere. Under given conditions of radiation, wind and moisture regime, plant physiognomy and plant density are the major determinants of plant temperature (e.g. Salisbury & Spomer, 1964; Gates *et al.* 1964). Körner & Cochrane (1983) have shown that plant structure can virtually overrule or exaggerate large altitudinal differences in air temperature. The

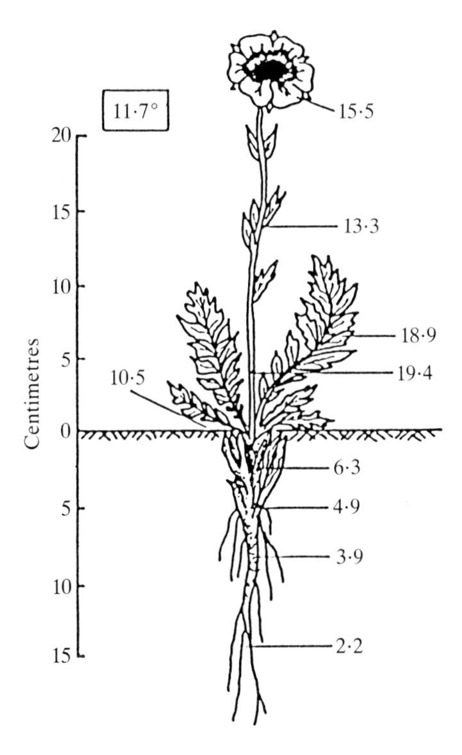

Fig. 1. Small scale variation in temperatures of a plant at an arctic habitat (after Tikhomirov, 1963).

examples shown in Table 2 illustrate that the elevational decline in mid-day leaf temperatures of trees can exceed that in air temperatures. Prostrate plants, on the other hand, may be warmer at high altitude than at low. Wilson, Grace & Allen (unpublished data cited by Grace, 1987) provide a good example for such life-form-dependent differences in leaf temperatures of *Pinus sylvestris* and the dwarf shrub *Arctostaphylos uva-ursi* in the Scottish mountains. Such microclimatic discrepancies among plant life forms will necessarily lead to diverging metabolic responses to temperatures, which will be discussed in a later section.

Numerous investigations have shown that low arctic and alpine plant canopies periodically exhibit temperatures equivalent to those of temperate or even tropical regions, as long as direct solar radiation occurs at moderate or high incidence angles. Reports of such accumulations of heat in low plant mats in the arctic and subarctic tundra are given for instance by Dahl (1951), Biebl (1968), Svoboda (1977) and Gauslaa (1984) and observations in alpine plants have been reviewed by Körner & Cochrane (1983). An excellent example is the annual course of leaf temperature in the low dwarf shrub *Loiseleuria procumbens* measured in the Central Alps by Cernusca (1976; Fig. 2). Mid-day canopy temperatures on most days without snowcover are in the range of 25 to 35 °C with peaks above 40 °C,

Table 2. *Comparison of climatic conditions in arctic, temperate–alpine and tropic–alpine tundra areas that support similar herb field, dwarf shrub or tussock grass vegetation (approximations from data compiled by Kessler, 1978; Larcher, 1980; Chapin, 1981; Körner & Mayr, 1981 and Körner* et al. *1983)*

	Arctic	Alpine				
		Temperate			Tropical	
		Oceanic	Semi-oceanic	Conti-nental	Humid	Mesic-dry
Area[a]:	A	B	C	D	E	F
Latitude	71° N	44° N	47° N	40° N	0°/5° S	8° N/10° S
Altitude (m)	5	1600	2600	3500	4400	4100
Altitude of upper forest line (m)	–	1200	2000	3400	3500	3200 (4100)
Mean length of growing season (d)	70	70–100	70–80	80	365	365
Mean daily global radiation in July (A,B,C,D)[a] and total year (E,F; 10^6 J m^{-2})[a]	18·1	*ca* 20	20·0	20·2	*ca* 15·0	*ca* 21·5
Growing season photoperiod (h d^{-1})	24	16	15·5	15	*ca* 12	*ca* 12
Mean air temperature in the warmest month (°C)	+4	+4	+5	+8·5	+3	+3/+4
Mean soil temperature in the warmest month (°C; 10–25 cm depth)	+2·4	+8/+10	+7	+13	*ca* 5	*ca* 5

[a] A: Alaska (Barrow); B, Southern Alps of New Zealand; C, Austrian Central Alps; D, Rocky Mountains (Niwot Ridge), Colorado; E, Mt Wilhelm, Papua New Guinea or Izombamba (3050 m, radiation only), Equador; F, Andes in Peru and Venezuela.

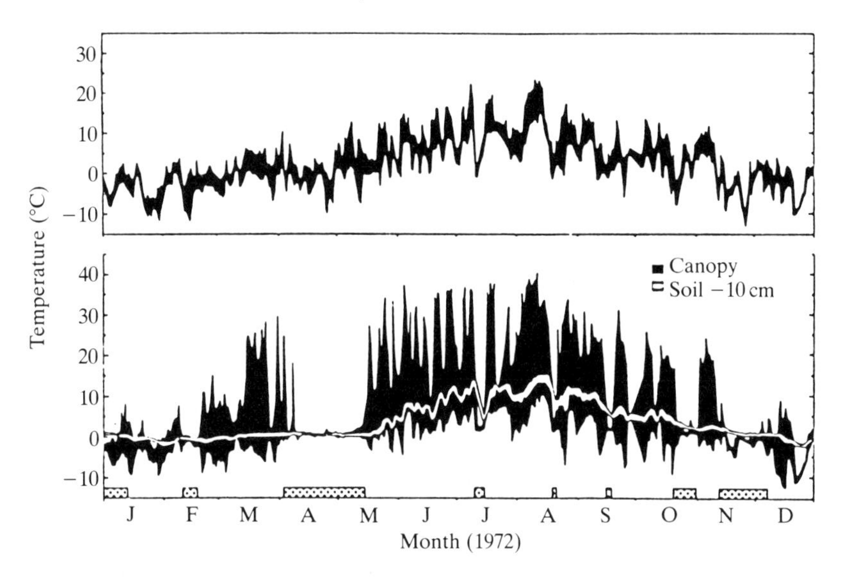

Fig. 2. Annual course of air temperature (upper diagram), leaf and soil temperature (lower diagram) in a wind-swept alpine dwarf shrub heath of *Loiseleuria procumbens* on Mt Patscherkofer (1950 m above sea level). The black area shows the range between minimum and maximum above ground temperatures, the white zone in the lower diagram illustrates the amplitude in soil temperature. Hatched bars at the bottom indicate periods with snow cover (from Cernusca, 1976).

while corresponding air temperatures at the same mid-day time range from 10 to 20°C. The mean midsummer (July) air temperature during this investigation was 7·5°C.

Compact, low rosette plants or seedlings on bare soil may occasionally be exposed to temperatures at or beyond their heat tolerance limit (Larcher, 1977, 1980; Larcher & Wagner, 1976, 1983; Körner & Cochrane, 1983; Gauslaa, 1984). Cushion plants are another life form that is particularly prone to heat accumulation both in arctic and alpine regions (e.g. Salisbury & Spomer, 1964; Svoboda, 1977; Körner & DeMoraes, 1979). In the tropics even less prostrate plants may accumulate substantial heat loads: leaf temperatures 15 K above ambient have been observed in tropic–alpine giant rosette plants (Larcher, 1975) and high temperatures are reported also for shrubby canopies under such conditions (Coe, 1969; Hedberg & Hedberg, 1979; Körner *et al.* 1983).

All these examples reflect the effects of strong radiant heating. However, full solar radiation is not required to elevate canopy temperatures above air temperature in low plant stands. Even under overcast sky conditions, longwave radiation can raise leaf temperatures above ambient. Fig. 3 illustrates the effect of incident radiation (here exemplified by quantum flux density in the visible range) on the temperatures within the herb layer of low and high altitude meadows in the Alps. In contrast to day-time conditions, night-time leaf temperatures of low vegetation tend to sink below air temperatures due to radiation losses, but differences are

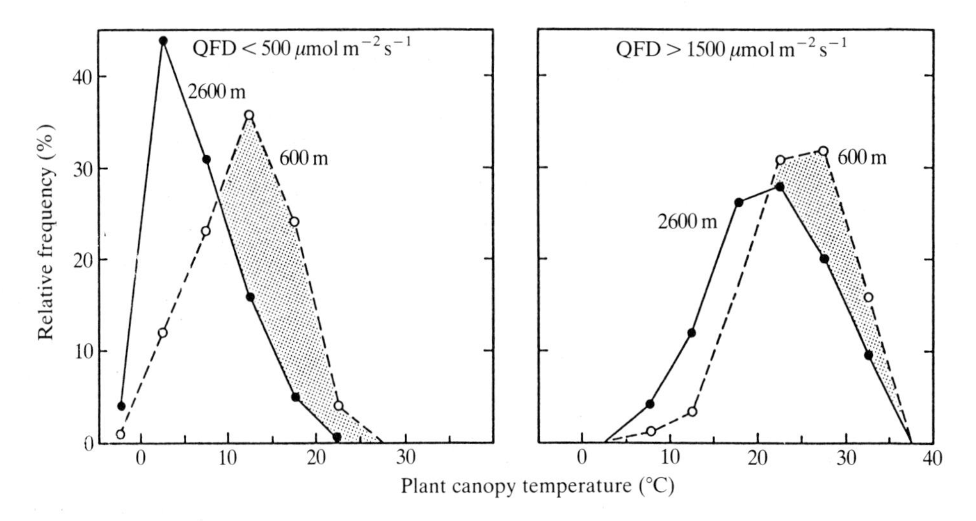

Fig. 3. Frequency distribution of hourly means of air temperature within dense herbaceous plant canopies for two ranges of incident quantum flux density (QFD) and two altitudes in the Alps (all hours with QFD > 30 μmol m^{-2} s^{-1} of the main growth period; from Körner & Diemer, 1987).

much smaller than during the day and will hardly exceed 3 to 5 K (cf. Sakai & Larcher, 1987).

In conclusion, low plant canopies in arctic and alpine regions are substantially warmer during the day but cooler during the night with respect to air temperature. Diurnal temperature oscillations are greater here than in the majority of vegetation types of the temperate lowlands or humid tropics. The 24-h amplitude in leaf temperature under clear sky conditions will normally be in the order of 25 K, but may reach extremes of 50 K in mountains of lower latitudes. Thus, an analysis of the temperature dependence of plant processes will yield different answers dependent on the type of life process (day-time or night-time activity) considered. Areas typically classified as cold climates, therefore, are not necessarily 'cold' for all plants and their respective functions.

Low temperature thresholds for survival of plants in cold climates

There is no doubt that frost plays a decisive role in global plant distribution (Larcher, 1981*a*; Larcher & Bauer, 1981; Grace, 1987; Sakai & Larcher, 1987; Woodward, 1987, 1988) and plants from cold regions will need to be particularly 'fit' to survive very low temperatures. However, it is rather difficult to rate the significance of extreme low temperatures for present day plant existence in cold climates. The extinction of a species in a certain area due to injurious low temperatures operates over long periods, and it can be assumed that persistence is due to resistance – at least on the population level. This is largely confirmed by available data (Sakai & Larcher, 1987). Yet resistance at the population level does not necessarily apply to individuals. Hence occasional lethal and sublethal injuries due to low temperatures are possible, which in turn alter productive contributions of respective species. The extent of such disturbances will depend on the leeway between temperatures which damage and those occurring in the real world.

In seasonal climates the winter period, i.e. the period with lowest temperatures, is unlikely to be a decisive factor with regard to survival of plant species native to cold regions, although winter frost resistance data are scarce for plants from the polar zones. Winter hardening provides sufficient protection, as revealed in laboratory experiments. Some hardened species tolerate even treatment with liquid nitrogen (Kainmüller, 1975; Larcher, 1980; Sakai & Larcher, 1987). The detection of individual plant extinction by frost in the field during winter is difficult, since the remnants of such plants will hardly permit the diagnosis of the cause of their death following snow melt.

If cold stress is to become effective for survival of native plants of cold climates, it is most likely to occur through low temperatures during the growing season in the temperate zone, and during clear nights in tropic-alpine regions. Deleterious effects occur as partial losses of plant parts rather than extinction of individuals (Sakai & Larcher, 1987). Hence, not the absolute annual minima of temperature recorded in an area, but the minima during the growth period or during periods lacking snow cover are important (Larcher, 1977, 1980).

Data presented by Pisek, Larcher & Unterholzner (1967), Larcher & Wagner (1976) and Yoshie & Sakai (1981) show that detectable leaf damage in arctic and alpine herbaceous plants and dwarf shrubs of humid regions first occurs between −3 and −7°C and 100% damage is caused by temperatures between −4 and −12°C. The low temperature threshold for damage of leaves from arid mountainous regions is approximately twice that low (around −12°C). Leaf frost resistance (TL_{50}) of megaphytes from tropical high mountains lies around −10°C (Goldstein *et al.* 1985; Beck, 1988) with adult *Lobelia* from Mt Kenya exhibiting resistance as low as −20°C. In many cases avoidance of cell damage is achieved by freezing point depression and supercooling. However, it has been shown that certain plants from high altitudes can also exhibit true freezing tolerance during the active life phase (Tyurina, 1957 and personal communication; Larcher & Wagner, 1983; Bodner & Beck, 1987). Seeds are generally less sensitive.

Thus, it appears that mountain plants can employ two resistance mechanisms to survive severe frost temperatures during the growth period (Sakai & Larcher, 1987): (1) by improving their ability to supercool and (2) by retaining a permanent tolerance of equilibrium freezing. The first of these mechanisms suffices where temperatures do not drop much below −12°C. This point appears to be close to the lowest limit for avoidance of freezing. Where temperatures below this critical threshold are experienced during the growth period (arid or semi-arid high mountains) the mechanism of freezing tolerance has to be brought into play. This mechanism is more effective and extends survival to lower temperatures. Freezing tolerance, however, requires specialized structures and biochemical adjustments for which a metabolic price has to be paid (Beck, 1988). Hence, these plants can be found frozen stiff in the morning, but remain entirely undamaged after thawing.

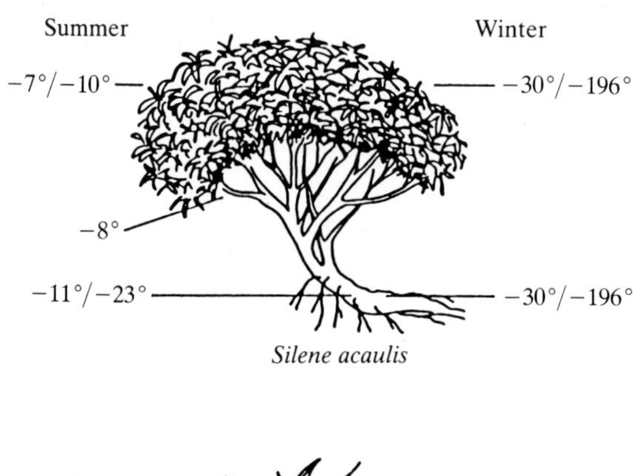

Silene acaulis

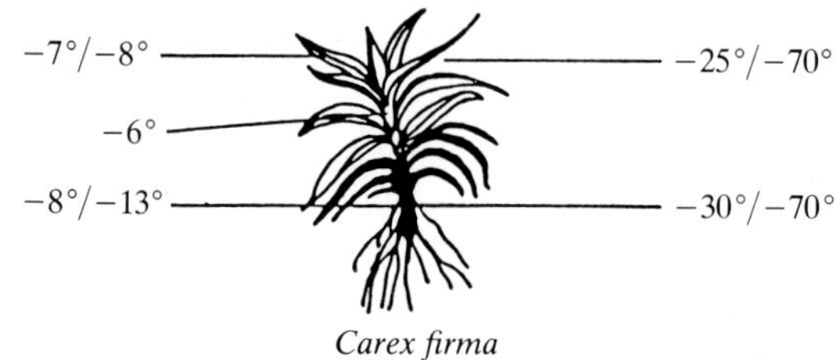

Carex firma

Soldanella alpina

Fig. 4. Frost resistance of high altitude plants. The paired temperature values indicate the minimum resistance after 3 to 5 days of warm temperature pretreatment (on the left) and maximum resistance after cold acclimation (on the right). The arctic–alpine element *Silene acaulis* grows on wind-exposed ridges, regularly lacking winter-time snow protection, in contrast to the typical snowbed plant *Soldanella alpina*, a southern European element. *Carex firma* holds an intermediate position with respect to snow cover (Larcher, 1980).

Fig. 4 and Tables 3 and 4 illustrate ranges of low temperature resistance in plants from various environments. Fig. 4 shows that frost resistance is not constant even under similar macroclimatic conditions, but depends also on microsite conditions, of which snow distribution is of major importance. Only tropical mountains exhibit altitudinal temperature gradients paralleled by resistance gradients. Extratropical mountains do not show such a correlation (Larcher, 1980, 1985).

The above conclusions illustrate that any estimates of resistance reserves ('margin of safety') will depend upon hardening state, plant organ, plant age etc. and will require relevant micrometeorological data. The absolute minimum air temperature in the snow free period and in areas where plants still grow in the

Table 3. *Frost resistance (lowest temperature in °C sustained without lethal injury) in alpine and arctic dwarf shrubs in summer (from Larcher & Bauer, 1981 and Yoshie & Sakai, 1981)*

Species	Leaf	Leaf bud	Stem	Root
Arctic plants[a]				
Arctostaphylos alpina	>−5	−15		−5
Empetrum nigrum	−15	−12·5		−7·5
Vaccinium vitis-idea	−15		−12·5	−10
Ledum palustre	−10	−10	−10	
Alpine plants[b]				
Arctostaphylos uva-ursi	−7 (−70)			
Empetrum nigrum	−8 (−70)		(−30)	(−30)
Vaccinium vitis-idea	−5 (−80)	(−30)	−8 (−30)	(−20)
Calluna vulgaris	−5 (−35)	(−30)	−5 (−30)	(−20)
Loiseleuria procumbens	−6 (−70)	(−40)	−10 (−60)	(−30)

[a] Canadian Tundra, N.W.T., Tuktoyaktuk, 7–12 August, 1980.
[b] Austrian Central Alps, peak growing season; maximum wintertime resistance in brackets.

Alps amounts to −10°C (Larcher, 1980). Radiant cooling at night may cause a further 2–4 K drop at the plant surface (Moser *et al.* 1977). Hence the lowest temperatures to which living plant tissue is possibly exposed in summer in the Alps range from −12 to −14°C. Such temperatures may approach or even surpass the low temperature threshold in hardened leaves in summer. If such extreme situations are considered it becomes evident that the margin of safety becomes negligible. This seems to be highly efficient with respect to the multiple direct and indirect costs of frost resistance.

Structural characteristics of plants from cold environments

Whole plant characteristics

It was shown by Nobel (1988) that plant morphology has a strong influence on the day-time temperature regime of plants. Here we ask the question whether plants from treeless low temperature regions have common structural features.

With few exceptions vascular plants from cold alpine and arctic environments are small, often prostrate and belong to four conspicuous growth forms: (1) herbaceous, mostly perennial rosette plants, (2) tussock graminoids, (3) cushion plants and (4) dwarf shrubs. Rosette forbs and cushion plants are typical for highest elevations, while tussocks and dwarf shrubs predominate lower alpine belts and polar tundras. Three of these growth forms are found in all cold environments, while true cushion plants are absent from certain regions (e.g. some humid tropical mountains), but predominant in others (Antarctic islands).

The cushion plant growth form, as defined by Rauh (1939), is genetically determined and is not a physiognomic modification under certain environmental influences. It has its advantages and disadvantages. A clear disadvantage of the

Table 4. *Frost resistance (TL_{50}) and supercooling ability (T_{sc}) of leaves of adult plants of tropical high mountains (from Goldstein* et al. *1985; Beck, 1988; after Sakai & Larcher, 1987)*

Plant species	Altitude (m)	Annual mean (absolute) minimum temperature of air (°C)	TL_{50} (°C)	T_{sc} (°C)
Venezuelan Andes				
Espeletia atropurpurea	2850	6·2 (4·0)	−5·9	−6·4
E. angustifolia	2850		−6·1	−6·6
E. lindenii	2850		−6·5	−7·5
E. jahnii	3100	4·8 (2·7)	−5·6	−5·7
E. atropurpurea	3100		−8·1	−7·3
E. marcana	3100		−8·0	−9·1
E. floccosa	3560	1·7 (−0·3)	−9·3	−8·5
E. schultzii	3560		−10·0	−10·8
E. spicata	4200	0·0 (−3·8)	−9·5	−10·0
E. lutescens	4200		−10·2	−10·5
E. schultzii	4200		−11·2	−10·0
E. moritziana	4200		−11·3	−10·6
Mt Kenya, Africa	4200	– (−13)		
Dendrosenecio keniodendron				
(Feb. 1983)	4200		−5 to −8	−7
(Mar. 1985)	4500		<−15	
Dendrosenecio brassica	4100		−10	−5·5
Lobelia keniensis	4100		−10 to −20	−4
Lobelia telekii	4200–4500		−14 to −20	−5
Happlocarpha rueppellii	4200		−13	
Ranunculus oreophytus	4200		−14	
Senecio purtschelleri	4250		−14	
Carduus chamaecephalus	4500		−15	

cushion morphotype is the limited growth potential caused by intrashoot-competition for space and the poor light harvesting potential caused by inheritable low leaf area indices (Körner & DeMoraes, 1979). Its major advantage is the high moisture and nutrient storage capacity in areas of poorly developed soils, not only in cold environments, but also in deserts including those at high elevations (Breckle, 1973; Agakhanyantz & Lopatin, 1978; Ruthsatz, 1978). The only unique environmental factor for all places where cushion plants naturally occur is strong wind. Retention of litter and nutrient recycling (Schinner, 1982*a*) may thus be the most beneficial effect of this growth form. We do not interpret the cushion morphotype as a primary adaptation to low temperatures.

The three remaining growth forms common to all cold areas fall into two groups: (1) persistent long lived graminoid tussocks and dwarf shrubs with predominantly vegetative propagation, and (2) non-graminoid herbaceous plants. The latter are almost exclusively sessile rosette plants, with shorter life cycles and the widest ecological amplitude of all growth forms. With respect to the area of land covered, the graminoid tussocks – largely *Carex* species and some genera from the *Poaceae* (e.g. *Deschampsia*, *Chionochloa*) are by far the most important growth form in cold environments.

In the following treatment we will try to quantify plant and leaf structures in cold climates in comparison to those in warm climates. Table 5 illustrates the great morphometric differences between related taxa of herbaceous perennial heliophytes from low and high altitude in the Alps. The mean summer air temperature decreases from 18 to 2 °C between the two sampling areas. Although day-time air temperatures differ much less (see climate section), shoots at high elevation are seven times shorter, while individual leaf area and total leaf area per individual are respectively ten and nine times smaller than at low elevation. These few numbers clearly show that plant exposure to cold climates causes dramatic physiognomic changes.

The question is, however, whether plants and their organs are just smaller in cold climates, or – in addition – exhibit fundamental alterations in their dry matter partitioning. A review of published data shows that there is no common pattern of dry matter partitioning to the different organs. For example, above- to below-ground dry matter ratios in individual plant species from arctic and alpine areas vary by two orders of magnitude from 0·05 to 7 (cf. Körner & Renhardt, 1987). Dwarf shrubs and tussock graminoids allocate most of their assimilates to below-ground organs, whereas most of the dry matter of real cushion plants from humid-cold regions is found above-ground. Herbaceous perennial plants of open habitats hold an intermediate position close to a 1:1 ratio of above- and below-ground dry matter, but variation among different species is very large.

It is important to note that the partitioning patterns in dwarf shrubs and tussock

Table 5. *Morphometric characteristics of representative herbaceous perennial plant species from 600 and 2600–3200 m altitude in the Alps at peak growth season (mean ± S.E., number of species in brackets; from Körner* et al. *1989)*

Altitude	600 m	2600–3200 m	Significance[a]
Base area (elliptic ground area occupied by one individual drawn through minimum and maximum horizontal extension (cm^2))	104 ± 34 (13)	20 ± 3 (25)	***
Height of inflorescence above ground (cm)	29 ± 4 (17)	4 ± 0·3 (31)	***
Number of basal leaves per adult individual	11 ± 2 (13)	16 ± 3 (17)	n.s.
Length of petiole (cm)	7 ± 2 (16)	1 ± 0·3 (24)	***
Maximum above-ground height extension of basal leaves (cm)	11 ± 2 (15)	2 ± 0·2 (23)	***
Single leaf area (cm^2	10 ± 2 (22)	1 ± 0·2 (33)	***
Total green leaf area per individual at peak vegetative development (cm^2)	143 ± 32 (17)	16 ± 2 (23)	***

[a] *** indicates significant difference among groups ($P < 0{\cdot}001$, *t*-test).

grasses are genotypic and possibly related to the predominant vegetative form of propagation, since they are also exhibited in warmer regions. For instance, tussocks and dwarf shrubs of lowland mires in the temperate zone still show this preference for below-ground allocation. The question whether increasingly cold environmental conditions invoke particular partitioning patterns was addressed by Körner & Renhardt (1987) in populations of perennial forbs along an elevational transect in the Alps.

Fig. 5 shows the mean dry matter fractions allocated to different plant compartments in 22 low- and 27 high-altitude taxa from the Alps. The mean leaf weight ratio does not change, and special storage organs retain similar dry matter fractions at both altitudes. However, due to alterations in the stem and fine root fractions, the relative portion of above-ground dry matter of total dry matter significantly decreases from 57 % at low to 42 % at high elevation (shoot/root ratio declines from 1·2 to 0·8; $P = 0{\cdot}02$). Interspecific variation, however, covers an order of magnitude at each elevation. Similar variations are reported by Scott & Billings (1964) for Rocky Mountain forbs and Nakhutsrishvili & Gamtsemlidze

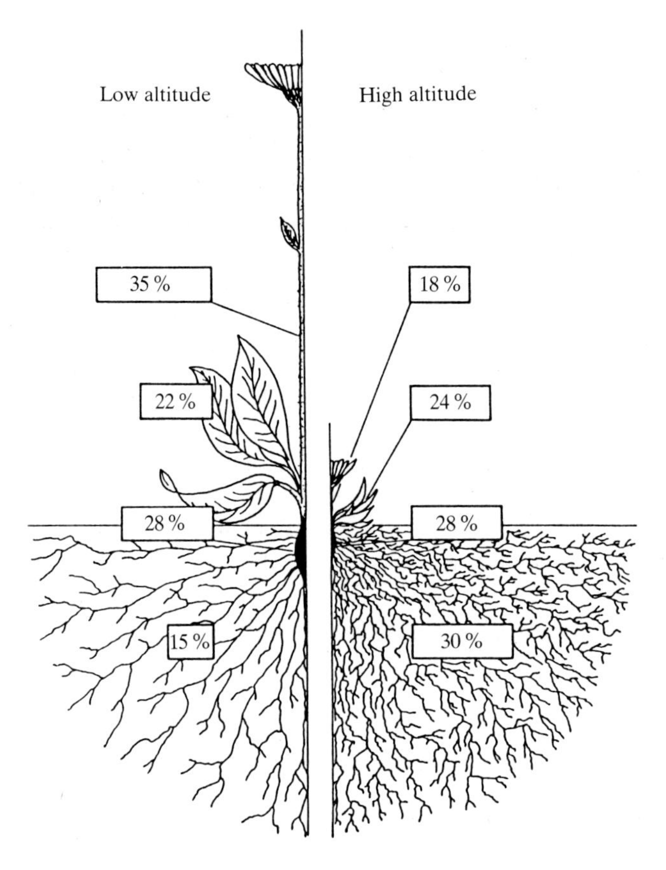

Fig. 5. A comparison of dry matter allocation in forbs from low (600 m) and high 2600 to 3200 m) altitude in the Alps. Means (±S.E.) for individual partitioning patterns in 22 species from low and 27 species from high altitude (from Körner & Renhardt, 1987).

(1984) for plants in the central Caucasus. Much more consistent and pronounced alterations are observed when root length/leaf area ratios are considered. Herbaceous plants from the highest places where vascular plants can grow in the Alps develop 4 to 5 times more root length per unit leaf area (Körner & Renhardt, 1987). The lack of effective mycorrhizal symbiosis at highest elevations (Haselwandter & Read, 1980) may explain this phenomenon.

Forbs from the subarctic belt of North Sweden exhibit similar partitioning patterns to those in the Alps (Körner & Sonesson, unpublished data). At low altitude the mean leaf weight ratios of five flowering forb species was 24 % of total dry matter, similar to the means obtained for all 49 species at low and high altitude in the Alps. At high altitude three species and one closely related pair of species found in both the subarctic belt and in the Alps have been compared. The three species represent high below-ground biomass types (*Ranunculus glacialis*, *Oxyria digyna* and *Polygonum viviparum*) and the two related species (*Cerastium uniflorum* and *C. alpinum*) represent the low below-ground biomass type. The above/below-ground ratios were not significantly different among the two cold areas (around 0·42 and 1·3 respectively for the two partitioning types). A wider sample of species indicated the same trend in root mass and root length with altitude as in the Alps.

Thus, there is a trend for increasing below-ground biomass in plants native to cold climates. However, this trend can only be detected at community level and within the same growth form. In both arctic and alpine areas herbaceous plants with rather low below-ground biomass investments have been found (e.g. *Cerastium* species, as shown above, and *Arabis* species). This indicates that high below-ground dry matter accumulation is not essential for plant performance in cold climates and, where it occurs, it is not necessarily coupled to annual carbon balance requirements. Aside from different functional advantages, there is also the possibility that large dry matter accumulation in older below-ground plant parts represents a functionally neutral side effect of retarded senescence and decreased tissue turnover due to low temperatures. In conclusion we may extend Billings's (1978) statement about alpine plants to all plants from cold regions: 'There is no single best adaptational strategy among plants species from cold climates'. Agakhanyantz & Lopatin (1978) arrived at the same conclusion for high-altitude arid conditions. This conclusion defies projections from individual plant responses (plasticity under experimental stress) to the real world of native plants and plant communities.

Leaf characteristics

The most striking morphological convergences observed among plant species from all cold areas of the globe are rolled nanophyllous, often ericoid leaves and narrow, often v-shaped graminoid leaves. They can be considered as typical elements of both high mountain and polar tundra ecosystems. However, as mentioned above, species of this type are less abundant near the higher extreme latitudinal and altitudinal limits of vascular plant existence. Here forbs and cushion plants predominate. Leaf width rarely exceeds 5 mm. The small size accompanied by strong development in the third dimension of these leaves may be advantageous with respect to mechanical stiffening, protection of stomates from wetting, buffering of short-term temperature variations, better utilization of diffuse radiation, pathogen resistance under snow, and litter capture for nutrient recycling.

Aside from the generally small size, leaves of plants from cold regions share some other structural features. In the following we try to compare quantitative leaf characteristics that reflect low temperature effects, not biased by other environmental factors like water stress. Among those, the most consistent feature is the decreased specific leaf area (SLA) in all cold regions. Table 6 shows SLAs of leaves from warm and cold sites of these various regions. Woodward's (1983) experiments with high and low-altitude *Phleum* species provide evidence that the low SLA of high-altitude plants is a heritable feature preserved under warm greenhouse conditions.

There is no simple explanation of why leaves of cold-climate plants contain more dry matter per unit area than leaves of comparable plants from warm climates. From the references in Table 6 it can be seen that leaf thickness increases by 10 to 40 % in both herbaceous and sclerophyllous species with decreasing temperature, but with no uniform temperature coefficient for different regions. Epidermal cell wall thickness increases by 20 to 200 % and cuticular thickness by 50 to 600 %. All these data refer to air temperature gradients of 8 to 15K, and to

Table 6. *Altitudinal differences of specific leaf area in plants from various climatic regions (herbaceous/sclerophyllous, $10^{-2}\,m^2\,g^{-1}$ d.w.; number of investigated species in brackets)*

	Low altitude	High altitude	Reference[a]
Austrian Alps (dicots only)	2·2 / – (21)	1·9 / – (28)	Körner *et al.* (1989)
Scottish mountains (grasses)	1·1 / – (2)	0·7 / – (2)	Woodward (1983)
New Zealand Alps (*Ranunculus, Nothofagus, Ericaceae*)	2·2 / 0·87 (2) (5)	1·2 / 0·64 (5) (5)	Körner *et al.* (1986)
Australian Snowy Mts (*Ranunculus, Viola, Eucalyptus*)	1·6 / 0·34 (2) (1)	1·5 / 0·28 (2) (1)	Körner & Cochrane (1985) and unpublished data
Mountains of New Guinea (mainly *Ranunculus, Ericaceae*)	1·8 / 0·56 (1) (4)	0·8 / 0·38 (6) (9)	Körner *et al.* (1983)
Different wet-tropical forest transects	– / 0·9– 1·3	– / 0·7	Grubb (1977)
Venezuelan Andes (*Espeletia*)	0·74 (1)	0·47 (1)	Baruch (1979)

[a] Comparison of the same or related species from different altitudinal ranges.

species that form vigorous canopies at high altitude. Conifer trees in the Krummholz belt in temperate zone mountains seem to make an exception, since some species were found to produce immature needles within the short growing season of unfavourable years and show reduced cuticular development near their upper altitudinal range (e.g. Baig & Tranquillini, 1976; Tranquillini, 1979).

In addition to the anatomical characteristics cited above, the internal/external leaf surface area ratio is substantially increased in plants native to extreme high elevations. This feature is important for CO_2 gas exchange, since it is correlated with photosynthetic capacity (Nobel & Walker, 1985). Au (1969) found that the internal/external area ratio is greater in alpine than in arctic populations of *Oxyria digyna*. Fig. 6 depicts a summary of leaf anatomical data from the Central Alps. Preliminary results from Northern Scandinavia exhibit a similar trend. Low-

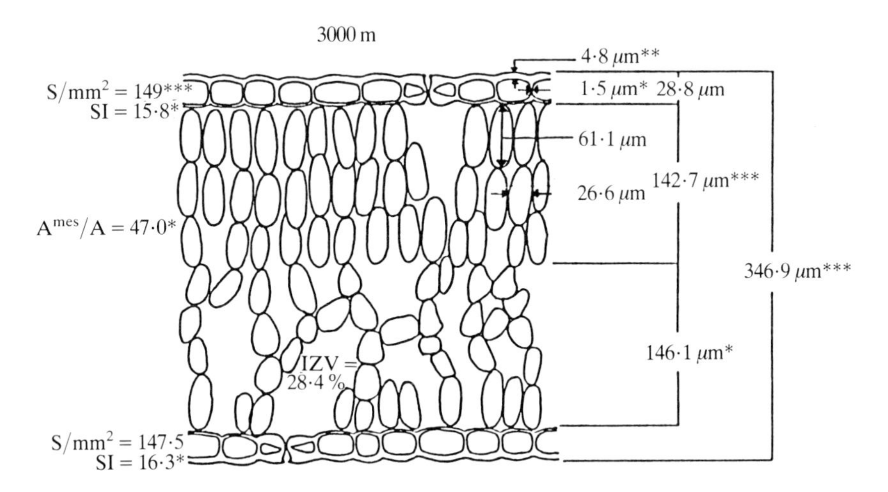

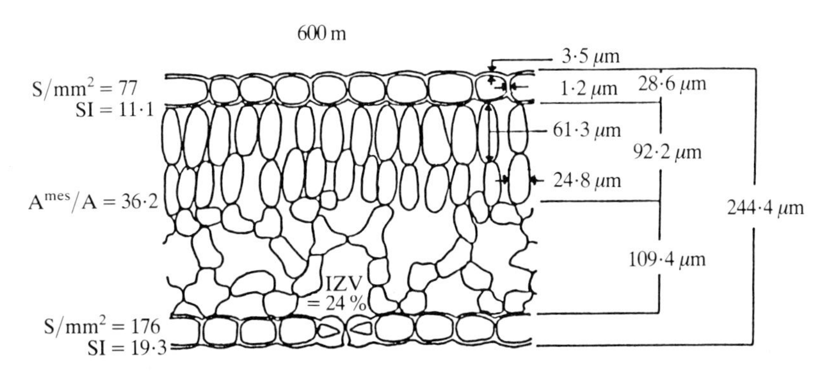

Fig. 6. Schematic cross-section of herbaceous leaves from low and high altitudes in the Alps. Numbers are means for 22 low- and 27 high-altitude species, (analysis of 5 representative leaves per species; Körner & Pelaez Mendes-Riedl, unpublished). Asterisks indicate significance levels of difference (*, **, ***, for $P < 0{\cdot}05$, 0·01, 0·001 respectively). S, stomata; SI, stomata index; A^{mes}/A, mesophyll surface area/projected leaf area; IZV, intercellular air space volume (% of total).

altitude (350 m) forbs have a mean leaf thickness of 229 μm compared to 375 μm close to the upper limit of vascular plant growth in this subarctic area (1150 m).

Summarizing the evidence from all these different regions and plant types, it appears that plants from cold environments develop thicker leaves with thicker cell walls, thicker epidermis and cuticula, lower specific leaf areas than comparable plants from warmer regions with high moisture supply. In other words the unit leaf area in cold environments tends to be more expensive in terms of dry matter than in warm areas.

The nutritional status of plants from cold environments

Low temperatures are usually assumed to limit mineral substrate weathering, ecosystem mineral nutrient recycling and microbial nitrogen fixation. Accordingly arctic tundra soils exhibit low levels of free mineral nutrients (e.g. Chapin *et al.* 1980; Chapin, 1980). Less is known about the nutritional status of soils in cold areas on high mountains. It has been shown that microbial activity drastically declines near the upper limits of plant existence (Schinner, 1982*b*; Haselwandter *et al.* 1983). Earlier studies by Rehder & Schäfer (1978) indicated that mineralization is also retarded in the lower alpine zone. A review of alpine soil characteristics by H. Franz (1980) also reveals comparatively low nutrient levels.

Mineral nutrient contents in plants from arctic and alpine areas show substantial interspecific variation, permitting no general rating of nutrient status. The crucial points are whether growth is limited to a greater extent by factors other than nutrient supply and whether nutrient demand exceeds nutrient supply? Furthermore, would increased supply be favourable for the present vegetation or would it create a different plant cover? These questions inevitably lead to an evaluation of optimality.

Two principal ways to examine this question have been adopted in the past, namely fertilization experiments and tissue assays of the mineral nutrient contents. These data were then compared to those of plants obtained from regions with supposedly high nutrient supply. However, none of these approaches is conclusive in an ecological context. Under almost all natural situations it is possible to push individual plants towards greater vegetative growth by fertilization. This is, however, by no means a justification for concluding that natural supply was suboptimal. If we were to define 'optimum supply' we need to agree upon the relevant level of complexity. Rather different definitions will be appropriate for individual species (that might be outcompeted by others) or whole communities. Even at the ecosystem level, it is not self-evident that increased productivity or higher rates of turnover will reflect a greater degree of optimality. So, in the case of persistence, for example, the opposite is true in most instances (Ellenberg, 1958; Bannister, 1976; Grime, 1977; Ernst, 1978; E. Franz, 1981; Körner, 1984; Rapport *et al.* 1985).

An example that illustrates the discrepancy between physiological optima for growth of individual species and their long-term success in a certain habitat was obtained in a five-year fertilization experiment in alpine dwarf shrubs on acidic soils (Körner, 1984). Similar to findings in the arctic tundra (e.g. Chapin *et al.* 1975), moderate amounts of additional mineral nutrients stimulated growth, altered shoot structure and thus canopy geometry and canopy climate, and changed leaf turnover. Phenological rhythms were disturbed, bud break accelerated and bud ripening delayed, causing increased frost damage, a consequence described also by Bell & Bliss (1980). The chemical composition of the plant reserves was also altered, with less fat and more starch. Perhaps related to all these changes, resistance to fungal pathogens, such as snow mould, declined, leading to biomass losses and final extinction of one species (*Loiseleuria procumbens*). Thus, initial growth stimulation by increased nutrient availability cannot be rated as an indication of nutrient limitation under natural conditions in these species. Responses of plants from higher altitudes, in particular by herbaceous plants from the highest peaks, where competitional barriers are almost absent, are unknown. This example illustrates the great difficulties encountered in the evaluation of nutrient shortage in wild plants, particularly under stress dominated, cold conditions. The following interpretations should therefore be treated with great caution.

Available data on mineral nutrient contents of tissues from many arctic and high altitude plants contrast the suspected supply limitation, particularly in perennial forbs and less so in shrubs and tussocks. A survey of the nutritional status of herbaceous and sclerophyllous plant species from low and high altitude in various mountain regions indicates no decline in nitrogen and phosphate contents per dry weight with increasing altitude. In several instances concentrations increase (literature reviewed by Körner *et al.* 1986). A comparison among 49 herbaceous plant species from low and high elevations in the Alps shows that mean nitrogen and phosphate contents of leaves increase by 17 and 50 % respectively over an elevational gradient from 600 to 3000 m. Due to declining specific leaf area, the mineral nutrient contents per unit leaf area increase even more. Fig. 7 shows the frequency distribution of nitrogen contents in plants from 600 and 3000 m in the Alps. Some of the species from uppermost sites were found to contain up to 5 % nitrogen, a value hardly reached by crop plants.

Similarly, Alaskan tundra plants often exhibit N-contents higher than in comparable plants from temperate regions (Chapin *et al.* 1975; Chapin & Oechel, 1983). The latter authors also showed that phosphate can be more limiting than nitrogen, and – at least in *Carex aquatilis* – phosphate acquisition is not impaired by low temperature effects on the uptake capacity of roots, but by the content and/or low rate of release from the mineral and dead organic substrate. Chapin *et al.* (1975) conclude that tundra plants are adapted to low nutrient availability by production of relatively small biomass with high nutrient contents and presumably high metabolic efficiency. Whether this is an 'adaptation' in the real sense, or a reflection of the predominance of other growth constraints operating in the same direction remains to be answered.

In seasonal mountain climates of the temperate zone, annual inputs of N by

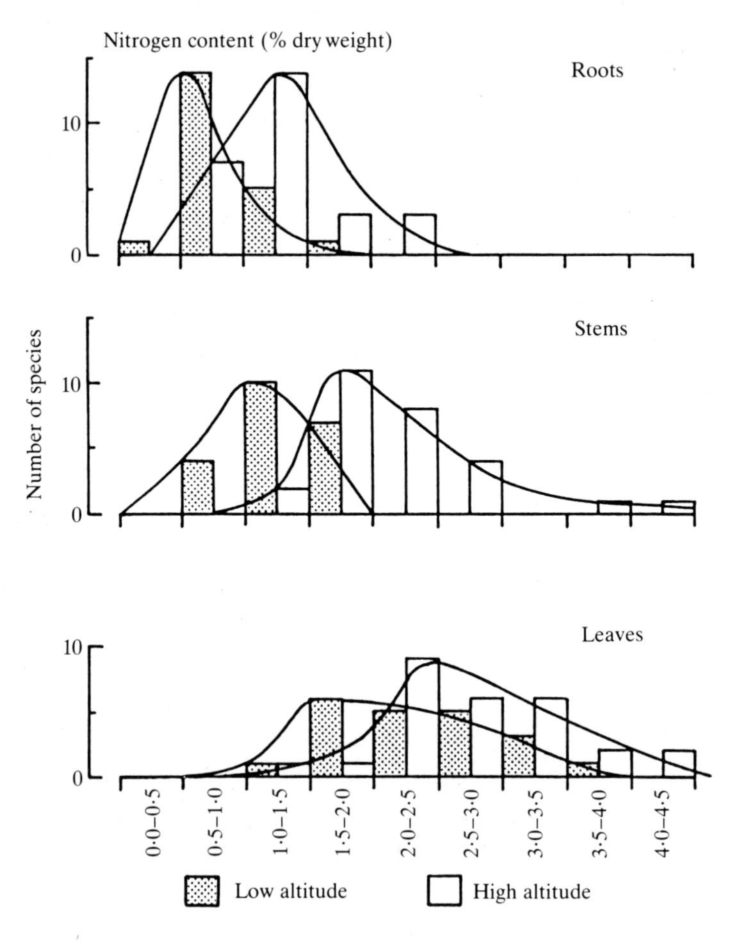

Fig. 7. Peak season nitrogen content of leaves, reproductive parts, storage organs plus fine roots in herbaceous plant species from low altitude (600 m) and high altitude (3000 m). (Körner & Renhardt, unpublished).

precipitation and mineralization may be sufficient to supply plants for a 2- to 3-month annual activity (Haselwandter *et al.* 1983). This may hold also for other nutrients, since it has been shown that annual aeolic mineral dust deposition at high altitudes in the Alps is substantial (cf. H. Franz, 1979). In this respect conditions are probably less favourable in cold regions with very low precipitation or little bare ground for dust abrasion and in high tropical mountains with year-round plant nutrient demand. In conclusion, mineral nutrient supply does not necessarily limit plant growth in cold environments. In fact, limited growth of plants from cold climates may prevent nutrient dilution within the plant (Chapin, 1983; Körner & Woodward, 1987).

The temperature response of metabolic processes in cold regions

Photosynthetic responses

Is carbon assimilation in cold regions limited by low temperature? In order to answer this question, one needs to know at least (1) the temperature response curves of CO_2-uptake under various quantum flux densities, (2) the relative limitations of CO_2-uptake by quantum supply, and (3) the frequency distribution of leaf temperatures under various light conditions in the field. If this information is incomplete, then predictions are impossible, a fact often overlooked when temperature adaptation is discussed. Discrepancies between light saturated temperature response and prevailing field leaf temperatures are irrelevant, when the plant is not light saturated under natural canopy situations, since the temperature response varies with light intensity. For high altitude plants the analysis provided for a predominant species of the central Alps, *Carex curvula*, serves as an example (Körner, 1982).

The annual carbon yield of this sedge was simulated using a simple statistical model of the light- and temperature-response, leaf angles and light extinction in the canopy, and real, as well as potentially optimal light- and temperature regimes in the canopy. The discrepancies between real and potential (= 100 %) yield are expressed as percentage 'losses'. Although this approach is rather simple and ignores the possibility that the plant may not be 'designed' to handle such a potential carbon yield in the long term, it is superior to mere comparisons of temperature optima and field temperature. Fig. 8 shows that carbon yield of this sedge within the snow-free period is mainly limited by irradiance. Suboptimal temperatures account for only 7·6 % 'losses' as compared to the theoretical

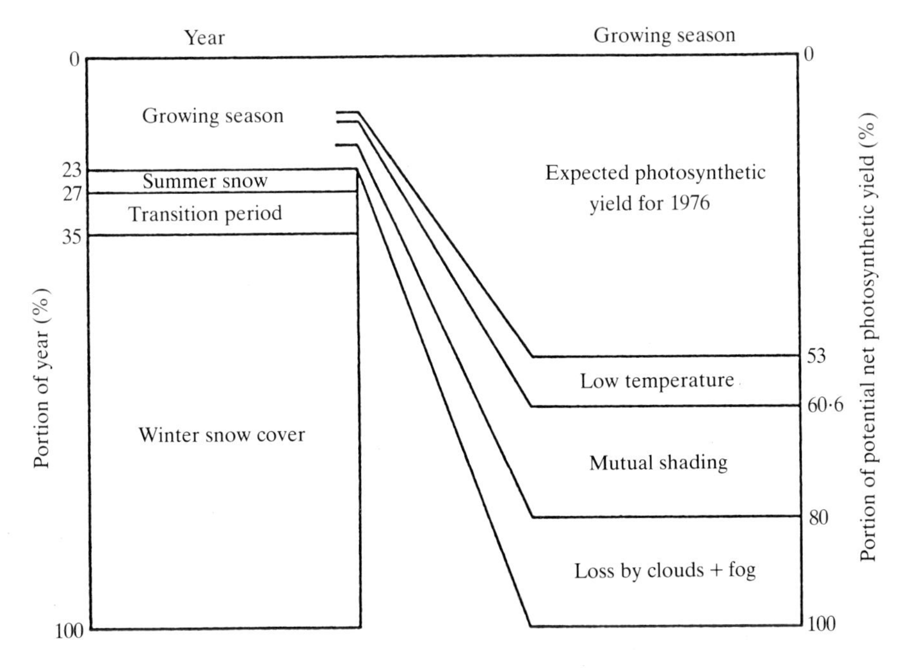

Fig. 8. Estimation of the relative limitations of the annual carbon yield of *Carex curvula* in 2300-m altitude in the Alps. Left: indirect effects of temperature by the duration of the growth period. Right: direct effects of temperature and light regime within the growth period (Körner, 1982).

maximum under the given light regime. The underlying temperature response of this sedge is such that saturating light conditions are optimally used (real = potential rate). Under light levels that permit only one tenth of maximum CO_2-uptake, however, prevailing leaf temperatures reduce rates by an additional 70%. The light saturated temperature optimum of *C. curvula* is 22·5°C compared to the mean air temperature for July and August of 4·5°C. Scott & Billings (1964) also concluded that temperature does not play a particularly limiting role to CO_2-uptake of alpine plants during the growth period. With respect to the high leaf temperatures under radiation conditions (see climate section) this is not surprising. On an annual basis (Fig. 8) the major overriding limitation is the duration of the growing season, an indirect effect of low temperature. An interesting analogy is provided by Nobel (1988) for hot desert plants. Nobel found that temperature is not a factor of direct significance for carbon uptake in *Agave* species in their natural environment.

Subsequent investigations in 19 herbaceous plant species, restricted either to high or low altitude in the Alps, revealed similar results (Körner & Diemer, 1987). As in *Carex curvula*, mean light saturated temperature optima reflect the warmest canopy temperatures under radiation conditions and are 23·9°C at low and 21·2°C at high altitude, thus differing by only 2·7 K but for sites at altitudes of 600 and 2600 m. Similar results were obtained by Moser *et al.* (1977) for the *in situ* temperature optimum of the highest ranging vascular plant of Europe, *Ranunculus glacialis*, which is 23°C. Again it needs to be emphasized that the mean air temperature in summer is 18°C at low and 2 to 5°C at high altitude. Temperature optima in arctic vascular plants tend to be lower and range between 15 and 20°C (Tieszen *et al.* 1981). The lowest optima are found in cryptogams such as arctic mosses. Lösch *et al.* (1983) report maximum light saturated rates of CO_2-uptake for snowbed bryophytes in northern Sweden between 6 and 11°C, and 60% of photosynthetic capacity have been recorded at 0°C in Alaskan mosses by Oechel & Collins (1976).

The responses for low or prostrate vascular species differ markedly from observations on tall plants with temperatures which are closely coupled to the free atmosphere. Examples are provided by the altitudinal change in the temperature optimum of photosynthesis in *Eucalyptus pauciflora* (Slatyer, 1978; Fig. 9; see also Fryer & Ledig, 1972). While the response of trees, such as *E. pauciflora* follows the elevational decline of mean maximum air temperature, low herbaceous plants, as explained above, do not show such a clear trend.

The low-temperature limit for net photosynthetic CO_2 uptake in freezing-intolerant leaves is very close to the low-temperature limit of leaf survival (see p. 31). However, freezing-tolerant leaves also show no CO_2 uptake after ice formation (Larcher & Bauer, 1981). High alpine forbs are able to produce a CO_2 gain down to about -5 ± 1°C (Pisek *et al.* 1967) and similar thresholds are found in arctic plants (Tieszen *et al.* 1981).

Little is known about *in situ* after-effects of low temperature stress on photosynthesis of plants from cold climates. Temporal depressions of photosyn-

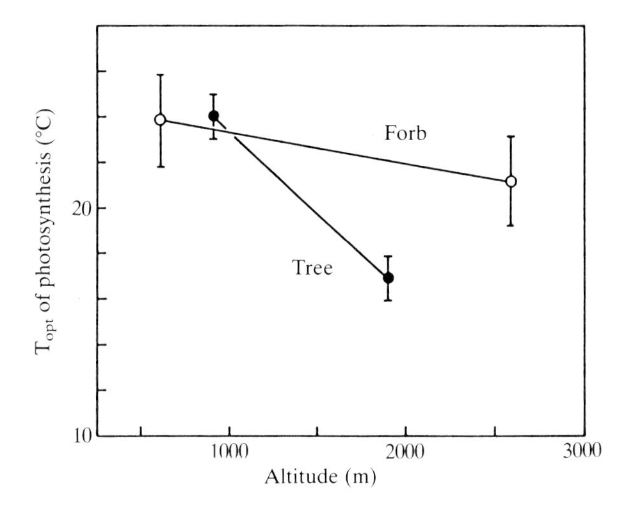

Fig. 9. The altitudinal variation of the temperature optimum of photosynthetic CO_2-uptake in two contrasting growth forms: The tree, in this case *Eucalyptus pauciflora* in the Australian Snowy Mountains (from Slatyer, 1978), and herbaceous perennial plants in the Alps (from Körner & Diemer, 1987).

thesis can result from both impairment of chloroplast reactions and stomatal closure (Larcher & Bauer, 1981, Öquist, 1983; Öquist & Martin, 1986). After-effects depend on the duration and severity of subfreezing temperatures and the hardening state, but tend to be most significant in trees, less pronounced in shrubs and often negligible in forbs from cold habitats. The time for full restoration of photosynthetic performance after non-injuring frost during the active growth period are in the order of several hours or less (Larcher, 1981*b*). Schulze *et al.* (1985) found that frozen leaves of afro-alpine giant rosettes regained full photosynthetic capacity immediately after thawing.

Photosynthetic capacity

Although alpine and arctic plants differ in many respects from taxonomically comparable plants from warmer regions, this is not the case for photosynthetic capacities, which are very similar under ambient CO_2 levels. Average rates for alpine and arctic perennial forbs and deciduous shrubs are in the order of $18 \pm 5\ \mu mol\ m^{-2}\ s^{-1}$ and 3 to $5\ \mu mol\ m^{-2}\ s^{-1}$ in evergreen shrubs (Grabherr, 1976; Larcher, 1977; Moser *et al.* 1977; Tieszen *et al.* 1981; Eckardt *et al.* 1982; Körner, 1982; Karlsson, 1985; Körner & Diemer, 1987). Since plants from high elevations live under lower partial pressure of CO_2 such high rates can only be achieved by higher efficiency of CO_2 uptake (Körner & Diemer, 1987). Again a different behaviour is reported for evergreen trees, which retain lower photosynthetic capacities at treeline than at lower altitudes (Slatyer & Morrow, 1977; Tranquillini & Havranek, 1985).

An important aspect for future plant growth in cold climates is the present increase of global CO_2. Sionit *et al.* (1981) demonstrated that elevated CO_2 may, to some extent, compensate for the adverse effects of low temperatures on growth. Körner & Diemer (1987) suggested that such effects are likely to be pronounced in high altitude plants, while no significant short term influences could be detected in whole stands of native arctic tundra plants in growth chambers (Billings *et al.* 1984) and field experiments (Tissue & Oechel, 1987).

Mitochondrial respiration

It is generally accepted that plants from cold climates operate with enhanced specific metabolic activities, as expressed by higher rates of mitochondrial respiration in arctic and alpine plants (Semikhatova & Gratchyeva, 1962; Semikhatova, 1965; Bliss, 1971; Stewart & Bannister, 1974; Ollerenshaw *et al.* 1976; Larcher, 1977, 1983; McNulty & Cummins, 1987). However, these experimental results for equal temperatures do not necessarily imply that plants from cold climates have higher absolute losses of carbon through respiration, since night-time and soil temperatures in the field are low. Warm acclimation of plants from cold habitats causes a rapid decline of temperature-specific mitochondrial respiration, leading to similar rates under largely differing growth conditions (Fig. 10; cf. Larcher, 1980). The mean night-time carbon losses of leaves of *Ranunculus nemorosus* at 600-m altitude (13 °C) and of *Ranunculus glacialis* from 2600 m (3 °C) in the Alps were similar, although specific respiration rates at 10 °C were almost twice as high in the alpine species. The same was found in thalli of the

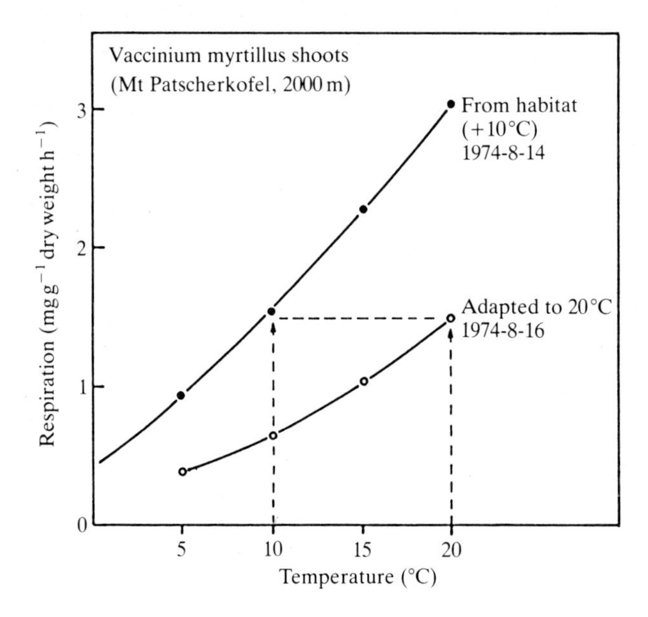

Fig. 10. Temperature response of mitochondrial respiration in *Vaccinium myrtillus*. ●, in its natural cool-adapted state in an alpine habitat; ○, after acclimation to higher temperatures, characteristic of a lowland site. Note similar rates at the temperature regimes characteristic of high and low altitudes (dashed lines; Larcher & Huber, unpublished).

Andean lichen *Dictyonema glabratum* from different altitudes (Larcher & Vareschi, 1988).

In conclusion, photosynthetic rates in plants from high altitudes are not significantly constrained by temperature. The same conclusion was drawn by Chapin (1983) in lowland, arctic tundra plants. Dark respiration rates show a homeostatic (i.e. compensatory) trend in response to low temperatures. In cold seasonal climates the major effect of low temperature on carbon assimilation on an annual basis is indirect, through reduction of the duration of the snow free period. This is also illustrated by Fig. 8.

Leaf growth

If we recall the conclusions of the previous sections, one might have problems understanding why plants in cold environments do not produce more biomass and are so small. Thermally the low vegetation is effectively uncoupled from low ambient temperatures during much of the day-time (particularly at mountains of lower latitudes), severe frost damage is unlikely, nutritional status is not so poor as often assumed, relative carbohydrate investments in assimilating tissues are similar to those in warmer regions (as long as similar growth forms are compared) and photosynthetic performance is excellent. What remains and what explains the observed slow rates of growth? What do we know about the growth process *per se*?

These questions represent the truly relevant points. Unfortunately, however, little is known about processes that lie between carbon assimilation and final biomass production. This is not only true for arctic and alpine plants (e.g. Monteith & Elston, 1983). Growth analysis provides some hints however. It was shown by Körner & Woodward (1987) that grass species at 3000-m altitude grow mainly during warm daylight hours. Normal, i.e. low, night-time temperatures inhibit leaf expansion. In contrast, low-altitude grasses were found to grow day and night. Unlike the photosynthetic response to temperature, the response of leaf expansion differs substantially between high- and low-altitude species (Fig. 11). Plant species from high altitudes exhibit inherently lower rates of leaf expansion than related taxa from low elevations, but are able to grow near 0°C (see also Billings, Peterson & Shaver, 1978). In both groups, low temperature thresholds for leaf extension are 6 to 8 K higher than photosynthetic thresholds (Woodward *et al.* 1986). Ollerenshaw *et al.* (1976), Woodward (1979) and Pollock & Eagles (1988) have found similar differences between grasses from cold and warm regions. In terms of Grime's (1977) concept of plant strategies, this behaviour is typical for plants from stress-dominated environments.

In Fig. 12, the light-saturated photosynthetic and growth responses to temperature of *Poa alpina* have been combined to illustrate the relative significancc of low temperatures on both processes. The combined effects of opposing exponential and rate-limiting saturation functions involved in the photosynthetic metabolism cause the bell-shaped temperature response. In both alpine and low-altitude forbs 95 % of maximum photosynthesis occurs over a 10 K wide range. If we consider the temperature range below 15°C it becomes clear that low temperature limitations of growth will always precede temperature limitations of the carbon fixation process, as long as light supply is sufficient (cf. Wardlaw, 1976). The different temperature response of growth and of photosynthesis is a well-known

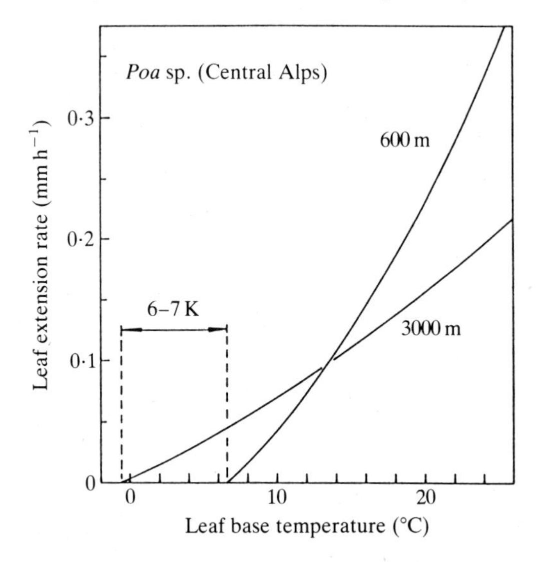

Fig. 11. Temperature response of leaf extension in *Poa* species at different altitudes (from Körner & Woodward, 1987).

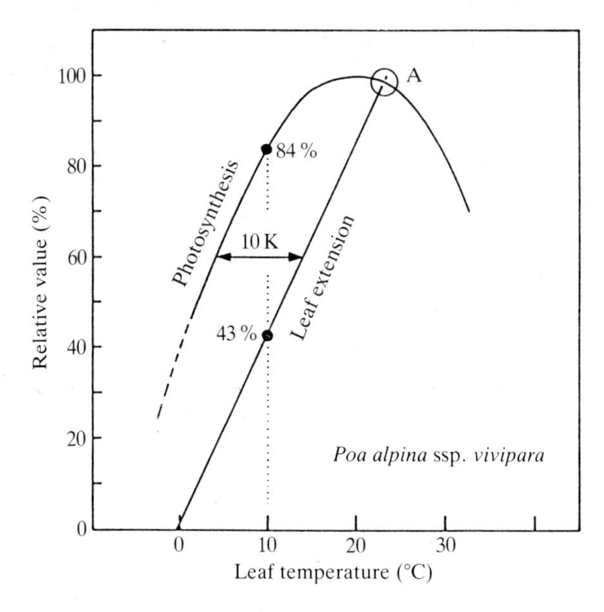

Fig. 12. *In situ* temperature response of photosynthesis and leaf extension growth in *Poa alpina* at high altitude in the Alps (from data of Körner & Woodward, 1987 and Körner & Diemer, 1987). Point A indicates the coincidence of maximum rate of photosynthesis and extension growth observed in the field. Further explanation in the text.

phenomenon. What makes this difference so significant for cold regions is its magnitude at temperatures just a few degrees above zero, which prevail for a substantial part of the life-time of plants.

At 0°C this difference becomes most obvious, when growth ceases but photosynthesis still occurs at 35 % of maximum in *Poa alpina* (Fig. 12). Leaf temperatures at high altitude will be more often around 10°C under rate-saturating radiation (*ca* half of full mid-day radiation). At this temperature, photosynthesis of *P. alpina* operates at 84 % of full capacity, while 43 % of the mean maximum growth rate that was observed in the field is reached. Only at 23°C do both processes operate at nearly 100 % rate (point A in Fig. 12). Here we ignore the rare instances under which leaf base temperature will exceed 23°C in 3000-m altitude and even greater growth rates could be achieved.

If we compare carbon assimilation and carbon investment rates in the leaf at point A (23°C), a great disparity between photosynthesis and growth is still found. For the purpose of this estimate the extension rates were tranformed to mass increments using measurements of leaf width and specific leaf area. This provides a maximum rate of leaf growth at 15 $\mu g\,h^{-1}$ per leaf. The photosynthetic rate of one leaf (2×30 mm) amounts to 144 μg CO_2 or 88 μg dry matter per hour. Hence, at 23°C an equivalent of only about 17 % of one leaf's carbon gain is used for growth of another leaf. At temperatures below 15°C this proportion will be even smaller, and approximately halved. The disparity is further increased by the ratio of mature to growing leaves in midsummer. Consequently, leaf carbon investment rates are substantially less than 10 % of the rates of leaf carbon gain and more than 90 % of assimilates need to be allocated to sinks other than instantaneous leaf dry matter increment (root growth, reserve pools, respiratory losses).

Since plants from cold climates show slow growth despite high photosynthetic carbon gain, other processes seem to exhibit superior control over growth. Experiments by Ollerenshaw & Baker (1981), Chapin & Oechel (1983) and Woodward *et al.* (1986) have also demonstrated that growth dynamics in herbaceous plants are genetically controlled. Similarly, forest geneticists have shown that inherently reduced growth is also a typical feature of conifer trees from cold areas (Langlet, 1971; Sakai & Larcher, 1987).

Cellular aspects of growth

Low growth rates may result from (1) low rates of cell division or expansion, (2) temporal cessation of cell production, (3) production of smaller cells (at equal rates of division) or (4) combinations of one or the other. Plants from cold environments, as well as their composite organs, are comparatively small (p. 35). Our results show that these size differences between cold and warm regions are attributable to cell number, rather than cell size. Anatomical data for leaves of related plant species from various cold and warm sites reveal no significant difference in the size of palisade and epidermis cells. In several instances species from colder regions had even larger cells. For instance *Ranunculus glacialis*,

Saxifraga rivularis and *Oxyria digyna*, which are typical herbaceous species of the coldest areas in the northern hemisphere, have extraordinarily large palisade cells, namely 88×31, 93×38 and 148×51 μm (length × width) compared to typical values of about 55×25 μm in most other lowland and high altitude herbaceous species. Fig. 13 illustrates the great uniformity of mean palisade cell size among plant populations from different altitudes and different latitudes.

Hence, cell multiplication seems to play a central role among the growth processes in plants from cold climates. It is still unclear whether the rate of cell division is the primary limiting factor, or whether slow expansion of newly formed cells feeds back negatively on the rate of mitosis. Whatever the cause, the number of cell divisions involved in the formation of a new leaf is drastically reduced in cold climates. Despite the ten times smaller size, the time required for a single leaf's development is equal or greater than in warm climates (Diemer & Körner, unpublished). Analysing the temperature dependency in horticultural plants, Paul (1985) arrived at the same conclusion, namely that inhibition of cell division is a limiting process in growth.

The rate of cell division is tightly regulated by temperature and depends in the size of the genome to be replicated (e.g. Francis & Barlow, 1988). Adaptations to growth in cold climates might therefore encounter positioning of meristems close to the warm soil surface – which is the case – and reduction of genome size. Bennett (1987) showed that genome size does decline in crop plants at their northern latitudinal limit. However, as shown above, plants native to cold regions may exhibit very large cells, suggesting larger genome size according to Grime (1983).

Unfortunately, developmental physiology is not a favoured research area within ecologically orientated plant biology (Larcher, 1983). A more fundamental

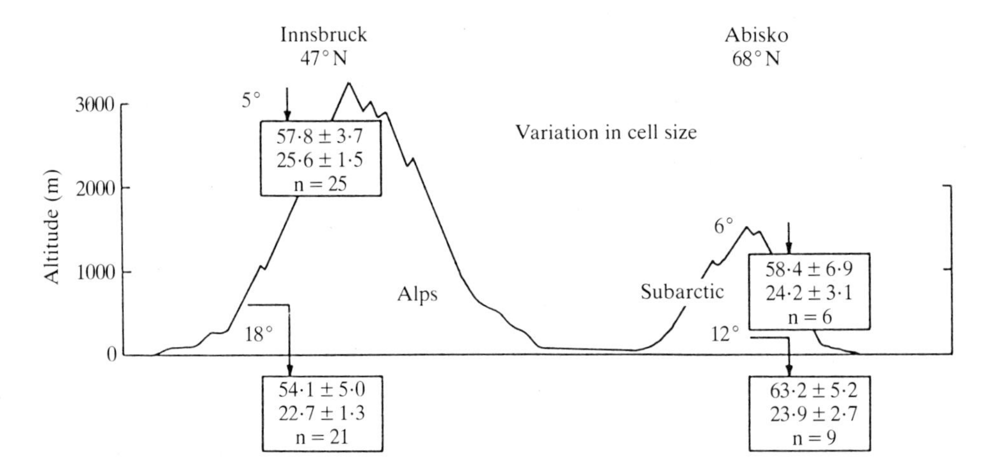

Fig. 13. The variation of cell size in plants from thermally diverse habitats in the Alps and northern Sweden (Körner & Pelaez Menendez-Riedl and Körner & Sonesson, unpublished). Length and width of cells (10^{-6} m), standard errors and number of observations (n), mean temperature of the warmest month (°C).

understanding of plant life in cold climates will depend on a better understanding of cellular processes in growth and development.

We suppose that plants from cold climates predominantly experience a disparity between high carbon assimilation rates and slow investment potential in growth. If this is true, effective mechanisms to channel energy overflow must exist and operate. One of these channels may be accumulation of fat. Another sink may be high respiratory consumption, particularly in roots. Lambers (1985) illustrated consistent negative correlations between respiratory losses and yield, particularly when the cyanide-resistant, alternative pathway is considered. This energetically wasteful pathway possibly plays a greater role in plants from cold climates (Lambers, 1985; McNulty & Cummins, 1987).

Since we are not used to accepting wasteful processes, particularly in an environment that presents serious constraints on survival, we may ask for possible functional benefits. One important point may be a higher buffering capacity to disturbances of all kinds, including direct low temperature effects (Huner, 1985; Lambers, 1985).

We propose that developmental processes play an equal or even more significant role at low temperatures than resource acquisition. More emphasis on developmental aspects of plant ecology is needed before we can leave the grounds of speculation regarding the control of slow growth and dwarf forms in cold climates.

The conceptual framework leading to this paper arose while the authors received generous support by the Austria Fonds zur Förderung der wissenschaftlichen Forschung (projects 637, 5597 and 6252). The unpublished work in the subarctic zone was supported by the Abisko Research Station of the Royal Swedish Academy of Science (Prof. M. Sonesson). We thank all those colleagues and students at the Botany Department in Innsbruck who have contributed to our understanding of mountain ecology over the past twenty years. The assistance by M. Diemer during the preparation of the manuscript and the helpful comments on the final text by F. I. Woodward are gratefully acknowledged.

References

Agakhanyantz, O. E. & Lopatin, I. K. (1978). Main characteristics of the ecosystems of the Pamirs, USSR. *Arctic Alpine Res.* **10**, 397–407.

Alexandrova, V. L. & Mateeva, N. V. (1979). *Arktichenskie tyndri i polariye pustyni Taimyra* (*Arctic Tundra and Polar Deserts of Taimyr*). Leningrad: Nauka.

Au, S. F. (1969). Internal leaf surface and stomatal abundance in arctic and alpine populations of *Oxyria digyna*. *Ecology* **50**, 131–134.

Baig, M. N. & Tranquillini, W. (1976). Studies on upper timber-line: morphology and anatomy of Norway spruce (*Picea abies*) and stone pine (*Pinus cembra*) needles from various habitat conditions. *Can. J. Bot.* **54**, 1622–1632.

Bannister, P. (1976). *Introduction to Physiological Plant Ecology*. Oxford, London, Edinburgh, Melbourne: Blackwell Sci. Publ.

Baruch, Z. (1979). Elevational differentiation in *Espeletia schultzii* (Compositae), a giant rosette plant of the venezuelan paramos. *Ecology* **60**, 85–98.

BECK, E. (1988). Cold tolerance. In *Tropical Alpine Environments, Plant Form and Function* (ed. P. W. Rundel), Berlin: Springer (in press).
BECK, E., SCHULZE, E.-D., SENSER, M. & SCHEIBE, R. (1984). Equilibrium freezing of leaf water and extracellular ice formation in afro-alpine 'giant' plants. *Planta* **162**, 276–282.
BELL, K. L. & BLISS, L. C. (1980). Autecology of *Kobresia bellardii*: Why winter snow accumulation limits local distribution. *Ecol. Monogr.* **49**, 377–402.
BENNETT, M. D. (1987). Variation in genomic form in plants and its ecological implications. *New Phytol.* **106**, 177–200.
BIEBL, R. (1968). Über Wärmehaushalt und Temperaturresistenz arktischer Pflanzen in Westgrönland. *Flora, B* **157**, 327–354.
BILLINGS, W. D. (1978). Aspects of the ecology of alpine and subalpine plants. In *High Altitude Revegetation Workshop* (ed. S. T. Kenny), Colorado Water Resources Res. Inst., Information 28, pp. 1–16. Ft. Collins: Env. Resources Center, Colorado State University.
BILLINGS, W. D. & MOONEY, H. A. (1968). The ecology of arctic and alpine plants. *Biol. Rev.* **43**, 481–529.
BILLINGS, W. D., PETERSON, K. M., LUKEN, J. O. & MORTENSEN, D. A. (1984). Interaction of increasing atmospheric carbon dioxide and soil nitrogen on the carbon balance of tundra microcosms. *Oecologia (Berlin)* **65**, 26–29.
BILLINGS, W. D., PETERSON, K. M. & SHAVER, G. R. (1978). Growth, turnover, and respiration rates of roots and tillers in tundra graminoids. In *Vegetation and Production Ecology of an Alaskan Arctic Tundra* (ed. L. L. Tieszen), Ecol. Studies 29, pp. 415–434. New York, Heidelberg, Berlin: Springer.
BLISS, L. C. (1971). Arctic and alpine plant life cycles. *A. Rev. Ecol. Syst.* **2**, 405–438.
BLISS, L. C., HEAL, O. W. & MOORE, J. J. (1981). *Tundra Ecosystems: a Comparative Analysis*. Int. Biol. Programme 25. Cambridge: Cambridge University Press.
BODNER, M. & BECK, W. (1987). Effect of supercooling and freezing on photosynthesis in freezing tolerant leaves of afro-alpine 'giant rosette' plants. *Oecologia (Berlin)* **72**, 366–371.
BRECKLE, S. W. (1973). Mikroklimatische Messungen und ökologische Beobachtungen in der alpinen Stufe des afghanischen Hindukusch. *Bot. Jb. Syst.* **93**, 25–55.
CERNUSCA, A. (1976). Energie- und Wasserhaushalt eines alpinen Zwergstrauchbestandes während einer Föhnperiode. *Arch. Met. Geoph. Biokl., Ser.* B **24**, 219–241.
CHAPIN, III F. S. (1980). The mineral nutrition of wild plants. *A. Rev. Ecol. Syst.* **11**, 233–260.
CHAPIN, III F. S. (1981). Field measurements of growth and phosphate absorption in *Carex aquatilis* along a latitudinal gradient. *Arctic Alpine Res.* **13**, 83–94.
CHAPIN, III F. S. (1983). Direct and indirect effects of temperature on arctic plants. *Polar Biol.* **2**, 47–52.
CHAPIN, III F. S., MILLER, P. C., BILLINGS, W. D. & COYNE, P. I. (1980). Carbon and nutrient budgets and their control in coastal tundra. In *An Arctic Ecosystem: The Coastal Tundra at Barrow, Alaska* (ed. J. Brown, P. C. Miller, L. L. Tieszen & F. L. Bunnell), pp. 458–490. Stoudsburg (PA): Dowden, Hutchinson & Ross.
CHAPIN, III F. S. & OECHEL, W. (1983). Photosynthesis, respiration, and phosphate absorption by *Carex aquatilis* ecotypes along latitudinal and local environmental gradients. *Ecology* **64**, 743–751.
CHAPIN, III F. S., VANCLEVE, K. & TIESZEN, L. L. (1975). Seasonal nutrient dynamics of tundra vegetation at Barrow, Alaska. *Arctic Alpine Res.* **7**, 209–226.
COE, M. J. (1969). Microclimate and animal life in equatorial mountains. *Zoologica Africana* **4**, 101–128.
DAHL, E. (1951). On the relation between summer temperatures and the distribution of alpine vascular plants in the lowlands of Fennoscandia. *Oikos* **3**, 22–52.
ECKARDT, F. E., HEERFORDT, L., JORGENSEN, H. M. & VAAG, P. (1982). Photosynthetic production in greenland as related to climate, plant cover and grazing pressure. *Photosynthetica* **16**, 71–100.
ELLENBERG, H. (1958). Bodenreaktion (einschließlich Kalkfrage). In *Handbuch Pflanzenphys. 4, Die mineralische Ernährung der Pflanze* (ed. W. Ruhland), pp. 638–708. Berlin: Springer.
ERNST, W. (1978). Discrepancy between ecological and physiological optima of plant species. A re-interpretation. *Oecol. Pl.* **13**, 175–188.
FRANCIS, D. & BARLOW, P. W. (1988). Temperature and the cell cycle. In *Plants and*

Temperature, Symp. Soc. Exp. Biol., vol. 42 (ed. S. P. Long & F. I. Woodward), pp. 181–202. Cambridge: The Company of Biologists Limited.

FRANZ, E. H. (1981). A general formulation of stress phenomena in ecological systems. In *Stress Effects on Natural Ecosystems* (ed. G. W. Barrett & R. Rosenberg), pp. 49–54. New York: John Wiley & Sons.

FRANZ, H. (1979). *Ökologie der Hochgebirge.* Stuttgart: Ulmer.

FRANZ, H. (1980). Die Gesamtdynamik der untersuchten Hochgebirgsböden. In *Untersuchungen an alpinen Böden in den Hohen Tauern 1974–1978, Stoffdynamik und Wasserhaushalt* (ed. H. Franz), 3, 277–286. Veröff. Österr. MaB-Hochgebirgsprogramm Hohe Tauern.

FRYER, J. H. & LEDIG, F. T. (1972). Microevolution of the photosynthetic temperature optimum in relation to the elevational complex gradient. *Can. J. Bot.* **50**, 1231–1235.

GATES, D. M., HIESEY, W. M., MILNER, H. W. & NOBS, M. A. (1964). Temperatures of *Mimulus* leaves in natural environments and in a controlled chamber. *Carnegie Instn Yb.* **63**, 418–426.

GAUSLAA, Y. (1984). Heat resistance and energy budget in different scandinavian plants. *Holarctic Ecology* **7**, 1–78.

GOLDSTEIN, G., RADA, F. & AZOCAR, A. (1985). Cold hardiness and supercooling along an altitudinal gradient in andean giant rosette species. *Oecologia (Berlin)* **68**, 147–152.

GRABHERR, G. (1976). Der CO_2-Gaswechsel des immergrünen Zwergstrauches *Loiseleuria procumbens* (L.)DESV. in Abhängigkeit von Strahlung, Temperatur, Wasserstress und phänologischem Zustand. *Photosynthetica* **11**, 302–310.

GRACE, J. (1987). Climatic tolerance and the distribution of plants. *New Phytol.* **106**, 113–130.

GRIME, J. P. (1977). Evidence for the existence of three primary strategies in plants and its relevance to ecological and evolutionary theory. *Am. Natural.* **111**, 1169–1194.

GRIME, J. P. (1983). Prediction of weed and crop response to climate based upon measurements of nuclear DNA content. *Aspects Appl. Biol.* **4**, 87–98.

GRUBB, P. J. (1977). Control of forest growth and distribution on wet tropical mountains: with special reference to mineral nutrition. *A. Rev. Ecol. Syst.* **8**, 83–107.

HASELWANDTER, K., HOFMANN, A., HOLZMANN, H. P. & READ, D. J. (1983). Availability of nitrogen and phosphorus in the nival zone of the Alps. *Oecologia (Berlin)* **57**, 266–269.

HASELWANDTER, K. & READ, D. J. (1980). Fungal associations of roots of dominant and subdominant plants in high-alpine vegetation systems with special reference to mycorrhiza. *Oecologia (Berlin)* **45**, 57–62.

HEDBERG, O.(1964). Features of afroalpine ecology. *Acta Phytogeographica Suecica* **49**, 8–89.

HEDBERG, I. & HEDBERG, O. (1979). Tropical-alpine life forms of vascular plants. *Oikos* **33**, 297–307.

HUNER, N. PA. (1985). Morphological, anatomical, and molecular consequences of growth and development at low temperatures in *Secale cereale* L. *cv. Puma. Am. J. Bot.* **72**, 1290–1306.

IVES, J. D. & BARRY, R. G. (1974). *Arctic and Alpine Environments.* London: Methuen.

KAINMÜLLER, CH. (1975). Temperaturresistenz von Hochgebirgspflanzen. *Anzeiger math.-naturwiss. Klasse Österr. Akad. Wiss.* **7**, 67–75.

KALLIO, P. (1984). The essence of biology in the north. *Nordia* **18**, 53–65.

KARLSSON, P. S. (1985). Photosynthetic characteristics and leaf carbon economy of a deciduous and an evergreen dwarf shrub: *Vaccinium uliginosum* L. and *V. vitis-idaea* L. *Holarctic Ecol.* **8**, 9–17.

KESSLER, A. (1978). Studien zur Klimatologie der Strahlungsbilanz unter besonderer Berücksichtigung der tropischen Hochgebirge und der kühltemperierten Zone der Südhalbkugel. In *Geoecological Relations between the Southern Temperate Zone and the Tropical Mountains* (ed. C. Troll & W. Lauer), pp. 49–61. Wiesbaden: F. Steiner Verlag GmbH.

KÖRNER, CH. (1982). CO_2 exchange in the alpine sedge *Carex curvula* as influenced by canopy structure, light and temperature. *Oecologia (Berlin)* **53**, 98–104.

KÖRNER, CH. (1984). Auswirkungen von Mineraldünger auf alpine Zwergsträucher. *Verhandl. Ges. Ökol.* **12**, 123–136.

KÖRNER, CH., ALLISON, A. & HILSCHER, H. (1983). Altitudinal variation in leaf diffusive conductance and leaf anatomy in heliophytes of montane New Guinea and their interrelation with microclimate. *Flora* **174**, 91–135.

KÖRNER, CH., BANNISTER, P. & MARK, A. F. (1986). Altitudinal variation in stomatal conductance, nitrogen content and leaf anatomy in different plant life forms in New Zealand. *Oecologia (Berlin)* **69**, 577–588.

KÖRNER, CH. & COCHRANE, P. (1983). Influence of plant physiognomy on leaf temperature on clear midsummer days in the Snowy Mountains, south-eastern Australia. *Acta Oecol., Oecol. Plant.* **4**, 117–124.

KÖRNER, CH. & COCHRANE, P. M. (1985). Stomatal responses and water relations of *Eucalyptus pauciflora* in summer along an elevational gradient. *Oecologia (Berlin)* **66**, 443–455.

KÖRNER, CH. & DEMORAES, J. A. P. V. (1979). Water potential and diffusion resistance in alpine cushion plants on clear summer days. *Oecol. Pl.* **14**, 109–120.

KÖRNER, CH. & DIEMER, M.(1987). *In situ* photosynthetic responses to light temperature and carbon dioxide in herbaceous plants from low and high altitude. *Functional Ecol.* **1**, 179–194.

KÖRNER, CH., NEUMAYER, M., PELAEZ MENENDEZ-RIEDL, S. & SMEETS-SCHEEL, A. (1989). Functional morphology of mountain plants. *Flora* (in press).

KÖRNER, CH. & RENHARDT, U. (1987). Dry matter partitioning and root length/leaf area ratios in herbaceous perennial plants with diverse altitudinal distribution. *Oecologia (Berlin)* **74**, 411–418.

KÖRNER, CH. & WOODWARD, F. I. (1987). The dynamics of leaf extension in plants with diverse altitudinal ranges. II. Field studies in *Poa* species between 600 and 3200 m altitude. *Oecologia (Berlin)* **72**, 279–283.

LAMBERS, H. (1985). Respiration in intact plants and tissues: Its regulation and dependence on environmental factors, metabolism and invaded organisms. In *Higher Plant Cell Respiration. Encyclopedia of Plant Physiology* vol. 18 (ed. R. Douce & D. A. Day), pp. 418–473. Berlin: Springer.

LANGLET, O. (1971). Two hundred years genecology. *Taxon* **20**, 653–722.

LARCHER, W. (1975). Pflanzenökologische Beobachtungen in der Paramostufe der venezolanischen Anden. *Anzeiger Math.-naturwiss. Klasse Österr. Akad. Wiss.* **11**, 194–213.

LARCHER, W. (1977). Ergebnisser des IBP-Projekts 'Zwergstrauchheide Patscherkofel'. *Sitzungsber. Österr. Akad. Wiss., Math.-naturwiss. Kl., Abt I* **186**, 301–371.

LARCHER, W. (1980). Klimastress im Gebirge – Adaptationstraining und Selektionsfilter für Pflanzen. *Rheinisch-Westfälische Akad. Wiss. Vorträge N* **291**, 49–88.

LARCHER, W. (1981*a*). Resistenzphysiologische Grundlagen der evolutiven Kälteakklimatisation von Sprosspflanzen. *Plant. Syst. Evol.* **137**, 145–180.

LARCHER, W. (1981*b*). Effects of low temperature stress and frost injury on plant productivity. In *Physiological Processes Limiting Plant Productivity* (ed. C. B. Johnson), pp. 253–269. London: Butterworths.

LARCHER, W. (1983). Ökophysiologische Konstitutionseigenschaften von Gebirgspflanzen. *Ber. dt. bot. Ges.* **96**, 73–85.

LARCHER, W. (1985). Winter stress in high mountains. In *Establishment and Tending of Subalpine Forest: Research and Management* (ed. H. Turner & W. Tranquillini). *Eidg. Anst. forstl. Versuchswesen, Ber.* **270**, 11–19.

LARCHER, W. & BAUER, H. (1981). Ecological significance of resistance to low temperature. In *Encyclopedia of Plant Physiology, New Series 12A, Physiological Plant Ecology I,* (ed. O. L. Lange, P. S. Nobel, C. B. Osmond & H. Ziegler), pp. 403–437. Berlin, Heidelberg: Springer.

LARCHER, W. & VARESCHI, V. (1988). Variation in morphology and functional traits of *Dictyonema glabratum* D. Hawksw. from contrasting habitats in the venezuelan Andes. *Lichenologist* **20**, 269–277.

LARCHER, W. & WAGNER, J. (1976). Temperaturgrenzen der CO_2-Aufnahme und Temperaturresistenz der Blätter von Gebirgspflanzen im vegetationsaktiven Zustand. *Oecol. Plant* **11**, 361–374.

LARCHER, W. & WAGNER, J. (1983). Ökologischer Zeigerwert und physiologische Konstitution von *Sempervivum montanum*. *Verh. Ges. Ökol. (Göttingen)* **11**, 253–264.

LÖSCH, R., KAPPEN, L. & WOLF, A. (1983). Productivity and temperature biology of two snowbed bryophytes. *Polar Biol.* **1**, 243–248.

MANI, M. S. (1978). *Ecology and Phytogeography of High Altitude Plants of the North-West Himalaya*. London: Chapman & Hall.
MCNULTY, A. K. & CUMMINS, W. R. (1987). The relationship between respiration and temperature in leaves of the arctic plant Saxifraga cernua. *Pl. Cell Environ.* **10**, 319–325.
MONTEITH, J. L. & ELSTON, J. (1983). Performance and productivity of foliage in the field. In *The Growth and Functioning of Leaves* (ed. J. E. Dale & F. L. Milthorpe), pp. 499–518. Cambridge: Cambridge University Press.
MOSER, W., BRZOSKA, W., ZACHHUBER, K. & LARCHER, W. (1977). Ergebnisse des IBP-Projekts Hoher Nebelkogel 3184 m. *Sitzungsber. Österr. Akad. Wiss., Mathem.-naturwiss. Kl, Abt I,* **186**, 387–419.
NAKHUTSRISHVILI, G. S. & GAMTSEMLIDZE, S. G. (1984). *Schisi rastenii b ekstremalnich uslowijach visokogornii, na primjeri zentralnovo Kafkasa.* (*Plant Life under Conditions of High Mountains, examplified by the Central Caucasus*). Leningrad: Isdatjelstwo Nauka.
NOBEL, P. S. (1988). Principles underlying the prediction of temperature in plants, with special reference to desert succulents. In *Plants and Temperature*, Symp. Soc. Exp. Biol., vol. 42 (ed. S. P. Long & F. I. Woodward), pp. 1–23. Cambridge: The Company of Biologists Limited.
NOBEL, P. S. & WALKER, D. B. (1985). Structure of leaf photosynthetic tissue. In *Photosynthetic Mechanisms and the Environment* (ed. J. Barber & N. R. Baker), pp. 502–536. London: Elsevier.
OECHEL, W. C. & COLLINS, N. J. (1976). Comparative CO_2 exchange patterns in mosses from two tundra habitats at Barrow, Alaska. *Can. J. Bot.* **54**, 1355–1369.
OLLERENSHAW, J. H. & BAKER, R. H. (1981). Low temperature growth in a controlled environment of Trifolium repens from northern latitudes. *J. appl. Ecol.* **18**, 229–239.
OLLERENSHAW, J. H., STEWART, W. S., GALLIMORE, J. & BAKER, R. H. (1976). Low-temperature growth in grasses from northern latitudes. *J. agric. Sci., Camb.* **87**, 237–239.
ÖQUIST, G. (1983). Effects of low temperature on photosynthesis. *Pl. Cell Environment* **6**, 281–300.
ÖQUIST, G. & MARTIN, B. (1986). Cold climates. In *Photosynthesis in Contrasting Environments* (ed. N. R. Baker & S. P. Long), pp. 237–293. Amsterdam: Elsevier.
PAUL, E. M. M. (1985). The inhibition of growth by sub-optimal temperatures and the potential for selection at the cellular level. *Euphytica* **34**, 467–473.
PISEK, A. (1960). Pflanzen der Arktis und des Hochgebirges. In *Handbuch der Pflanzenphysiologie*, Bd. 5 (ed. W. Ruhland), pp. 377–413. Berlin, Göttingen, Heidelberg: Springer.
PISEK, A., LARCHER, W. & UNTERHOLZNER, R. (1967). Kardinale Temperaturbereiche der Photosynthese und Grenztemperaturen des Lebens der Blätter verschiedener Spermatophyten. I. Temperaturminimum der Nettoassimilation, Gefrier- und Frostschadensbereiche der Blätter. *Flora Abt.* B **157**, 239–264.
POLLOCK, C. J. & EAGLES, L. F. (198). Low temperature and the growth of plants. In *Plants and Temperature*, Symp. Soc. Exp. Biol., vol. 42 (ed. S. P. Long & F. I. Woodward), pp. 157–180. Cambridge: The Company of Biologists Limited.
RAPPORT, D. J., REGIER, H. A. & HUTCHINSON, T. C. (1985). Ecosystem behavior under stress. *Am. Natural.* **125**, 617–640.
RAUH, W. (1939). Über polsterförmigen Wuchs. Ein Beitrag zur Kenntnis der Wuchsformen der hoheren Pflanzen. *Nova Acta Leopoldina (Halle/Saale)* **7**, 271–505.
REHDER, H. & SCHÄFER, A. (1978). Nutrient turnover studies in alpine ecosystems. IV Communities of the central Alps and comparative survey. *Oecologia (Berlin)* **34**, 309–327.
RUTHSATZ, B. (1978). Las plantas en cojinde los semi-desiertos andinos del Noroeste Argentino. *Darwinia* **21**, 492–539.
SAKAI, A. & LARCHER, W. (1987). Frost survival of plants. In *Ecol. Studies 62* (ed. W. D. Billings, F. Golley, O. L. Lange, J. S. Olson & H. Remmert). Berlin: Springer.
SALISBURY, F. B. & SPOMER, G. G. (1964). Leaf temperatures of alpine plants in the field. *Planta* **60**, 497–505.
SCHINNER, F. (1982*a*). CO_2-Freisetzung, Enzymaktivitäten und Bakteriendichte von Böden unter Spaliersträuchern und Polsterpflanzen in der alpinen Stufe. *Acta Oecologica* **3**, 49–58.
SCHINNER, F. (1982*b*). Soil microbial activities and litter decomposition related to altitude. *Pl. Soil* **65**, 87–94.

SCHRÖTER, C. (1905). *Das Pflanzenleben der Alpen*. Zürich: Raustein.
SCHULZE, E.-D., BECK, E., SCHEIBE, R. & ZIEGLER, P. (1985). Carbon dioxide assimilation and stomatal response of afroalpine giant rosette plants. *Oecologia (Berlin)* **65**, 207–213.
SCOTT, D. & BILLINGS, W. D. (1964). Effects of environmental factors on standing crop and productivity of an alpine tundra. *Ecol. Monogr.* **34**, 243–270.
SEMIKHATOVA, O. A. (1965). O dykhanii vysokogorykh rastenii (Respiration in high mountain plants). Waprosi biologii i fisiologii rastenii w uslowijach wisoko gorii. *Problemi botaniki (Isda telstwo Nauka, Moskwa-Leningrad)* **7**, 142–158.
SEMIKHATOVA, O. A. & GRATCHYEVA, G. I. (1962). O dykhanii vysokogornykh rastenii sadadnowo (On the respiration of high mountain plants in the western Caucasus). *Trudi tebjerdniskowo gosudarstwennowo sadobednika (RSFSR)* **3**, 155–170.
SHIBATA, O. (1985). Altitudinal Botany (jap.). Tokio: Uchida Rokakuho.
SIONIT, N., STRAIN, B. R. & BECKFORD, H. A. (1981). Environmental controls on the growth and yield of okra. I. Effects of temperature and of CO_2 enrichment at cool temperature. *Crop. Sci.* **21**, 885–888.
SLATYER, R. O. (1978). Altitudinal variation in the photosynthetic characteristics of snow gum, *Eucalyptus pauciflora* Sieb. ex Spreng. VII. Relationship between gradients of field temperature and photosynthetic temperature optima in the Snowy Mountains area. *Aust. J. Bot.* **26**, 111–121.
SLATYER, R. O. & MORROW, P. A. (1977). Altitudinal variation in the photosynthetic characteristics of snow gum, *Eucalyptus pauciflora* Sieb. ex Spreng. I. Seasonal changes under field conditions in the Snowy Mountains area of south-eastern Australia. *Aust. J. Bot.* **25**, 1–20.
SONESSON, M. (1980). Ecology of a subarctic mire. *Ecol. Bull.* 30. Stockholm: NFR Editorial Services.
STEWART, W. S.& BANNISTER, P. (1974). Dark respiration rates in *Vaccinium* spp. in relation to altitude. *Flora* **163**, 415–421.
SVOBODA, J. (1977). Ecology and primary production of raised beach communities, Truelove Lowland. In *Truelove Lowland, Devon Island, Canada: A High Arctic Ecosystem* (ed. L. C. Bliss), pp. 185–216. Edmonton: University of Alberta Press.
TIESZEN, L. L., LEWIS, M. C., MILLER, P. C., MAYO, J., CHAPIN, III F. S. & OECHEL, W. (1981). An analysis of processes of primary production in tundra growth forms. In *A Comparative Analysis of Tundra Ecosystems* (ed. L. C. Bliss, J. B. Cragg, D. W. Heal & J. J. Moore), pp. 285–356. International Biological Programme 25, Cambridge: Cambridge University Press.
TIKHOMIROV, B. A. (1963). *Ocherki po biologii rastenii Arktiki* (*Investigations on the Biology of Arctic Plants*). Izdatestvo Akad. nauk SSSR, Leningrad.
TISSUE, D. T. & OECHEL, W. C. (1987). Response of *Eriophorum vaginatum* to elevated CO_2 and temperature in alaskan tussock tundra. *Ecology* **68**, 401–410.
TRANQUILLINI, W. (1979). Physiological ecology of the alpine timberline. In *Ecological Studies* vol. 31 (ed. W. D. Billings, F. Golley, O. L. Lange & J. S. Olson). Berlin: Springer.
TRANQUILLINI, W. & HAVRANEK, W. M. (1985). Influence of temperature on photosynthesis in spruce provenances from different altitudes. In *Establishment and Tending of Subalpine Forest: Research and Management* (ed. H. Turner & W. Tranquillini), pp. 41–51. Ber. Eidgen. Anst. Forstl. Versuchswesen 270.
TURESSON, G. (1925). The plant species in relation to habitat and climate. Contributions to the knowledge of genecological units. *Hereditas* **6**, 147–236.
TYURINA, M. M. (1957). *Issledovanie morozostoikosti rastenii v usloviakh vysokogorii pamira* (*Investigations of Frost Resistance in Plants of the High Pamirs*). Izdat. Nauk Tadzhik. SSR, Stalinabad.
WARDLAW, I. F. (1976). Assimilate partitioning: Cause and effect. In *Transport and Transfer Processes in Plants* (ed. I. F. Wardlaw & I. B. Passioura), pp. 381–391. New York: Academic Press.
WARDLE, P. (1974). Alpine timberlines. In *Arctic and Alpine Environments* (ed. J. D. Ives & R. G. Barry), pp. 371–402. London: Methuen.
WIELGOLASKI, F. E. (1975). Fennoscandian tundra ecosystems. I. Plants and microrganisms. *Ecol. Studies* 16. Berlin-Heidelberg-New York: Springer.

WOODWARD, F. I. (1979). The differential temperature responses of the growth of certain plant species from different altitudes. I. Growth analyses of *P. alpinum* L., *Phleum bertolonii* D. C., *Sesleria albicans* Kit. and *Dactylis glomerata* L. *New Phytol.* **82**, 397–405.
WOODWARD, F. I. (1983). The significance of interspecific differences in specific leaf area to the growth of selected herbaceous species from different altitudes. *New Phytol.* **95**, 313–323.
WOODWARD, F. I. (1987). *Climate and Plant Distribution. Cambridge: Cambridge University Press.*
WOODWARD, F. I. (1988). Temperature and the distribution of plant species and vegetation. In *Plants and Temperature,* Symp. Soc. Exp. Biol. vol. 42 (ed. S. F. Long & F. I. Woodward), pp. 59–75. Cambridge: The Company of Biologists Limited.
WOODWARD, F. I., KÖRNER, CH. & CRABTREE, R. C. (1986). The dynamics of leaf extension in plants with diverse altitudinal ranges. I. Field observations on temperature responses at one altitude. *Oecologia (Berlin),* **70**, 222–226.
YOSHIE, F. & SAKAI, A. (1981). In *Water Status and Freezing Resistance of Arctic Plants* (ed. S. Kinosita), [Joint studies on physical and biological environments in the permafrost North Canada.] Inst. Low Temperature Sci. Contrib. **2427**, 95–110.

TEMPERATURE AND THE DISTRIBUTION OF PLANT SPECIES

F. I. WOODWARD

Department of Botany, University of Cambridge, Downing Street, Cambridge CB2 3EA, UK

Summary

An understanding of the mechanisms by which temperature influences the distribution of species and vegetation has been attempted by modelling population growth and establishing those stages of the plant life cycle which, when diminished by extremes of temperature, for example, may have the greatest impact on plant survival. This analysis suggests that the heat sum of the growing season, measured as day-degrees, controls the distribution of annual vegetation. For perennial vegetation both the heat sum of the growing season and the annual, absolute minimum temperature are critical.

Climatic correlations and experimental analyses indicate that, in northern Europe, the northern latitudinal and upper altitudinal limits of lowland and southern vegetation are directly controlled by climate. In contrast, the southern and lower altitudinal limits of upland and northern vegetation are likely to be controlled by temperature-sensitive competition with southern or lowland species.

Many of the temperature-sensitive processes of plant growth and development, such as the non-linearity of extension growth and variations in the threshold temperatures of processes, may increase the realized heat sum at a particular geographical location. However, in more northerly climates, photoperiodic control is crucial in avoiding precocious development in the highly variable climatic conditions of early spring.

Introduction

The global patterns of vegetation are well described (e.g. Cain, 1944; Walter, 1979) and clearly correlate with global patterns in temperature and rainfall (Schimper, 1898). However, these correlations provide no clear indication of the physiological mechanisms which may be involved in defining these patterns of distribution.

Woodward (1987) has attempted to use known physiological mechanisms, in particular minimum temperature resistance and the physiological control of evapotranspiration, to define the global patterns of vegetation on a mechanistic basis. The basic approach was confined to mature plants with no attempt to consider other stages of the life cycle. This paper includes a population approach for defining that period of the life cycle which is likely to be the most sensitive to temperature and which has the greatest potential for controlling plant distribution.

Nobel (1988) has described a wide range of processes which are temperature-sensitive and which may be crucial in controlling plant distribution. However, establishing which of these processes are most critical in controlling plant distribution must be determined by predictions which can be tested experimentally. This chapter aims to combine considerations of plant life cycles and ecophysiology to predict the mechanisms which may be involved in the control of plant distribution by temperature.

Life cycles

The life cycle of a flowering plant may vary from a few days to at least 8000 years (Friday & Ingram, 1985). This range of longevities also defines the range of vegetation types from annual vegetation to perennial forest. The need for annual plants to complete the life cycle to seed production in one growing season contrasts with the ability of very long-lived species of trees to exist in a forest for many years without significant seed production or regeneration from seed (Pigott & Huntley, 1978, 1980, 1981).

For summer annuals in the temperate zone, winter temperatures may be irrelevant to plant survival if the overwintering seed is buried in the soil and also insulated from very low temperatures by an additional blanket of snow (Ylimaki, 1962). In contrast the survival of evergreen trees from Mediterranean vegetation may be critically dependent on the absolute minimum temperatures during the winter growing season (Larcher & Mair, 1969).

The definition of these climatically sensitive periods of the life cycle has been investigated by establishing matrix models of population growth (e.g. Sarukhan & Gadgil, 1974; Hartshorn, 1975) for the most extreme lengths of life cycle, annual herbs and perennial trees, using published life cycle data (Harper, 1977). The schematics of the two life cycles are shown on Fig. 1 and the selected probabilities of the different stages for establishing a population at equilibrium with a growth rate of zero, are shown on Table 1. The probabilities of seed bank survival, germination, establishment, dispersal and fecundity are the same for the annual and perennial cycles; while high probabilities of survival are characteristic of later stages of the perennial life cycle.

A more realistic simulation of population growth includes density-dependent probabilities of survival (Law *et al.* 1977; Woodward & Jones, 1984). These probabilities regulate population development towards an equilibrium, with close to zero growth. However, details of these probabilities are not readily available. This lack of information was avoided by selecting life cycle probabilities which would lead to an equilibrium population, with zero growth, the end result of density-dependent effects.

The sensitivity of population growth to one-off reductions in the survival probabilities of each stage of the life cycle was then tested. This was achieved by reducing just one probability, e.g. germination or winter survival, for one year and to a level which just allowed subsequent population recovery. This technique was designed to test the sensitivity of the life cycle to an extreme climatic event. Rare extreme events, which occur for a rather short period of time, and probably independently of average climate (Woodward, 1987), can have catastrophic influences on plant survival (e.g. Woodward, 1987) through lethal effects at a

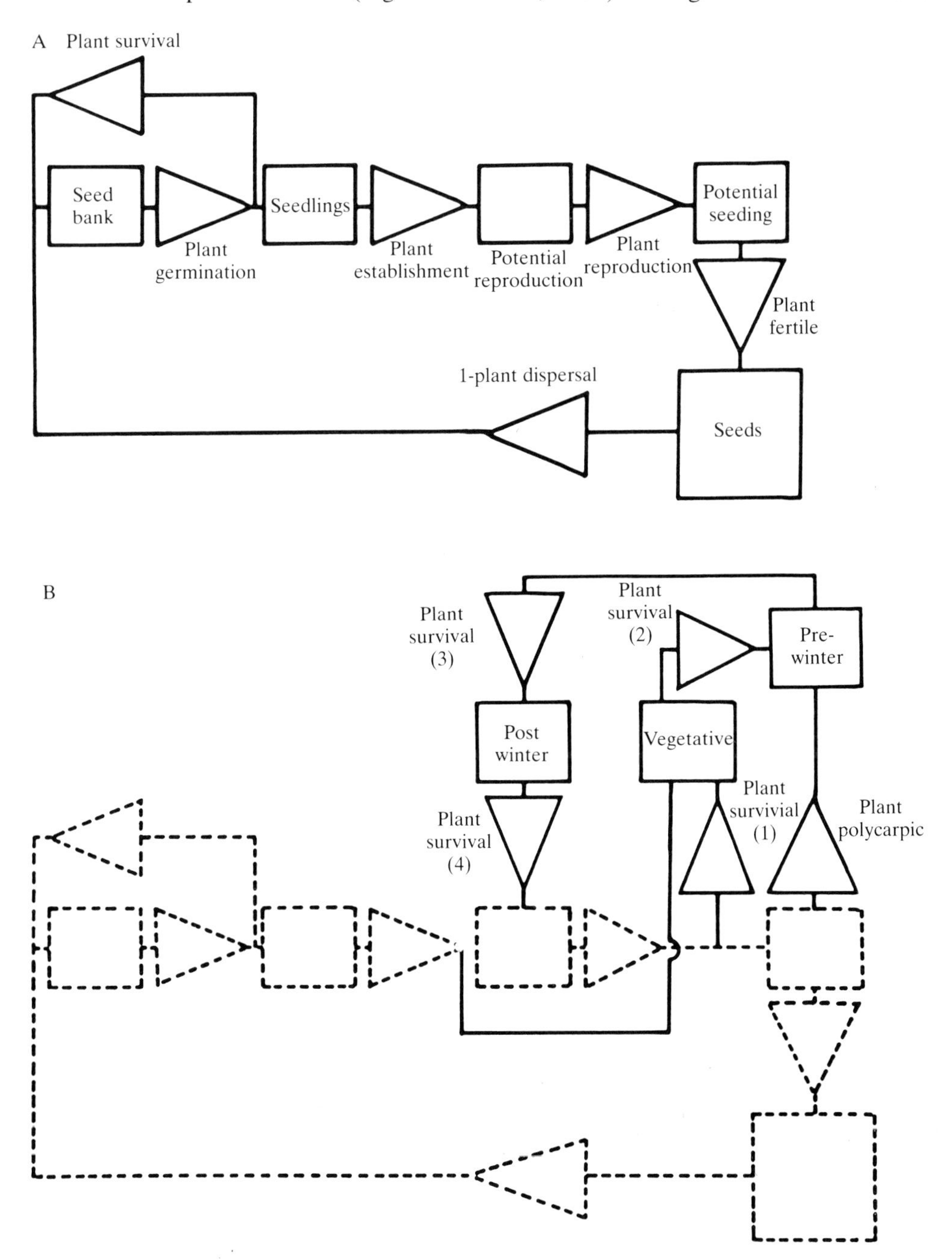

Fig. 1. Life cycle schematics. (A) Annual life cycle. (B) Perennial life cycle.

Table 1. *Selected life-cycle probabilities for stable population establishment*

Stage	Probabilities	
	Annual cycle	Perennial cycle
1. Survival in seed bank	0·15	0·15
2. Germination	0·6	0·6
3. Establishment	0·2	0·2
4. Reproduction	0·4	0·7
5. Fecundity	200	200
6. 1-probability	0·1	0·1
7. Non-reproductive survival (1) to vegetative stage	–	0·2
8. Vegetative survival (2) to prewinter stage	–	0·75
9. Winter survival (3)	–	0·8
10. Polycarpy	–	0·9
11. Postwinter survival (4) to reproductive stage	–	0·88

particular stage of the life cycle. Such effects have been shown for *Verbena officinalis*, a species which was killed by an extreme winter frost (Yaqub, 1981) which has not been repeated over a period of 10 years. The extent to which an extreme event at one stage of the life cycle influences population survival is measured by the attenuation coefficient (Fig. 2). The attenuation coefficient measures the reduction in population growth following an extreme event, and the greater the coefficient the greater the sensitivity of population growth to an extreme event.

For vegetation consisting of annual species, all stages of the life cycle, except for the survival of seeds in the seed bank, are very sensitive to attenuation. Population recovery would be limited to, at the most, a 40 % attenuation of a particular stage of the life cycle.

The sensitivity of the perennial life cycle is quite distinct from the annual cycle, with three particular stages expected to be most sensitive to attenuation: the probabilities of becoming reproductive, winter survival and post-winter establishment. The early stages of seed germination and establishment are not crucial in the short term because of the continued survival of mature plants, with the ability to set seed from one year to the next. Replacement of mature plants, by establishment from seed, occurs with a low probability and so the winter and post-winter survivals of the perennial vegetation become particularly sensitive stages of the life cycle.

These two tests of life-cycle sensitivity indicate that, for annual vegetation, temperatures in the growing season will be most critical in controlling distribution. In the case of perennial vegetation, the temperatures during the period of nil, or low growth, as in the case of winter survival in temperate zones, will be crucial. This will be in addition to the length of the growing season, which will influence both pre- and post-winter survival (Figs 1 and 2) and also the probability of reproductive success.

Distribution and temperature

The tests of resilience on the life cycle models have indicated the periods of the year when the life cycle characteristics should be most sensitive to perturbation by temperature. The predictions are that annual vegetation should be controlled by heat sum during the growing season, while perennial vegetation should be sensitive to temperatures in both the growing season and the dormant season. These predictions will be tested by comparing the distributions of annual and perennial species and vegetation with spatial variations in temperature over the area of northern Europe. This region possesses a wide range of vegetation types from tundra to deciduous forest (Good, 1974) and includes the northern limits of deciduous forests and the southern limits of boreal forests.

Four types of vegetation have been selected for investigation. The most northerly perennial type is coniferous forest, with a distribution represented by *Pinus sylvestris* (Fig. 3A). The more southerly, winter deciduous forests are represented by *Tilia cordata* (Fig. 3B). Permanently annual vegetation is not naturally extensive in Europe and annual species are very rare in cold climates. However, *Koenigia islandica* (Fig. 3C) provides an example of a tundra annual from areas of sparse and naturally disturbed vegetation. The activities of man in

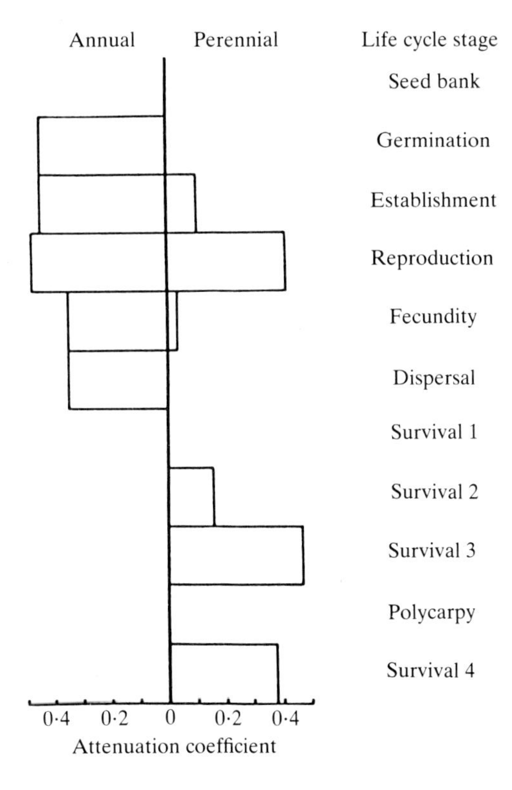

Fig. 2. Sensitivity of population growth to a one-cycle attenuation of each step of the life cycle. The attenuation coefficient is the maximum attenuation from which population recovery can occur.

the warmer areas of Europe provide continuous disturbance, which encourages the development of annual vegetation; *Bromus sterilis* (Fig. 3D) is a characteristic example of such vegetation.

The European distributions of these vegetation types can be correlated with the spatial variation in climate recorded at the network of meteorological stations in Europe (Müller, 1982). Many features of the thermal climate have been used in the past for explaining the distribution of plants (Box, 1981) but, in this case, the sensitivity analysis of the life cycle dynamics indicates that two measurements should be sufficient for explaining the distribution of perennial plants and just one for annual plants. Woodward (1987) has argued that the absolute minimum temperature is crucial in explaining the distribution of plants, because this is a limiting temperature that must be endured for survival, from chilling temperatures in tropical regions to extreme freezing temperatures in the boreal zone. The physiological mechanisms of survival are well established (Levitt, 1980), although all of the processes concerned still require further exploration. The absolute

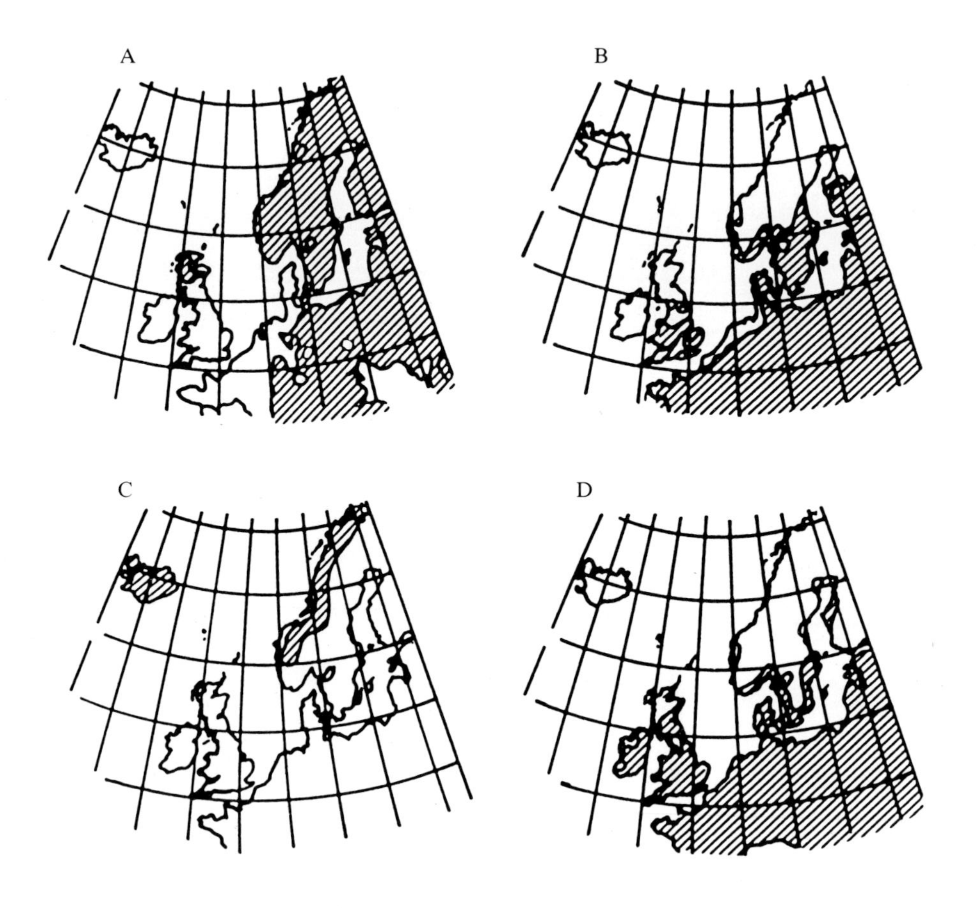

Fig. 3. Distribution of species in northern Europe. (A) *Pinus sylvestris*, (B) *Tilia cordata*, (C) *Koenigia islandica*, (D) *Bromus sterilis*. From: Dahl, 1963; Fitter, 1978; Fitter, Fitter & Farrer, 1984.

annual minimum temperature can therefore be selected as one of the temperature axes.

A measure of the growing season is perhaps not as simple. However, Solomon *et al.* (1981) have demonstrated that the product of the period of the year in which growth is possible and the mean temperature of this period – the heat sum as day-degrees – is an adequate measurement. The growing season has been defined from observations on *T. cordata* and *P. sylvestris* planted in the European network of phenological gardens (Schnelle *et al.* 1984) and is used as the other axis for correlation with observations of distribution. Day-degrees have been taken from an arbitrary threshold of 0 °C.

The absolute minimum temperatures and growing season day-degrees for 113 meteorological stations in northern Europe (the area defined by Fig. 3) have been used as the axes of temperature for establishing correlations with the distributions of the four vegetation types represented by the four species on Fig. 3.

Coniferous vegetation, as represented by *P. sylvestris* (Fig. 4), occurs in the coldest regions and is confined to regions colder than an absolute minimum temperature of −22 °C and a heat sum of 2800 day-degrees. Many species of boreal conifer can survive when established, by man, in Botanic Gardens in warmer

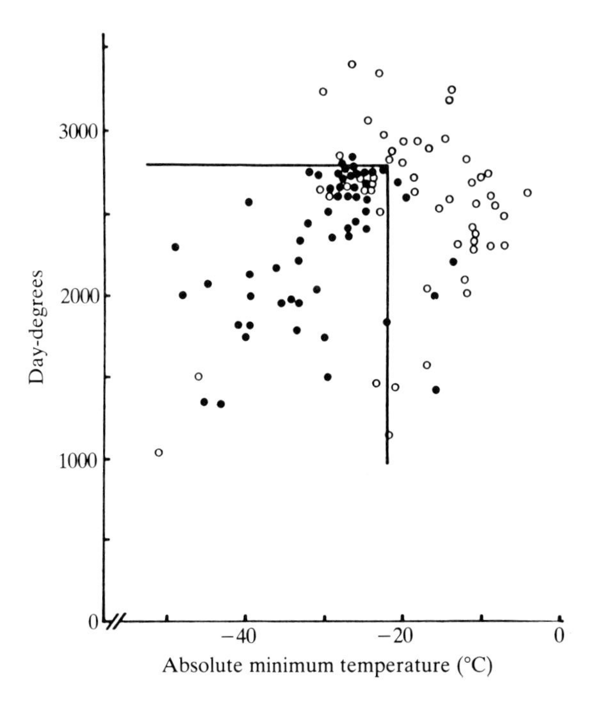

Fig. 4. Climatic correlation for *Pinus sylvestris*. ●, present; ○, absent. The vertical and horizontal limits were determined as the positions along the axes at which a χ^2 contingency test indicated a significant difference between the overall mean probability of occurrence, for all sites, and the average occurrence over an interval of either 4 °C or 200 day-degrees.

climates than the limits shown on Fig. 4 (Carlisle & Brown, 1968). This indicates that the relationship between temperature and distribution is not simple.

T. cordata (Fig. 5) contrasts with *P. sylvestris*, occurring in climates warmer than a minimum temperature of −42°C and a heat sum of 2000 day-degrees. Extensive work by Pigott (Pigott, 1981; Pigott & Huntley, 1978, 1980, 1981) has clearly demonstrated that the cold limit of 2000 day-degrees operates by limiting or preventing seed set, with no regeneration occurring beyond this boundary. The minimum temperature limit coincides closely with the temperature at which supercooled water may spontaneously nucleate, causing freezing and death in overwintering buds (Sakai, 1978). In the case of temperate deciduous vegetation, as demonstrated by *T. cordata*, the temperature limits to the geographical distribution, as shown on Fig. 5, can be explained by observed mechanisms.

K. islandica (Fig. 6) occurs in cold climates with a heat sum of less than 1800 day-degrees but, as predicted above, there is no correlation with absolute minimum temperature. Experiments by Löve & Sarkar (1957) indicate that *K. islandica* can survive high temperatures to at least 40°C, showing that an inability to survive high temperatures is not the simple explanation of the distribution limit at 1800 day-degrees. Growth trials (Löve & Sarkar, 1957) have demonstrated that the species can develop from seed to seed in about 700 day-degrees.

The distribution of *B. sterilis* is confined to climates with a heat sum of 2000 day-degrees or greater (Fig. 7). No relationship emerged with the absolute minimum temperature. In the northern part of its range the species is summer green,

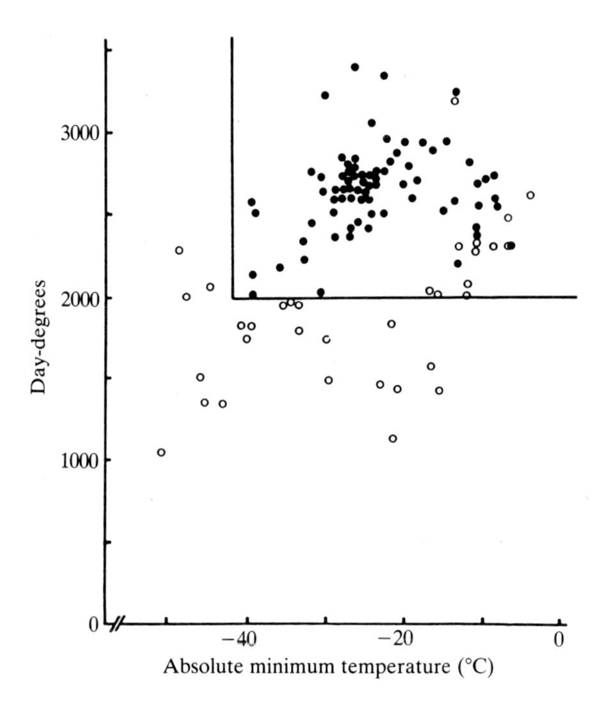

Fig. 5. Climatic correlation for *Tilia cordata* (details as for Fig. 4).

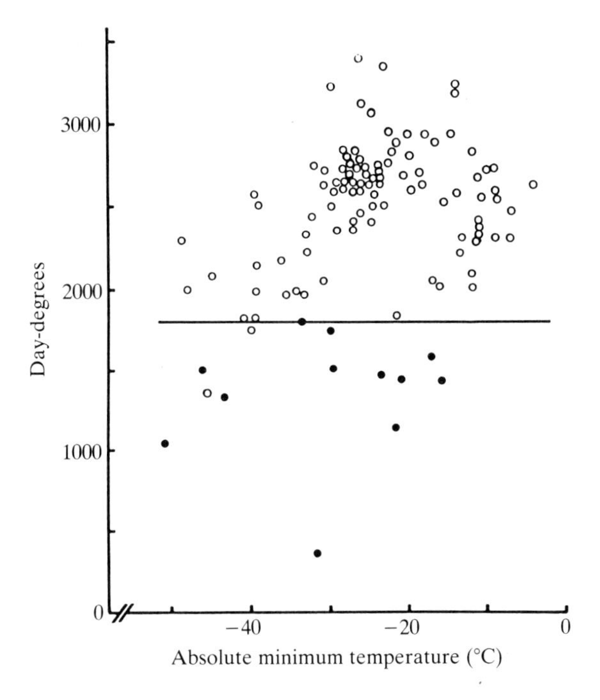

Fig. 6. Climatic correlation for *Koenigia islandica* (details as for Fig. 4).

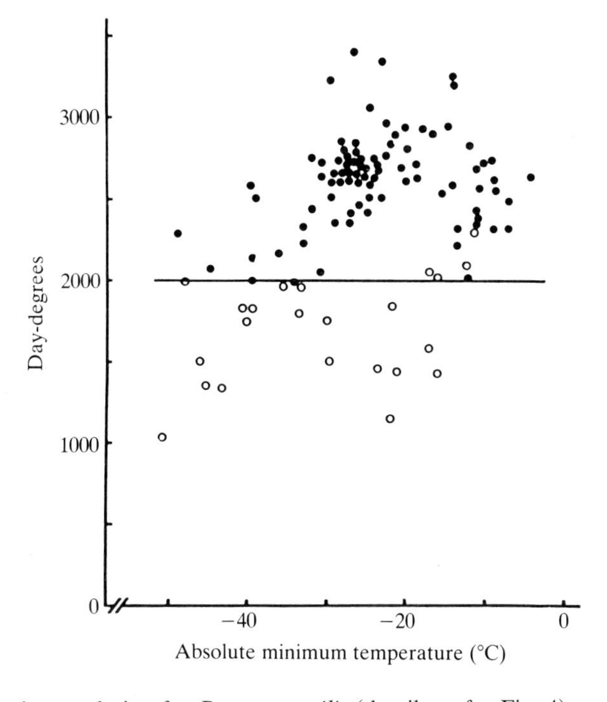

Fig. 7. Climatic correlation for *Bromus sterilis* (details as for Fig. 4).

however it can be winter green in southern France, indicating that the winter climate may be crucial for the distribution of the species, at least in drier, Mediterranean climates. Such climates are represented on Fig. 7 by high values of heat sums, indicating that the potential heat sum may not necessarily equal the realized heat sum, particularly if an additional climatic feature, such as drought, is limiting.

Figs 4 to 7 present a range of climatic correlations which may be tested by experiment, such as is possible for *T. cordata*. In addition it is possible to use the relationships which have emerged from the correlations to predict the likely distribution of the species in another area, outside this data set of northern Europe. Such a prediction has been attempted for *B. sterilis* in North America, an area in which the species is an introduced alien. The data presented on Fig. 7 indicate a cold limit to the distribution of *B. sterilis* at 2000 day-degrees. The 2000 day-degree isotherm has been established from the meteorological stations of North America (Müller, 1982). This may then be compared with the northern distribution limits of *B. sterilis*, which have been established from a range of state floras (Fig. 8). Away from areas of drought the predicted and observed limits to the distribution agree closely, indicating the general predictability and relevance of the heat sum limit.

A similar approach has been successfully applied by Booth *et al.* (1987), for

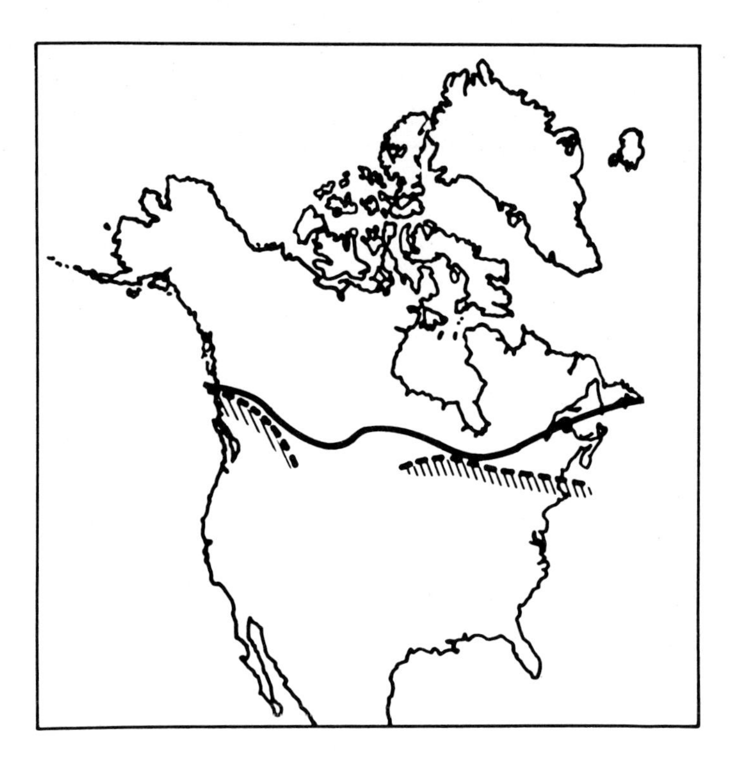

Fig. 8. Comparison between the 2000 day-degree limit (——) and the northern distribution of *Bromus sterilis* (shaded area) in north America.

predicting the areas of the globe which have the appropriate climatic conditions for the introduction of alien species of trees for forestry.

Implications for mechanisms of distribution control

Low temperature tolerance

The climatic correlations for the species with a more southerly range (Figs 5 and 7) may be explained in terms of mechanisms which are directly associated with the chosen axes of climate, i.e. the heat sum to complete the life cycle and, for *T. cordata*, the absolute winter minimum temperature. Vegetation from southern regions, as represented by *T. cordata* (Fig. 5) and *B. sterilis* (Fig. 7) cannot be extended (Schnelle *et al.* 1984; Woodward, 1987) beyond the minimum heat sums and, for the deciduous forest, the absolute winter minimum temperature defined in the climatic correlations (Figs 5 and 7). In contrast, species of the boreal coniferous forest may be grown in Botanic Gardens with warmer climates (Schnelle *et al.* 1984) than the limits defined for *P. sylvestris* (Fig. 4). Similarly *K. islandica* (Fig. 6) may also be grown in warmer climates than native populations (Gauslaa, 1984).

The likely explanation for the high-temperature distribution limits for the species from cold climates is competitive exclusion by species from warmer climates. This hypothesis was tested by a long-term study of the field growth trials established by Woodward (Woodward & Pigott, 1975) at a range of altitudes. In these trials plants of *Sedum rosea*, a species restricted to high altitudes in England, were grown in a competitive array with *Sedum telephium*, a species restricted to low altitudes. These trials were established in 1971 and were then revisited at intervals for up to 15 years. In addition, plants of each species were established as a monoculture at the same sites.

After a period of three years the survival patterns of the species became clear-cut (Fig. 9). In monoculture, *S. rosea* establishes and survives in heat sums from

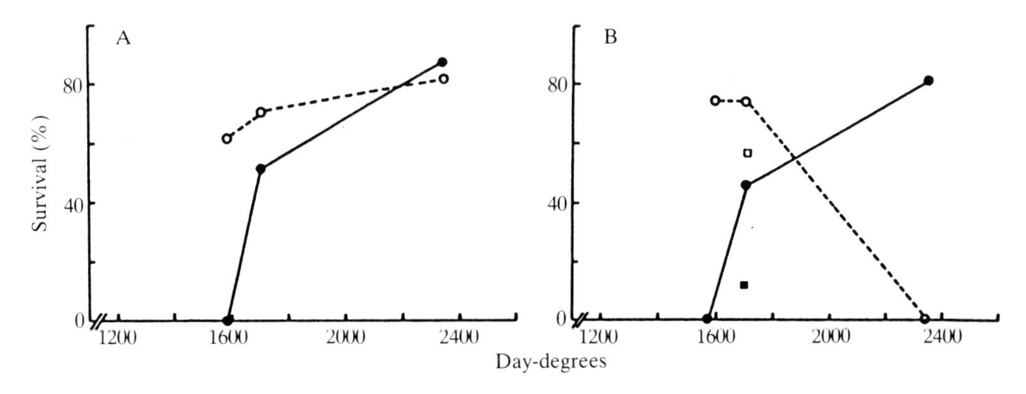

Fig. 9. Survival of *Sedum rosea* (○——○) and *S. telephium* (●——●) after 3 year's growth at different heat sums (Woodward & Pigott, 1975). (A) Growth in monoculture. (B) Growth with interspecific competition. Survival in competition after 15 years, *S. rosea* (□), *S. telephium* (■).

1585 day-degrees (at an altitude of 640 m) to 2346 day-degrees (an altitude of 46 m). In contrast, *S. telephium* becomes extinct at the highest altitude. The response curve for *S. telephium* is not affected by growth in competition with *S. rosea*. However, this is not the case for *S. rosea* (Fig. 9B) which is excluded from the lowest altitude, with the greatest heat sum. The survival of *S. rosea* is not affected by competition from *S. telephium* at higher altitudes.

Both species survived together at a heat sum of 1700 day-degrees, an altitude of 550 m which is 150 m higher than the observed, natural limit for *S. telephium* in England. However, after 15 years *S. telephium* shows poor survival, in contrast to *S. rosea*, suggesting that the species may ultimately become extinct.

These data do conform to the hypothesis that the high temperature limit to the distribution of the cold climate species may be explained by temperature-sensitive competition, often over an appreciable time scale. The question therefore emerges of whether the greater capacity of the cold climate perennials to survive low winter temperatures is at the expense of competitive ability. This relationship does not appear to have been investigated directly, however evidence for such a correlation may be seen for the winter growth of different provenances of *Festuca arundinacea* (Robson & Jewiss, 1968). In this case, growth is negatively correlated with frost tolerance (Fig. 10) and could lead to poorer annual growth, and perhaps competitive ability, by the most frost resistant provenances in warm climates. It seems likely that the ability of perennial species to tolerate low temperatures is

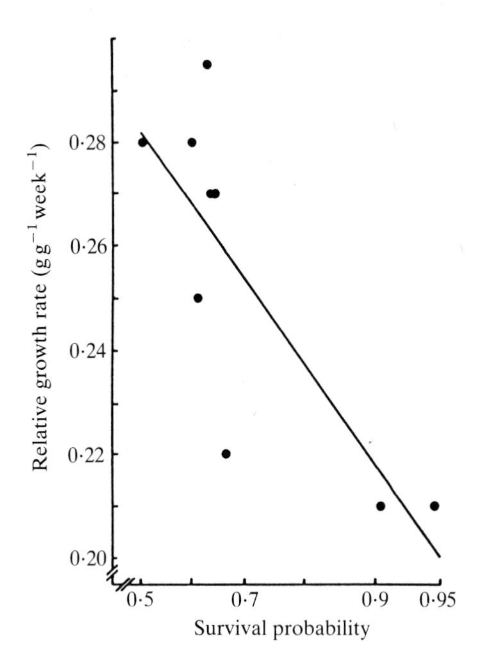

Fig. 10. The relationship between winter growth and low temperature resistance of different provenances of *Festuca arundinacea* (from Robson & Jewiss, 1968). The *x* axis is an inverse sine transformation, r = −0·814 for the regression of the transformed data.

controlled by many biochemical and developmental processes (Levitt, 1980; Woodward, 1987), processes which may prove to be sinks for otherwise growth productive investment.

Heat sum

The temperate climate of Europe shows considerable year-to-year variation (e.g. Lamb, 1982; Woodward, 1987) and in some years the heat sum may be too small for life cycles to be completed. This has important consequences for annual vegetation, the survival of which depends to some considerable degree on current seed production. For perennial vegetation a complete failure of seed production in any one year is unlikely to be crucial. However, a growing season with a low heat sum may be followed by an early or an exceptionally cold winter (Lamb, 1982), which may kill plants or leaves which are incompletely developed (Tranquillini, 1979).

In such marked and seasonally variable climates, plants have evolved photoperiodic controls of development which are not readily overriden by periods of ameliorated conditions (Larcher & Wagner, 1976). In *P. sylvestris*, for example, short days cause the onset of dormancy and low temperature hardening but inhibit significant growth, even in warm conditions (Wareing, 1950; Carlisle & Brown, 1968). Photoperiodic curtailment of the potential growing season may also be enhanced towards the colder limits of the distribution of *P. sylvestris* (Eiche, 1966), where early autumns and winters may be more frequent (Hagner, 1965).

Photoperiodic thresholds can therefore limit the realized heat sum. However, a number of plant responses effectively increase the realized heat sum. The manner in which this can be achieved can be demonstrated by *Poa alpina*, a species which occurs at mid-altitudes in the Austrian Alps, and *P. alpina* ssp. *vivipara*, the sub-species which occurs at higher altitudes (Woodward *et al.* 1986). Field measurements of leaf extension have shown the characteristic temperature responses shown on Fig. 11. For *P. alpina*, from warmer environments than the sub-species *vivipara*, no extension occurs below 3 °C (Woodward *et al.* 1986), therefore defining a low temperature threshold for the effective heat sum. In the case of *P. alpina* ssp. *vivipara*, no threshold was detected to 0 °C, effectively increasing the range of temperatures over which leaf extension occurs. For *P. alpina* and all of the other species observed in the field, there was a predictable relationship between the threshold temperature for extension and the absolute rates of extension at higher temperatures. It appears that the capacity to extend at low temperatures is at the expense of high rates of extension at high temperatures. At 25 °C the extension rate of *P. alpina* exceeds that of *P. alpina* ssp. *vivipara* by over fourfold. In addition, the temperature response of leaf extension is non-linear and markedly so for *P. alpina*. An equal number of day-degrees at 10 °C and 20 °C leads to a 22 % increase in total extension at 20 °C, for *P. alpina* ssp. *vivipara* but a 100 % increase for *P. alpina*. This amplification was most marked for the lowland species that were observed in the field (Woodward *et al.* 1986).

Increase of the realized heat sum may also result from radiant heating of the vegetation above ambient temperature, on days of sunny weather. It has been shown that leaf temperatures may exceed air temperature by up to 30°C in prostrate, alpine vegetation (Dahl, 1963; Salisbury & Spomer, 1964; Körner & Cochrane, 1983), thereby increasing the potential rates of processes with positive temperature coefficients. However, enhanced leaf temperatures during periods of very high radiation may cause plant temperatures to reach a lethal threshold (Dahl, 1963). Gauslaa (1984) has shown that even plants of low altitude tundra exhibit very high lethal threshold temperatures. In *K. islandica*, for example, the threshold for leaves is 44°C, equal to that of *Quercus suber* in southern Spain (Lange & Lange, 1963).

For species with overwintering leaves which are not covered by snow, such as *P. sylvestris*, the increase of the realized heat sum by radiant heating will increase the probability that leaf growth will be completed, with a fully developed cuticle which minimizes winter desiccation (Tranquillini, 1979; Wilson *et al.* 1987).

Maximizing the leaf temperature by radiant heating is dependent on days of clear, sunny weather, an uncommon and unreliable feature on mountains, particularly those with a dominant, maritime air stream (Barry, 1981). In addition such days may be followed by nights of clear skies with intense radiant cooling and a high probability of frosts (Geiger, 1965), a feature which may cause damage during periods of active growth (Carlisle & Brown, 1968). Plants may, therefore, utilize the available heat sum more efficiently by variations in the threshold and amplification responses of particular processes.

Heat storage may also ameliorate the plant microclimate. Heat stored in the soil beneath prostrate vegetation will conduct to the plant above when the temperature gradient is towards the plant (Geiger, 1965). Such heat storage extends the available period for growth and development. This storage may be complemented by storage of carbohydrates, during periods of low temperatures when the growth

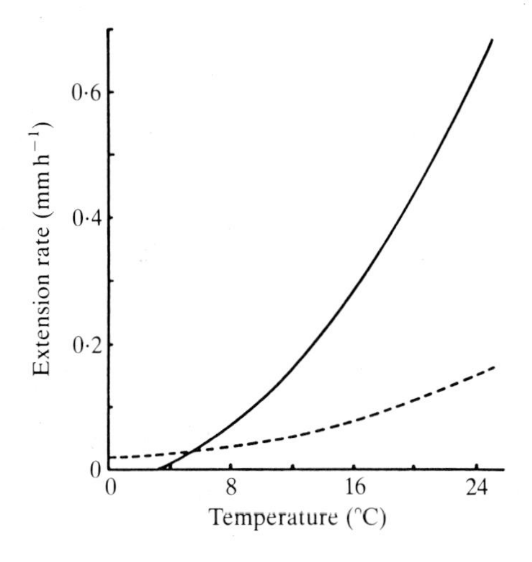

Fig. 11. The temperature responses of leaf extension in *Poa alpina* (——) and *P. alpina* ssp. *vivipara* (- - -), from Woodward *et al.* (1986).

rate is low but the photosynthetic rate is high (Mooney & Billings, 1960). These carbohydrates may be exported to areas of growth during periods of higher temperatures (Mooney, 1972). Such a storage system depends on the plant's ability to acclimate, so that photosynthesis proceeds at lower temperatures than growth (Pearcy, 1977; Borland & Farrar, 1987).

Conclusion

The observed patterns of global vegetation and their correlations with climate are well established facts but an understanding of the mechanisms involved in controlling plant distribution is still in its infancy. A population-based approach indicates that for annual species and vegetation the mechanisms relate in the main to the growing season climate. For perennial species at their northern (cold) limits, winter climate appears as, or more crucial than the climate of the growing season. For species at the southern (warm) limits to their range, the growing season becomes more crucial, through competitive interactions with more vigorous species.

I am grateful for comments on the manuscript by A. D. Friend and A. C. Newton.

References

Barry, R. G. (1981). *Mountain Weather and Climate*. London: Methuen & Co. Ltd.
Booth, T. H., Nix, H. A., Hutchinson, M. F. & Busby, J. R. (1987). Grid matching: a new method for homoclime analysis. *Agric. For. Meteorol.* **39**, 241–255.
Borland, A. M. & Farrar, J. F. (1987). The influence of low temperature on diel patterns of carbohydrate metabolism in leaves of *Poa annua* L. and *Poa* × *jemtlandica* (Almq.) Richt. *New Phytol.* **105**, 255–263.
Box, E. O. (1981). *Macroclimate and Plant Forms: An Introduction to Predictive Modeling in Phytogeography*. The Hague: Junk.
Cain, S. A. (1944). *Foundations of Plant Geography*. New York: Harper.
Carlisle, A. & Brown, A. H. F. (1968). Biological Flora of the British Isles. *Pinus sylvestris* L. *J. Ecol.* **56**, 269–307.
Dahl, E. (1963). On the heat exchange of a wet vegetation surface and the ecology of *Koenigia islandica*. *Oikos* **14**, 190–211.
Eiche, V. (1966). Cold damage and plant mortality in experimental provenance plantations with Scots pine in Northern Sweden. *Studia for. suec.* **36**, 5–219.
Fitter, A. (1978). *An Atlas of the Wild Flowers of Britain and Northern Europe*. London: Collins.
Fitter, R., Fitter, A. & Farrer, A. (1984). *Collins Guide to the Grasses, Sedges, Rushes and Ferns of Britain and Northern Europe*. London: Collins.
Friday, A. & Ingram, D. S. (ed.) (1985). *The Cambridge Encyclopedia of Life Sciences*. Cambridge: Cambridge University Press.
Gauslaa, Y. (1984). Heat resistance and energy budget in different Scandinavian plants. *Holarctic Ecology* **7**, 1–78.
Geiger, R. (1965). *The Climate near the Ground*. 4th. edn. Cambridge, Mass: Harvard University Press.
Good, R. (1974). *The Geography of the Flowering Plants*. 4th edn. London: Longman.
Hagner, S. (1965). Cone crop fluctuations in Scots Pine and Norway Spruce. *Stud. For. Suec. Skogshogski* **33**, 1–21.

HARPER, J. L. (1977). *Population Biology of Plants.* London: Academic Press.
HARTSHORN, G. S. (1975). A matrix model of tree population dynamics. In *Tropical Ecological Systems* (ed. F. B. Golley & E. Medina), pp. 41–51. New York: Springer-Verlag.
KÖRNER, CH. & COCHRANE, P. (1983). Influence of plant physiognomy on leaf temperature on clear midsummer days in the Snowy Mountains, south-eastern Australia. *Acta Oecol./Oecol. Pl.* **4**, 117–124.
LAMB, H. H. (1982). *Climate, History and the Modern World.* London: Methuen.
LANGE, O. L. & LANGE, R. (1963). Untersuchungen über Blattemperaturen, Transpiration und Hitzeresistenz an Pflanzen mediterraner Standorte (Costa Brava, Spanien). *Flora* **153**, 387–425.
LARCHER, W. & MAIR, B. (1969). Die Temperaturresistenz als ökophysiologisches Konstitutionsmerkmal: 1. *Quercus ilex* und andere Eichenarten des Mittelmeergebietes. *Oecol. Pl.* **4**, 347–376.
LARCHER, W. & WAGNER, J. (1976). Temperaturgrenzen der CO_2-Aufnahme und Temperaturresistenz der Blätter von Gebirgspflanzen im vegetationsaktiven Zustand. *Oecol. Pl.* **11**, 361–374.
LAW, R., BRADSHAW, A. D. & PUTWAIN, P. D. (1977). Life history variation in *Poa annua. Evolution* **31**, 233–246.
LEVITT, J. (1980). *Responses of Plants to Environmental Stresses*, vol. 1. *Chilling, Freezing and High Temperature Stresses.* 2nd. edn. New York: Academic Press.
LÖVE, A. & SARKAR, P. (1957). Heat tolerances of *Koenigia islandica. Bot. Notiser.* **110**, 478–481.
MOONEY, H. A. (1972). The carbon balance of plants. *A. Rev. Ecol. Syst.* **3**, 315–346.
MOONEY, H. A. & BILLINGS, W. D. (1960). The annual carbohydrate cycle of alpine plants as related to growth. *Am. J. Bot.* **47**, 594–598.
MÜLLER, M. J. (1982). *Selected Climatic Data for a Global Set of Standard Stations for Vegetation Science.* The Hague: Junk.
NOBEL, P. S. (1988). Principles underlining the prediction of temperature in plants, with special reference to desert succulents. In *Plants and Temperature*, Symp. Soc. Exp. Biol. 42 (ed. S. P. Long & F. I. Woodward), pp. 1–23. Cambridge: The Company of Biologists Limited.
PEARCY, R. W. (1977). Acclimation of photosynthetic and respiratory carbon dioxide exchange to growth temperatures in *Atriplex lentiformis* (Torr.) *Wats. Pl. Physiol.* **59**, 795–799.
PIGOTT, C. D. (1981). Nature of seed sterility and natural regeneration of Tilia cordata near its northern limit in Finland. *Ann. Bot. Fennici* **18**, 255–263.
PIGOTT, C. D. & HUNTLEY, J. P. (1978). Factors controlling the distribution of *Tilia cordata* at the northern limits of its geographical range. I. Distribution in north-west England. *New Phytol.* **81**, 429–441.
PIGOTT, C. D. & HUNTLEY, J. P. (1980). Factors controlling the distribution of *Tilia cordata* at the northern limits of its geographical range. II. History in north-west England. *New Phytol.* **84**, 145–164.
PIGOTT, C. D. & HUNTLEY, J. P. (1981). Factors controlling the distribution of *Tilia cordata* at the northern limits of its geographical range. III. Nature and causes of seed sterility. *New Phytol.* **87**, 817–839.
ROBSON, M. J. & JEWISS, O. R. (1968). A comparison of British and north African varieties of tall fescue (*Festuca arundinacea*). II. Growth during winter and survival at low temperatures. *J. appl. Ecol.* **5**, 179–190.
SAKAI, A. (1978). Freezing tolerance of evergreen and deciduous broad-leaved trees in Japan with reference to tree regions. *Low Temp. Sci. Ser. B* **36**, 1–19.
SALISBURY, F. B. & SPOMER, G. G. (1964). Leaf temperatures of alpine plants in the field. *Planta* **60**, 497–505.
SARUKHAN, J. & GADGIL, M. (1974). Studies on plant demography; *Ranunculus repens* L., *R. bulbosus* L. and *R. acris* L. III. A mathematical model incorporating multiple modes of reproduction. *J. Ecol.* **62**, 921–936.
SCHIMPER, A. F. W. (1898). *Pflanzengeographie auf physiologischer Grundlage.* Jena.
SCHNELLE, F., BAUMGARTNER, A. & FREITAG, E. (1984). *Arboreta Phaenologica* **28**.
SOLOMON, A. M., WEST, D. C. & SOLOMON, J. A. (1981). Simulating the role of climatic change

and species immigration in forest succession. In *Forest Succession Concepts and Applications.* (ed. D. C. West, H. H. Shugart & D. B. Botkin), pp. 154–177. New York: Springer-Verlag.

TRANQUILLINI, W. (1979). *Physiological Ecology of the Alpine Timberline.* Berlin: Springer-Verlag.

WALTER, H. (1979). *Vegetation of the Earth and Ecological Systems of the Geo-Biosphere.* 2nd. edn. New York: Springer-Verlag.

WAREING, P. F. (1950). Growth studies in woody species. II. Effect of day-length on shoot-growth in *Pinus sylvestris* after the first year. *Physiologia Pl.* **3**, 300–314.

WILSON, C., GRACE, J., ALLEN, S. & SLACK, F. (1987). Temperature and stature: a study of temperatures in montane vegetation. *Functional Ecology* **1**, 405–413.

WOODWARD, F. I. (1987). *Climate and Plant Distribution.* Cambridge: Cambridge University Press.

WOODWARD, F. I. & JONES, N. (1984). Growth studies of selected plant species with well-defined European distributions. I. Field observations and computer simulations on plant life cycles at two altitudes. *J. Ecol.* **72**, 1019–1030.

WOODWARD, F. I., KÖRNER, CH. & CRABTREE, R. C. (1986). The dynamics of leaf extension in plants with diverse altitudinal ranges. I. Field observations on temperature responses at one altitude. *Oecologia* **70**, 222–226.

WOODWARD, F. I. & PIGOTT, C. D. (1975). The climatic control of the altitudinal distribution of *Sedum rosea* (L.) Scop. and *S. telephium* L. I. Field observations. *New Phytol.* **74**, 323–334.

YAQUB, M. (1981). The Implications of Climate in the Control of the Distribution of *Verbena officinalis* and *Eupatorium cannabinum*. Ph.D. Thesis, University of Wales.

YLIMAKI, A. (1962). The effect of snow cover on temperature conditions in the soil and overwintering of field crops. *Ann. Agric. Fenn.* **1**, 192–216.

INTERACTION OF TEMPERATURE AND OTHER ENVIRONMENTAL VARIABLES INFLUENCING PLANT DISTRIBUTION

J. A. TEERI

The University of Michigan Biological Station, Ann Arbor, Michigan 48109, USA

Summary

Statistical analyses of the relationships between the patterns of distribution of plants with C_4 photosynthesis and Crassulacean acid metabolism (CAM) have revealed a variety of correlations with environmental variables. The worldwide abundance of C_4 grasses, relative to C_3 grasses, is highly positively correlated with growing season temperature. However, microscale analyses have revealed that C_4 grasses are more abundant than C_3 grasses in habitats with high levels of solar irradiance and low moisture availability. There are numerous exceptions to these generalizations. C_4 dicots generally are more abundant in habitats characterized by high rates of evaporation. Species possessing CAM occur in habitats having low levels of soil moisture store. In the Cactaceae such habitats also have very high potential rates of evaporation and the CAM pathway is the primary mechanism of the uptake of atmospheric CO_2. In contrast many species of the Crassulaceae grow in habitats with lower potential rates of evaporation and the CAM pathway is less important or not used at all in the uptake of atmospheric CO_2.

Introduction

Biologists have attempted for centuries to discover the causal mechanisms that result in the variety of geographical patterns of distribution and abundance of organisms on Earth. In some cases single environmental variables are observed to have a major effect on organism presence or absence. Examples include the availability of water in deserts, the lack of heat energy at high latitudes and elevations, and the lack of photosynthetically active radiation in the understorey of dense evergreen forests. However, it is a widely recognized principle of ecology that all environmental variables can potentially influence the performance of an organism. For the past several decades a considerable effort has been expended in attempting to understand what are the relative magnitudes of influence of different environmental variables on plant performance. Apart from its intrinsic value, such an understanding can have profound importance for decisions about how best to manage the native and agricultural plant communities of the earth. In particular, the relationship of temperature to other climatic variables has received much attention.

The utilization of the energy balance equation to assess the causes of variation in microclimate (Geiger, 1965, and earlier editions; Nobel, 1988) represented a major conceptual advance. This approach permits an objective evaluation of all of the processes that can potentially result in gains or dissipations of energy from environmental and organismal surfaces. Units of energy are used to quantify all of the considered variables. This permits a quantitative comparison of the effects of variation in radiant energy that is received, reflected or emitted by a surface, as well as variation in energy transferred by convection, conduction, evaporation/condensation and metabolism. Rashke (1956) first used this approach to analyse how climatic variables interact with each other and with the performance of plant leaves. The energy balance approach has been used effectively by many authors to analyse and interpret mechanisms of interactions of plants and climatic variables. Complex and often subtle interaction variables such as air temperature, humidity, wind speed, light level, the optical properties of plant organs, transpiration rates and metabolism are now better understood particularly at the individual plant level (see Nobel, 1988). Temperature is intimately related to all terms of the energy balance equation. In this chapter I will explore recent studies that have attempted to relate geographical patterns of abundance of plant species to geographical patterns of variation in climatic variables. The emphasis will be on plant families that possess C_4 photosynthesis and families that possess Crassulacean acid metabolism (CAM). I will consider the relationship of temperature to other correlated variables.

C_4 grasses

Soon after the discovery of C_4 photosynthesis in grasses it was observed that many genera possessing this pathway were tropical in origin. There were no observations of the C_4 pathway in grasses native to arctic regions. Based on a laboratory-derived understanding of the functional significance of the C_4 pathway, a number of environmental circumstances were suggested in which possession of the C_4 trait might be advantageous relative to possession of the C_3 trait. These circumstances include habitats characterized by high growing-season temperatures, high levels of solar irradiance, low levels of soil moisture availability and environments conducive to high rates of transpiration. Initially, there was a tendency to treat the above variables as being correlated, leading to the assumption that the C_4 trait should be considered as an adaptation to hot desert environments. However, subsequent biogeographical surveys revealed that C_4 grasses, while relatively abundant in hot deserts, were also abundant in mesic prairies, savannahs, and many other vegetation types. In fact, the absolute number of C_4 grass species was found to be greater in moist regions than in dry regions. These observations resulted in studies in a number of laboratories with the goal of attempting to establish the extent to which temperature and other environmental variables are correlated with patterns of distributions of C_4 species. A wide range of methods have been used in these investigations.

The various studies have attracted attention because of the striking, often near linear, positive correlations observed between change in growing season tempera-

Table 1. *The ranking of correlation coefficients of 18 climatic variables with the percent C_4 grass species in North America*

r[a]	Climatic variable
0·97[b]	Normal July minimum temperature
0·95[b]	Log mean annual degree days
0·92	Mean annual degree days
0·92	Normal July average temperature
0·89	Mean number days per year with maximum temperature ≥32 °C
0·87	Average annual potential evapotranspiration
0·82	Log mean number days per year with maximum temperature ≥32 °C
0·78	Normal July maximum temperature
0·74	Average annual temperature
0·70	Mean July dewpoint
0·68	Mean annual pan evaporation
0·60	Annual mean daily solar irradiance
0·60	Mean length annual freeze-free period
0·59	Log mean length annual freeze-free period
0·41	Normal July precipitation
0·36	Average precipitation 10 driest summers
0·08	Normal annual precipitation
−0·23	July mean daily solar irradiance

[a] Data adapted from Teeri & Stowe, 1976.
[b] These two correlation coefficients are significantly larger than any of the other correlation coefficients.

ture and change in percent C_4 grass species. In a study of North American C_4 grasses, the two variables that were most highly correlated with the percent C_4 grasses were normal July minimum temperature and log mean annual degree-days (Table 1). Both of these variables were significantly more highly correlated with percent C_4 than any of the other studied variables. There was a high correlation ($r = 0{\cdot}97$) between the normal July minimum temperature and log mean annual degree-days. However, many other variables exhibited relatively high levels of correlation with percent C_4 grasses (Table 1). The five variables most highly correlated with percent C_4 are all temperature variables that are either direct or indirect measures of growing season temperature. The sixth-ranked variable annual average potential evapotranspiration is calculated using temperature in the equation. In addition, average annual potential evapotranspiration is itself highly correlated with three temperature variables (normal July minimum temperature $r = 0{\cdot}84$; mean number of days per year with maximum temperature ≥32°C $r = 0{\cdot}90$; and mean annual degree-days $r = 0{\cdot}97$). The variables ranked 7–10 are also direct or indirect estimates of growing season temperature.

Studies conducted in separate geographical regions of the Earth have frequently identified different combinations of variables as being highly correlated with the abundance patterns of C_4 grasses (Table 2). Temperature, usually expressed or inferred to be growing season temperature, is the only variable that has been found consistently to be associated with C_4 abundance. Other variables reported

Table 2. *Environmental variables found to be highly correlated with the geographic patterns of abundance of C_4 grass species*

Region	Environmental variables	Reference
North America	Average July minimum temperature and log mean annual degree-days	Teeri & Stowe, 1976
Australia	Average summer temperature and summer precipitation	Hattersley, 1983
Costa Rica	Annual precipitation and growing season temperature	Chazdon, 1978
Sinai, Negev and Judean Deserts	Annual precipitation and growing season temperature	Vogel *et al.* 1986
South Africa	Average growing season temperature	Vogel *et al.* 1978
Southwest Africa/ Namibia	Growing season temperature[a]	Ellis *et al.* 1980
Montane tropical Africa	Temperature[b]	Livingstone & Clayton, 1980
Hawaiian Islands	Temperature, precipitation and irradiance	Rundel, 1980
Montane southeastern Arizona	Temperature, moisture availability and irradiance	Wentworth, 1983

[a] Annual rainfall was also correlated (+) with malate-forming C_4 species and (−) with aspartate-forming C_4 species.

[b] Tieszen *et al.* (1979) suggested that, in addition to temperature, soil moisture availability and level of irradiance may also be important in montane tropical Africa; further study is needed.

to be highly correlated include precipitation and irradiance. Thus the available surveys of C_4 grasses in many different regions of the world appear to have confirmed the predictions based on laboratory investigations of the physiology of C_3 and C_4 grasses. In certain ecological circumstances the possession of C_4 physiology is associated with increased abundance relative to C_3 plants in habitats that are hotter, drier, or exposed to higher levels of solar irradiance. However, as Hattersley (1983) pointed out, Australian C_4 grass species are more abundant in hot moist regions than they are in hot, dry regions. This is also true in the United States where the absolute number of C_4 grass species in the moist sub-tropical environments of southern Florida is 14 % greater than the number of C_4 species in the Sonoran Desert (Teeri, unpublished).

In those studies that place a greater emphasis on mesoscale (kilometres) and microscale (e.g. centimetres to tens of metres) variation in the environment (e.g. Gurevitch, 1986; Rundel, 1980; Wentworth, 1983) there appears to be an increase in the number of variables that are detected as important. Competitive interactions between species occur on scales of centimetres to metres. In a southern Arizona grassland, Gurevitch (1986) observed that the perennial C_3 grass *Stipa neomexicana* was restricted to dry ridge crests and about 40 other grass species, all of which were C_4, occurred on slopes and other sites that were often more moist. In this case, the restriction of the C_3 grass to the dry exposed site was found to be the result of competitive exclusion from the more favourable sites by the C_4

grasses. In the favourable sites, following precipitation the C_4 grasses appear to outcompete the C_3 grass for water.

Consideration of appropriate scales of analysis also includes taxonomic scale. For example, the C_4 grasses are comprised of three sub-types: NADP-malic enzyme, NAD-malic enzyme and PEP (phosphoenolpyrurate) carboxykinase. An analysis conducted in the Sinai, Negev and Judean deserts (Vogel *et al.* 1986) led to the conclusion that these subtypes exhibit different distribution patterns. While there is considerable overlap in their distributions, Vogel *et al.* (1986) suggest that NADP-malic enzyme grasses are more abundant in habitats where availability of moisture is not severely low. The NAD-malic enzyme species can occur in locations that are extremely dry. The PEP-carboxykinase species appear to be intermediate in their association with dry soils. In Australia, Hattersley (1983) also found that in comparison to the NAD-malic enzyme and PEP-carboxykinase types, the NADP-malic enzyme grasses were more abundant in hot moist habitats. Thus, at a finer taxonomic scale that permits discrimination among the biochemical sub-types of C_4 grasses, substrate moisture availability emerges as an important correlate of geographical distribution.

I suspect that there are several reasons why virtually all studies have identified temperature as the sole variable, or as one of several variables, correlated with C_4 abundance. The first is that temperature does have a direct effect on the biology of C_4 grasses. For example, the quantum yield of C_4 plants does not vary with changing leaf temperature. In contrast, the quantum yield of C_3 plants exhibits a pronounced temperature dependence. The quantum yields of C_3 plants become increasingly greater than those of C_4 plants as leaf temperatures decline from *ca* 30°C. In contrast the quantum yields of C_3 plants become increasingly lower than those of C_4 plants as leaf temperatures rise above 30°C. Using a computer simulation, Ehleringer (1978) showed that these differences could potentially account for a differential temperature sensitivity of the daily net photosynthesis of C_3 and C_4 grass canopies. At higher temperatures C_4 performance becomes increasingly more efficient than C_3 performance. A second reason is that temperature, per se, is the environmental variable that is usually highly correlated with measures of growing season moisture availability and levels of solar irradiance. As a result of mesoscale and microscale heterogeneity in measures of moisture availability and irradiance the importance of these variables may be less readily detectable by the approaches used in macroscale studies. Supporting evidence for this suggestion was provided by a microscale analysis of the local habitat preferences of C_3 and C_4 grasses of North America. In that study (Teeri & Stowe, 1976; Teeri, 1979), the C_4 grasses were observed to be significantly more abundant in microsites characterized by high levels of irradiance and low levels of moisture availability.

While all studies have identified a positive, near-linear, correlation between growing season temperature and C_4 grass abundance, there is considerable variation in the absolute relationship of temperature to abundance. The floristic crossover temperature is the growing season temperature at that location along an

Table 3. *The observed floristic crossover temperatures for C_3 and C_4 grasses in different geographic regions of the earth*

Crossover temperature[a] (°C)	Region	Reference
18	North America	Teeri & Stowe, 1976
14–15	Australia	Hattersley, 1983
13	Southeastern Arizona	Wentworth, 1983
9	Hawaii	Rundel, 1980
8	Kenya	Rundel, 1980

[a] Mean minimum temperature of the warmest month at the geographic location where the flora contains equal numbers of C_3 and C_4 grass species.

environmental gradient where the grass flora is composed of equal numbers of C_3 and C_4 species. With decreasing latitude the floristic crossover temperature becomes substantially lower (Table 3). Rundel (1980) suggested that the greater abundance of C_4 grasses at lower temperatures in the tropics is in part due to the smaller annual variation in temperature at low latitudes. The C_4 photosynthetic pathway evolved in tropical groups of grasses. Presumably, the long evolutionary history of the groups in tropical environments has resulted in many other adaptations that may permit these grasses to outperform non-tropical grasses in their native habitats, even at reduced growing-season temperatures. Regardless of the cause, the data in Table 3 clearly indicate that there is a great amount of quantitative variation in the relationship between temperature and the relative abundances of C_3 and C_4 grasses.

A similar analysis of the distributions of C_4 species in the Cyperaceae (Teeri *et al.* 1980) also resulted in the finding that the square of July minimum daily temperature ($r = 0{\cdot}93$) was most highly correlated with C_4 species abundance. However, in this case there were six other variables that were also highly correlated with percent C_4 Cyperaceae. Two of these variables were measures of the tendency for evaporation (log potential evapotranspiration $r = 0{\cdot}92$; potential evapotranspiration $r = 0{\cdot}91$). The remaining four variables were measures of temperature (July minimum daily temperature $r = 0{\cdot}93$; mean annual degree-days $r = 0{\cdot}91$; number of days per year with maximum temperature $\geqslant 32°C$ $r = 0{\cdot}91$; July average daily temperature $r = 0{\cdot}90$). In the stepwise multiple regression equation for the C_4 Cyperaceae two moisture variables (mean annual pan evaporation; June–August precipitation) were substantially more important than any temperature variables. Future research should attempt to determine the underlying causes of the differences in environmental correlates between the C_4 grasses and the C_4 sedges.

C_4 dicotyledonous plants

There have been fewer attempts to analyse the relationships of the geographical

Table 4. *The climatic variables found to be most highly correlated with the percent C_4 or C_3 species in eight C_4 dicot families in North America*[a] *(Data from Teeri & Stowe, 1978)*

r	Climatic variable
C_4 species as percent Spermatophyte flora	
0·95	Mean May–October pan evaporation
0·93	Mean annual pan evaporation
0·93	Mean annual dryness ratio
C_3 species as percent Spermatophyte flora	
0·89	Mean May–October pan evaporation
C_4 species as percent C_3 and C_4 species in eight families	
0·77	Log mean May–October pan evaporation

[a] The eight families are: Aizoaceae, Amaranthaceae, Capparidaceae, Chenopodiaceae, Euphorbiaceae, Nyctaginaceae, Portulacaceae and Zygophyllaceae.

patterns of distribution of C_4 species of the Dicotyledonae. Doliner & Jolliffe (1979) compared the habitat preferences of two sub-samples of C_3 and C_4 species native, respectively, California and central Europe. They lumped monocot and dicot species in their analysis; however, the great majority of both C_3 and C_4 species were in dicot families. On both continents the C_4 species were found to be statistically more abundant in habitats characterized by indices of high temperature and low moisture availability.

An analysis of the climatic distributions of C_4 species in eight dicot families of North America revealed that indices of the level of summer evaporation were the variables most highly positively correlated with the abundance of C_4 species (Table 4). The C_4 species of the eight families were lumped and analysed as a percent of the Spermatophyte flora of different regions. The three variables most highly correlated with percent C_4 were summer pan evaporation, annual pan evaporation and annual dryness ratio. The dryness ratio is '...the ratio of the annual net radiation to the heat energy required to evaporate the mean annual precipitation' (Stowe & Teeri, 1978).

The strength of the correlations and the fact that all three variables were measures of the tendency for evaporation led to the conclusion that on average the climatic preferences of C_4 dicots were different from those of C_4 grasses. A second analysis of the data was performed, expressing the C_4 species as a percent of the total number of C_3 and C_4 species in the eight families for each geographical region. Again (Table 4), summer pan evaporation emerged as the variable most highly correlated with C_4 abundance. A final analysis compared the relative abundance of C_3 species in the eight C_4 families (Table 4). The C_3 species were also more highly correlated with summer pan evaporation than other studied variables. Thus, unlike the case with the C_4 grasses, both C_3 and C_4 species in the eight dicot families are more abundant in hot deserts, and the variable summer pan evaporation best describes this relationship.

Turitzin & Drake (1981) observed that the erect architecture of a C_4 grass canopy permits the plants to intercept high levels of sunlight with an associated high level of canopy net photosynthesis. When such canopies were naturally or artificially flattened both light interception and canopy net photosynthesis were substantially reduced. It remains to be determined to what extent the average difference in canopy architecture between C_4 grasses and C_4 dicots is related to their different strength of correlation to temperature and evaporation.

Crassulacean acid metabolism

It has long been recognized the Crassulacean acid metabolism (CAM) is a type of photosynthesis that has the potential to confer on plants a great reduction in transpirational water loss during growth. In addition photosynthetic organs that are capable of CAM are succulent, often massively so, and capable of storing large quantities of water, particularly in comparison to the photosynthetic organs of C_3 and C_4 plants. Two widespread CAM families in North America are the Cactaceae and the Crassulaceae. The Cactaceae are primarily stem succulents and the Crassulaceae primarily leaf succulents.

Not surprisingly, for both CAM families the environmental variables (Table 5) found to be most highly correlated with patterns of abundance were measures of environmental moisture availability (Teeri *et al.* 1978). The species of the Cactaceae achieve their greatest abundance, expressed as percent of the total Spermatophyte flora, in the warm deserts of the southwestern United States and northwestern Mexico.

The abundance of species of the Cactaceae was most highly correlated (negatively) with the log of the January coefficient of humidity, an estimate of soil moisture availability. According to this correlation Cactus species become increasingly abundant as soil moisture storage declines. The coefficient of

Table 5. *The ranking of correlation coefficients of climatic variables with the abundance of species in the Cactaceae and Crassulaceae. (Data from Teeri* et al. *1978, Table 3)*

r	Climatic variable
Cactaceae	
−0·93	Log January coefficient of humidity
0·87	Annual pan evaporation
0·87	Summer pan evaporation
0·87	March daily temperature range
0·85	Mean annual lake evaporation
Crassulaceae	
−0·74	July coefficient of humidity
−0·68	Summer precipitation
−0·68	Average precipitation in 10 driest summers
−0·68	Warm season precipitation
0·63	July mean daily irradiance

humidity was not significantly more highly correlated with Cactus abundance than the other four listed variables in Table 5. Three of these variables are measures of the tendency for evaporation and the fourth is the March daily temperature range. The multiple regression equation (Teeri *et al.* 1978) that best predicted cactus species' abundance also included the log of the January coefficient of humidity, the March daily temperature range, autumn (September–November) precipitation, annual precipitation, and July coefficient of humidity. Intuitively, the strong negative correlation with soil moisture storage makes sense for the Cactaceae which, instead of utilizing water stores in the soil for growth, utilize water stored in their succulent organs and tissues. Daily temperature amplitude has been demonstrated to be an important determinant of rates of carbon gain by CAM plants conducting dark uptake of atmospheric CO_2 during growth. Cactus species in the southwestern United States frequently grow following winter rains. The daily temperature range for March, and other winter months (Teeri *et al.* 1978) was more highly correlated with species abundance than were the daily temperature ranges for May, July or October.

The Crassulaceae of the United States and Canada exhibit their greatest abundance in the Spermatophyte floras of the mountain ranges along the Pacific coast from Canada to southern California. The variable most highly correlated (negatively) with Crassulaceae species' abundance was also an estimate of soil moisture storage, July coefficient of humidity (Table 5). Three precipitation variables were all negatively correlated with abundance and July mean daily irradiance was positively correlated with abundance. In the multiple regression equation that best predicted abundance of the Crassulaceae the important variables were July coefficient of humidity (negative), winter precipitation (positive) and dryness ratio (positive). Taken together, the evidence suggests that lack of available moisture in the summer and dependable winter precipitation are important correlates of abundance of the Crassulaceae.

The environmental correlations of the Cactaceae and Crassulaceae differ in several respects. The correlations with measures of soil moisture storage and rates of evaporation are high for the Cactaceae. The correlations with measures of soil moisture storage and summer precipitation are, respectively, lower for the Crassulaceae. Growing season daily temperature range was important for the Cactaceae but not for the Crassulaceae. These observations led to the suggestion that the control of transpiration rate via dark uptake of atmospheric CO_2 might be of greater importance in determining the abundance patterns of species of the Cactaceae than those of the Crassulaceae. A subsequent survey of field biomass $\delta^{13}C$ values (Teeri, 1982) revealed that most, but not all, species of the Crassulaceae in the United States and Canada have very negative (*ca* -27 ‰) $\delta^{13}C$ values strongly indicating that atmospheric CO_2 is captured in the daylight via C_3 photosynthesis. In contrast, many studies have shown that most Cactus species have less negative $\delta^{13}C$ values (e.g. -12 ‰) strongly suggesting that dark CO_2 uptake by CAM is the major contributor to growth.

It is now clear that there are many genetically determined CAM phenotypes.

Future research can exploit these variant types to improve the understanding of how different patterns of day and night CO_2 uptake contribute to fitness in different environments.

Discussion

It must be re-emphasized (e.g. Teeri & Stowe, 1976) that statistical analyses of the correlations between species distribution patterns and environmental variables can never establish the underlying causal relationships. Analyses such as those discussed here are useful in identifying the possible ranking of the importance of environmental variables as determinants of species distribution patterns. However, the limitations of statistical analyses (e.g. Stowe & Teeri, 1978; Hattersley, 1983) must be understood.

Choice of scale is a critical factor in these studies. For example, numerous laboratory and field measurements have consistently demonstrated that most C_4 plants exhibit net photosynthetic rates that do not become light saturated in maximum natural sunlight. It would be reasonable to expect that irradiance would emerge as a variable highly correlated with C_4 species distribution patterns. However, several marcoscale statistical analyses (Hattersley, 1983; Doliner & Jolliffe, 1979; Teeri & Stowe, 1976) did not identify the studied measures of irradiance to be well correlated with C_4 distributions. However, at the microscale several investigators have found that, on average, C_4 grass species occur in microhabitats that have significantly higher levels of irradiance than do C_3 grass species.

Statistical analyses have been helpful in identifying environmental variables to which high priority should be given in initial attempts to discover underlying causal relationships. Statistical analyses have also been important in revealing the great variability in the relative importance of different environmental variables in different ecological settings. These analyses have also revealed that not all C_4s are similar in environmental requirements or tolerances. This strongly suggests that both the C_4 trait and the CAM trait may confer many different kinds of adaptative significance in different groups and in different settings.

From the perspective of the energy balance of a plant, a change in the magnitude of any term in the energy balance equation has the potential for changing the amount of sensible heat storage and thus changing the temperature of the organism. In this sense, organismal temperature integrates all of the energy exchange processes of the organism. From the perspective of the organism's local habitat, changes in the magnitude of any term in the energy balance of the ground surface has the potential for changing the amount of energy available to heat the air above the soil surface. Such a change is likely to be accompanied by a change in air temperature.

It is trite but true that there is always an interaction of temperature with other environmental variables in determining the geographic patterns of distribution of plant species. The question is the extent to which any individual variable, or a particular combination of variables, predominate in shaping the pattern of distribution.

In the case of the C_4 dicots the evidence strongly suggests that environmental aridity, as measured by summer pan evaporation, is more important than temperature, per se, in characterizing environments favoured by these species. However, the C_3 species in the eight studied C_4 dicot families were also relatively more abundant in environments characterized by high levels of summer pan evaporation. This raises the possibility that other characteristics in the evolutionary background of the C_4 dicots are equally or more important than photosynthetic type in determining their presence in arid regions.

The type of photosynthetic physiology possessed by a plant is only one of a very large number of physiological, anatomical, morphological and developmental characteristics that ultimately determine its growth, likelihood of survival and fitness. Plants with different types of photosynthesis have the potential to have different rates of light harvesting, CO_2-uptake, water loss and sensitivity to temperature. Statistical analyses of the correlation of climatic variables to distribution have revealed that, on average, these functional differences often are reflected in broad geographic patterns of abundance. However, the many other phenotypic properties of these plants are also involved in determining plant performance. Exceptions to average correlations are abundant. Many C_4 plants grow in wet habitats (Robichaux & Pearcy, 1984; Teeri, 1979; Hattersley, 1983), other C_4 species occur in cool, deeply shaded habitats (Robichaux & Pearcy, 1980). There are many C_3 species that grow in hot, dry habitats. Some CAM plants occur in relatively mesic regions with very cold winters and other CAM plants grow as submerged aquatics (Keeley, 1981). The task that remains is to understand how natural selection acts to produce the various combinations of characteristics that are combined to result in populations of organisms that are adapted to various kinds of environmental circumstances.

Future directions of research

Patterson (1983) has summarized a series of careful investigations comparing how differences at the physiological, anatomical, morphological and developmental levels can all contribute to whole plant response to variation in temperature, light level, soil moisture availability, and atmospheric water vapour pressure deficit. Using such an integrative approach, it is possible to develop predictions as to what types of environments will be more or less favourable to a particular species as it is presently genetically constituted. The increasing availability of easily portable, battery-powered instrumentation such as infrared gas analysers, steady state porometers, dew point hygrometers and environmental measurement instruments offers the opportunity to extend such studies to a wide range of species and habitats. It is now time to develop testable hypotheses based on the correlations produced in the statistical analyses. Such an approach will probably combine the energy balance perspective, growth analysis, an understanding of competitive interactions and the determinants of fitness under field conditions.

References

CHAZDON, R. L. (1978). Ecological aspects of the distribution of C_4 grasses in selected habitats of Costa Rica. *Biotropica* **10**, 265–269.

DOLINER, L. J. & JOLLIFFE, P. A. (1979). Ecological evidence concerning the adaptive significance of the C_4 dicarboxylic acid pathway of photosynthesis. *Oecologia* **38**, 23–34.

EHLERINGER, J. R. (1978). Implications of quantum yield differences on the distributions of C_3 and C_4 grasses. *Oecologia* **31**, 255–267.

ELLIS, R. P., VOGEL, J. C. & FULS, A. (1980). Photosynthetic pathways and the geographical distribution of grasses in South West Africa/Namibia. *S. Afr. J. Sci.* **76**, 307–314.

GEIGER, R. (1965). *The Climate near the Ground.* Transl. by Scripta Technica, Inc., pp. 1–611. Cambridge: Harvard University Press.

GUREVITCH, J. (1986). Competition and the local distribution of the grass *Stipa neomexicana*. *Ecology* **67**, 46–57.

HATTERSLEY, P. W. (1983). The distribution of C_3 and C_4 grasses in Australia in relation to climate. *Oecologia* **57**, 113–128.

KEELEY, J. E. (1981). *Isoetes howellii*: a submerged aquatic CAM plant? *Am. J. Bot.* **68**, 420–424.

LIVINGSTONE, D. A. & CLAYTON, W. D. (1980). An altitudinal cline in tropical African grass floras and its paleoecological significance. *Quat. Res.* **13**, 392–402.

NOBEL, P. S. (1988). Principles underlying the prediction of temperature in plants, with special reference to desert succulents. In *Plants and Temperature*, Symp. Soc. Exp. Biol., vol. 42 (ed. S. P. Long & F. I. Woodward), pp. 1–23. Cambridge: Company of Biologists.

PATTERSON, D. T. (1983). Phenotypic and physiological comparisons of field and phytotron grown plants. In *Crop Reactions to Water and Temperature Stresses in Humid, Temperate Climates* (ed. C. D. Raper, Jr & P. J. Kramer), pp. 299–314. Boulder, Colorado: Westview Press.

RASCHKE, K. (1956). Über die physikalischen Beziehungen zwischen Wärmeübergangszahl, Strahlungsaustausch, Temperatur and Transpiration eines Blattes. *Planta* **48**, 200–238.

ROBICHAUX, R. H. & PEARCY, R. W. (1980). Photosynthetic responses of C_3 and C_4 species from cool shaded habitats in Hawaii. *Oecologia* **47**, 106–109.

ROBICHAUX, R. H. & PEARCY, R. W. (1984). Evolution of C_3 and C_4 plants along an environmental moisture gradient: patterns of photosynthetic differentiation in Hawaiian *Scaevola* and *Euphorbia* species. *Am. J. Bot.* **71**, 121–129.

RUNDEL, P. W. (1980). The ecological distribution of C_4 and C_3 grasses in the Hawaiian Islands. *Oecologia* **45**, 354–359.

STOWE, L. G. & TEERI, J. A. (1978). The geographic distribution of C_4 species of the Dicotyledonae in relation to climate. *Am. Naturalist* **112**, 609–623.

TEERI, J. A. (1979). The climatology of the C_4 photosynthetic pathway. In *Topics in Plant Population Biology* (ed. O. T. Solbrig, S. Jain, G. B. Johnson & P. H. Raven), pp. 356–374. New York: Columbia Univ. Press.

TEERI, J. A. (1982). Carbon isotopes and the evolution of C_4 photosynthesis and Crassulacean acid metabolism. In *Biochemical Aspects of Evolutionary Biology* (ed. M. H. Nitecki), pp. 93–130. Chicago and London: The University of Chicago Press.

TEERI, J. A. & STOWE, L. G. (1976). Climatic patterns and the distribution of C_4 grasses in North America. *Oecologia* **23**, 1–12.

TEERI, J. A., STOWE, L. G. & LIVINGSTONE, D. (1980). The distribution of C_4 species of the Cyperaceae in North America in relation to climate. *Oecologia* **47**, 307–310.

TEERI, J. A., STOWE, L. G. & MURAWSKI, D. A. (1978). The climatology of two succulent plant families: Cactaceae and Crassulaceae. *Can. J. Bot.* **56**, 1750–1758.

TIESZEN, L. L., SENYIMBA, M. M., IMBAMBA, S. K. & TROUGHTON, J. H. (1979). The distribution of C_3 and C_4 grasses and carbon isotope discrimination along an altitudinal and moisture gradient in Kenya. *Oecologia* **37**, 337–350.

TURITZIN, S. N. & DRAKE, B. G. (1981). Canopy structure and the photosynthetic efficiency of a C_4 grass community. In *Photosynthesis VI. Photosynthesis and Productivity, Photosynthesis and Environment* (ed. G. Akoyunoglou), pp. 73–80. Philadelphia: Balaban International Science Services.

VOGEL, J. C., FULS, A. & ELLIS, R. P. (1978). The geographical distribution of Kranz grasses in South Africa. *S. Afr. J. Sci.* **74**, 209–215.

VOGEL, J. C., FULS, A. & DANIN, A. (1986). Geographical and environmental distribution of C_3 and C_4 grasses in the Sinai, Negev, and Judean deserts. *Oecologia* **70**, 258–265.

WENTWORTH, T. R. (1983). Distributions of C_4 plants along environmental and compositional gradients in southeastern Arizona. *Vegetatio* **52**, 21–34.

TEMPERATURE AS A DETERMINANT OF PLANT PRODUCTIVITY

JOHN GRACE

Department of Forestry and Natural Resources, University of Edinburgh, Edinburgh EH9 3JU, UK

Summary

Quite small variations in temperature such as those that exist between the north and south of Britain, or on a local scale between north and south aspects of a hill, may cause large variations in plant productivity. An attempt is made to assess the effect of a 1°C increase in the mean temperature of the growing season by using published data from various sources. The conclusion reached is that a 1°C increase in a north temperate climate may be expected to increase plant productivity by about 10%, providing that other factors like water or nutrients do not become limiting. But, for annual crops, a negative effect of temperature has sometimes been observed. This may be because temperature speeds up the development of the crop and thus reduces the duration of photosynthesis. In the natural environment variations in solar radiation and wind speed may often exert an effect on growth because they influence the energy balance, altering the temperature of meristems and other plant tissues in subtle ways which have a significant impact on the pattern of carbon utilization. A more precise understanding of the relationship between growth and weather variables is unlikely to be achieved unless models are developed which enable integration of component processes of growth with what is known about the spatial and temporal variation in the weather variables.

Introduction

Temperature is one of the most important components of the plant environment. Within the British Isles, regional variations in mean air temperature seem rather slight yet they are biologically highly significant. The difference in the annual mean of air temperature between the north of Scotland and the south of England is only 3°C but the former has an impoverished flora and lies within a different bioclimatic zone (Table 1). Altitudinal variation is more than this (6–10°C per km above sea level), and topographic variation may be highly significant for plants: Rorison *et al.* (1986) found about a 1°C difference between the annual mean of air temperature at a north- versus a south-facing slope in a Derbyshire dale. Year-to-year variation at any one site is of a similar magnitude (Table 1), and longer term trends are thought to be in the range 0·1–0·4°C per decade (Clark, 1982).

Microclimatological variation is likely to be of particular significance within

Table 1. *Variation in mean temperature near sea level over the British Isles*

Location	Latitude (degrees and minutes N)	Mean annual temperature (°C)	Mean July temperature (°C)
Lerwick	60 09	7·1 ± 0·5	11·7 ± 0·8
Stornaway	58 12	8·3 ± 0·4	12·9 ± 0·7
Kew	54 24	10·6 ± 0·5	17·5 ± 1·0
Plymouth	50 23	10·7 ± 0·5	15·9 ± 0·8

short vegetation, where vertical gradients may be as high as 1°C mm^{-1} in sunshine; and in the surface layers of exposed soil.

The purpose of this article is to assess the sensitivity of plant production to such variations in temperature, making use of data from a variety of sources.

Data

In the field

On a global scale it is self-evident that temperature and water availability are major determinants of biological productivity. Cold places contain fewer species, with a biological spectrum displaced towards dwarf life forms, and the standing crop and productivity are very low (Raunkiaer, 1934; Rosenzweig, 1968; Lieth, 1972; Box, 1981). Field measurements of community productivity have now been made in natural and semi-natural vegetation world-wide, many of them as part of the International Biological Programme. Lieth (1972) attempted to correlate these data on net primary productivity with temperature and rainfall and, on the basis of the relationships, draw a world map of productivity. Accurate field measurements of net primary productivity are notoriously difficult to achieve because of sampling problems, especially when considering the below-ground fraction; but perhaps this is the best place to begin our survey of data for, despite difficulties, the data were collected in natural vegetation, as opposed to in agricultural systems or controlled environment rooms. However, in reviewing this work it is as well to remember the limitations of the data: as a result of sampling errors each determination has a precision of perhaps only ±20%, and (more seriously) there may well be a systematic error introduced by various assumptions concerning the growth of the below-ground fraction. The data suggest that in the range of annual temperatures prevailing over much of the earth's inhabited surface (0–30°C) the relationship between annual dry matter production and temperature can be considered to be linear (Fig. 1), although Lieth (1972) fitted a sigmoid curve. At mean temperatures equal to those of the south of England (about 10°C) the effect of a one degree shift in temperature, on a linear basis, would be to increase productivity by 10% or about 140 g m^{-2}. Another representation of the data is possible: curve A forms the upper limit of an envelope which encloses all the data and suggests what might be the effect of temperature if other variables were optimal. Using curve A, we obtain a figure of only 3%.

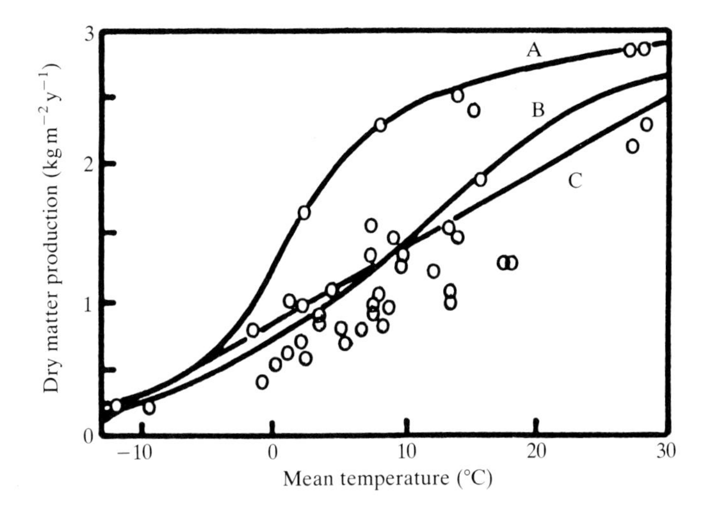

Fig. 1. The effect of mean annual temperature on annual dry matter production. Data points from Lieth (1972). Lines represent: A, hand-drawn envelope representing possible maximum production, that might be obtained if all other factors were optimal; B, curve fitted by Lieth (1972); C, linear regression using least squares.

An alternative approach to studying the variation in productivity over a large area would be to examine the variation at a single site in relation to the annual variations in temperature over a long period of time. In agricultural crops there have been several attempts to do this. In Britain the most notable data set in existence is perhaps that of the long record of wheat yields at the Broadbalk field, Rothamsted. At this site, wheat was harvested in most years starting from 1853. The data have been analysed by Fisher (1924), Tippett (1926) and Buck (1961). It should be borne in mind that the data here are not those of rates of total dry matter accumulation, but rather the yield of grain after a growing period which is likely to have varied with the weather. Moreover, Rothamsted is in one of the drier parts of the British Isles, and it would be unwise to generalize beyond this region or, for that matter, beyond this crop. The method of analysis is of considerable historical interest: for each climatological variate a fifth-order polynomial was fitted to describe the variate's trend over the year. This was used to define month-by-month estimates of each independent variate. Then, taking the entire data set (values of final harvest each year), correlations were sought between monthly values of, e.g., rainfall and yield. The conclusion was that rainfall often accounted for a large part of the year-to-year variation in yield (Fisher, 1924), hours of sunshine sometimes did too (Tippett, 1926) but temperature never did (Buck, 1961).

Monteith (1981*a*) drew attention to a *negative correlation* between the yield of wheat crops in certain parts of southern England and the mean temperature of the period May, June and July. He proposed an interesting explanation of this phenomenon, which may apply to many annual crops, as follows. The rate at which the crop develops from sowing, through phenological stages to flowering, is often determined by temperature (or a combination of temperature and photoperiod). Thus, in warm years, the *duration* of the crop is less and consequently the total photosynthesis and therefore dry-matter production is also reduced, even though the rate of photosynthesis per area of leaf per unit of time may be enhanced. Other explanations are possible: warm years tend to be dry years and crop growth in southern England is probably more limited by water stress in such years.

It might be fruitful to make comparisons of such crop–weather relationships in other parts of the country, including northern and western regions where water stress is less important. An alternative approach is to utilize a natural altitudinal gradient to obtain a range of experimental conditions.

Several weather variables change with altitude, but the most important may be temperature, and so this work furnishes an estimate of the effect of temperature on grass growth. Attempts have been made to estimate the effect of altitude on the performance of the grass crop in the Scottish uplands (Hunter & Grant, 1971). Water deficits are generally slight in these conditions. The authors found that floral development was delayed by 0·04 day m^{-1} and yield by 0·1 % m^{-1}. Since air temperatures decline by 7–10 °C km^{-1}, the implied rate of change of yield with respect to temperature is of the order of 10 % $°C^{-1}$. Prince (1976) grew barley at high altitude in the English Pennines and observed that it produced seeds one month later than at lowland sites, the grains being reduced in weight.

In the case of trees, a rich source of data on the effect of temperature on productivity comes through the analysis of radial increments obtained by taking cores from the stem. Productivity is in this case radial trunk growth and not yield as in the studies just referred to. Over northern Europe, the growth of Scots pine *Pinus sylvestris* has been shown to be strongly correlated with summer temperature (Mikola, 1962; Hughes *et al.* 1984). The correlation between temperature and increment becomes progressively stronger as one approaches the altitudinal or latitudinal limits of the species. In Finland (Mikola, 1962), the correlation coefficient between July temperature and stem increment was about 0·85 in the case of one data set (Fig. 2), and the relationship is linear throughout the entire range of July temperatures occurring naturally. The warmest years in Finland have a July temperature rather like the mean for central Britain, and here a 1 °C decline in temperature is likely to cause a 20 % reduction in the width of stems. However, total timber production is probably not simply related to radial increment: for one thing, the narrow increments formed at low temperatures have a higher density of latewood (Schweingruder, Braker & Schar, 1979); and for another, the height increment may not be well related to girth increment. Nevertheless, the information which can be obtained from tree stems is potentially very important, enabling the relative importance of weather variables to be assessed at widely-scattered sites over an entire continent (Schweingruber *et al.* 1979). So far, this information has been exploited mainly by dendrochronologists in the reconstruction of past climates, and only to a limited extent in the assessment of plant sensitivity to the weather. Other approaches to the analysis of tree growth have

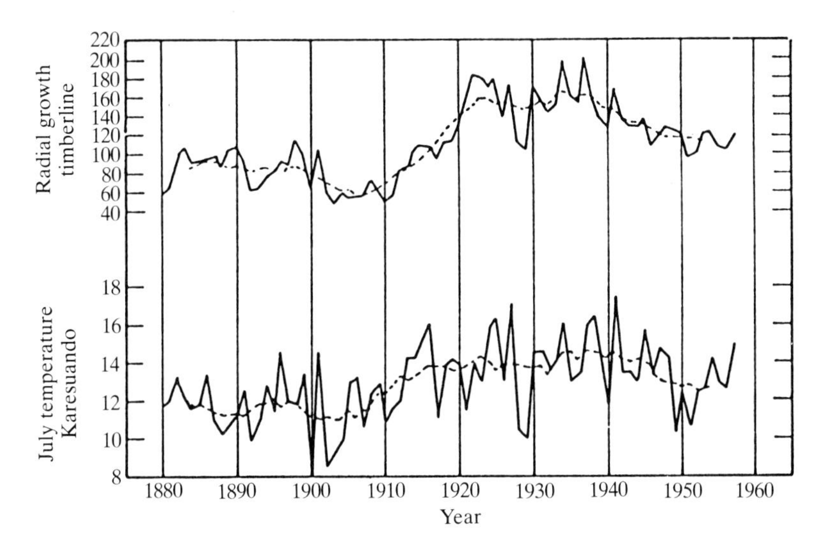

Fig. 2. Radial growth of *Pinus sylvestris* in Finland, and the July temperatures at a nearby site. Reproduced with permission from Mikola (1962).

been employed. The day-to-day or week-to-week measurement of height growth has often been related to temperature (Millar, 1965; White, 1972, 1974), though there are difficulties in the analysis of data of this type, as discussed by Ford & Milne (1981).

The strong and nearly linear relationship between temperature and growth of *Pinus sylvestris*, suggested by Mikola's data, has been recently confirmed in a controlled environment study (Juntilla, 1986).

In controlled environments

In controlled environment rooms the temperature, light or humidity regime can be varied whilst other variables are held constant. Plants are generally grown without mutual interference and they can be well watered and well supplied with nutrients, but for technical reasons the photon flux density has often been extremely low. Plants are usually grown for a few weeks during which they are in the vegetative part of their life cycle. Many authors have assessed plant performance as relative growth rate (RGR), and so it is possible to compare the results from different laboratories (Fig. 3). The result shows clearly that over the temperature range typical of the north temperate summer (6–18 °C) the relative growth rate increases sharply with temperature and in most cases the relationship can be approximated by a straight line. The relative growth rate of wild or semi-wild plants increased by about 10 % °C^{-1} at 15 °C, and that of C_3 crop plants by rather less (7 % °C^{-1}). As RGR is related exponentially to plant weight, these differences would result in a very large variation in weight after several weeks.

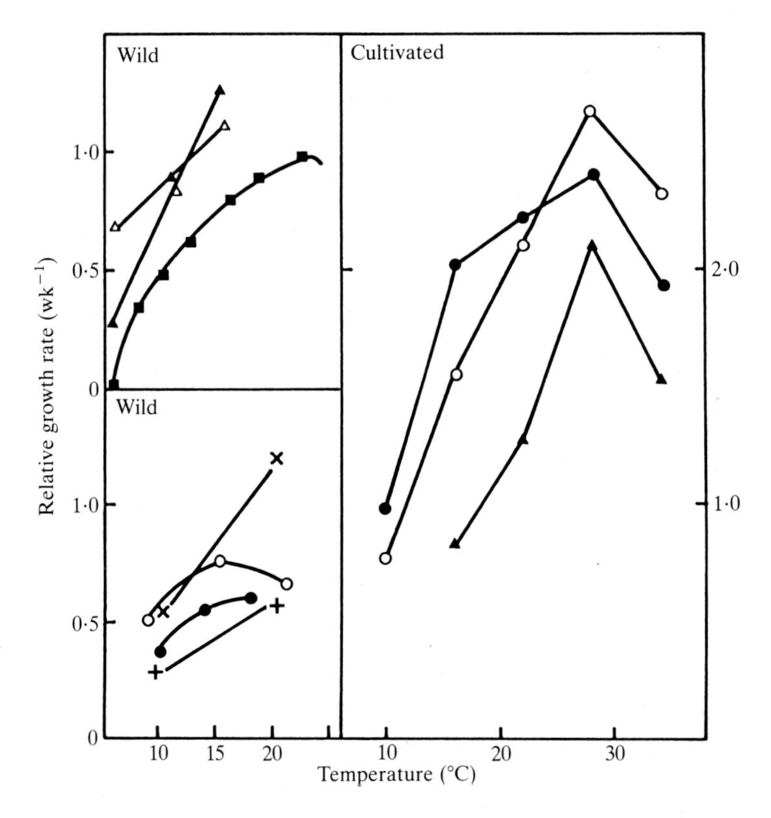

Fig. 3. Relative growth rates of wild plants and cultivated crops plotted against temperature. Wild plants: ●, *Plantago maritima* (Arnold, 1974); ○, *Poa pratensis* (Hay & Heide, 1983); ×, *Dactylis glomerata* (Woodward, 1979); +, *Sesleria albicans* (Woodward, 1979); △, *Geum urbanum* (Graves & Taylor, 1986); ▲, *Geum rivale* (Graves & Taylor, 1986); ■, *Urtica dioica* (Furness, 1978), Cultivated plants: ●, rape; ○, sunflower; ▲, maize (Warren-Wilson, 1966).

Temperatures of plant organs

Coupling

One of the difficulties in establishing relationships between temperature and productivity is that organs of the plant in the natural environment occur along a microclimatological gradient, which in some cases is very steep. Consequently, the temperature of the air above the vegetation may bear only a weak relationship to that of the meristems and other tissues.

It is only relatively recently that attention has been focussed on meristem rather than foliar temperatures. This follows work by Watts (1971) and Peacock (1975) which demonstrated that the temperature of basal meristems was more important than that of leaves in determining extension growth of grasses. It is now clear that cell division and extension are generally more sensitive to temperature than is the rate of photosynthesis, and that growth is less dependent on photosynthesis than was previously thought.

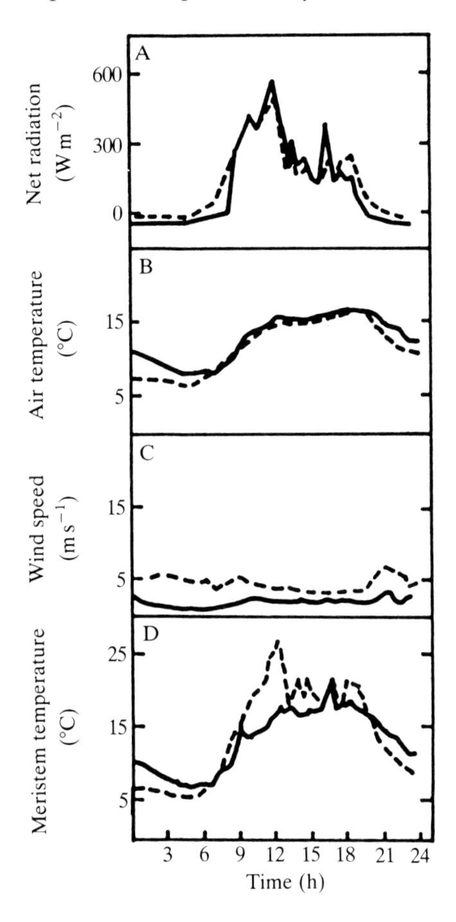

Fig. 4. Diurnal trends in climatological variables and meristem temperatures measured at a montane site in tall (16 m) and dwarf (0·1 m) vegetation: (——), (*Pinus sylvestris*) forest at 450 m above sea-level; (– – –), *Arctostaphylos uva-ursi* dwarf shrub at 650 m above sea-level. From Wilson *et al.* (1986).

It may be that the temperatures of meristems are more closely affected by fluctuations in radiation or wind speed than they are by air temperature, so any attempt to correlate growth and screen temperature would be fruitless. This situation has been examined in short montane vegetation by Wilson *et al.* (1987). The meristem temperatures of the terminal shoots were measured using fine thermocouples attached within dwarf vegetation to *Arctostaphylos uva-ursi* and, simultaneously, to the corresponding organs in a forest of *Pinus sylvestris* (Fig. 4). In the tall vegetation, meristem temperature was coupled to air temperature: during the day it was never more than a few degrees higher than air temperature and always peaked at the same time as air temperature. In contrast, the meristem temperatures of short vegetation correlated strongly with net radiation, displaying peaks whenever the sun shone, when the absolute temperatures became considerably warmer than those of the air (at night, net radiation is negative and the meristems of the short vegetation may be a few degrees colder than during the day). Thus, even though the dwarf vegetation occurs at a higher altitude than the tall vegetation, the meristems spend much of their day at higher temperatures than those of the pine trees (see also Körner & Larcher, 1988). The difference in the nature of this coupling appears to be entirely due to the difference in stature *per se*. Short vegetation is aerodynamically smooth, and so turbulent mixing between the air at the surface and that in the atmosphere as a whole, is correspondingly poor.

Very high temperatures have often been observed in dwarf alpine vegetation on sunny days and are remarkable because they are sometimes so close to the lethal point (Larcher, 1980).

What structural features of the vegetation determine the extent of coupling of meristem temperatures to air temperatures? In general, coupling to air temperature requires the boundary layer surrounding the organ to be relatively thin, with efficient mixing between the air in the canopy and that of the atmosphere as a whole. Tight coupling to air temperature occurs when leaves are small (low aerodynamic resistance) and canopies are rough (large-scale turbulent eddies). The concept of coupling was introduced by Monteith (1981*b*) and later developed by McNaughton & Jarvis (1984) who introduced the omega factor as an index of coupling between the plant and the regional climate.

In general, broad-leaved herbaceous vegetation is not particularly well coupled to air temperature, especially in bright sunshine and very cold or very hot environments, a point made many years ago by Linacre (1964, 1967). In many cases, plants growing in hot environments display leaf temperatures which are up to ten degrees cooler than air temperatures as a result of high rates of transpiration (e.g. Lange, 1959; Smith, 1978; Althawadi & Grace, 1986). Similarly, plants in arctic, alpine and montane environments can be up to 20 °C warmer than air (Cernusca, 1976; Körner & Cochrane, 1983; Larcher, 1980), displaying temperatures which fluctuate in relation to radiation and wind speed.

Much information concerning the influence of temperature upon growth has been obtained from controlled environment rooms where it has been the air temperature that has been controlled. In most cases, the temperatures of shoot and root meristems have not been recorded yet alone controlled. In older controlled environment rooms the radiation regime has been very low ($<100\,W\,m^{-2}$) so tissue temperatures may not have departed from air temperatures by more than 5 °C. In modern installations, metal halide lamps are often used to provide realistically-high radiation fluxes, and here it is very important to measure tissue temperatures and to provide air speeds in the cabinet which reduce the aerodynamic resistances as much as possible without causing unwanted physical damage to the plants.

In glasshouses the tissue temperatures are often more than 5 °C different from air temperatures as a result of low air speeds, and considerable variation may occur from place to place according to distance from the ventilator fans.

Models and interpretations

Elucidating a quantitative, mechanistic relationship between temperature and yield, or moving towards 'a biological explanation of yield' remains a major goal for plant scientists. Statistical relationships between yield and temperature, as we have seen, are often weak and consequently of limited practical value. The problem of predicting plant productivity is in two parts. The first part is to predict the rate of phenological development from germination (or budbreak in woody perennials), through leaf development to the production of flowers (or the onset of dormancy in woody perennials). The second part of the problem is to predict the carbon assimilation and distribution during the growth phase. In general, the first part of the problem is easier than the second.

Heat sum and development

One of the first authors to investigate the quantitative relationship between temperature and plant performance was Réaumur, who soon after inventing his thermometer used it to explore the effect of temperature on the dates of harvest of grapes and wheat in France (Réaumur, 1735). He proposed the 'heat-unit' law, that a fixed 'quantity of heat' is needed for a plant to reach a particular stage of development. This 'law', though frequently violated, has proved useful in many instances, leading to the widespread use of degree-days (or degree-days above a certain threshold temperature) as an index of the extent of the growing season (e.g. Hopkins, 1938; Manley, 1952; Nuttonson, 1955; Chandler & Gregory, 1976). The law may be stated in relation to any time interval but the most common is degree-days:

$$S = \sum_{\text{day}=1}^{\text{day}=j} (\bar{t} - t_0)$$

where S is the sum of degree-days, also called thermal time, t_0 is the threshold temperature for the onset of development also called the base temperature and $\bar{t}$ is the mean temperature of the day. The utility of this approach is illustrated with reference to data of Ong (1983*a*): when development of pearl millet from early and late sowings (corresponding to two quite different temperature regimes) was plotted against time, two distinct progressions were observed. However, when the independent variate was degree-days the two disparate data sets fell on the same line (Fig. 5). Crop physiologists have found excellent linear relationships between state of development and accumulated temperature in a wide range of species, including north temperate trees (Sucoff, 1971; Hari, 1972; Landsberg, 1974; Cannell & Smith, 1984), temperate crops (Milford & Riley, 1980; Russell *et al.* 1982), tea bushes in Malawi (Squire, 1979) and tropical graminaceous crops (Ong, 1983*a*,*b*; Hadley *et al.* 1983; Roberts, Hadley & Summerfield, 1985). The developmental stages of the crop have been variously defined, for example germination → floral initiation → anthesis → maturity. A natural objection to modelling development as a function of accumulated temperature is that the approach

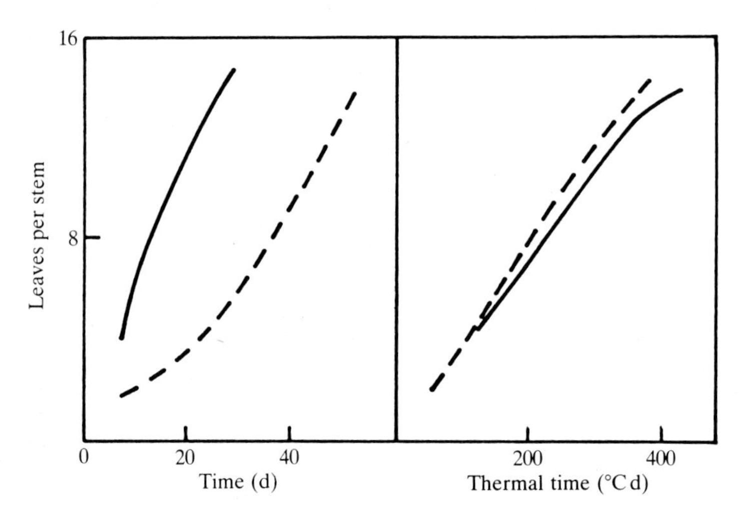

Fig. 5. Illustrating the utility of accumulated temperature or thermal time, using data from Ong (1983*a*) on pearl millet grown at high (——) and low (---) temperatures.

neglects the role of other climatic variables which interact with temperature and genotype (Aitken, 1974). However, simple quantitative models involving photoperiod as well as temperature can often be fitted to the data (Nuttonson, 1955; Robertson, 1968; Porter & Delecolle, 1988).

Growth room studies have been particularly useful in elucidating the separate roles of temperature and photoperiod in the progression of development towards flowering. When tropical legumes were grown in factorial combinations of photoperiod and day/night temperature the reciprocal of the time from sowing to first flowering ($1/f$) was a linear function of mean temperature $\bar{t}$ (Hadley *et al.* 1983, 1984; Roberts, Hadley & Summerfield, 1985):

$$1/f = a + b\bar{t}$$

where a and b are empirical coefficients determined from regression analysis (this representation implies that flowering occurs when a thermal sum of $1/b$ above a base temperature of $-a/b$ has been attained). In those genotypes that were sensitive to photoperiod, and in the range of photoperiod over which sensitivity occurred, an analogous relationship could be found to relate flowering time to photoperiod at a given temperature. Overall, a general equation was proposed to enable an estimate of the time of flowering to be made from a knowledge of photoperiod p and mean temperature:

$$1/f = a + bp + c\bar{t}$$

The model may be particularly useful in the screening of genotypes to determine their suitability for use in different climatic zones. The coefficients determined in the growth room may not be directly applicable in the field, where the environment is different in several respects, and where more variables come into play. However, from similar work in the field on other types of crop grown at

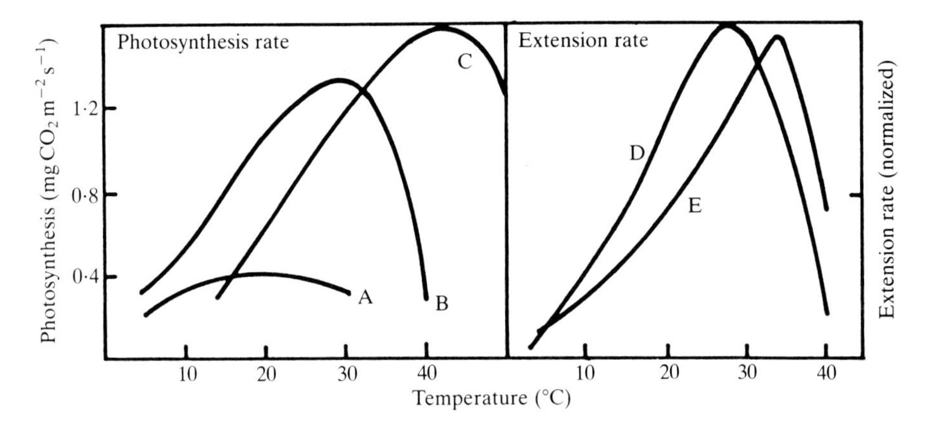

Fig. 6. Rates of photosynthesis of A, a C_3 grass of north temperate distribution (*Sesleria caerulea*, Lloyd & Woolhouse, 1976); B, a north temperate C_4 grass (*Spartina anglica*, Long & Woolhouse, 1978) and C, a C_4 plant of the hot desert in North America (*Tidestromia oblongifolia*, Bjorkman, 1981). Also, rates of extension growth of D, a C_3 grass (*Lolium perenne*, Peacock, 1975) and E, a C_4 grass (*Zea mays*, Watts, 1971). These rates have dimensions of length time^{-1} but have been normalized so that the maximum value is unity.

different latitudes it seems likely that this general form of equation may indeed be useful in predicting the time of flowering (Nuttonson, 1955; Porter & Delecolle, 1988).

Temperature and vegetative growth

The linearity between state of development and accumulated temperatures can hold only if rates development (e.g. rate of leaf production) are themselves linear with respect to temperature. For this to hold, it seems probable that rates of physiological processes would also have to be linear functions of temperature. Rates of leaf extension and photosynthesis have now been measured in many species and it is clear that the linearity applies approximately within a restricted range, often 5–20 °C (Fig. 6). Tissue temperatures often exceed this range and so the good linear fit of the model is surprising.

The component processes of growth may not all have the same sensitivity to temperature. Inspection of published temperature responses suggests, for example, that the rate of photosynthesis is much less sensitive to temperature than is the rate of leaf extension growth and may have a lower threshold temperature (Thorne *et al.* 1967). In many plants, a reduction in temperature with the onset of winter causes an increase in the pool sizes of soluble carbohydrates, presumably a consequence of these differences (Eagles, 1967). There is also strong evidence from classical growth analysis to suggest that the effect of temperature on growth is greater than the effect of temperature on photosynthesis. For example, Warren-Wilson (1966) shows that the specific leaf area is the variable most sensitive to temperature with the ratio of area to weight increasing with temperature.

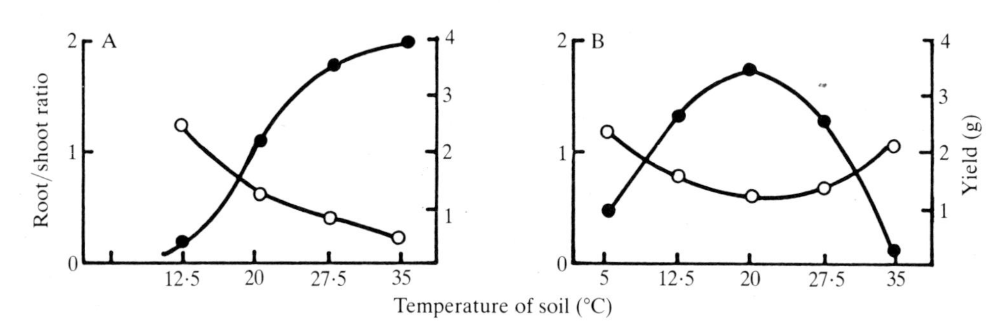

Fig. 7. The influence of soil temperature on the root/shoot ratio (○) and final yield (●) of pasture species (Davidson, 1969). A, subtropical species; B, temperate species.

Recently, a similar but even more pronounced trend has been shown in pearl millet and ground nut (Overseas Development Administration, 1987).

The growth rate of a plant is bound to depend on the partitioning of assimilate between root and shoot, and it is notable that this is also very sensitive to environmental fluctuations, especially root temperature. In experiments on 12 pasture species, in which leaf temperature fluctuated diurnally whilst root temperature was varied independently, very pronounced sensitivity of the root:shoot ratio was observed (Fig. 7, from Davidson, 1969). At the optimum temperature for growth, dry weight partitioning to the roots was generally at a minimum.

In view of the different temperature sensitivities of physiological processes, and the importance of partitioning in determining the rate of growth, there is an urgent need to construct detailed models of plant growth as a vehicle for the integration of knowledge and as an aid in exploring plant response to the weather. Progress in this field has been surprisingly slow, not because of a lack of models but because the experimental and observational data required to test them over a wide range of conditions are hard and expensive to obtain (e.g. Cooper & Thornley, 1976; Lang & Thorpe, 1983; Johnson, 1985; Johnson & Thornley, 1985*a,b*).

As an illustration of this modelling approach, a scheme of the carbon and nitrogen balance is shown in Fig. 8. The influence of temperature on the process of phloem transport is rather small, as far as can be judged from experiments in which stems and petioles have been chilled whilst observations have been made on translocation of marker substances (Geiger & Sovonick, 1970; Lang, 1974; Watson, 1975; Helms & Wardlaw, 1977). Thus, interest centres on the relationship between the activity of each organ as a 'sink' for carbon, and the supply of carbon (the 'source'). Difficulty in treating flows of carbon from source to sink as simply diffusional (as we do in the case of water transport) arises because of the chemical transformations of carbon from one chemical form to another, and because of the complex compartmentalization within the tissue. Were it not for this complication, it might be possible to model the utilization of substrate for structural growth using classical enzyme Michaelis–Menten kinetics.

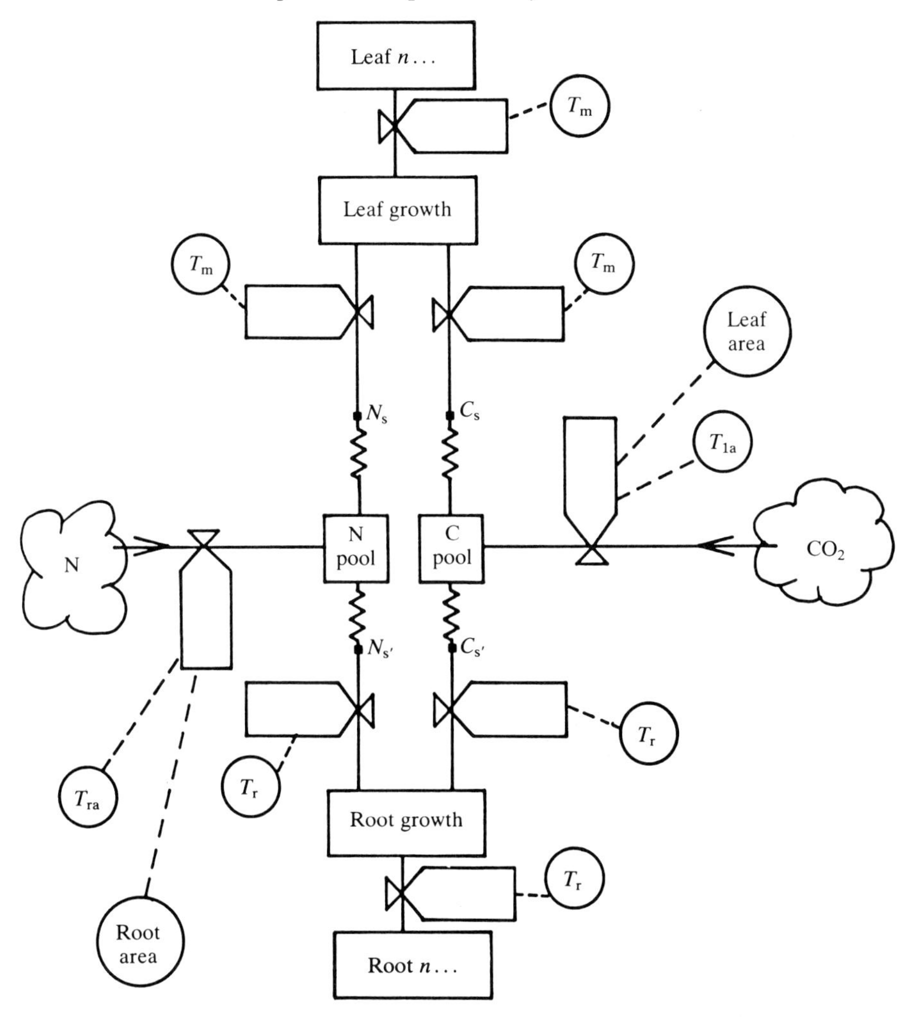

Fig. 8. Model of plant growth to show the points of influence of temperature. Developing leaves and roots take up carbon and nitrogen from substrates at concentrations C_s and N_s respectively, at rates determined by leaf meristem temperatures T_M and root meristem temperatures T_R. Supply of C and N from the external environment depends on the temperature of the leaf area T_{LA} and root area T_{RA} respectively. Transport resistances (-WW-) are only weakly dependent on temperature.

Conclusions

At northern latitudes there is no doubt that small variations in temperature may influence the distribution of species and the performance of those crops which are near the limits of their climatological range. In the majority of cases warmer years or warmer microclimates give rise to higher yield. In annual crops there is some evidence of a rather weak or even negative relationship between yield and temperature. It has been suggested that this occurs because higher temperatures speed up development so that less time is available for photosynthesis before seeds are produced.

The thermal time concept has been widely used to predict the progress through phenological stages to flowering. In many cases a photoperiod term is required as well, and there may be scope for the addition of other terms to improve the accuracy of the prediction in the field.

Productivity and yield depend on the expansion and maintenance of leaf area, and the rate of net photosynthesis of the leaf area. This is more difficult to predict from climatological data as models which carry out an hour-by-hour or day-by-day integration are called for.

I wish to acknowledge the useful discussion I had with Paul Hadley regarding the possible negative effects of temperature on yield.

References

AITKEN, Y. (1974). Flowering time, climate and genotype. Melbourne: Melbourne University Press.

ALTHAWADI, A. M. & GRACE, J. (1986). Water use by the desert cucurbit *Citrullus colocynthis* L. Schrad. *Oecologia* **70**, 475–480.

ARNOLD, S. M. (1974). The relationship between temperature and seedling growth of two species which occur in Upper Teesdale. *New Phytol.* **73**, 333–340.

BJÖRKMAN, O. (1981). The response of photosynthesis to temperature. In *Plants and their Atmospheric Environment* (ed. J. Grace, E. D. Ford & P. G. Jarvis), pp. 273–301. Oxford: Blackwell Scientific Publications.

BOX, E. O. (1981). *Macroclimate and Plant Forms: an introduction to predictive modelling in phytogeography*. The Hague: Junk.

BUCK, S. F. (1961). The use of rainfall, temperature, and actual transpiration in some crop-weather investigations. *J. agric. Sci., Camb.* **57**, 355–365.

CANNELL, M. G. R. & SMITH, R. I. (1984). Spring frost damage on young *Picea sitchensis*. II. Predicted dates of budburst and probability of frost damage. *Forestry* **57**, 177–197.

CERNUSCA, A. (1976). Bestandesstruktur, Bioklima und Energiehaushalt von alpinen Zwergstrauchbeständen. *Oecologia Plantarum* **11**, 71–102.

CHANDLER, T. J. & GREGORY, S. (1976). *The Climate of the British Isles*. London: Longman.

CLARK, W. C. (1982). *Carbon Dioxide Review: 1982*. Oxford: Clarendon.

COOPER, A. J. & THORNLEY, J. H. M. (1976). Responses of dry matter partitioning, growth and carbon and nitrogen levels in the tomato plant to changes in root temperature: experiment and theory. *Ann. Bot.* **40**, 1139–1152.

DAVIDSON, R. L. (1969). Effect of root/leaf temperature differentials on root/shoot ratios in some grasses and clover. *Ann. Bot.* **33**, 561–569.

EAGLES, C. F. (1967). Variation in the soluble carbohydrate content of climatic races of *Dactylis glomerata* (cocksfoot) at different temperatures. *Ann. Bot.* **31**, 645–651.

FISHER, R. A. (1924). The influence of rainfall on the yield of wheat at Rothamsted. *Phil. Trans. R. Soc.* **213**, 89–142.

FORD, E. D. & MILNE, R. (1981). Assessing plant response to the weather. In *Plants and their Atmospheric Environment* (ed J. Grace, E. D. Ford & P. G. Jarvis), pp. 333–362. Oxford: Blackwell Scientific Publications.

FURNESS, S. B. (1978). Effects of temperature upon growth. Annual Report of the Unit of Comparative Plant Ecology (NERC), 1978, pp. 5–6. Sheffield: University of Sheffield.

GEIGER, D. R. & SOVONICK, S. A. (1970). Temporary inhibition of translocation velocity and mass transfer rates by petiole cooling. *Pl. Physiol.* **46**, 847–849.

GRAVES, J. D. & TAYLOR, K. (1986). A comparative study of *Geum rivale* L. and *G. urbanum* L.

to determine those factors controlling their altitudinal distribution. 1. Growth in controlled and natural environments. *New Phytol.* **104**, 681–691.

Hadley, P., Roberts, E. H., Summerfield, R. J. & Minchin, F. R. (1983). A quantitative model of reproductive development in cowpea (*Vigna unguiculata* (L.) Walp.) in relation to photoperiod and temperature and implications for screening germplasm. *Ann. Bot.* **51**, 531–543.

Hadley, P., Roberts, E. H., Summerfield, R. J. & Minchin, F. R. (1984). Effects of temperature and photoperiod on flowering in soyabean (*Glycine max* (L.) Merrill): a quantitative model. *Ann. Bot.* **53**, 669–681.

Hari, P. (1972). Physiological stage of development in biological models of growth and maturation. *Ann. Bot. Fenn.* **9**, 107–115.

Hay, R. K. M. & Heide, O. M. (1983). Specific photoperiodic stimulation of dry matter production in a high-latitude cultivar of *Poa pratensis. Physiologia Pl.* **57**, 135–142.

Helms, K. & Wardlaw, I. F. (1977). Effect of temperature on symptoms of tobacco mosaic virus and movement of photosynthate in *Nicotiniana glutinosa. Phytopathology* **67**, 344–350.

Hopkins, A. D. (1938). Bioclimates: a science of life and climate relations. Miscellaneous Publications of the U.S. Department of Agriculture, no. 280, Washington DC.

Hughes, M. K., Schweingruber, F. H., Cartwright, D. & Kelly, P. M. (1984). July–August temperature at Edinburgh between 1721 and 1975 from tree-ring density and width data. *Nature* **308**, 341–344.

Hunter, R. F. & Grant, S. A. (1971). The effect of altitude on grass growth in east Scotland. *J. appl. Ecol.* **8**, 1–20.

Hustich, I. (1947). Climate fluctuation and vegetation growth in northern Finland during 1890–1939. *Nature* **160**, 478–479.

Johnson, I. R. (1985). A model of the partitioning of growth between the shoots and roots of a vegetative plant. *Ann. Bot.* **55**, 421–431.

Johnson, I. R. & Thornley, J. H. M. (1985*a*). Temperature dependence of plant and crop processes. *Ann. Bot.* **55**, 1–24.

Johnson, I. R. & Thornley, J. H. M. (1985*b*). Dynamic model of the response of a vegetative grass crop to light, temperature and nitrogen. *Plant, Cell Envir.* **8**, 485–499.

Juntilla, O. (1986). Effects of temperature on shoot growth in northern provenances of *Pinus sylvestris* L. *Tree Physiology* **1**, 185–192.

Körner, Ch. & Cochrane, P. (1983). Influence of plant physiognomy on leaf temperature on clear midsummer days in the Snowy Mountains of south-eastern Australia. *Acta Oecologia/Oecologia Plantarum* **4**, 117–124.

Körner, Ch. & Larcher, W. (1988). Plant life in cold climates. In *Plants and Temperature*, Symp. Soc. Exp. Biol., vol. 42 (ed. S. P. Long & F. I. Woodward), pp. 25–57. Cambridge: Company of Biologists.

Landsberg, J. J. (1974). Apple fruit bud development and growth: analysis and an empirical model. *Ann. Bot.* **38**, 1013–1023.

Lang, A. (1974). The effect of petiolar temperature on the translocation rate of ^{137}Cs in the phloem of *Nymphoides peltata. J. exp. Bot.* **25**, 71–80.

Lang, A. & Thorpe, M. R. (1983). Analysing partitioning in plants. *Pl. Cell Envir.* **6**, 267–274.

Lange, O. L. (1959). Untersuchungen über Wärmehaushalt und Hitzeresistenz mauretanischer Wüsten- und Savannenpflanzen. *Flora* **147**, 595–651.

Larcher, W. (1980). *Physiological Plant Ecology*, 2nd Edition. Berlin: Springer-Verlag.

Lieth, H. (1972). Modelling the primary productivity of the world. *Tropical Ecology* **13**, 125–130.

Linacre, E. T. (1964). A note on a feature of leaf and air temperatures. *Agric. Met.* **1**, 66–72.

Linacre, E. T. (1967). Further notes on a feature of leaf and air temperatures. *Arch. Met. Geophys. Bioklim.* **15**, 422–436.

Lloyd, N. D. H. & Woolhouse, H. W. (1976). The effect of temperature on photosynthesis and transpiration in populations of *Sesleria caerulea. New Phytol.* **77**, 553–559.

Long, S. P. & Woolhouse, H. W. (1978). The response of net photosynthesis to light and temperature in *Spartina townsendii* (*sensu lato*), a C_4 species from a cool temperate climate. *J. exp. Bot.* **29**, 803–814.

McNaughton, K. G. & Jarvis, P. G. (1984). Using the Penman-Monteith equation predictively. In *Evapotranspiration from Plant Communities* (ed. M. L. Sharma), pp. 263–278. Amsterdam: Elsevier.

MANLEY, G. (1952). *Climate and the British Scene*. London: Collins.

MIKOLA, P. (1962). Temperature and tree growth near the northern timber line. In *Tree Growth* (ed. T. T. Kozlowski), pp. 256–274. New York: Ronald Press.

MILFORD, G. F. R. & RILEY, J. (1980). The effects of temperature on leaf growth of sugar beet varieties. *Ann. appl. Biol.* **94**, 431–443.

MILLAR, A. (1965). The effect of temperature and day length on the height growth of birch *Betula pubescens* at 1900 feet in the northern Pennines. *J. appl. Ecol.* **2**, 17–29.

MONTEITH, J. L. (1981*a*). Climatic variation and the growth of crops. *J. R. Met. Soc.* **107**, 749–774.

MONTEITH, J. L. (1981*b*). Coupling of plants to the atmosphere. In *Plants and their Atmospheric Environment* (ed. J. Grace, E. D. Ford & P. G. Jarvis), pp. 1–29. Oxford: Blackwell Scientific Publications.

NUTTONSON, M. Y. (1955). *Wheat-climate relationships and the use of phenology in assertaining the thermal and photo-thermal requirements of wheat*. Washington, DC: American Institute of Crop Ecology.

ONG, C. K. (1983*a*). Response to temperature in a stand of pearl millet (*Pennisetum typhoides* S. & H.). 1. Vegetative development. *J. exp. Bot.* **34**, 322–326.

ONG, C. K. (1983*b*). Response to temperature in a stand of pearl millet (*Pennisetum typhoides* S. & H.). 4. Extension of individual leaves. *J. exp. Bot.* **34**, 1731–1739.

OVERSEAS DEVELOPMENT ADMINISTRATION (1987). *Microclimatology in Tropical Agriculture*, Volume 1. London: ODA.

PEACOCK, J. M. (1975). Temperature and leaf growth in *Lolium perenne* II. The site of temperature perception. *J. appl. Ecol.* **12**, 115–124.

PEACOCK, J. M. & SHEEHY, J. E. (1974). The measurement and utilisation of some climatic resources in agriculture. In *Climatic Resources and Economic Activity* (ed. J. A. Taylor), pp. 87–107. London: David & Charles.

PORTER, J. R. & DELECOLLE, R. (1988). Interaction of temperature with other environmental factors in controlling the development of plants. In *Plants and Temperature*, Symp. Soc. Exp. Biol., vol. 42 (ed. S. P. Long & F. I. Woodward), pp. 133–156. Cambridge: Company of Biologists.

PRINCE, R. D. (1976). The effect of climate on grain development in barley at an upland site. *New Phytol.* **76**, 377–389.

RAUNKIAER, C. (1934). *The Life Forms of Plants and Statistical Plant Geography*. (Translation from 1909 original). Oxford: Oxford University Press.

REAUMUR, R. A. F. (1735). Observations du thermomètre faites à Paris pendant l'année 1735, comparées avec celles qui ont été faites sous la ligne, à l'Isle de France, à Alger, et an quelques-unes de nos iles de l'Amérique. Pp. 545–575, Memoires de l'Academie Royal des Sciences, Paris (cited by Aitken, 1974).

ROBERTS, E. H., HADLEY, P. & SUMMERFIELD, R. J. (1985). Effects of temperature and photoperiod on flowering in chickpeas (*Cicer arietinum* L.). *Ann. Bot.* **55**, 881–892.

ROBERTSON, G. W. (1968). A biometeorological time scale for a cereal crop involving day and night temperatures and photoperiod. *Int. J. Biomet.* **12**, 191–223.

RORISON, I. H., SUTTON, F. & HUNT, R. (1986). Local climate topography and plant growth in Lathkill Dale NNR 1. A twelve-year summary of solar radiation and temperature. *Pl. Cell Envir.* **9**, 49–56.

ROSENZWEIG, M. L. (1968). Net primary productivity of terrestrial communities: predictions from climatological data. *Am. Nat.* **102**, 67–74.

RUSSELL, G., ELKS, R. P., BROWN, J., MILBOURN, G. M. & HAYTER, A. M. (1982). The development and yield of autumn- and spring-sown barley in south east Scotland. *Ann. appl. Biol.* **100**, 167–178.

SCHWEINGRUBER, F. H., BRAKER, O. U. & SCHAR, E. (1979). Dendroclimatic studies on conifers from central Europe and Great Britain. *Boreas* **8**, 427–452.

SMITH, W. K. (1978). Temperatures of desert plants: another perspective on the adaptability of leaf size. *Science* **201**, 614–616.

Squire, G. R. (1979). Weather, physiology and seasonality of tea (*Camellia sinensis* L.) yields in Malawi. *Exp. Agric.* **15**, 321–330.
Sucoff, E. (1971). Timing and rate of bud formation in *Pinus resinosa*. *Can. J. Bot.* **49**, 1821–1832.
Thorne, G. N., Ford, M. A. & Watson, D. J. (1967). Effects of temperature variation at different times on growth and yield of sugar beet and barley. *Ann. Bot.* **31**, 71–101.
Tippett, L. H. C. (1926). On the effect of sunshine on wheat yield at Rothamsted. *J. Agric. Sci., Camb.* **16**, 159–165.
Warren-Wilson, J. (1966). Effect of temperature on net assimilation rate. *Ann. Bot.* **30**, 753–761.
Watson, B. T. (1975). The influence of low temperatures on the rate of translocation in the phloem of *Salix viminalis* L. *Ann. Bot.* **39**, 889–900.
Watts, W. R. (1971). Role of temperature in the regulation of leaf extension in *Zea mays*. *Nature* **229**, 46–47.
White, E. J. (1972). Orthogonalized regressions of height increments on meteorological variables. Research Papers in Forest Meteorology: an Aberystwyth Symposium (ed. J. A. Taylor), pp. 109–125. Aberystwyth: Cambrian News.
White, E. J. (1974). Multivariate analysis of tree height increment on meteorological variables, near the altitudinal tree limit in northern England. *Int. J. Biomet.* **8**, 199–210.
Wilson, C., Grace, J., Allen, S. & Slack, F. (1987). Temperature and stature: a study of temperatures in montane vegetation. *Functional Ecology* **1**, 391–397.
Woodward, F. I. (1979). The differential temperature response of the growth of certain plant species from different altitudes. 1. Growth analysis of *Phleum alpinum* L., *P. bertolonii* D.C., *Sesleria albicans* Kit. and *Dactylis glomerata* L. *New Phytol.* **82**, 385–395.

Printed in Great Britain © *Society for Experimental Biology 1988*

TEMPERATURE AND SEED GERMINATION

E. H. ROBERTS

Department of Agriculture, University of Reading, Earley Gate, P.O. Box 236, Reading RG6 2AT, UK

Summary

Temperature can affect the percentage and rate of germination through at least three separate physiological processes.

1. Seeds continuously deteriorate and, unless in the meanwhile they are germinated, they will ultimately die. The rate of deterioration depends mainly on moisture content and temperature. The Q_{10} for rate of loss of viability in orthodox seeds consistently increases from about 2 at −10°C to about 10 at 70°C.
2. Most seeds are initially dormant. Relatively dry seeds continuously lose dormancy at a rate which is temperature-dependent. Unlike enzyme reactions, the Q_{10} remains constant over a wide range of temperature at least up to 55°C, and typically has a value in the region of 2·5–3·8. Hydrated seeds respond quite differently: high temperatures generally reinforce dormancy or may even induce it. Low temperatures may also induce dormancy in some circumstances, but in many species they are stimulatory (stratification response), especially within the range −1°C to 15°C. Small, dormant, hydrated seeds are usually also stimulated to germinate by alternating temperatures which typically interact strongly and positively with light (and often also with other factors including nitrate ions). The most important attributes of alternating temperatures are amplitude, mean temperature, the relative periods spent above and below the median temperature of the cycle (thermoperiod) and the number of cycles.
3. Once seeds have lost dormancy their rate of germination (reciprocal of the time taken to germinate) shows a positive linear relation between the base temperature (at and below which the rate is zero) and the optimum temperature (at which the rate is maximal); and a negative linear relation between the optimal temperature and the ceiling temperature (at and above which the rate is again zero). The optimum temperature for germination rate is typically higher than that required to achieve maximum percentage germination in partially dormant or partially deteriorated seed populations.

None of the sub-cellular mechanisms which underlie any of these temperature relations are understood. Nevertheless, the temperature responses can all be quantified and are fundamental to designing seed stores (especially long term for genetic conservation), prescribing germination test conditions, and understanding seed ecology (especially that required for the control of weeds).

Introduction

There are two main attributes of germination which can be affected by temperature – how many and how quickly the seeds germinate (i.e. the percentage and rate of germination). Temperature can affect both attributes in three distinct ways – through its effect on seed deterioration (ageing), through its effect on loss of dormancy, and through its effect on the germination process itself.

Effects of temperature on seed deterioration culminating in loss of viability

So far as deteriorative processes and storage characteristics are concerned, seeds may be divided into two categories – orthodox and recalcitrant (Roberts, 1973*b*). Orthodox seeds can be dried to low moisture contents, typically to 5 % or less, without damage. Between this value and a critical moisture content, which is typically 15–18 % (all moisture contents are expressed on a wet weight basis in this paper) in oily seeds or 26–28 % in non-oily seeds (Roberts, 1986), the lower the moisture content the greater the longevity. It is probable that the critical moisture content is equivalent to a water activity of about 0·95 (i.e. in equilibrium with a relative humidity of 95 %) (Roberts, 1986), or a water potential in the region of −14·5 MPa (Roberts, 1988).

In contrast, recalcitrant seeds – which are typically large and fleshy and often (but not always) produced by woody species – are immediately damaged even by slight desiccation. Because they cannot be dried, neither can they be cooled to freezing temperatures without being killed, because of ice formation within the tissues. Furthermore, most tropical species of recalcitrant seeds – e.g. rubber [*Hevea brasiliensis* (Willd ex Adr. Juss) Muell, Arg.] and cocoa (*Theobroma cacao* L.) – are also subject to 'chilling injury' at temperatures of about 10°C and below. Except for these limitations, in general the lower the temperature the greater the lifespan. Even under the best storage conditions, however, – i.e. in moist, aerobic storage in cool temperatures – seldom can recalcitrant seeds be stored for more than a few weeks or months (King & Roberts, 1980; Roberts & King, 1980, 1982). Because of the difficulty of working with these species the relation between temperature and lifespan has not yet been defined.

In the case of orthodox seeds the relation between temperature and longevity is well defined at moisture contents below the critical value; and indeed it appears to be similar, and possibly identical, in all species in this category which have so far been examined. The relation between temperature, moisture content and longevity was first fully defined in barley (*Hordeum vulgare* L.) (Ellis & Roberts, 1980*a*,*b*) (Fig. 1), and the same type of relationship has since been confirmed in other species – e.g. in onion (*Allium cepa* L.) (Ellis & Roberts, 1981), in soyabean [*Glycine max* (L.) Merr.], chickpea (*Cicer arietinum* L.), cowpea [*Vigna unguiculata* (L.) Walp.] (Ellis *et al.* 1982) (Fig. 2), and lettuce (*Lactuca sativa* L.) (Kraak & Vos, 1987) (Fig. 3). In all these papers it was shown that, under a wide range of storage conditions, percentage viability could be accurately predicted by the equation

$$v = K_i - p/10^{K_E - C_W \log m - C_H t - C_Q t^2} \qquad \text{(Eqn 1)}$$

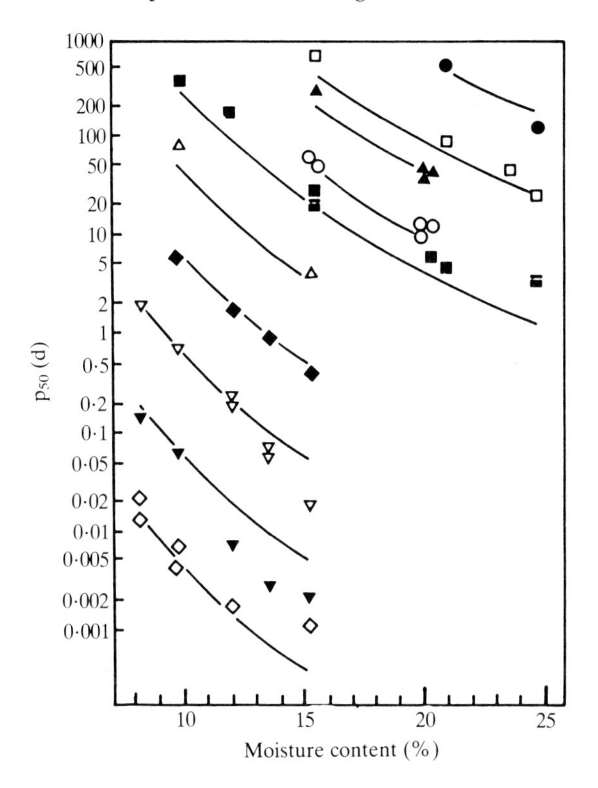

Fig. 1. The relation between moisture content, temperature and time taken for seed viability to fall to 50% (p_{50}) (log scale) in barley. Curves fitted according to Eqn 1. Temperatures: 3°C (●); 20°C (□); 25°C (▲); 35°C (○); 40°C(■); 50°C (△); 60°C (◆); 70°C (▽); 80°C (▼); 90°C (◇). (From Ellis & Roberts, 1980*b*.)

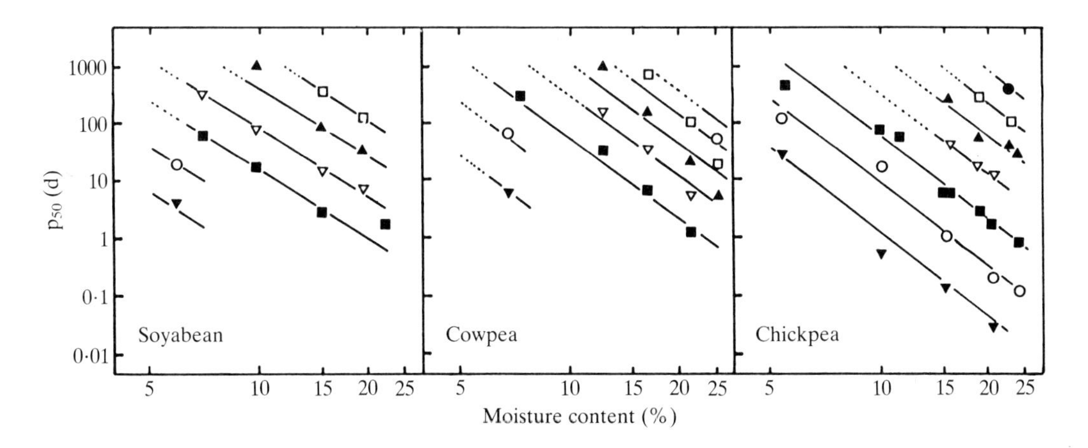

Fig. 2. The relation between moisture content (log scale), temperature and time taken for seed viability to fall to 50% (p_{50}) (log scale) in soyabean, cowpea and chickpea. Temperatures: 10°C (●); 20°C (□); 30°C (▲); 40°C (▽); 50°C (■); 60°C (○); 70°C (▼). (From Ellis *et al.* 1982.)

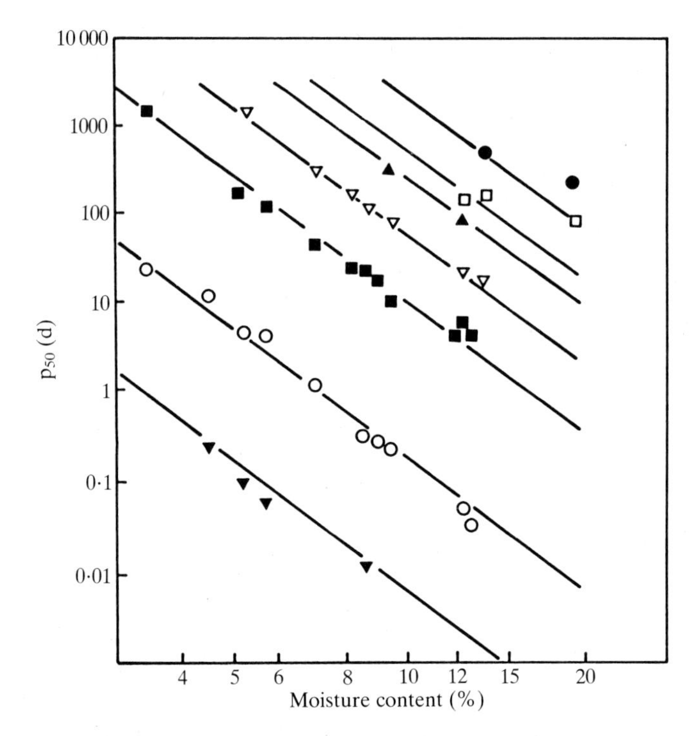

Fig. 3. The relation between moisture content (log scale) and time taken for seed viability to fall to 50% (p_{50}) (log scale) in lettuce. Temperature: 5°C (●); 15°C (□); 20°C (▲); 30°C (▽); 40°C (■); 60°C (○); 75°C (▼). (From Kraak & Vos, 1987.)

in which v is probit of percentage viability after any storage period, p (days), at any temperature, t (°C), and moisture content, m (%), within the constraints previously mentioned. The seed-lot constant (K_i), is the intercept of the survival curve at zero time and is equivalent to the probit of percentage initial viability, and K_E, C_W, C_H and C_Q are the species constants. K_i is a measure of the initial quality of the seed lot: it indicates potential longevity which is subsequently modified by the storage conditions. Whereas the value of K_i varies amongst seed lots according to their quality, the values of the other four species constants are invariant within a species.

Two of the species constants, C_H and C_Q, define the temperature relation. When experiments are carried out over a restricted range of temperatures it is not always possible to show the advantage of incorporating the quadratic term, C_Qt^2 (e.g. Roberts & Abdalla, 1968; Tompsett, 1984; Dickie & Bowyer, 1985; Dickie *et al.* 1985). But a careful consideration of all the evidence, especially when the data are obtained over a wide temperature range, shows that there is little doubt that the entire temperature term, $C_Ht - C_Qt^2$, is required. Furthermore the values of both constants may well be common to all orthodox species, the best current estimates being 0·04 for C_H and 0·000428 for C_Q (Ellis, 1988). The inclusion of the

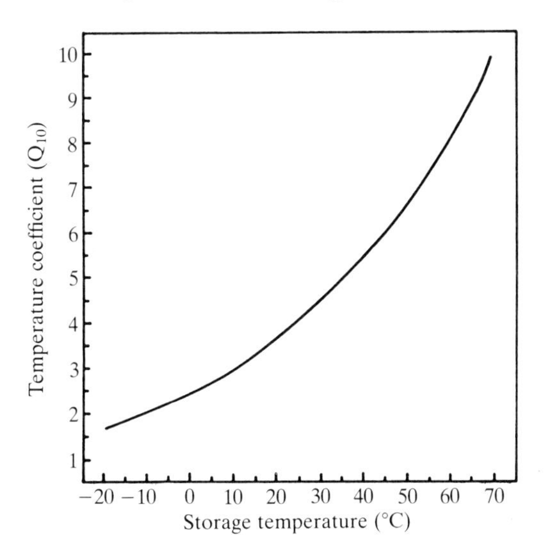

Fig. 4. The relation between storage temperature and temperature coefficient (Q_{10}) for rate of loss of seed viability. The curve was calculated for barley but is probably appropriate for other species. (From Ellis, 1987.)

quadratic term means that the temperature coefficient (Q_{10}) for rate of loss of viability increases from about 2 at −10°C to about 10 at 70°C (Fig. 4).

Although seeds of different species survive for quite different periods under the same conditions, the common relationship with temperature suggests that the underlying cause(s) of deterioration may be very similar. Because so many different sub-cellular systems are affected at the same time (Roberts, 1972*a*, 1979, 1983; Roberts & Ellis, 1982) it is not possible to attribute loss of viability to a single event or type of damage. It seems unlikely that the deteriorative processes are the result of enzyme activity for the following reasons: (1) the smooth relation between temperature and loss of viability over a very wide range in which the Q_{10} *increases* (rather than decreases) with increase in temperature; (2) the very high Q_{10} values (high activation energy) at high temperatures; and (3) the temperature relation which is identical over a very wide range of moisture contents, including values lower than 5 % at which enzyme activity seems improbable.

In orthodox seeds below the critical moisture content, the evidence suggests that damage to cellular components occurs at a rate dependent on temperature and moisture content; and it continuously accumulates since enzymatic repair or replacement of the vital components is not possible at sub-critical moisture contents. Accordingly, at these moisture contents the presence of oxygen is not an advantage since it is not required for respiration to support maintenance of repair activity; indeed, the presence of oxygen is somewhat deleterious to seed survival at moisture contents less than the critical value (Ibrahim *et al.* 1983; Roberts, 1988). It may be that all the damage observed may simply reflect the gradual denaturation of macromolecules which proceeds at a rate which is dependent mainly on temperature and moisture content but can also be exacerbated by oxidative reactions.

In orthodox seeds the relationship described by Eqn 1 does not continue above the critical moisture content. Above it, and providing oxygen is present, longevity *increases* with increase in moisture content reaching a maximum value at full hydration (which is typically about 50 % moisture content in most seeds). In the absence of oxygen there is a further decline in longevity with increase in moisture content above the critical value (Ibrahim *et al.* 1983). Above the critical moisture content the relation with temperature is similar to that described for moisture contents below the critical value except that, at any given temperature, the Q_{10} is slightly less (Ibrahim & Roberts, 1983).

It is clear that repair mechanisms can occur in fully hydrated seeds and this may counteract the damage which otherwise accumulates – an hypothesis first postulated by Villiers (Villiers, 1975; Villiers & Edgecumbe, 1975). It now seems probable that oxygen is necessary for respiratory activities to support the repair and that some repair may start at about the critical moisture content, but its efficacy is probably dependent on water potential so that it does not become fully active until the seeds are fully hydrated (Ibrahim *et al.* 1983). Longevity above the critical moisture content may therefore represent the outcome of the balance between non-enzymatic accumulation of damage and its enzymatic repair.

During seed deterioration sub-cellular damage ultimately reaches catastrophic proportions so that the seed becomes incapable of germination (loses viability). But in view of the sub-cellular damage which accumulates, it is not surprising that most measures of seed performance (which collectively comprise seed vigour) – e.g. ability to germinate under stressful conditions, rate of germination, ability to produce morphologically normal seedlings – gradually decline before the seed ultimately loses viability (Ellis & Roberts, 1980*c*). One aspect of this ageing syndrome is that partially deteriorated seeds lose their ability to germinate over a wide range of temperatures so that, for example, the optimum temperature for maximum germination of low-vigour seed lots in barley is about 10°C whereas non-dormant seeds of high vigour can germinate equally well over a range of temperatures from 2 to 30°C (Fig. 5), and the optimum temperature for rate of germination (of non-dormant seeds) is about 27°C (Fig. 8) (Ellis & Roberts, 1981; Ellis *et al.* 1987*a*).

The effect of temperature on the germination of non-dormant seeds

There are several different definitions of germination but, for many physiological studies, it is convenient to consider that the process starts when seeds are set to imbibe water and that it is complete when the radicle emerges. It is not possible to measure directly how rapidly the process is occurring in an individual seed (i.e. the rate of germination), but it is possible to time it. By analogy with other processes, e.g. the speed of a vehicle or the rate of an enzyme reaction, the rate of germination may be calculated by taking the reciprocal of the time taken to complete it.

This approach is proving to be generally useful in developmental physiology since it is commonly found that the rate of development is a linear function of temperature, e.g. in flowering (Hadley *et al.* 1983; Roberts & Summerfield, 1987). Labouriau (1970) was probably the first to show the value of the approach in seed germination (of *Vicia graminea* Sm) when he demonstrated that at sub-optimal constant temperatures there is a positive linear relation between germination rate (reciprocal of the time taken) between the base temperature (T_b), at which the rate is zero, and the optimum temperature (T_o), at which seeds germinate most rapidly; and that at supra-optimal constant temperatures there is another, but negative, linear relationship between the optimum temperature and the ceiling temperature (T_c), when the rate is again zero. This pattern was later confirmed in seeds of *Dolichos biflorus* L. (Labouriau & Pacheco, 1979).

This early work in South America apparently escaped the attention of European workers who subsequently used a similar approach. For example, Hegarty (1973) showed the positive linear relation in carrot seeds (*Daucus carota* L.) at constant temperatures between 8 and 24°C. Then Bierhuizen & Wagenvoort (1974) showed a similar positive relation in 31 vegetable species, albeit over a relatively restricted temperature range, as did Washitani & Saeki (1986) in *Pinus densiflora* Sieb. et Zucc. and Gummerson (1986) in *Beta vulgaris* L.

Both the positive and negative relations were confirmed by Garcia-Huidobro *et al.* (1982*a*) in pearl millet (*Pennisetum typhoides* S. & H.) and in several grain legumes (Covell *et al.* 1986; Ellis *et al.* 1986*b*; Ellis *et al.* 1987*b*). In summary, this

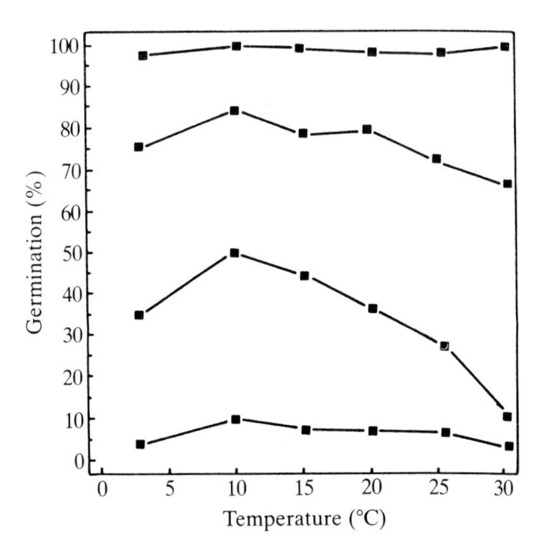

Fig. 5. The effect of temperature during the germination test on the percentage germination of four seed lots of barley (cv. Luke) which had suffered different amounts of deterioration. (From Ellis & Roberts, 1981.)

latter work has culminated in two equations which describe the mean germination rates of all seeds within a given seed lot by incorporating the seed-to-seed variation in germination rates. To explain this, it is necessary first to consider the relation for any given cumulative percentage of seeds (G) to germinate, when the basic relation at sub-optimal temperatures ($T_b < T < T_o$) can be described as

$$1/t(G) = [T - T_b(G)]/[\Theta_1(G)] \qquad \text{(Eqn 2)}$$

in which T is temperature, T_b is the base temperature at which the germination rate is zero, and Θ_1 is the thermal time in day-degrees above the T_b required to accumulate the given percentage germination (G).

At supra-optimal temperatures ($T_o < T < T_c$)

$$1/t(G) = [T_c(G) - T]/[\Theta_2(G)] \qquad \text{(Eqn 3)}$$

in which T_c is the ceiling temperature at which the germination rate is zero and Θ_2 is the thermal sum.

At sub-optimal temperatures the base temperature, T_b, is the same for all seeds within a population, and so all the variation in germination rate has to reside in variation in Θ_1 which determines the slope of the relation (as can be seen in Fig. 6). Θ_1 is normally distributed. Accordingly Eqn 2 can be developed when $T_b < T < T_o$ to give

$$1/t(G) = [T - T_b]/\{[\text{probit }(G) - K]\sigma\} \qquad \text{(Eqn 4)}$$

where σ is the standard deviation of Θ_1 and K is a constant.

At supra-optimal temperatures it is found that Θ_2, i.e. the slope of the relation, remains constant, while the seed-to-seed variation in germination rate is accounted for by a normal distribution in the ceiling temperature, T_c. So that at supra-optimal temperatures ($T_o < T < T_c$), Eqn 3 can be developed to give

$$1/t(G) = \{[K_s - \text{probit }(G)]\sigma\} - T/\Theta_2 \qquad \text{(Eqn 5)}$$

in which σ is the standard deviation of T_c and K_s is a constant (Ellis *et al.* 1987*b*).

The importance of these rather complicated equations is that they allow the prediction of rate of germination (or time taken to germinate) of every seed in a population at all temperatures from germination tests carried out at only four different temperatures – two sub-optimal and two supra-optimal. Such tests can provide a simple screen for investigating or selecting genetic variants in the cardinal temperatures which define the limiting temperatures, T_b and T_c, and optimal temperature, T_o, for germination, and also variants in the rates of germination at sub-optimal temperatures, which depend on T_b and Θ, and the rates at supra-optimal temperatures, which depend on T_c and Θ_c (Ellis *et al.* 1987).

The effect of fluctuating temperatures may also be accommodated within these concepts, but Garcia-Huidobro *et al.* (1982*b*) showed that wide temperature alterations may themselves lead to increases in germination rate in some circumstances. However, the extra stimulation they observed on rate of germination may have been associated with an effect on some residual dormancy since, as will be discussed below, alternating temperatures can have a profound effect on seed dormancy in many species.

Before leaving this topic, attention should be drawn to an entirely different

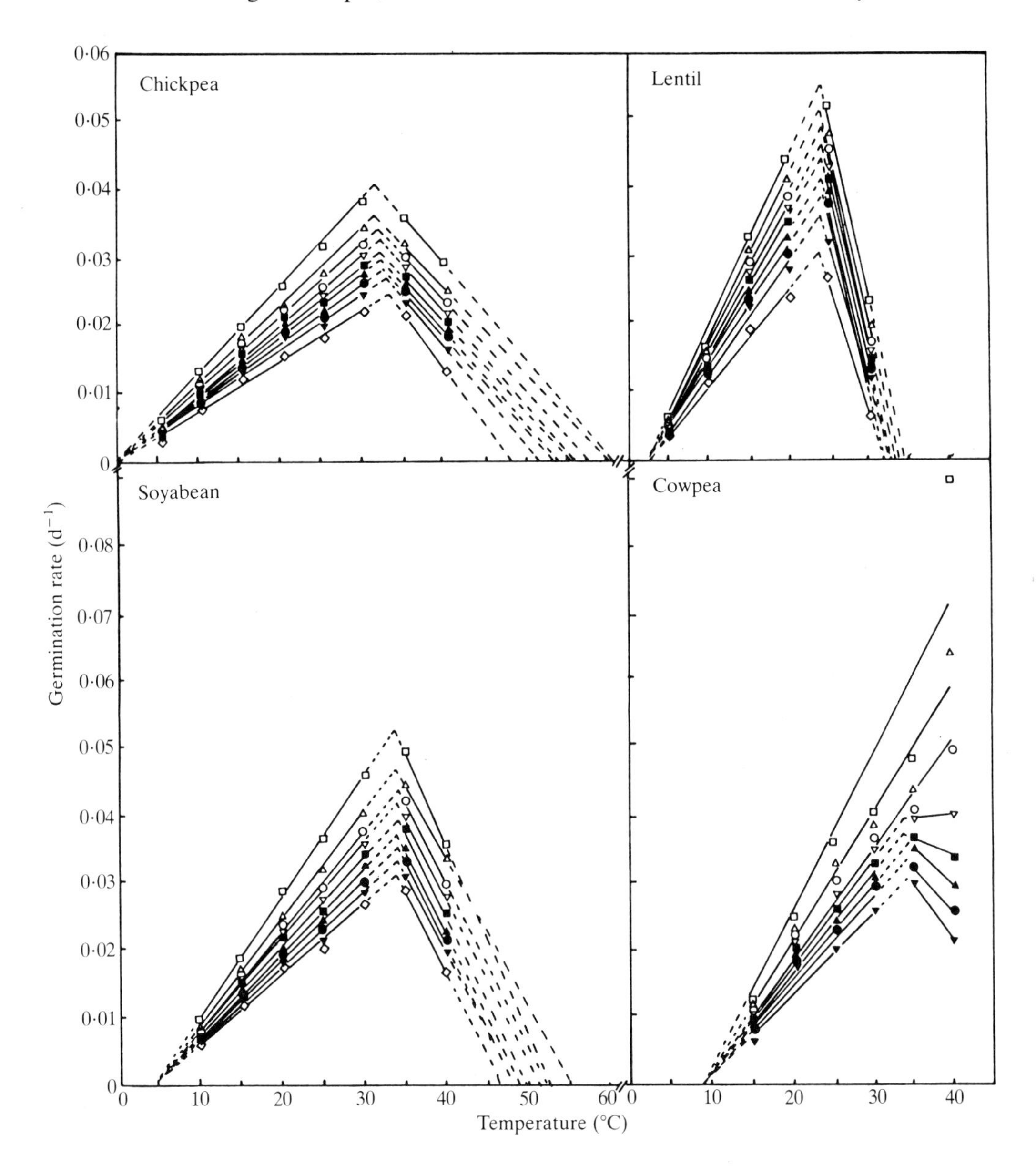

Fig. 6. The relation between temperature and rate of progress to germination to 10 % (□), 20 % (△), 30% (○), 40% (▽), 50% (■), 60% (▲), 70% (●), 80% (▼) and 90% (◇) for four legume species. Broken lines represent extrapolation beyond experimental values. The supra-optimal temperature data in cowpea are complicated by the fact that the first cohorts to germinate were not deleteriously affected by warm temperatures because germination was complete before the seeds were damaged. (From Covell *et al.* 1986.)

approach which has been made to the relation between temperature and the rate of germination, in which the logarithm of the rate of germination (reciprocal of the time taken to germinate) is related to the reciprocal of the absolute temperature. The rationale of this approach is that, in essence, it represents an Arrhenius plot and, therefore, may be more closely related to the effect of temperature on the underlying chemical reactions which influence germination. However, the plots are seldom linear over the entire sub-optimal temperature range, and the curves which have been published have generally been construed as two or more straight lines of different slopes. Physiological significance has often been attached to the temperature at which the break is perceived and to the activation energies calculated above and below any break. Simon (1979) has been an advocate of this approach, and he and others have speculatively associated the high activation energies calculated at cool temperatures with protein denaturation. However, others would argue that the statistics of 'bent-stick' curve fitting are problematical and that smooth, but non-linear, curves would fit many of the Arrhenius plots equally well. If Eqn 4 applies, then one would expect this to translate into a continuous curve on the basis of an Arrhenius plot and, as Hegarty (1973) saw, this implies a Q_{10} which decreases continuously from high values at low temperatures to low values at warmer temperatures.

Certainly the simpler approach, in which the relation between the rate of germination (not log rate of germination) and temperature (not reciprocal of absolute temperature) is assumed to be linear, seems to be of very general application. In many earlier experiments the results were presented as 'germination character curves', i.e. the time taken to achieve, say, 50 % germination was plotted as a function of temperature. The result is generally an asymmetric U-shaped curve (e.g. Thompson, 1970*a*,*b*,*c*,*d*, 1973, 1974). But if the results are re-plotted as rates they generally become asymmetric pyramids, as would be expected if Eqns 4 and 5 apply.

It is not at all clear why there should be a linear relation between temperature and rate of development between the base and the optimum temperature (Monteith, 1981) or, indeed, between the optimum and ceiling temperature. One speculation concerning the sub-optimal temperature range is as follows. The underlying processes are indeed of the Arrhenius type, but it would not matter whether one uses temperature, T (°C), or the reciprocal of absolute temperature, $[1/(273+T)]$ as the independent variable, because, over the temperature ranges considered, the relation between both measures of temperature is not detectably different from linear. Therefore over the limited range, $T_b < T < T_c$, the logarithm of the chemical reaction rate would be an approximately linear function of T. Then, it is well-known that for many growth and development processes, the relation between the concentration of the substance (e.g. growth hormone or limiting mineral ion) and rate is only linear if the rate is plotted as a function of the logarithm of the concentration of the substance. If these two arguments are combined then it would not necessarily be surprising that the rate of development is a linear function of temperature (Roberts & Summerfield, 1987).

The effect of temperature on seed dormancy

There are four main ways in which temperature can affect seed dormancy. (1) Most seeds become innately dormant as soon as they mature on the mother plant and then, after they have dried, they begin to lose dormancy. At ambient temperatures this process, often referred to as dry after-ripening, may take weeks or months, but it occurs much more rapidly at higher temperatures. (2) Once the seed has been rehydrated the effects of temperature are quite different. Many seeds respond to stratification, i.e. low temperature stimulates loss of dormancy. (3) Either warm temperatures or cool temperatures, which prevent germination, may also induce dormancy which is not immediately reversible. (4) Many hydrated seeds also lose dormancy when subjected to alternating temperatures.

Dry after-ripening

The quantitative relation between temperature and loss of dormancy in dry seeds was first shown in rice (*Oryza sativa* L. and *O. glaberrima* Steud.) (Roberts, 1962, 1965). In any seed, the time in dry storage until the point is reached when the seed is capable of germination may be referred to as the dormancy period. In rice, as in many other species, it is found that the dormancy periods within a given seed lot (population) are normally distributed. This gives rise to a cumulative normal distribution if seeds are stored under constant conditions and the percentage capable of germinating are plotted against time (Fig. 7A). From studies of this type the mean dormancy period can be calculated (or estimated from the time taken to reach a capability of 50 % germination). It can then be shown that, over a wide range of temperatures, the relationship can be described as

$$\log D = a + bT \qquad \text{(Eqn 6)}$$

where D is the mean dormancy period, T is the temperature and a and b are constants (Fig. 7B). This relation was shown to extend from 27 °C to 57 °C in rice (Roberts, 1965) and from 0° to 30 °C in *Dactylis glomerata* L. (Stoyanova & Kostov, 1983) and *Festuca arundinacea* Schreb. (Stoyanova *et al.* 1984). It seems reasonable to assume, therefore, that the relationship holds over a very wide range of temperatures. Although dormancy is gradually lost at very low temperatures, at −75 °C for example in *Dactylis glomerata*, it is so slow as to be undetectable after 110 weeks (Probert, Smith & Birch, 1985*b*).

The form of the relationship is such that the temperature coefficient (Q_{10}) does not vary with temperature. In rice the value appears to be the same, irrespective of the genotype: i.e. the slope constant, b, is identical in all cultivars (Fig. 7B). However, different Q_{10} values have been calculated for different species viz: rice 3·1 (Roberts, 1965), barley 3·9 (Roberts & Smith, 1977), *Dactylis glomerata* and *Festuca arundinacea* between 2 and 3 (Stoyanova & Kostov, 1983; Stoyanova *et al.* 1984).

The relationship can be used to devise practical techniques for physiologically removing seed dormancy as a plant breeding tool: a treatment at 47 °C for a week is appropriate for most genotypes in the common rice species, *Oryza sativa*

(Carpenter & Roberts, 1962). However, in African rice, *O. glaberrima*, which is more dormant, more complex treatments are needed, e.g. pre-treatments in dilute nitric acid and hydrogen peroxide followed by germination in 2-mercapto-ethanol in diurnal alternating temperatures of 16 h at 34 °C and 8 h at 11 °C (Ellis *et al.* 1983). This alternative is necessary because longer treatments of dry after-ripening

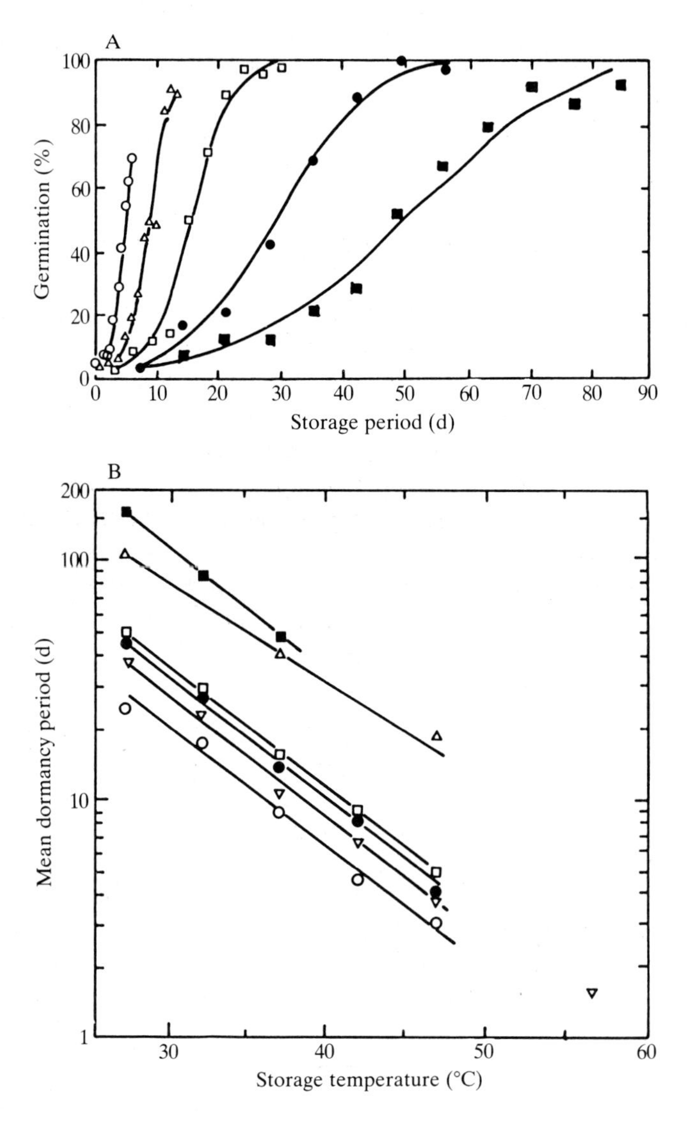

Fig. 7. The relation between storage temperature and loss of seed dormancy during dry after-ripening in rice. A. Results of germination tests after different periods of storage of cv. Lead 35 at different temperatures: 27 °C (■), 32 °C (●), 37 °C (□), 42 °C (△), 47 °C (○). B. The effect of storage temperature on mean germination period on cvs Masalacci A4 (■), Bayawuri (△), Lead 35 (□), India Pa Lil 46 (●), Nam Dawk Nai (▽), Mas 2401 (○). The anomalous regression line for Bayawuri is questionable since it is based on extrapolation from relatively few results. (From Roberts, 1965.)

at warm temperatures are required for *O. glaberrima*, and prolonged warm treatments can damage viability before dormancy is completely lost.

From experiments carried out on rice over a relatively narrow range of moisture contents (12·0–14·5 %) it was originally thought that the effect of temperature on dry after-ripening was independent of moisture content (Roberts, 1962). However, subsequent work carried out over the range 4·9–11·8 % moisture content showed that, over this range at least, the effect of temperature on loss of dormancy is greatly dependent on moisture content – the lower the moisture content the less effective the temperature treatment (Ellis *et al.* 1983). Nevertheless, it is still appropriate to use the term dry after-ripening since it occurs at moisture contents typical of normal storage. Although the relations have only been quantified in a few graminaceous species there is considerable evidence that similar relationships probably apply in many other species, e.g. in *Draba verna* (Baskin & Baskin, 1979) and in *Avena fatua* L. (Quail & Carter, 1969).

Because of these after-ripening relationships, special problems arise in devising appropriate germination tests for assessing viability in seeds stored for genetic conservation. In order to extend their longevity, such seeds are typically stored at −20 °C and about 5 % moisture content and, under these conditions, it is clear that dry after-ripening is exceedingly slow; and so special methods of removing dormancy in gene banks have to be developed in which the temperature regime during the germination test is often crucial (Ellis *et al.* 1985*a*,*b*).

Stratification and stimulation of loss of dormancy at cool temperatures

Vegis (1964) drew attention to the fact that, as seed populations lose dormancy, so the temperature range over which they are capable of germinating increases. The most common form of this behaviour in many species, including temperate cereals, is that seeds are initially capable of germinating only at cool temperatures but gradually become capable of germinating also at warmer temperatures. In barley, for example, it could easily be concluded that no dormancy exists if seeds were always tested at relatively low temperatures between 2 and 15 °C. Above this range dormancy is clearly expressed, although it is gradually lost in the germination test itself so that ultimately, even in warmer temperatures, the seeds will germinate after 25–30 days (Roberts & Smith, 1977; Ellis *et al.* 1987*a*). Careful analysis shows, however, that there is some dormancy in the cooler temperatures, but this is lost within a matter of only one or two days in the germination test (Roberts & Smith, 1977) and would often go unnoticed unless specifically sought.

The pattern of loss of the dormancy in which dormancy is only obvious in warmer temperatures is illustrated in Fig. 8. This shows the results of an experiment in which seeds were after-ripened at 40 °C and 15·2 % moisture content for various periods up to 23 days (Ellis *et al.* 1988). Initially the rate of germination up to 17 °C is as one would expect from Eqn 4 but, above this temperature, there is a discontinuity so that the germination of seeds is considerably delayed. After the seeds had been after-ripened for 24 h the range of temperature in which germination is rapid had extended up to 20 °C. After 10 days all signs of dormancy

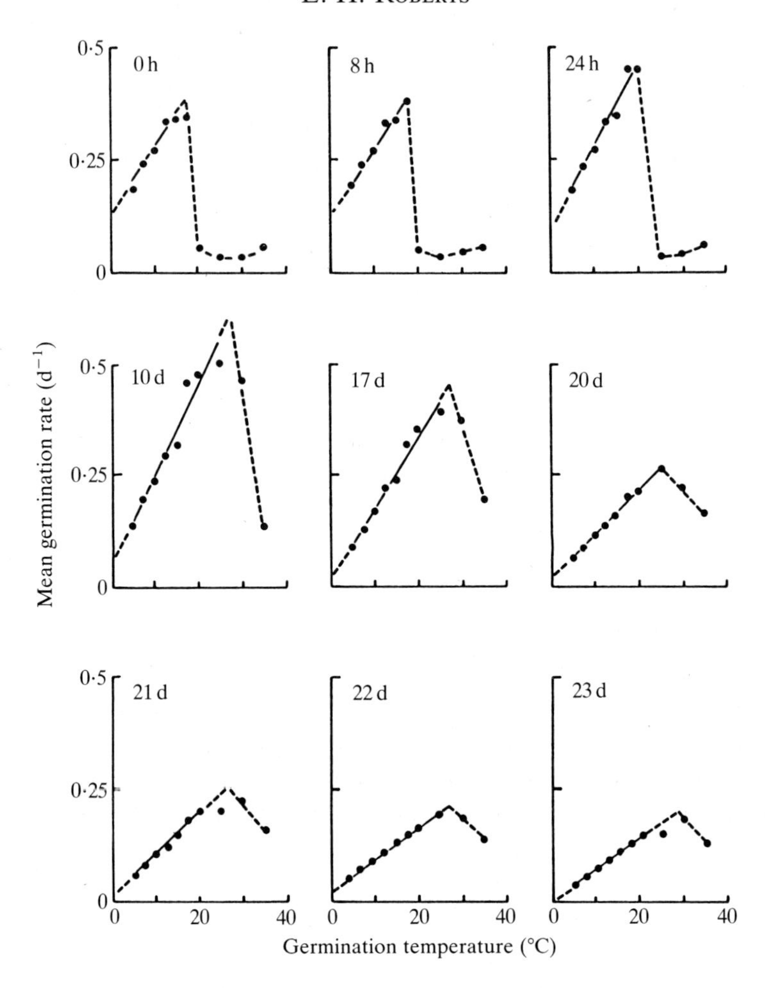

Fig. 8. The effect of temperature on rate of germination in barley (cv. Proctor) after various periods of dry after-ripening at 40°C and 15·2 % moisture contents. (From Ellis *et al.* 1987*a*.)

had disappeared and the standard pyramidal pattern for non-dormant seeds is seen: in this case Eqn 4 applies up to an optimum temperature of about 27°C and Eqn 5 above this temperature. After 17 days, signs of seed deterioration are apparent so that the rate of germination has declined at all temperatures (the height of the pyramid has decreased), and deterioration continues with increase in storage period until the end of the experiment.

Seeds of many species respond to a stratification treatment in which seeds are imbibed and maintained under cool conditions before being transferred to a warmer temperature when germination takes place. This is similar in many respects to the case outlined above for temperate cereals, except that the cereal seeds do not have to be transferred to warmer temperatures for germination to take place. The optimum temperature for stratification is often thought to be about 3–5°C, but effects are also found from −5° to 15°C and, depending on species, the period of treatment required may vary from a few days to a few months (Bewley & Black, 1982). However, some care is needed in interpreting the optimum temperature because, in some cases at least, the optimum temperature may vary with time of treatment owing to the simultaneous induction of secondary dormancy during the stratification treatment, as discussed below.

The relation between temperature and the induction of dormancy

'Some seeds are born dormant, some seeds achieve dormancy and some have dormancy thrust upon them.' So Harper (1959) neatly paraphrased Shakespeare to define three main types of dormancy – innate, induced and enforced. Many other different terms have been used, most of which have been carefully compared by Bewley & Black (1982): e.g. primary is an alternative to innate but also sometimes includes enforced, while secondary is often used as an alternative to induced. For the most part this discussion has been concerned with innate dormancy (that which the seed is born with) but sometimes also with enforced because, in the case of temperate cereals, for example, it could be said that dormancy is enforced by warm temperatures but this usually disappears immediately on transfer to cooler temperatures. However, in other circumstances, seeds kept under conditions which prevent germination for any length of time often develop induced or secondary dormancy; this condition is diagnosed when the dormancy remains after transfer to non-inhibitory conditions. Factors which might induce dormancy in some circumstances include light (high irradiation or far-red), solutions of high osmotic potential, anaerobic conditions, or excessively warm or cool temperatures (Bewley & Black, 1982).

In *Rumex crispus* L. and *Rumex obtusifolius* L. there is evidence that stratification treatments can decrease primary dormancy while secondary dormancy is being induced at the same time (Totterdell & Roberts, 1979). Fig. 9A shows the result of an experiment in which seeds of *R. obtusifolius* were hydrated in the dark for various periods at 1·5 °C, 10 °C or 15 °C before being transferred to a temperature of 25 °C for 4 weeks. It will be seen that maximum germination was achieved after prolonged treatment at 1·5 °C, but when either of the warmer temperatures were used for stratification there was a lower and transitory peak of germination. This pattern is most easily explained by assuming that any temperature between 1·5 °C and 15 °C has a stratification effect but that, at the same time, secondary dormancy is induced at constant temperatures at a rate dependent on temperature (Fig. 9B). As Bewley & Black (1982) remark, it is possible that this type of response may be widespread.

If this interpretation is correct it may be related to the observations where dormancy is so transitory as to go unnoticed at temperatures less than 17 °C in temperate cereals. The main differences between temperate cereals and *Rumex* are that only a day or two at cool temperatures are required to remove dormancy in cereals instead of a much longer period, that a subsequent transfer to a warmer temperature is not necessary in cereals as it appears to be in *Rumex* (Roberts &

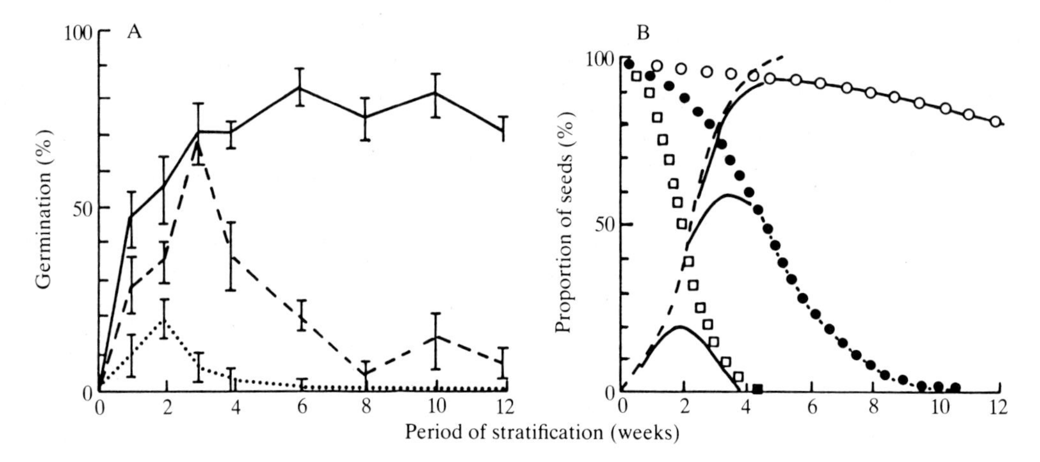

Fig. 9. A. Germination of *Rumex obtusifolius* after 4 weeks at 25°C in the light following various periods of stratification at 1·5°C (——), 10°C (---) or 15°C (·····). B. Expected germination if loss of dormancy due to stratification (---) is independent of temperature over the range investigated, while the induction of dormancy increases with increase in temperature so that the proportion of seeds with no induced dormancy is shown as follows: 1·5°C (○), 10°C (●), 15°C (□). Predicted germination is the product of (proportion of seeds with no induced dormancy) × (proportion of seeds which have lost innate dormancy through stratification) (——). It is assumed (arbitrarily) that seed-to-seed variability is similar for both processes with a coefficient of variation of 50 %, and that log of the mean time taken to induce dormancy is linearly related to temperature. Note the similarity of the predicted curves with those in A. (From Totterdell & Roberts, 1979.)

Totterdell, 1981), and that secondary dormancy is not easily induced in cultivated cereals.

The characteristics of alternating temperatures which remove dormancy

It has been pointed out that most small seeds require mechanisms which prevent germination when the seed is more than a centimetre or so below the soil surface; otherwise germination will result in self-destruction since the food reserves will run out before emergence occurs and photosynthesis becomes possible. It will also often be an advantage for a seed to delay germination if the ground is covered by a dense foliar canopy, otherwise it may not be able to compete sufficiently after emergence to reach maturity and reproduce. It may also be an advantage to restrict germination to those seasons when the environment is not inimicable to early seedling growth (Roberts, 1982). This often means avoiding the winter in temperate latitudes, and the dry season in tropical latitudes. A common feature of most small-seeded plants which facilitates these adaptations is that dormancy (primary, induced or enforced) is lost in response to a combination of white light and alternating temperatures. These factors typically interact positively and strongly so that little germination occurs unless both factors are experienced. Either or both may also interact with nitrate ions which are at higher concentrations near the soil surface and especially at the beginning of the growing season

– i.e. in the spring in temperate latitudes and at the start of the rainy season in the tropics. In temperate species there may also be a stratification response so that a preceding period of cool temperature is a pre-requisite which helps to ensure that seeds do not germinate in the winter (Roberts, 1972*b*, 1973, 1981). Accordingly first-, second-, and third-order positive interaction between these factors are the norm rather than the exception (e.g. Thompson, 1969; Vincent & Roberts, 1977; Bostock, 1978; Roberts & Benjamin, 1979; Totterdell & Roberts, 1979, 1980; Probert *et al.* 1985*a*,*b*; Goedert & Roberts, 1986; Probert & Smith, 1986; Probert *et al.* 1987).

So far as the light component is concerned, the phytochrome system provides a mechanism which enables the seed to perceive and respond to its position in the soil profile so that a seed only germinates if it is at or near the soil surface. Not only this but, because of the high far-red/red ratios beneath foliar canopies, phytochrome provides a sensor by which some seeds remain dormant if they find themselves beneath a dense canopy. Again, the decreased amplitude in temperature alternations beneath a canopy may be an important reinforcement of the signal to delay germination (Roberts, 1981).

It is not easy to define which attributes of alternating temperatures stimulate loss of dormancy. It has been pointed out that there are at least nine characteristics of diurnal temperature alternation which could be involved, and it is not possible to alter any one of them experimentally without confounding it with a change in at least one of the other characteristics (Roberts & Totterdell, 1981). Since then we (A. J. Murdoch and myself) have added to the list, further analysed the various attributes, and classified them into primary, secondary and tertiary characteristics (Fig. 10). Each primary characteristic can be separately changed experimentally but that change will affect at least three characteristics of higher order. If it is intended to hold any of the secondary characteristics constant, then it is not possible to alter any of the primary characteristics which determine it without confounding it with a change in at least one of the other primary characteristics. For example, most experiments involve a cycle duration (which is a secondary characteristic) of 24 h. If this duration is held constant, then it is not possible to alter either of the primary characteristics which determine it, without confounding it with a change in the other. Another example is that, if the amplitude is held constant, it is not possible to alter the minimum temperature without confounding it with an alteration in the maximum temperature.

For these reasons it is difficult to devise experiments which unequivocally identify those characteristics of alternating temperatures which affect dormancy. One approach is to apply a very large number of treatments – for which a two-dimensional thermo-gradient plate is almost a practical necessity (e.g. Ellis *et al.* 1982; Ellis *et al.* 1983; Goedert & Roberts, 1986). These results can then be used to postulate the simplest models which can be discerned to account for the results.

The first steps in the analysis of the results obtained by these means may be to produce a diagram such as that shown in Fig. 11, in which isopleths of equal germination can be related to several scales, e.g. minimum temperature, maxi-

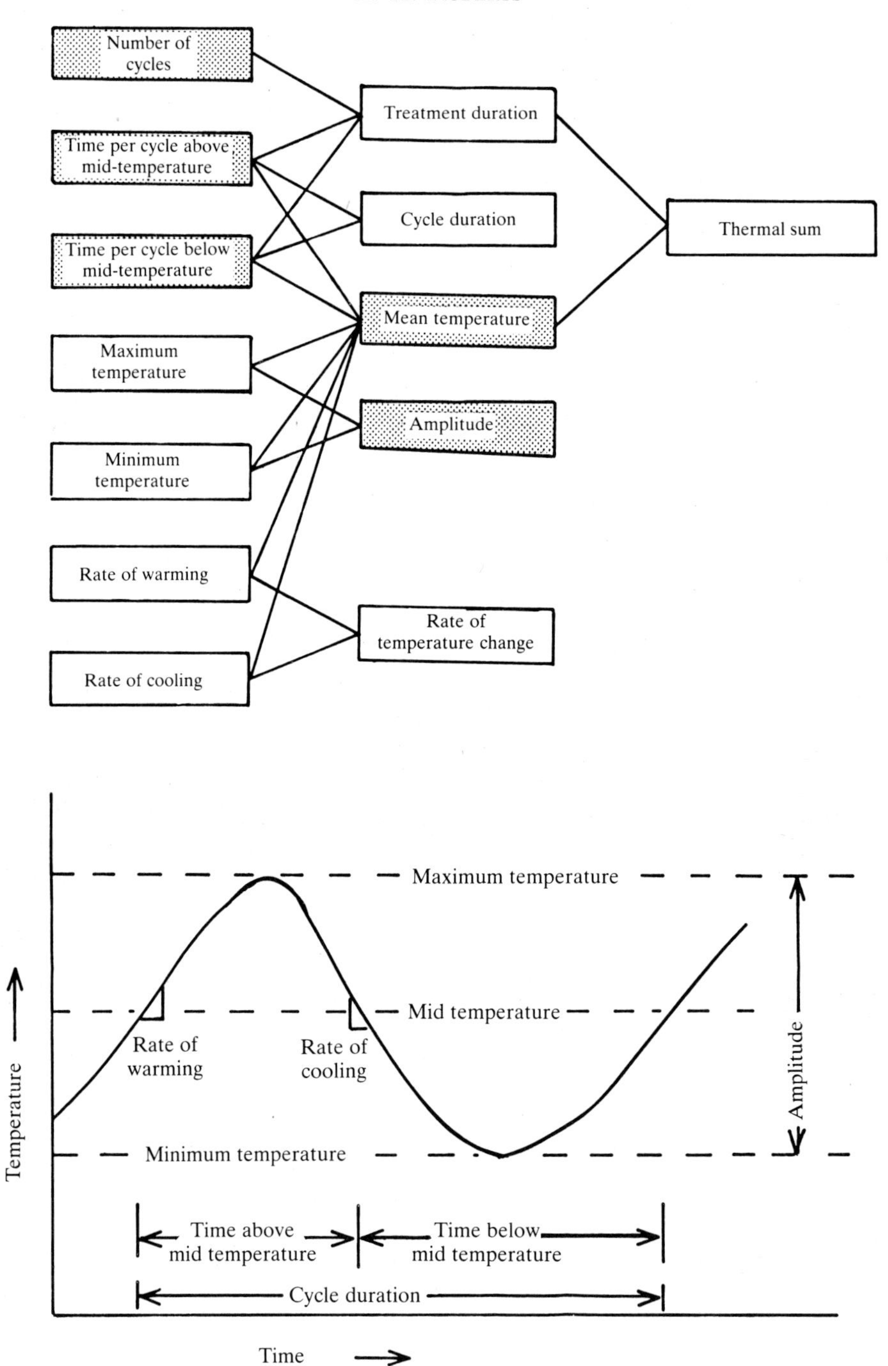

Fig. 10. Characteristics of alternating-temperature regimes which might affect loss of dormancy. The characteristics have been divided into primary (left-hand column) which can be independently controlled, and secondary characters which depend on two or more primary characters. Stippling indicated those characteristics which the evidence suggests are of greatest significance. The graph explains some of the terminology.

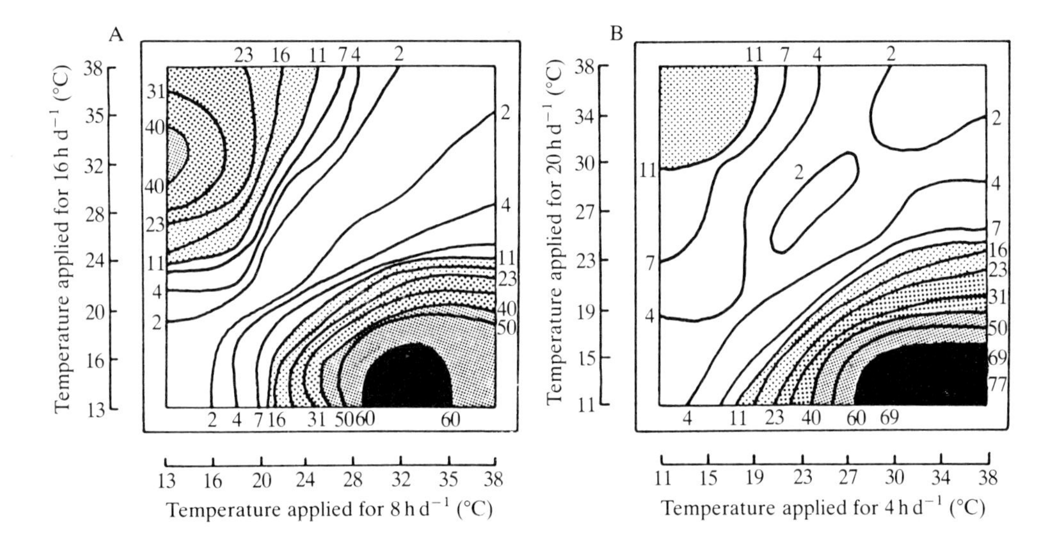

Fig. 11. A. Isopleths of percentage germination of *Brachiaria humidicola* after 40 days in thermoperiods of 8 h/16 h. Percentage germination of the isopleths are marked on the perimeter of the square. B. As in A, but in thermoperiods of 4 h/20 h. (From Goedert & Roberts, 1986.)

mum temperature, and temperature amplitude and mean temperature. The four scales can be shown simultaneously (Ellis *et al.* 1982) but only the first two are shown in Fig. 11. In this case it is clear that in the tropical grass *Brachiaria humidicola* (Rendle) Schweickerdt, little germination takes place at any constant temperature, that short periods (4 h day^{-1}) at the warmer temperature in diurnal cycles are more stimulatory than longer periods and that the optimum minimum and maximum temperatures in that thermoperiod are approximately 13 °C and 34 °C, respectively. These diagrams also establish that if temperatures never exceed 20 °C or are continuously above 25 °C, very little germination takes place. Although the rules for effective temperature alternations begin to emerge from this type of analysis, it is difficult to quantify the responses using this approach.

Another subsequent approach, then, is to postulate a number of simple quantitative models and then test them against the data. Fig. 12 is the end result of a process of this type. In general models can often be linearized by transforming percentage germination values to probits. This is because many dormancy characteristics are normally distributed with respect to time (e.g. Fig. 7A) or to the logarithm of the dose of stimulatory factors such as light (e.g. Bartley & Frankland, 1982; Ellis *et al.* 1986*b*,*c*; Probert *et al.* 1987), or a number of temperature alternations (Probert & Smith, 1986; Probert *et al.* 1986; Probert *et al.* 1987). Accordingly, probit percentage germination is shown in Fig. 12 as a function of mean temperature at various temperature amplitudes for seeds of *Chenopodium album* L. at two thermoperiods. It will be seen that at constant

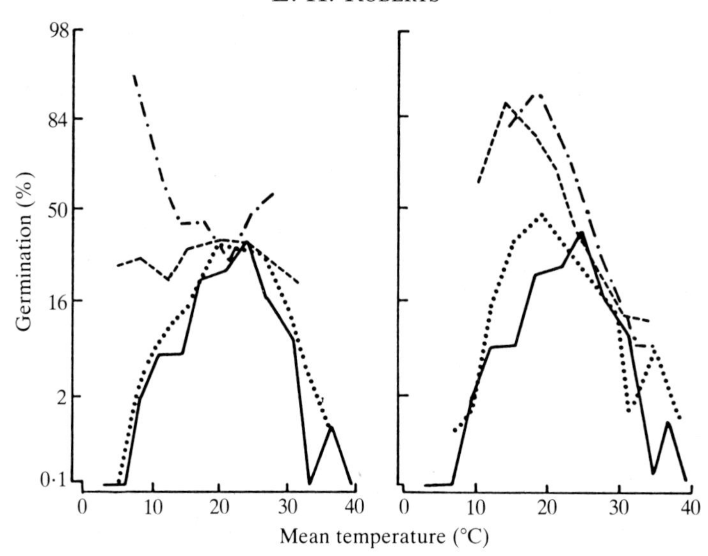

Fig. 12. Germination of *Chenopodium album* as a function of mean temperature after 28 days in 8 h/16 h thermoperiods when either 8 h day^{-1} in each cycle was at the warmer temperature (left) or 16 h day^{-1} at the warmer temperature (right). Temperature amplitude: 0°, i.e. constant temperature, (——); 6·2 °C (·····); 12·3 °C (– – –); 18·5 °C (–·–). (A. J. Murdoch & E. H. Roberts, previously unpublished data.)

temperatures, probit of percentage germination seems to be more-or-less linearly related to temperature – positively from a base temperature of about 4 °C up to an optimum of 24 °C, and negatively from the optimum to a ceiling temperature of 40 °C. (Note, as mentioned earlier, these cardinal temperatures for percentage germination are not necessarily the same as for rate of germination.) At sub-optimal mean temperatures, alternating temperatures stimulate germination and the effect increases with increase in amplitude up to 18·5 °C. Furthermore, the optimum mean temperature of alternating-temperature regimes decreases with increase in amplitude. At mean temperatures greater than the optimum constant temperature, alternating temperatures have no effect. A temperature alternation in which the warmer temperature is applied for 16 h in the 24-h cycle is more stimulatory than one in which it is applied for 8 h. Although it cannot be expressed in Fig. 12, it has been shown that in this, as in many other species, in any given regime the response is proportional to the number of cycles experienced. In *Dactylis glomerata*, for example, the probit of the germination percentage is linearly related to the logarithm of the number of cycles (Probert *et al.* 1986).

From these examples and other data, it is becoming clear that the stimulatory effect of alternating temperature depends on amplitude, mean temperature, thermoperiod and the number of cycles. Other experiments (e.g. Goedert, 1984) suggest that other characteristics such as rate of temperature change are probably unimportant. Unpublished work in our laboratory has shown that the general response pattern illustrated in Fig. 12 is typical of several other species, including tropical grasses such as *Panicum maximum* Jacq. Maximum germination of dormant seeds of many species, therefore, is achieved at a relatively low mean temperature of large amplitude. Species vary with respect to the optimum thermoperiod.

This analysis shows that it is now becoming feasible to describe the response to alternating temperatures in terms of a few parameters which can be quantified; but clearly more work is required before satisfactory quantitative models of alternating temperature effects are produced. These should be of considerable value in helping to predict ecological behaviour and defining the characteristics that cellular physiology will need to explain.

References

Bartley, M. R. & Frankland, B. (1982). Analysis of the dual role of phytochrome in the photo-inhibition of seed germination. *Nature, Lond.* **300**, 750–752.

Baskin, J. M. & Baskin, C. C. (1979). Effect of relative humidity on after-ripening and viability in seeds of the winter annual *Draba verna*. *Bot. Gaz.* **140**, 284–287.

Bewley, J. D. & Black, M. (1982). *Physiology and Biochemistry of Seeds in Relation to Germination*, vol. 2, 375 pp. Berlin: Springer-Verlag.

Bierhuizen, J. F. & Wagenvoort, W. A. (1974). Some aspects of seed germination in vegetables. I. The determination and application of heat sums and minimum temperature for germination. *Scientia Horticulturae* **2**, 213–219.

Bostock, S. J. (1978). Seed germination strategies of five perennial weeds. *Oecologia* **36**, 113–126.

Carpenter, A. J. & Roberts, E. H. (1962). Some useful techniques in speeding up rice breeding programmes. *Empire J. Exp. Agric.* **30**, 127–131.

Covell, S., Ellis, R. H., Roberts, E. H. & Summerfield, R. J. (1986). The influence of temperature on seed germination rate in grain legumes. I. A comparison of chickpea, lentil, soyabean and cowpea at constant temperatures. *J. exp. Bot.* **37**, 705–715.

Dickie, J. B. & Bowyer, J. T. (1985). Estimation of provisional seed viability constants for apple (*Malus domestica* Borkh. cv. Greensleeves). *Ann. Bot.* **56**, 271–275.

Dickie, J. B., McGrath, S. & Linington, S. H. (1985). Estimation of provisional seed viability constants for *Lupinus polyphyllus* Lindley. *Ann. Bot.* **55**, 147–151.

Ellis, R. H. (1988). The viability equation, seed viability nomographs and practical advice on seed storage. *Seed Sci. & Technol.* **16**, 29–50.

Ellis, R. H., Covell, S., Roberts, E. H. & Summerfield, R. J. (1986*a*). The influence of temperature on seed germination. II. Intraspecific variation on chickpea (*Cicer arietinum* L.) at constant temperatures. *J. exp. Bot.* **37**, 1503–1515.

Ellis, R. H., Hong, T. D. & Roberts, E. H. (1982). An investigation of the influence of constant and alternating temperature on the germination of cassava seed using a two-dimensional temperature gradient plate. *Ann. Bot.* **49**, 241–246.

Ellis, R. H., Hong, T. D. & Roberts, E. H. (1983). Procedures for the safe removal of dormancy from rice seed. *Seed Sci. & Technol.* **11**, 77–112.

Ellis, R. H., Hong, T. D. & Roberts, E. H. (1985*a*). *Handbook of Seed Technology for Genebanks. I. Principles and Methodology*, pp. 1–210. Rome: International Board for Plant Genetic Resources.

Ellis, R. H., Hong, T. D. & Roberts, E. H. (1985*b*). *Handbook of Seed Technology for Genebanks. II. Compendium of Specific Germination Information and Test Recommendations*, pp. 211–667. Rome: International Board for Plant Genetic Resources.

Ellis, R. H., Hong, T. D. & Roberts, E. H. (1986*b*). Quantal response of seed germination in *Brachiaria humidicola, Echinochloa turnerana, Eragrostis tef* and *Panicum maximum* to photon dose for the low energy reaction and the high irradiance reaction. *J. exp. Bot.* **179**, 742–753.

Ellis, R. H., Hong, T. D. & Roberts, E. H. (1986*c*). The response of seeds of *Bromus sterilis*

L. & *Bromus mollis* L. to white light of varying photon flux density and photoperiod. *New Phytol.* **104**, 485–496.
Ellis, R. H., Hong, T. D. & Roberts, E. H. (1987*a*). Comparison of cumulative germination and rate of germination of dormant and aged barley seed lots at different constant temperatures. *Seed Sci. & Technol.* **15**, 717–727.
Ellis, R. H., Osei-Bonsu, K. & Roberts, E. H. (1982). The influence of genotype, temperature and moisture content on longevity in chickpea, cowpea and soyabean. *Ann. Bot.* **50**, 69–82.
Ellis, R. H. & Roberts, E. H. (1980*a*). Improved equations for the prediction of seed longevity. *Ann. Bot.* **45**, 13–30.
Ellis, R. H. & Roberts, E. H. (1980*b*). The influence of temperature and moisture on seed viability period in barley (*Hordeum distichum* L.). *Ann. Bot.* **45**, 31–37.
Ellis, R. H. & Roberts, E. H. (1980*c*). Towards a rational basis for testing seed quality. In *Seed Production* (ed. P. D. Hebblethwaite), pp. 605–635. London: Butterworth.
Ellis, R. H. & Roberts, E. H. (1981). The quantification of ageing and survival in orthodox seeds. *Seed Sci. & Technol.* **9**, 373–409.
Ellis, R. H., Simon, G. & Covell, S. (1987*b*). The influence of temperature on seed germination rate in grain legumes. III. A comparison of five faba bean genotypes at constant temperatures using a new screening method. *J. exp. Bot.* **38**, 1033–1043.
Garcia-Huidobro, J., Monteith, J. L. & Squire, G. R. (1982*a*). Time, temperature and germination in pearl millet (*Pennisetum typhoides* S. & H.). *I. Constant temperature. J. exp. Bot.* **33**, 288–296.
Garcia-Huidobro, J., Monteith, J. L. & Squire, G. R. (1982*b*). Time, temperature and germination of pearl millet (*Pennisetum typhoides* S. & H.). II. Alternating temperature. *J. exp. Bot.* **33**, 297–302.
Goedert, C. O. (1984). *Seed Dormancy of Tropical Forage Grasses and Implications for the Conservation of Genetic Resources.* PhD Thesis, University of Reading.
Goedert, C. O. & Roberts, E. H. (1986). Characterization of alternating-temperature regimes that remove seed dormancy in seeds of *Brachiaria humidicola* (Rendle) Schweickerdt. *Plant Cell Env.* **9**, 521–525.
Gummerson, R. J. (1986). The effect of constant temperatures and osmotic potentials on the germination of sugar beet. *J. exp. Bot.* **37**, 729–741.
Hadley, P., Summerfield, R. J. & Roberts, E. H. (1983). Effects of temperature and photoperiod on reproductive development of selected grain legume crops. In *Temperate Grain Legumes: Physiology, Genetics and Nodulation* (ed. D. G. Jones & D. R. Davies), pp. 19–41. London: Pitman.
Harper, J. L. (1959). The ecological significance of dormancy and its importance in weed control. *Proc. 4th Int. Congress in Crop Protection*, Hamburg 1957, vol. I, pp. 415–420. Braunschweig.
Hegarty, T. W. (1973). Temperature relations of germination. In *Seed Ecology* (ed. W. Heydecker), pp. 411–432. London: Butterworth.
Ibrahim, A. E. & Roberts, E. H. (1983). Viability of lettuce seeds. I. Survival in hermetic storage. *J. exp. Bot.* **34**, 620–630.
Ibrahim, A. E., Roberts, E. H. & Murdoch, A. J. (1983). Viability of lettuce seeds. II. Survival and oxygen uptake in osmotically controlled storage. *J. exp. Bot.* **34**, 631–640.
King, M. W. & Roberts, E. H. (1980). Maintenance of recalcitrant seeds in storage. In *Recalcitrant Crop Seeds* (ed. H. F. Chin & E. H. Roberts), pp. 53–89. Kuala Lumpur: Tropical Press.
Kraak, H. L. & Vos, J. (1987). Seed viability constants for lettuce. *Ann. Bot.* **59**, 343–349.
Labouriau, L. G. (1970). On the physiology of seed germination in *Vicia graminea* Sm. *Ann. Acad. brasil. Ciênc.* **42**, 235–262.
Labouriau, L. G. & Pacheco, A. (1979). Isothermal germination rates in seeds of *Dolichos biflorus* L. *Bol. Soc. Venez. Ciênc. Nat. Caracas* **34**, 73–112.
Monteith, J. L. (1981). Climatic variation in the growth of crops. *Q. Jl Royal Meteorol. Soc.* **107**, 749–754.
Probert, R. J., Gajjar, K. H. & Haslam, I. K. (1987). The interactive effects of phytochrome, nitrate and thiouroa on the germination response to alternating temperatures in seeds of *Ranunculus sceleratus* L.: a quantal approach. *J. exp. Bot.* **38**, 1012–1025.

PROBERT, R. J. & SMITH, R. D. (1986). The joint action of phytochrome and alternating-temperatures in the control of seed germination in *Dactylis glomerata. Physiologia. Pl.* **67**, 299–304.

PROBERT, R. J., SMITH, R. D. & BIRCH, P. (1985*a*) Germination responses to light and alternating temperatures in European populations of *Dactylis glomerata* L. II. The genetic and environmental components of germination. *New Phytol.* **99**, 317–322.

PROBERT, R. J., SMITH, R. D. & BIRCH, P. (1985*b*). Germination responses to light and alternating temperatures in European populations of *Dactylis glomerata* L. IV. The effects of storage. *New Phytol.* **101**, 521–529.

PROBERT, R. J., SMITH, R. D. & BIRCH, P. (1986). Germination responses to light and alternating temperatures in European populations of *Dactylis glomerata* L. V. The principle componenets of the alternating temperature requirements. *New Phytol.* **102**, 133–142.

QUAIL, P. H. & CARTER, O. G. (1969). Dormancy in seeds of *Avena ludoviciana* and *A. fatua. Aust. J. agric. Res.* **20**, 1–11.

ROBERTS, E. H. (1962). Dormancy in rice seed. III. The influence of temperature, moisture and gaseous environment. *J. exp. Bot.* **13**, 75–94.

ROBERTS, E. H. (1965). Dormancy in rice seed. IV. Varietal responses to storage and germination temperatures. *J. exp. Bot.* **16**, 341–349.

ROBERTS, E. H. (1972*a*). Cytological, genetical, and metabolic changes associated with loss of viability. In *Viability of Seeds* (ed. E. H. Roberts), pp. 253–306. London: Chapman and Hall.

ROBERTS, E. H. (1972*b*). Dormancy: a factor affecting seed survival in the soil. In *Viability of Seeds* (ed. E. H. Roberts), pp. 321–359. London: Chapman and Hall.

ROBERTS, E. H. (1973*a*). Oxidative processes and the control of seed germination. In *Seed Ecology* (ed. W. Heydecker), pp. 189–218. London: Butterworths.

ROBERTS, E. H. (1973*b*). Predicting the viability of seeds. *Seed Sci. & Technol.* **1**, 499–514.

ROBERTS, E. H. (1979). Seed deterioration and loss of viability. *Adv. Res. Technol. Seeds* **4**, 25–42.

ROBERTS, E. H. (1981). The interaction of environmental factors controlling loss of dormancy in seeds. *Ann. appl. Biol.* **98**, 552–555.

ROBERTS, E. H. (1983). Loss of seed viability during storage. *Adv. Res. Technol. Seeds* **8**, 9–34.

ROBERTS, E. H. (1986). Quantifying seed deterioration. In *Physiology of Seed Deterioration* (ed. M. B. McDonald & C. J. Nelson), pp. 101–123. Madison, WI: Crop Science Society of America.

ROBERTS, E. H. (1988). Seed aging – the genome and its expression. In *Senescence and Aging in Plants* (ed. L. D. Noodén & A. C. Leopold), pp. 465–498. New York: Academic Press.

ROBERTS, E. H. & ABDALLA, F. H. (1968). The influence of temperature, moisture and oxygen on period of seed viability in barley, broad beans, and peas. *Ann. Bot.* **32**, 97–117.

ROBERTS, E. H. & BENJAMIN, S. K. (1979). The interaction of light, nitrate and alternating temperature on the germination of *Chenopodium album, Capsella bursa-pastoris* and *Poa annua* before and after chilling. *Seed Sci. & Technol.* **7**, 379–392.

ROBERTS, E. H. & ELLIS, R. H. (1982). Physiological, ultrastructural and metabolic aspects of seed viability. In *The Physiology and Biochemistry of Seed Development, Dormancy and Germination* (ed. A. A. Khan), pp. 465–485. Amsterdam: Elsevier Biomedical Press.

ROBERTS, E. H. & KING, M. W. (1980). The characteristics of racalcitrant seeds. In *Recalcitrant Crop Seeds* (ed. H. F. Chin & E. H. Roberts), pp. 1–5. Kuala Lumpur: Tropical Press.

ROBERTS, E. H. & KING, M. W. (1982). Storage of recalcitrant seeds. In *Crop Genetic Resources – the Conservation of Difficult Material* (ed. J. T. Williams & L. A. Withers), pp. 39–48. Paris: International Union of Biological Sciences.

ROBERTS, E. H. & SMITH, R. D. (1977). Dormancy and the pentose phosphate pathway. In *The Physiology and Biochemistry of Seed Dormancy and Germination* (ed. A. A. Khan), pp. 385–411. Amsterdam: North-Holland.

ROBERTS, E. H. & SUMMERFIELD, R. J. (1987). Measurement and prediction of flowering in annual crops. In *Manipulation of Flowering* (ed. J. G. Atherton), pp. 17–50. London: Butterworths.

ROBERTS, E. H. & TOTTERDELL, S. (1981). Seed dormancy in *Rumex* species in response to environmental factors. *Pl. Cell Envir.* **4**, 97–106.

SIMON, E. W. (1979). The Role of the Membrane. In *Low Temperature Stress in Crop Plants* (ed. J. M. Lyons, D. Graham & J. K. Raison), pp. 37–45. London: Academic Press.
STOYANOVA, S. & KOSTOV, K. (1983). Effect of storage temperature on the post harvest dormancy period of *Dactylis glomerata* L. *Plant Sci., Sofia* **20**, 94–100 [Russian: English Summary].
STOYANOVA, S., KOSTOV, K. & ANGELOVA, A. (1984). Influence of storage temperature on the post-harvest dormancy period of *Festuca arundinacea* Schreb. *Plant Sci., Sofia* **21**, 99–108.
THOMPSON, P. A. (1969). Germination of *Lycopus europaeus* L. in response to fluctuating temperatures and light. *J. exp. Bot.* **20**, 1–11.
THOMPSON, P. A. (1970*a*). A comparison of the germination character of species of Caryophyllaceae collected in central Germany. *J. Ecol.* **58**, 699–711.
THOMPSON, P. A. (1970*b*). Germination of Caryophyllaceae in relation to their geographical origin in Europe. *Ann. Bot.* **34**, 427–449.
THOMPSON, P. A. (1970*c*). Characterization of the germination response to temperature of species and ecotypes. *Nature, Lond.* **225**, 827–831.
THOMPSON, P. A. (1970*d*). Germination of species of Caryophyllaceae to their geographical distribution in Europe. *J. exp. Bot.* **34**, 427–449.
THOMPSON, P. A. (1973). Effects of cultivation on the germination character of the corn cockle (*Agrostemma githago* L.). *Ann. Bot.* **37**, 133–154.
THOMPSON, P. A. (1974). Characterization of the germination responses to temperature of vegetable seeds. I. Tomatoes. *Scientia Horticulturae* **2**, 35–54.
TOMPSETT, P. B. (1984). The effect of moisture content and temperature on the seed storage life of *Araucaria columnaris*. *Seed Sci. & Technol.* **12**, 801–816.
TOTTERDELL, S. & ROBERTS, E. H. (1979). Effects of low temperature on the loss of innate dormancy and the development of induced dormancy in seeds of *Rumex obtusifolius* L. and *Rumex crispus* L. *Pl. Cell Envir.* **2**, 131–137.
TOTTERDELL, S. & ROBERTS, E. H. (1980). Characteristics of alternating temperatures which stimulate loss of dormancy in seeds of *Rumex obtusifolius* L. and *Rumex crispus* L. *Pl. Cell Envir.* **3**, 3–12.
VEGIS, A. (1964). Dormancy in higher plants. *A. Rev. Pl. Physiol.* **15**, 185–224.
VILLIERS, T. A. (1975). Genetic maintenance of seeds in imbibed storage. In *Crop Genetic Resources for Today and Tomorrow* (ed. O. H. Frankel & J. G. Hawkes), pp. 465–485. Cambridge: Cambridge University Press.
VILLIERS, T. A. & EDGECUMBE, D. J. (1975). On the cause of seed deterioration in dry storage. *Seed Sci. & Technol.* **3**, 761–774.
VINCENT, E. M. & ROBERTS, E. H. (1977). The interaction of light, nitrate and alternating temperature in promoting the germination of dormant seeds of common weed species. *Seed Sci. & Technol.* **5**, 659–670.
WASHITANI, I. & SAEKI, T. (1986). Germination responses of *Pinus densiflora* seeds to temperature, light and interrupted imbibition. *J. exp. Bot.* **37**, 1376–1387.

Printed in Great Britain © Society for Experimental Biology 1988

INTERACTION OF TEMPERATURE WITH OTHER ENVIRONMENTAL FACTORS IN CONTROLLING THE DEVELOPMENT OF PLANTS

J. R. PORTER[1] *and R. DELECOLLE*[2]

[1] Department of Agricultural Sciences, University of Bristol, Institute of Arable Crops Research, Long Ashton Research Station, Long Ashton, Bristol BS18 9AF, UK
[2] Station de Bioclimatologie, Institut National de la Recherche Agronomique, Domaine Saint-Paul, Montfavet 84140, France

Summary

Development is the ordered sequence of changes in plant form and phenology that occurs through time. As such it is distinguished from that of growth and includes the succession of ontogenetic stages as well as the initiation of leaves, shoots and roots.

It is more useful to look at rates of development as functions of environmental influences rather than, for example, total number of days to a given stage. It is often found that linear, or close to linear, relationships are found between rates and temperature, photoperiod and the duration of cold temperatures. These ideas are treated theoretically and illustrated with data from an experiment where wheat was grown at ten sites in Britain. We conclude that temperature is the most important factor governing differences in developmental rates between sites and sowing dates, especially for later stages of development. Variation in the timing of earlier stages depends on either exposure to photoperiods of different lengths or on the degree of vernalization of the plants. Calculation is made of the maximum effectiveness of these environmental factors in shortening the vegetative phases in a range of wheat cultivars and the basis of their interaction is also explored. For wheat it seems that they act multiplicatively in their effect on developmental rate.

The rate at which plants produce their leaves is important in canopy development. The sensitivity of the rate of leaf emergence, measured in thermal time, has been linked to the rate at which daylength changes ($d\phi/dt$). A possible mechanism may involve plants responding to the ratio of light in the red (R, *ca* 660 nm) and far-red (FR, *ca* 730 nm) wavebands (R/FR). This ratio decreases when $d\phi/dt$ is becoming more negative and vice versa. Linkage can be found between the change in R/FR at twilight and commensurate changes in calculated phytochrome photoequilibria. Thus, the possibility exists that phytochrome may be involved in modulating the leaf production rate in response to different sowing dates.

Alternatively, variation in rate of leaf production may be through changes in the base temperature, in response to the 'spectrum' of temperatures the plants experience during development. Thus, any involvement of photoperiod in this response may be spurious. Controlled environment studies of leaf emergence tend to support this latter interpretation as does analysis of experiments from three European sites.

Introduction

In studying the biology of plants, at the scale of the community or that of the whole plant, we recognize two dependent, yet separate, components, growth and development. For many crops dry matter accumulates at a rate that is determined by the amount of intercepted radiation (Gallagher & Biscoe, 1978) but other climatic factors affect the duration of growth. Thus, on a world-wide basis, plants are restricted to geographical zones and to particular seasons of growth. Whilst factors such as excessive salinity or the seasonality of the water supply limit the extent of these areas, the growth cycle of plants is also constrained to fit into a range of local climatic conditions (Bunting, 1975; Roberts & Summerfield, 1987; Summerfield & Roberts, 1987). This paper considers how environmental factors interact with temperature in controlling plant development which, in turn, is the basis for predicting one important response of plants to their growing conditions (Porter, 1985).

Although factors such as water shortage and different nutrient availabilities appreciably affect the rates of dry matter accumulation (Legg *et al.* 1979) they are of secondary importance in their effects on development (Monteith, 1981; Thorne & Wood, 1987) and their interaction with temperature will not be discussed. It is even plausible that shortages (or excesses) of water and nutrients exert their 'effects' on development via changes induced in the local plant or crop microclimate which then initiates a temperature-based developmental response. Recently, a conceptual review of the temperature dependence of many plant and crop processes has been made by Johnson & Thornley (1985).

A dramatic example of the different ways in which dry matter accumulation, measured by yield, and development respond to temperature is shown in Fig. 1. In his experiment on spring wheat, Robertson (1983) compared the effect of maximum temperature during the period from jointing to ear emergence on rate of development as opposed to the rate of growth. Whilst the relative increase in development rate is linear with increasing temperature, that of yield is parabolic with an optimum temperature, under these experimental conditions, of just over 20°C. This effect of increasing temperature is referred to as the positive effect of temperature on development to distinguish it from low temperature vernalization responses. Higher temperatures favour faster development (although there are upper limits; see Roberts, 1988). Temperatures, up to about 25°C for wheat, also have a positive effect on growth rate (Penning de Vries *et al.* 1979) but the decreased duration of growth at high temperatures means that total yield will be reduced. Also, high temperatures and high amounts of incident radiation, to which the rate of evapo-transpiration is related directly, are often coincidental. Therefore, an equally important determinant of the yield/temperature relationship is the effect of reduced water availability on growth rate.

Robertson's experiment is one in a long tradition of attempts to understand the relationship between the environment and the development of plants. It is just over 250 years since modern experimentation began into the response of plant development to temperature (de Réaumur in Paris in 1735, cited in Kolbe, 1974). The reasons for such an ongoing and abiding interest are not difficult to understand. The effect of the weather on plant development is of interest not just to plant physiologists but also for the growers of crops and their advisers. The relative proportions of the life cycle spent in pre-, as opposed to, post-flowering growth will influence the relative sizes of the components of yield (Halloran & Pennell, 1982). Also, certain developmental stages are used as markers in a crop's life as the time for effective use of growth regulators, herbicides or fertilizers with minimum risk to the crop.

In his work on development, de Réaumur postulated that one developmental stage will succeed another when a certain summation of daily temperature has been reached and that the base temperature above which increments may contribute positively to a developmental progression may be above 0 °C. In other words, he concluded that, between given temperature limits, the rate of plant development is linear with temperature. Such early insight is remarkable. It is humbling to realize that we still do not know why responses to temperature tend to be linear for whole plants whereas, at the cellular and sub-cellular levels, rate constants show exponential responses to absolute temperature.

Following de Réaumur, studies into plant phenology developed quickly in the eighteenth and nineteenth centuries; the term phenology was coined by the Belgian botanist Morren in about 1850. Korniche & Werner (1885) (quoted in Kolbe, 1974) provided thermal time totals for the life cycles of different crops with about 1500 °Cd required for the maturation of spring barley and about 4000 °Cd for rice, both above base temperatures of 0 °C.

These early studies, perhaps for technical reasons, were not able to shed light on how development might be affected by a combination of temperature with other environmental factors. Thus, it was not until the 1920s that Garner & Allard (1920) showed that certain cereals were either developmentally responsive to changing daylengths, above or below critical values, or they were neutral in their response. Oats, spring barley, wheat and rye were found to be long-day plants (LDP), whereas maize was classified as a short-day plant (SDP) with winter barley as day neutral. As Roberts & Summerfield (1987) have pointed out, the terms long and short-day do not refer to the length of effective photoperiods but to whether a critical night-length has or has not been exceeded for obligate SDP or LDP, respectively. Plants with facultative or quantitative responses to photoperiod (Vince-Prue, 1975) are distinguished by rates of development which proceed at faster rates when the photoperiod is increased or reduced. Whilst we commonly talk in terms of photoperiod being the stimulus it is, of course, the length of the dark period that is perceived by plants.

By the end of the 1920s it had become clear that a third environmental factor was necessary for the more complete understanding of the control of development especially for plants grown in northern latitudes. This was the effect of cold temperatures on the process, originally referred to as Jarowisation by Lysenko and his colleagues, but now known as vernalization. This effect was first noted in the 1857 Yearbook for the Ohio State Board of Agriculture, USA (see Kolbe, 1974). It was found, in experiments by Klippart, that it was possible to 'transform winter wheat into summer wheat...by softening the seed and allowing it to germinate, then freezing it and keeping it in the frozen state until it is sown in the spring'. The previously held idea was that spring and winter cereal cultivars had fixed, genetically controlled developmental phases of different lengths whereas, we now recognize, it is the requirement for a cold temperature experience that, in the main, separates winter and spring types of cereals; the length of the vegetative phase in winter cereals can approach that of spring varieties provided vernalization conditions have been met.

Interactions

The interactions between temperature, photoperiod and vernalization are poorly understood. Temperature and photoperiod have been reported to act both synergistically and independently in their effects on development. Ford *et al.* (1981) concluded, on the basis of a combination of controlled-environment and field experiments for northern temperate cultivars of wheat, that photoperiod and temperature effects did not interact. Halse & Weir (1970) found vernalization responses only for a specific range of temperatures, which implies that vernalization is an effect independent of the generally continuous response of winter wheat to temperature.

Marcellos & Single (1971) maintained that photoperiod and temperature effects were linear with respect to development, but correlated. Following a review of experiments, mostly from controlled environments and for a variety of LDP and SDP, Roberts & Summerfield (1987) held that there were 'no interactions between mean temperature (up to an optimum) and photoperiod'. They were, however, considering the whole developmental period from emergence to flowering as a single unit. For cereals at least, emergence to anthesis covers, at a minimum, two phenological phases; from emergence to terminal spikelet (Fig. 2) and from terminal spikelet to flowering. It is known that successive phenological phases may respond differently to temperature. For example, Angus *et al.* (1981*a*,*b*) found higher base temperatures for later stages in the life cycle of several spring wheats. Porter *et al.* (1987) found, similarly, that apparent base temperatures increased with later stages as did also the length of the photoperiods effective in forwarding development. The influence of photoperiod in controlling the developmental rate is generally smaller (at least in wheat) for later stages than for earlier ones.

Before considering data on temperature, photoperiod and vernalization interactions in field-grown crops it is worth discussing how responses to these factors can be analysed. At their simplest, typical data from a controlled environment experiment in which plants are grown in a constant temperature, but different

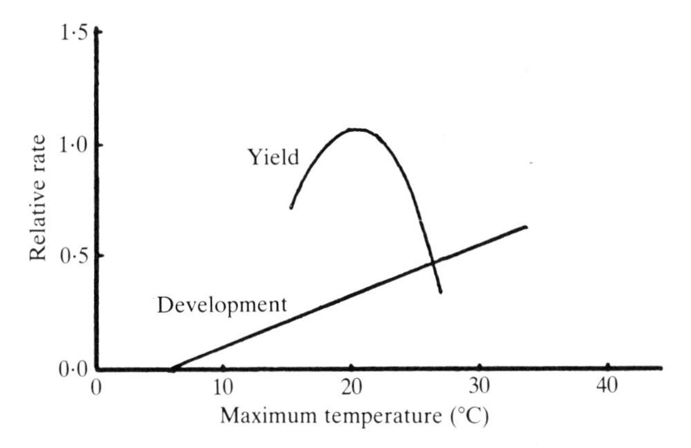

Fig. 1. Comparison of the effect of maximum temperature during the phase from jointing to heading on the relative rate of development and yield for spring wheat. (Used with permission; Robertson, 1983.)

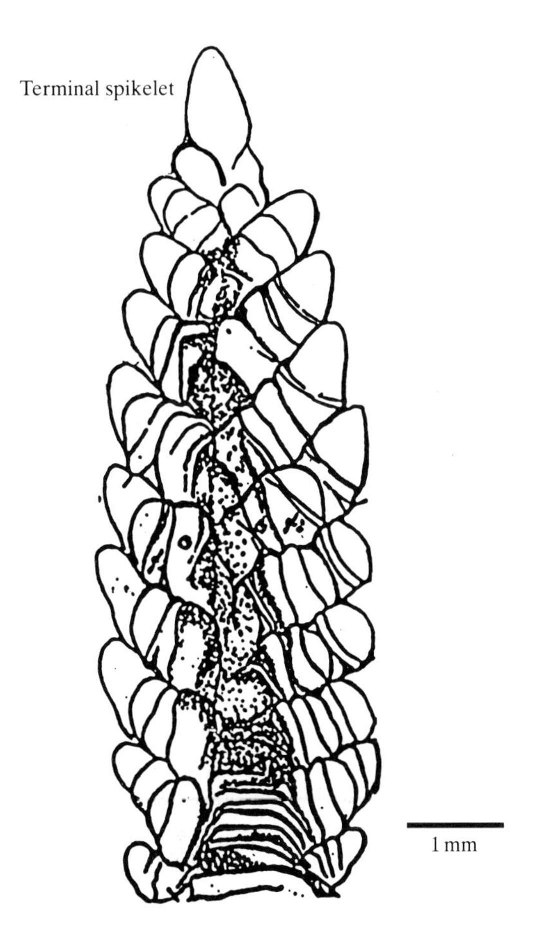

Fig. 2. Apex of wheat at the terminal spikelet stage.

photoperiods, might look like Fig. 3 (Roberts & Summerfield, 1987). For a typical SDP, in this case, we see that as photoperiod shortens so the number of days to flowering also decreases but in a hyperbolic fashion. Also of note is the minimum observed time and the maximum possible time to flowering, i.e. there is a photoperiod at which the time to flowering becomes infinite.

It is much more useful, and usual, to plot these data as their reciprocals, to provide an estimate of the relationship between, in this case, photoperiod and the *rate* of development, with rate defined as the reciprocal of the duration of a phenological phase. Strictly speaking, the measured rate is the average rate of development between the stages and should, perhaps, be thought of as the fractional progress between one phenological event and another. The relationship between the rate of development and photoperiod is often linear (Fig. 4) and the value of the photoperiod at which the rate of development is zero, the base photoperiod, can be found from the extrapolated line. Base values for environmental variables can never be observed, only estimated by extrapolation, and their method of estimation has important implications, as will be seen later.

There are many data that illustrate the above principles of development in crops. An example is the work of Robertson (1983) who linked the length of the period from heading to maturity for Marquis wheat to the average mean daily temperatures, during the phase, from eight stations in Canada. The results (Fig. 5) show a negatively sloping curvilinear relationship. As temperatures increase, so the duration of the phase decreases non-linearly. The relationship is made linear by using the reciprocal of the duration, i.e. the rate, for the *y*-axis (Fig. 6). This gives an extrapolated threshold temperature of about 6°C, a value close to that found by other workers for this phase (Vos, 1981; Weir *et al.* 1984). However, there are situations in which the relationship between rate and temperature may

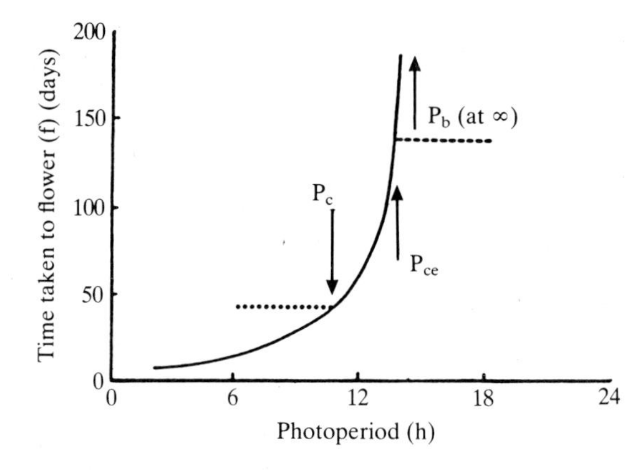

Fig. 3. Typical photoperiodic response of a short-day plant showing time taken to flower (f) as a function of photoperiod. P_b, the extrapolated base photoperiod, above which time to flowering is infinite; P_c, the critical photoperiod at or below which time to flowering is minimal; P_{ce}, the critical photoperiod, below which the time to flowering is progressively shortened down to P_c. (Used with permission; Roberts & Summerfield, 1987.)

not be linear or in which it may not be possible to distinguish between a linear and non-linear form (Skjelvag, 1981*a*,*b*). One possible reason has already been mentioned; the length of the phase may be too long and may comprise more than a single homogeneous phenological period. In addition, daily mean temperature may not be a sufficiently sensitive measure of the environment; there may be different developmental responses to day and night temperatures or apical temperatures may differ from ambient to a significant extent. However, it has been shown that mean air temperature, for a site in central England, is a good approximation to mean temperature near the soil surface (Gallagher, 1976). For cereals, prior to stem extension, the apex is at a depth of 20–40 mm.

Experimentation

To disentangle the interaction of environmental factors and measure their relative importance in the control of plant development requires careful experimentation. Roberts & Summerfield (1987) concluded from comprehensive experiments using controlled environment facilities that rates of progress towards flowering are, almost without exception, linear functions of temperature or photoperiod or both. The daily contribution of the effect of environment on progress towards flowering acts as if successive days *add* their respective increments rather than advance phenology in any other functional manner, for instance, multiplicatively. This is not to say that interactions between environmental factors are also additive but their combined effect on development does seem to be one of promoting an arithmetic progression from one stage to the next. Of the six species looked at by Roberts and his co-workers barley has the most complex response to a combination of environmental factors. It shows differential responsiveness to photoperiod depending on the current phenological phase

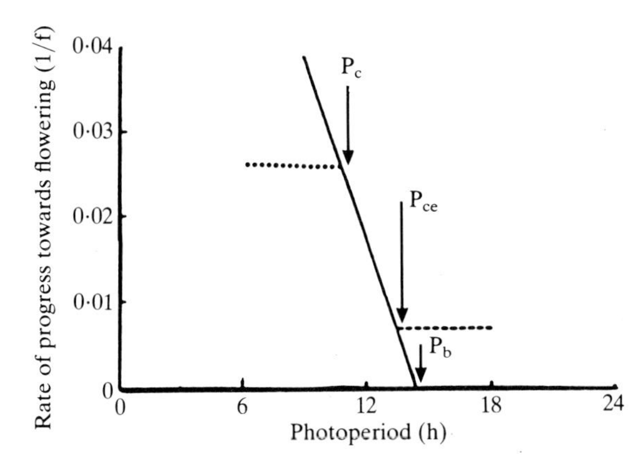

Fig. 4. Photoperiodic response of a short-day plant showing the rate of progress (1/f) as a function of photoperiod. P_b, the extrapolated base photoperiod, above which the rate of flowering is zero; P_c, the critical photoperiod at or below which the rate of flowering is maximal; P_{ce}, the critical photoperiod, below which the rate of flowering is progressively increased up to P_c. (Used with permission; Roberts & Summerfield, 1987.)

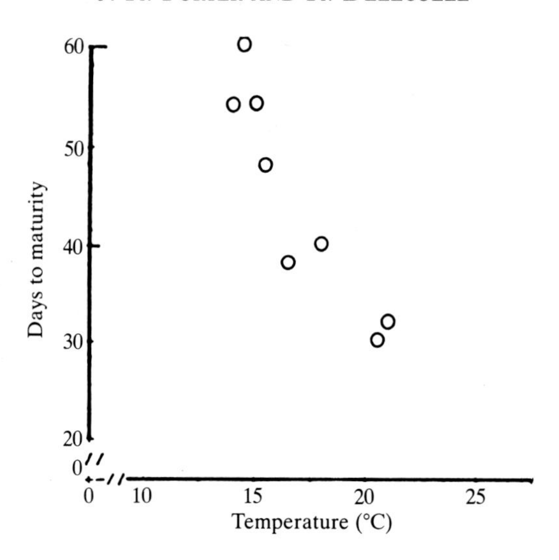

Fig. 5. Number of days from ear appearance to maturity for wheat grown at eight sites in Canada and the mean daily temperature (°C) during the phase. (Used with permission; Robertson, 1983.)

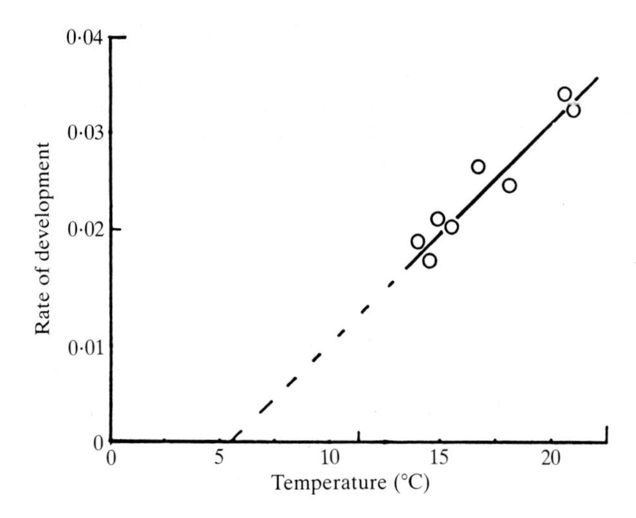

Fig. 6. Average rate of development from ear emergence to maturity for wheat grown at eight sites in Canada and the mean daily temperature (°C) during the phase. (Used with permission: Robertson, 1983.)

together with the ability of low-temperature vernalization to be partially substituted for by short days (8–10 h). Long days, following ear initiation, stimulated the rate of development (Roberts *et al.* 1988).

Whilst not allowing the same precise, manipulative control over experimental stimuli as that obtained under controlled environments, experiments designed to look at rates of morphological development in the field are important in

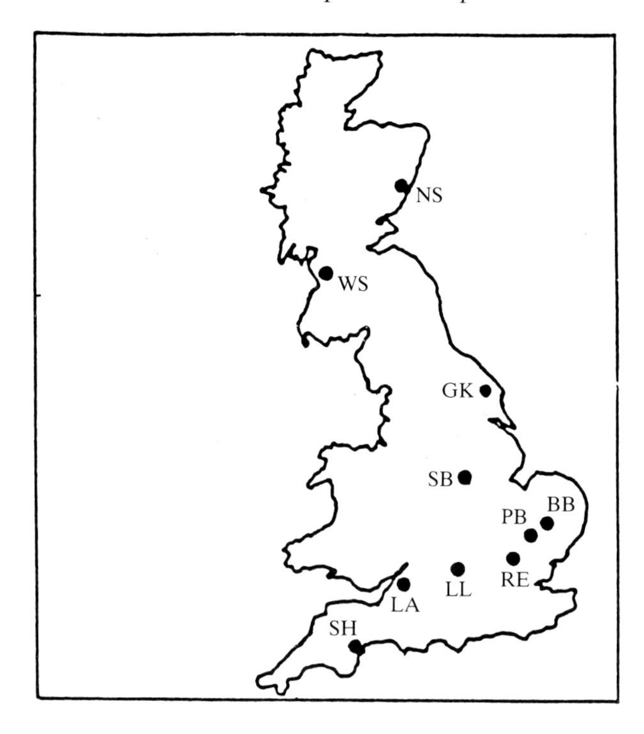

Fig. 7. Location of UK sites for a survey of the development of *cv.* Avalon wheat. NS, Aberdeen; WS, Auchincruive; GK, Driffield; SB, Sutton Bonington; BB, Higham; PB, Cambridge; RE, Harpenden; LL, Letcombe Regis; LA, Long Ashton; SH, Newton Abbott. (Used with permission; Porter *et al.* 1987.)

complementing the work mentioned above and similar studies. We need to be sure that the conclusions reached in the laboratory apply when temperature varies diurnally and seasonally and when the end-of-day change is not an abrupt one.

Accordingly, in 1983–84, an experimental analysis of the development of winter wheat (*cv.* Avalon) was performed at ten sites in England and Scotland (Fig. 7). The sites covered 7° of latitude and included much of the main cereal growing areas of Britain. Details of the experiment are given in Porter *et al.* (1987) and Kirby *et al.* (1987). At each site Avalon wheat was sown in mid-September (S1), mid-October (S2) and mid-November (S3) 1983. Throughout the following growing season plants were monitored for their stages of apical development (Kirby & Appleyard, 1981), primordium initiation and leaf production (Stern & Kirby, 1979).

Development of the mid-September sowing (Fig. 8) shows the differences between sites for the length of the period from sowing to emergence (maximum 20 d, minimum 9 d) but far greater variation was found for the phase emergence to double ridges. For example, the largest difference between the earliest and the latest occurrence of any stage for any sowing date, at the ten sites, was seen in double ridges for the September sowing. This stage occurred 46 days later at Aberdeen than at Newton Abbot for S1 but only 28 days later and 16 days later for

S2 and S3, respectively. Also the 46-day difference between the geographical extremes for double ridges had been reduced to a 17-day difference by anthesis. Similar north–south differences were found for anthesis from S2 and S3. Generally, but not uniformly, the length of the grain-filling period increased in a northerly direction, with the longest grain filling being at Aberdeen for S3 (61 d) and equal shortest at Newton Abbott for S3 and Cambridge for S2 (39 d).

If we consider the phase which is most variable in length, that is emergence to double ridges, a plot of the data from all sites and sowing dates shows no simple relationship between rate of development and temperature (Fig. 9). For S1 and S2 the rate was correlated with mean temperature, but the slopes and intercepts differed. Data for S3 showed no significant correlation but pointed generally to faster rates of development than S2 for similar mean temperatures.

One way of examining the effect of photoperiod on the rate of development is to express the duration of the emergence to double ridge phase in thermal time by summing the mean daily temperatures during the period. The reciprocal of thermal time, equivalent to a rate, may then be plotted against mean photoperiod. If significantly more of the overall variation is accounted for than for the simple

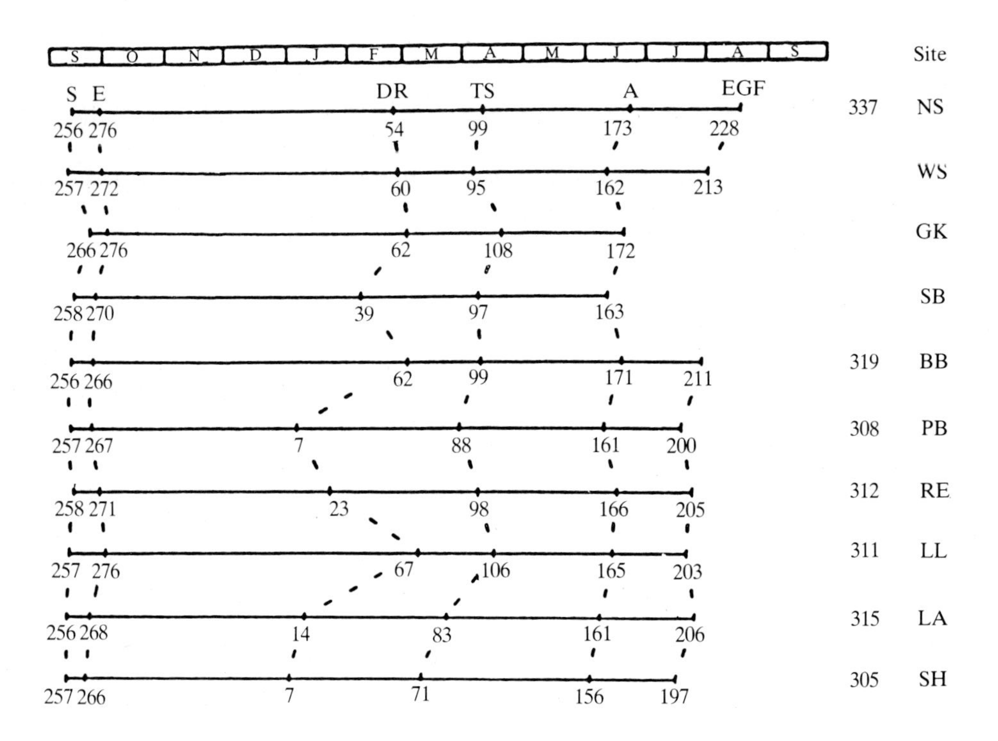

Fig. 8. Timing (as day number from 1 January) of phenological events for wheat crops sown in mid-September 1983 at ten sites in Britain. Sites are arranged in north–south order down the figure. S, sowing; E, emergence; DR, double ridge; TS, terminal spikelet; A, anthesis; EGF, end of grain filling for most sites. Months are shown at the top of the figure and the number of days from sowing to the end of grain filling is shown, for most sites, to the right. (Used with permission; Porter *et al.* 1987).

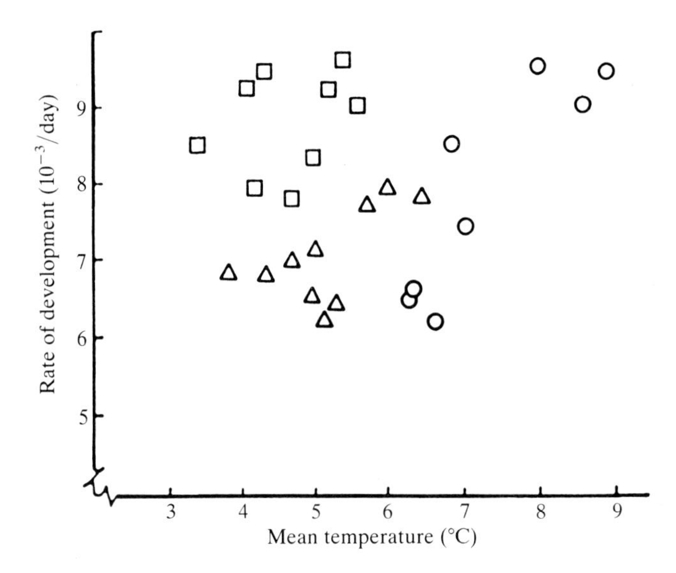

Fig. 9. Rate of development from emergence to double ridges for *cv.* Avalon wheat crops grown at ten sites in Britain plotted against mean temperature during the phases. ○, sowing 1; △, sowing 2; □, sowing 3. (Used with permission; Porter *et al.* 1987.)

rate against temperature we conclude that photoperiod is having an extra and significant influence on development.

Mean photoperiod did not vary appreciably between sites or sowing dates but the relation between the thermal rate of development and mean photoperiod suggests that variation in site-to-site development rate for S3, in contrast to S1 and S2, was explained by photoperiodic effects (Fig. 10). For S1 and S2 there was a negative correlation between thermal rate and photoperiod.

The contribution of vernalization to the control of development is assessed by a method analogous to that described above. In this case the phase duration is calculated in photo-thermal time and its reciprocal plotted against an estimate of the degree of vernalization of the plants. Photo-thermal is thermal time modified in its accumulation by developmentally effective photoperiods. The degree of vernalization of the plants is assessed by the number of days that the mean temperature is between certain limits (see Porter *et al.* (1987) for details). It is difficult to define the form of the vernalization response because the relationship of the degree of vernalization to temperature is non-linear, dependent on a plant's previous exposure to effective temperatures. The sensitivity of rate of development to vernalizing temperatures decreases as vernalization is progressively satisfied (Chujo, 1966; Marcellos & Single, 1971; Weir *et al.* 1984). The degree of vernalization, on a scale between 0 and 1·0, is shown as parameter Fv in Fig. 11.

The mean value of the vernalization parameter for plants from S2 and especially S3 was near to 1·0 for all sites, indicating they were completely vernalized throughout the whole emergence to double ridge phase. However, for S1 the degree of estimated vernalization significantly accounted for variation in the

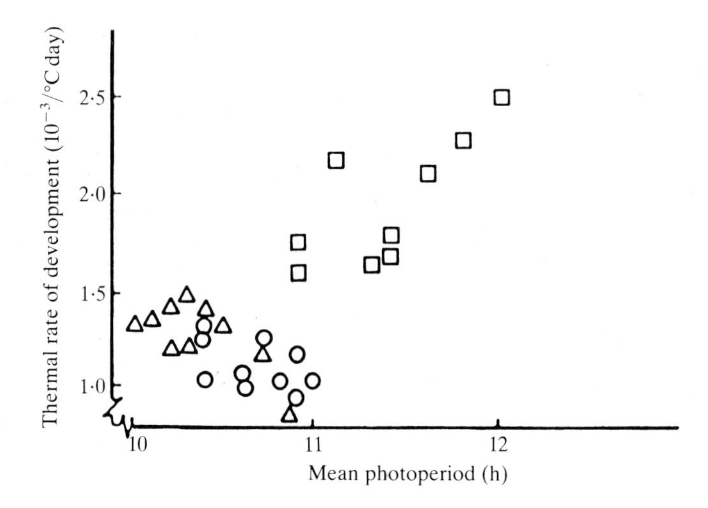

Fig. 10. Thermal rate of development from emergence to double ridges for *cv*. Avalon wheat crops grown at ten sites in Britain plotted against mean photoperiod during the phase. ○, sowing 1; △, sowing 2; □, sowing 3. (Used with permission; Porter *et al.* 1987.)

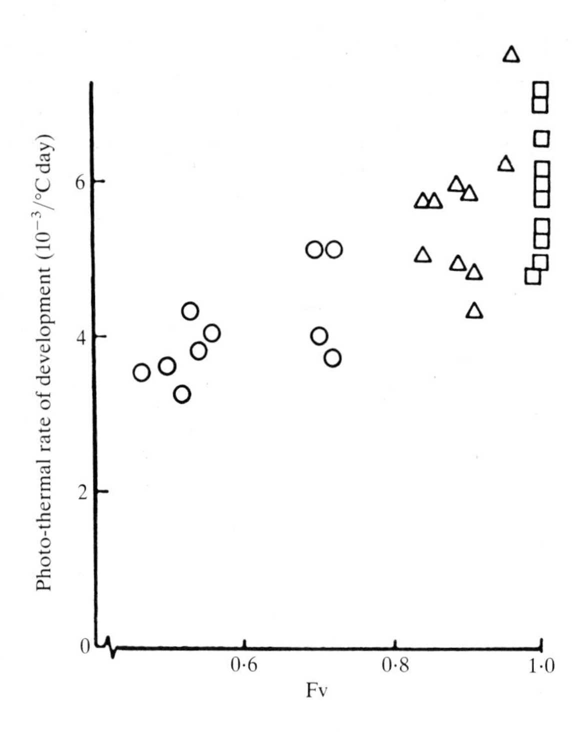

Fig. 11. Photo-thermal rate of development from emergence to double ridges for *cv*. Avalon wheat crops grown at ten sites in Britain plotted against a mean vernalization factor (Fv) during the phase. See text for explanation of variables. ○, sowing 1; △, sowing 2; □, sowing 3. (Used with permission; Porter *et al.* 1987.)

photo-thermal rate of development (Fig. 11). Over all sowings, there was a highly significant linear relationship between photo-thermal rate of development and the degree of vernalization. Comparison of the spread of points in Figs 9 and 11 shows, for S1, that positive temperature effects accounted for as much variation as vernalization effects in determining the rate of development. However, the variance explained by vernalization is additional to that explained by positive temperature effects because developmental rates have been successively calculated to embrace the influence of temperature and photoperiod. In contrast, photoperiod varied little from emergence to double ridges for S1 and accounted for an insignificant amount of variation in thermal development rate.

The overall conclusion from this work is that temperature is the over-riding environmental factor that controls development rate, and photoperiodic and vernalization factors operate with decreasing effectiveness as the plants age and, for early stages, with a relative importance dependent on their variability. Site and sowing-date effects on development in winter cereals are achieved by exposing plants to particular ranges, or combination of ranges, of climatic conditions.

The strength of vernalization and photoperiodic effects

Following the preceding analysis, two questions present themselves; (i) if temperature is the over-riding environmental control on development then what is the contribution of photoperiod and vernalization to development in its early stages?; (ii) since we are primarily concerned with interactions in the control of development what form of interaction best accounts for our observations?

For the phase during which photoperiod and vernalization exert their major effects it is possible to estimate their contribution to developmental progress using the data of Rahman (1980). In his collection of 30 wheat varieties gathered between latitudes 64°N and 42°S, most cultivars had low vernalization requirements coupled with medium-strength photoperiodic responses. Based on their response to vernalizing conditions of 49 or 0 days at 4°C or photoperiods of 10- or 16-h cultivars were grouped according to the degree of shortening of the phase from emergence to floral initiation (Stern & Kirby, 1979). For those with a 'high' vernalization response the mean time to floral initiation from emergence for vernalized plants was 20 d as against 41 d for high response non-vernalized plants, a 50 % faster rate of development. A similar calculation for photoperiod gave an increase in rate of about 36 % for highly photoperiodically sensitive plants. The majority (23 of 30) of the other varieties had smaller responses than these and would reflect more closely the findings reported above. It is recognized that Rahman's sample was biased towards mid-latitude Australian cultivars (*ca* 33°S) and it is possible that varieties exist that would surpass these estimates.

Form of interaction

An important feature of the analysis used in assessing the effects of three environmental factors on crop development is that the interplay between these components has been viewed as multiplicative but without separate interaction terms. Formally, the model can be written;

$$\frac{dF}{dt} = f(T \times P \times V)$$

where F represents the change from one stage into the next and temperature T, photoperiod, P, and vernalization, V, are viewed as having their separate effects multiplied. This has one result, in particular, that extreme values of temperature do not disproportionately influence the developmental rate as would be the case if the influences were added. Another reasonable model would be one that postulated the rate of development to be a function of a single minimum or maximum value of one of the components.

The evidence is, from formal optimization of the results from the Avalon survey (Travis, personal communication), that neither additive nor limiting factor models offered any improvement over the simple multiplicative approach although sometimes they were no worse. The inclusion of additional interaction terms was also unnecessary. Non-linear responses to temperature and photoperiod were also tried in the analysis but none were found to fit significantly better than linear responses. It is from the results of experiments and analyses such as these that computer models aimed at predicting crop development in the field have been based (Porter, 1983, 1985, 1987).

Leaf production in winter wheat

Over the past ten years a number of computer simulation models have attempted to simulate the growth and development of wheat (Groot, 1987; Penning de Vries & van Laar, 1982; Ritchie, 1985; Weir *et al.* 1984). The most difficult, least well understood and least well modelled component of the crop development/environment interaction is how the crop canopy develops over time so that radiation is intercepted and dry matter produced and partitioned to plant parts. In the model, AFRCWHEAT (Weir *et al.* 1984), canopy development was modelled in detail by considering the contribution of separate groups of shoot and leaves to the total leaf area. Such cohorts of leaves and shoots had life cycles whose demographic characteristics were linked to the density and age structure of the crop. The net result of the birth, development and death of shoots and leaves was the Green Area Index (GAI) of the crop on any day. Similar ideas for modelling the changes in tiller number in barley have been suggested by Gandar *et al.* (1984) and Gandar & Bertaud (1984). Such approaches are in contrast to most other crop simulation models where, to calculate GAI, an equivalence is made between the calculated dry weight of leaf material on a day and its area via the specific leaf area (SLA), an input into such models. The parameter value taken by SLA can change with the phenological stage of the crop. This idea offers simplicity in model parameterization but it has the problem that the increase in GAI is very sensitive to the value chosen for SLA. If such a value it too high then serious over-estimation of GAI can occur; this increases the estimate of leaf weight which increases the estimate of leaf area and an unstable positive feed-back loop is generated leading to potential over-estimates of dry-matter production.

The rate at which leaves are produced is fundamental to a model of canopy development. In AFRCWHEAT (Porter, 1984) the predicted rate of leaf emergence, for a given crop emergence date, is calculated from an experimentally determined relationship (Baker *et al.* 1980), linking the rate of leaf production in thermal time (i.e. the reciprocal of the thermal time interval between successive leaves) to the rate at which daylength is changing at the time of crop emergence (Fig. 12). Thus, there appears to be a clear example of an interaction between the rate of development of the vegetative component of a crop, temperature and another environmental factor, in this case, the rate of change of daylength ($d\phi/dt$). Such a relationship goes some way towards explaining how the rate of leaf emergence in wheat can be about 50% slower for cereal crops sown in September in the UK compared with those sown in the spring.

If $d\phi/dt$ is a cue to predicting the rate of future leaf production in wheat then it would be interesting to know with which other environmental signal it may be correlated and possibly substituting. At face value, $d\phi/dt$ says that plants are, in some way, able to compare the durations of two or three consecutive days, 'memorize' the result and refer to it for the rest of the vegetative part of the life cycle in order to calculate how fast leaves should be produced. What other environmental signals might be sensitive enough to provide the initial rate fixing with sufficiently accurate information?

The physiology of how plants detect the transition from light to dark has been researched intensively over the years (Vince-Prue, 1975, 1981; Salisbury, 1981). What is important about such investigations is that they should be supported by field-based studies designed to test laboratory findings (Smith, 1986; Smith & Holmes, 1977).

One hypothesis to explain how plants detect the twilight signal that marks the end of the day is that plants perceive the coming darkness by changes in the

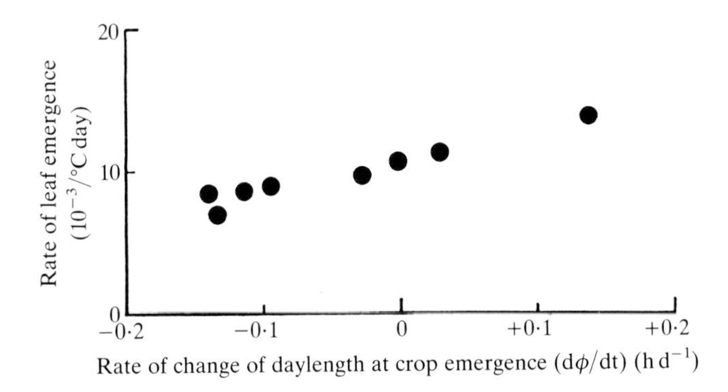

Fig. 12. Mean rate of leaf appearance in thermal time plotted against the rate of change of daylength ($d\phi/dt$) at crop emergence for eight crops of *cv.* Maris Huntsman wheat grown at Sutton Bonington, UK. Sowing dates ranged from 4 October 1975 to 6 March 1978. (Redrawn from Baker *et al.* 1980.)

spectral composition of light, particularly in the wavebands centred about 660 nm and 730 nm. These correspond to light in the red (R) and the far-red (FR) part of the spectrum and during twilight the ratio (R/FR) of photon flux densities in these bands decreases from a value of about 1·2 to about 0·7 during a period of 60–90 min (Smith & Morgan, 1981). It has been suggested that the fall provides a cue for detecting the onset of darkness. We wish to present circumstantial evidence both in favour of and against this hypothesis operating in the field.

Few measurements of the annual cycle of R/FR have been made. An exception is shown in Fig. 13 (Gorski, 1980) where the ratio was measured every 30 days at a site in Poland (51 °N) for three solar angles. Baker *et al.* (1980) performed their experiments at 53 °N. The observations by Gorski can be used to predict changes in R/FR around twilight (i.e. at low solar angles). Fig. 14 shows the correlation between R/FR and $d\phi/dt$ for 5° solar elevation throughout the year. The overall correlation is negative but not significantly so ($0{\cdot}10 > P > 0{\cdot}05$). For times of the year when daylength is increasing (day numbers 30, 60, 90 and 120, counting from 1 January) there is a (non-linear) increase in R/FR. This is matched by a decrease in R/FR when daylength is falling (day numbers 210, 240, 270 and 300). Thus, at the time of year when daylength is changing R/FR also changes in a consistent direction.

There is general agreement (Smith, 1982, 1986) that plant photomorphogenesis is regulated by the equilibrium conditions of the reversible phytochrome system between Pfr, the active from of phytochrome and the inactive form, Pr. Ptotal is the sum of Pr and Pfr and the red to far-red ratio of light modulates this equilibrium.

If the R/FR ratio, or any other environmental signal, is important in providing end-of-day signals and thus substituting for $d\phi/dt$, three conditions have to be met. First, any change in R/FR and consequent change in the phytochrome equilibrium (defined as the ratio Pfr/Ptotal) should show what engineers call good 'gain'. Thus, the magnitude of the change in the stimulus (R/FR) should be matched by, at least, a similar change in the response. Secondly, that twilight (in

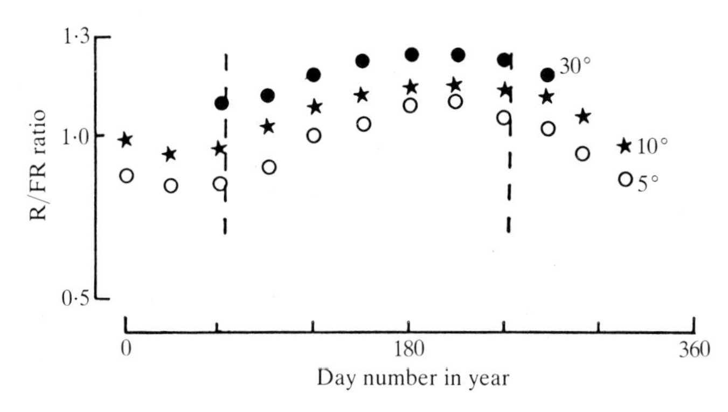

Fig. 13. Mean annual cycle of R/FR ratio in the direct solar radiation measured at three angles of solar elevation for Pulaway, Poland during 1975–1978. (Redrawn from Gorski, 1980.)

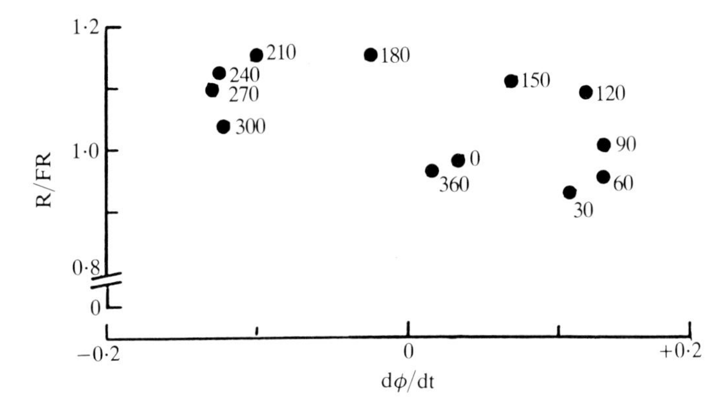

Fig. 14. Annual relationship between the change in R/FR ratio and the rate of change of daylength ($d\phi/dt$) at 30-day intervals for Pulaway, Poland. Numbers are day numbers in the year (1 January = 1).

this case) should be the time when the stimulus changes most over a 24-h cycle and, thirdly, that the duration of any change in the stimulus should allow sufficient time for equilibrium conditions to be set up in the response. Smith (1982) provides evidence to satisfy the first two conditions. Fig. 15 shows that during the twilight period there is an absolute reduction of 0·5 in R/FR for about a 0·1 reduction in Pfr/Ptotal and that compared with any diurnal change in R/FR the maximum change occurs at the end of the day. It is worth noting that phytochrome equilibrium changes induced by changes in R/FR detected under canopies dwarf those at twilight by a factor of four times although the rates of change are very similar for the two situations.

Using hypocotyls from *Cucurbita pepo*, Smith & Holmes (1977) showed that a Pfr/Ptotal equilibrium, the photo-equilibrium, was set up after 20 s of light from a mixed spectral source following saturation with either R or FR light (Fig. 16). The equilibrium value differed depending on whether the samples were pre-irradiated with R or FR but it is clear that if 20 s is the time in which a stable Pfr/Ptotal is established then, except for one example from Baker *et al.* (1980), for which $d\phi/dt$ is close to zero, the magnitude of $d\phi/dt$ is sufficient for a particular photo-equilibrium condition to the established. What is unknown is whether different phytochrome equilibria may be set up for different rates of change of daylength and the extent to which changes in photo-equilibrium at dusk compare with changes in phytochrome levels throughout the rest of the dark period. If the former are small compared to the latter it would be necessary to rethink this hypothesis.

The above statements are merely hints and possibilities about the involvement of R/FR, modulated by phytochrome metabolism, in governing the rate of leaf production in wheat. Naturally, at this level of (un)certainty there is also contrary evidence. It is difficult to see how a change in R/FR can give information about temporal position in a year if the amplitude of annual changes in the ratio (Fig. 13)

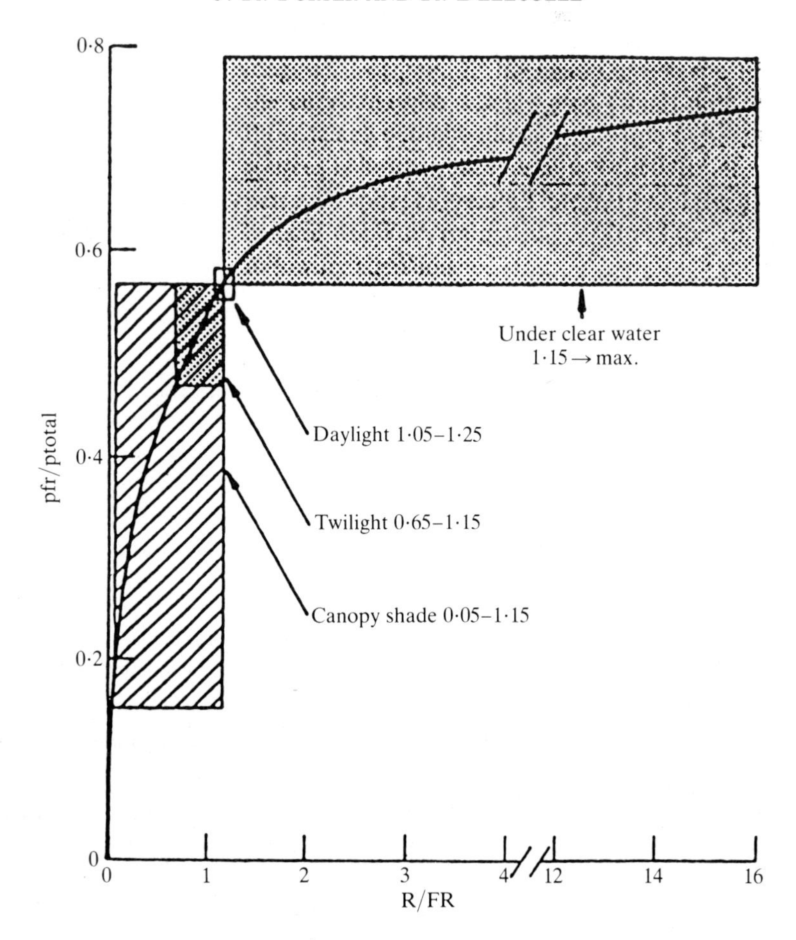

Fig. 15. Relationship between R/FR and calculated phytochrome photoequilibrium (Pfr/Ptotal). The blocked areas indicate the observed variation in R/FR under certain environmental conditions. (Used with permission; Smith, 1982.)

are so similar to that observed on a diurnal timescale (Fig. 15). In this case how would a plant tell the duration of a day from that of a year? Experimentation is needed to shed further light onto this subject. Although technically difficult it would be a direct test of the above hypothesis to isolate phytochrome from plants that have emerged at different times of the year, with the hope that the Pfr/Ptotal value would vary in a systematic way with emergence date. A simpler experiment would be to alter the R/FR environment of plants emerging on the same date, testing the prediction that this would alter the rate of leaf production. Emerging plants under an already existing canopy would be one way of achieving this objective since it is known that R/FR spectral ratios change more under canopy shade than during the day or at twilight (Holmes & Smith, 1977).

An alternative to the involvement of R/FR in substituting for the $d\phi/dt$ signal is that plants detect the end-of-day by a tenfold decrease in total light level (Salisbury, 1981). Interestingly, human beings require a 10^7 change in illuminance

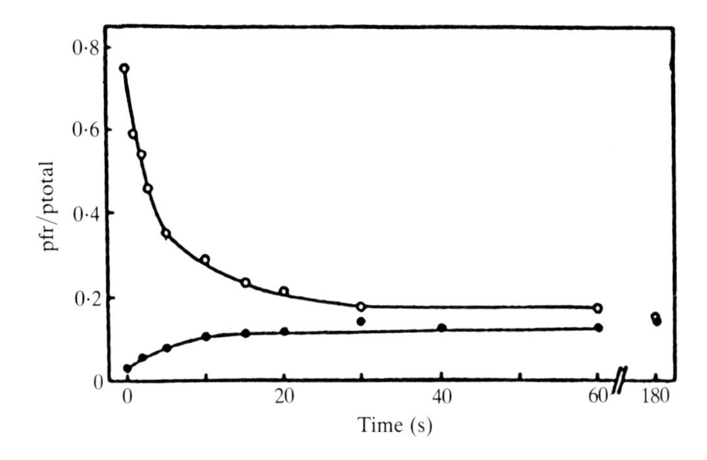

Fig. 16. Phytochrome phototransformation in *Cucurbita pepo* hypocotyl hooks under a mixed light spectrum providing a photoequilibrium of Pfr/Ptotal = 0·15. Samples were pre-irradiated with red (○) or far-red (●) light. (Used with permission; Smith & Holmes, 1977.)

in order to register such a change. Significant changes in Pfr/Ptotal equilibrium have been correlated with the changes in irradiance around dusk (Frankland, 1986). Thus, it would be wise to clarify the importance of irradiance as a signal for the end of the day.

Any attempt to include the influence of light quality or irradiance as a mechanism governing the rate of production of leaves is dependent on the original assumption that the base temperature for all emergence dates was 0 °C. If this assumption is not justified then the differences in leaf production rate for different sowing dates may not be linked to photoperiod at all. Any relationship implying this linkage may simply derive from the general correlation between daylength and temperature that bedevils attempts to partition the effects and understand the interactions between environmental factors. Fig. 17 shows an idealized curve, based on observed rates of leaf emergence and temperature from emergence to floral initiation for Avalon wheat, during the season 1984–85 at three sites (Delecolle, unpublished). A curve of very similar shape but shifted to the right on the horizontal axis has been published for maize by Warrington & Kanemasu (1983). From such a relationship it is apparent that any extrapolated estimation of the base temperature is strongly dependent on the range in temperature experienced by the crop during the period of leaf production. Such a range would be expected to alter with sowing date. For example, if the range of temperatures experienced by the plant during development was from 10° to 15 °C then linear extrapolation to the abscissa gives an intercept, at zero rate, of about 7·5 °C; for a range of 10° to 5 °C the estimate of base temperature falls by 2·5 °C. Such apparent base temperatures, which vary with sowing and emergence date, can thus be calculated for the rate of leaf emergence at the different sites and sowing dates used in this experiment (Fig. 18). Base temperatures were higher both for

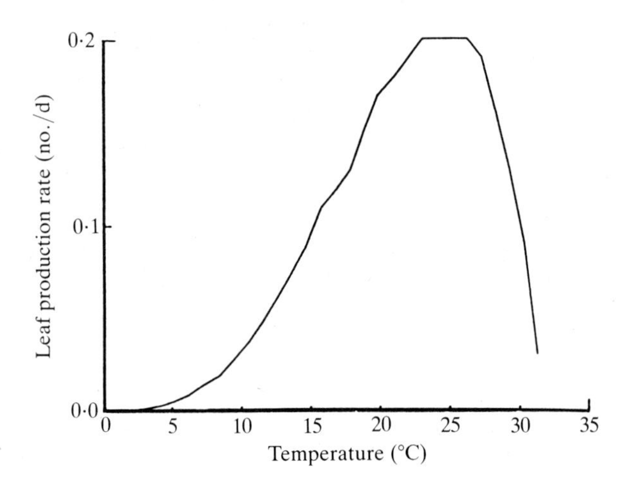

Fig. 17. Relationship between leaf production rate and temperature for wheat.

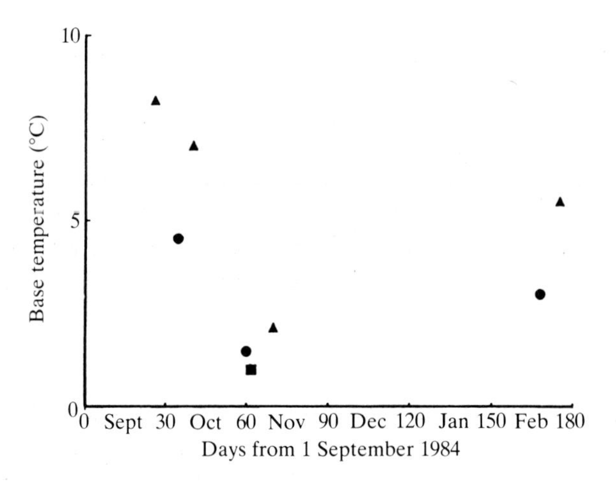

Fig. 18. Apparent base temperatures for the rate of leaf appearance and sowing date at three sites; ▲, Avignon, France; ●, Auchincruive, UK; ■, Mons, France.

September- and early-October-sown crops and also the spring sowings of this winter variety.

Compared with the assumption that the base temperature is zero, such increases in base temperatures mean that rates of leaf emergence calculated for crops that emerge in the early autumn are likely to have been under-estimated previously, those in the sping to have been over-estimated and at points when both the rate of change of daylength and the base temperature approach true zero, to have been calculated correctly. For the results of Baker *et al.* this means that the rate of leaf emergence varies less between sowing dates and is independent of photoperiodic influence, be it the rate of change of daylength or photoperiod itself (Jones & Allen, 1986; Wright & Hughes, 1987). Controlled environment experiments are

Table 1. *The response of processes involved in canopy formation to important environmental factors. Responses to an* increase *in the factor are;* +++, (−−−) *a very strong increase (decrease) in response;* ++, (−−) *a moderate increase (decrease) in response, often species or cultivar related;* +, (−) *a slight influence, direction uncertain. The degree of uncertainty of the relationship is:* ? *more work needed on the details of any response;* ?? *much more work needed on the fundamental response itself.* T_b *base temperature for process;* T_o *optimal temperature for the process. (After Monteith, 1987.)*

	Response			
Process	Temperature (T_b to T_o)	Soil dryness	Air dryness	Daylength
Emergence				
rate	+++	−−− ??		
survival	++ ?	−−− ??		
Initiation				
rate	+++	−− ?	−− ?	
duration	+ ?			++ ?
leaf number	+++	−− ?	−− ?	++ ?
Expansion				
rate	+++	−−− ?	−− ?	
duration	−− ?			+ ??
leaf area index	+++	−−− ?	−−	+ ??

reported to show that photoperiod does not affect the rate of leaf initiation (Ritchie, personal communication).

Conclusions

In a report to the Overseas Development Agency describing the research completed by his group at Sutton Bonington into tropical crop production, Monteith (1987) ends with an overview of the effect of an increase in a given environmental factor on various processes related to canopy formation (Table 1). Together with the plusses and minuses which indicate the force of a particular environmental factor on the process involved there is also a liberal spreading of question marks indicating a lack of knowledge at an empirical, descriptive level and thus, also, in our ability to understand how effects come about. Were one to ask question about how environmental *interactions* influence plant growth and development it is clear that the question marks would easily outnumber the answers.

There is a substantial body of evidence, developed from both field and controlled-environment studies, within specified ranges, that rates of development are linear with increasing temperature. Our best analyses from field experiments suggest that any interaction with photoperiod and/or specific cold temperatures can be viewed as multiplicative, although other interpretations seem not to be inferior. Furthermore, as plants move closer to the flowering stage, it appears that the influence of other factors besides temperature is reduced. Whilst some circumstantial support can be found for the involvement of photoperiodic effects interacting with temperature in controlling the rate of leaf production in wheat, and other cereals (Porter, unpublished), the initial hypothesis is looking more doubtful. We have reached the point which as scientists we should relish. We are unsure and badly in need of critical work although some relatively simple experiments suggest themselves. This should also be an area where integration of cellular and sub-cellular events to this important aspect of plant behaviour could be achieved.

References

ANGUS, J. F., CUNNINGHAM, R. B., MONCUR, M. W. & MACKENZIE, D. H. (1981*a*). Phasic development in the field. I. Thermal response in the seedling stage. *Field Crops Res.* **3**, 365–378.

ANGUS, J. F., MACKENZIE, D. H., MORTON, R. & SCHATER, C. A. (1981*b*). Phasic development in field crops. II. Thermal and photoperiodic responses of spring wheat. *Field Crops Res.* **4**, 269–283.

BAKER, C. K., GALLAGHER, J. N. & MONTEITH, J. L. (1980). Daylength change and rate of leaf emergence in wheat. *Pl. Cell. Environ.* **3**, 285–287.

BUNTING, A. H. (1975). Time, phenology and the yield of crops. *Weather* **30**, 312–325.

CHUJO, H. (1966). Difference in vernalisation effect in wheat under various temperatures. *Proc. Crop Sci. Soc. Japan* **35**, 177–186.

FORD, M. A., AUSTIN, R. B., ANGUS, W. J. & SAGE, G. C. M. (1981). Relationships between the responses of spring wheat genotypes to temperatures and photoperiodic treatment and their performance in the field. *J. agric. Sci., Camb.* **96**, 623–634.

FRANKLAND, B. (1986). Perception of light quantity. In *Photomorphogenesis in Plants* (ed. R. E. Kendrick & G. H. M. Kronenberg), pp. 219–235. Dordrecht: Martinus Nijhoff.

GALLAGHER, J. N. (1976). *The growth of cereals in relation to weather*. Ph.D. thesis. University of Nottingham.

GALLAGHER, J. N. & BISCOE, P. V. (1978). Radiation absorption, growth and yield of cereals. *J. agric. Sci., Camb.* **91**, 103–116.

GANDAR, P. W. & BERTAUD, D. S. (1984). Modelling tillering and yield formation in spring-sown Mata barley. *Proc. Agron. Sci. N.Z.* **14**, 89–94.

GANDAR, P., BERTAUD, D. S., CLEGHORS, J. A., WITHERS, N. J. & SPRIGGS, T. W. (1984). Modelling tillering and yield formation in spring-sown Karamu wheat. *Proc. Agron. Soc. N.Z.* **14**, 83–88.

GARNER, W. & ALLARD, A. H. (1920). Effect of the relative length of the day and night and other factors of the environment on growth and reproduction of plants. *J. agric. Res.* **18**, 553–606.

GORSKI, T. (1980). Annual cycle of the red and far-red radiation. *Int. J. Biometeor.* **24**, 361–365.

GROOT, J. J. R. (1987). *Simulation of Crop Growth and Nitrogen Dynamics in a Winter Wheat-Soil System.* Wageningen: CABO-TT Simulation Reports. 98 pp.

HALLORAN, G. & PENNELL, A. L. (1982). Duration and rate of development phases in wheat in two environments. *Ann. Bot.* **49**, 115–121.

HALSE, N. J. & WEIR, R. N. (1970). Effects of vernalisation, photoperiod and temperature on physiological development and spikelet number of Australian wheat. *Aust. J. Agric. Res.* **21**, 383–393.

HOLMES, M. G. & SMITH, H. (1977). The function of phytochrome in the natural environment. II. The influence of vegetation canopies on the spectral energy distribution of natural daylight. *Photochem. Photobiol.* **25**, 539–545.

JOHNSON, I. R. & THORNLEY, J. H. M. (1985). Temperature dependence of plant and crop processes. *Ann. Bot.* **55**, 1–24.

JONES, J. L. & ALLEN, E. J. (1986). Development in barley (*Hordeum sativum*). *J. agric. Sci., Camb.* **107**, 187–213.

KIRBY, E. J. M. & APPLEYARD, M. (1981). *Cereal Development Guide.* Stoneleigh, Warwickshire, UK: NAC Cereal Unit.

KIRBY, E. J. M., PORTER, J. R., DAY, W., ADAM, J. S., APPLEYARD, M., AYLING, S., BAKER, C. K., BELFORD, R. K., BISCOE, P. V., CHAPMAN, A., FULLER, M. P., HAMPSON, J., HAY, R. K. M., MATTHEWS, S., THOMPSON, W. J., WEIR, A. H., WILLINGTON, V. B. A. & WOOD, D. W. (1987). An analysis of primordium initiation in Avalon wheat crops with different sowing dates and at nine sites in England and Scotland. *J. agric. Sci., Camb.* **109**, 123–134.

KOLBE, W. (1974). Seasonal course of cereal growth stages in relation to annual weather and crop protection measures. *Pflanz. Nach. Bayer* **27**, 312–363.

KORNICHE, F. & WERNER, H. (1885). *Handbuch des Getreidebaues.* Bonn: Verlag von Emil Strauss.

LEGG, B. J., DAY, W., LAWLOR, D. W. & PARKINSON, K. J. (1979). The effects of drought on barley growth: models and measurements. *J. agric. Sci., Camb.* **92**, 703–716.

MARCELLOS, A. & SINGLE, W. V. (1971). Quantitative responses of wheat to photoperiod and temperature in the field. *Aust. J. Agric. Res.* **22**, 343–357.

MONTEITH, J. L. (1981). Climatic variation and the growth of crops. *Quart. J. R. Met. Soc.* **107**, 749–774.

MONTEITH, J. L. (1987). *Microclimatology in Tropical Agriculture*, vol. 1. London: ODA.

PENNING DE VRIES, F. W. T., WITLAGE, J. M. & KREMER, D. J. (1979). Rates of respiration and of increase in structural dry matter in young wheat, ryegrass and maize plants in relation to temperature, to water stress and to their sugar content. *Ann. Bot.* **44**, 591–609.

PENNING DE VRIES, F. W. T. & VAN LAAR, H. H. (1982). *Simulation of Plant Growth and Crop Production.* Wageningen: Pudoc.

PORTER, J. R. (1983). Modelling stage development in winter wheat. In *Influence of Environmental Factors on Herbicide Performance and Crop and Weed Biology* (ed. A. Walker), pp. 449–455. *Aspects of Applied Biology* **4**, IHR Wellesbourne, Warwick: AAB.

PORTER, J. R. (1984). A model of canopy development in winter wheat. *J. agric. Sci., Camb.* **102**, 383–392.

PORTER, J. R. (1985). Models and mechanisms in the growth and development of wheat. *Outlook Agric.* **14**, 190–196.

PORTER, J. R. (1987). Modelling crop development in wheat. *Span* **30**, 19–22.

PORTER, J. R., KIRBY, E. J. M., DAY, W., ADAM, J. S., APPLEYARD, M., AYLING, S., BAKER, C. K., BEALE, P., BELFORD, R. K., BISCOE, P. V., CHAPMAN, A., FULLER, M. P., HAMPSON, J., HAY, R. K. M., HOUGH, M. N., MATTHEWS, S., THOMPSON, W. J., WEIR, A. H., WILLINGTON, V. B. A. & WOOD, D. W. (1987). An analysis of morphological development stages in Avalon winter wheat crops with different sowing dates and at ten sites in England and Scotland. *J. agric. Sci., Camb.* **109**, 107–121.

RAHMAN, M. S. (1980). Effect of photoperiod and vernalisation on the rate of development and spikelet number per ear in 31 varieties of wheat. *J. Aust. Inst. Agric. Res.* **46**, 68–70.

REAUMUR, M. DE (1735). Observations du thermometre, faites a Paris pendant l'annee. *Mem. l'Acad. roy. sci., Paris* 737–754.

RITCHIE, J. T. (1985). A user-orientated model of the soil water balance in wheat. In *Wheat Growth and Modelling* (ed. W. Day & R. K. Atkin), pp. 293–305. New York: Plenum Press.

ROBERTS, E. H. (1988). Temperature and seed germination. In *Plants and Temperature*, Symp. Soc. Exp. Biol., vol. 42 (ed. S. P. Long & F. I. Woodward), pp. 109–132. Cambridge: The Company of Biologists Ltd.

ROBERTS, E. H. & SUMMERFIELD, R. J. (1987). Measurement and prediction of flowering in annual crops. In *Manipulation of Flowering* (ed. J. G. Atherton), pp. 17–50. Proceedings of the 45th Easter School, Faculty of Agricultural Science, University of Nottingham. London: Butterworths.

ROBERTS, E. H., SUMMERFIELD, R. J., COOPER, J. P. & ELLIS, R. H. (1988). Environmental control of flowering in barley (*Hordeum vulgare* L.). I. Photoperiod limits to long-day responses, photoperiod-insensitive phases and effects of low temperature and short-day vernalisation. *Ann. Bot.* In the Press.

ROBERTSON, G. W. (1983). Weather-based mathemetial models for estimating development and ripening of crops. *Technical Note No.* **180**, Geneva: WMO.

SALISBURY, F. B. (1981). Twilight effect: Initiating dark measurement in photoperiodism of *Xanthium. Plant Physiol.* **67**, 1230–1238.

SKJELVAG, A. O. (1981*a*). Effects of climatic factors on the growth and development of the field bean (*Vicia faba* L. var. *minor*). I. Phenology, height growth and yield in a phytotron experiment. *Acta Agric. Scand.* **31**, 358–371.

SKJELVAG, A. O. (1981*b*). Effects of climatic factors on the growth and development of the field bean (*Vicia faba* L. var. *minor*). II. Phenological development in outdoor experiments. *Acta Agric. Scand.* **31**, 372–381.

SMITH, H. (1982). Light quality, photoperception and plant strategy. *Ann. Rev. Plant Physiol.* **33**, 481–518.

SMITH, H. (1986). The perception of light quality. In *Photomorphogenesis in Plants* (ed. R. E. Kendrick & G. H. M. Kronenberg), pp. 187–217. Dordrecht: Martinus Nijhoff.

SMITH, H. & HOLMES, M. G. (1977). The function of phytochrome in the natural environment. III. Measurement and calculation of phytochrome photoequilibria. *Photochem. Photobiol.* **25**, 547–550.

SMITH, H. & MORGAN, D. C. (1981). The spectral characteristic of the visible radiation incident upon the surface of the earth. In *Plants and the Daylight Spectrum* (ed. H. Smith), pp. 3–20. London: Academic Press.

STERN, W. R. & KIRBY, E. J. M. (1979). Primordium initiation at the shoot apex in four contrasting varieties of spring wheat in response to sowing date. *J. agric. Sci., Camb.* **93**, 203–215.

SUMMERFIELD, R. J. & ROBERTS, E. H. (1987). Effects of illuminance on flowering in long- and short-day legumes: a reappraisal and unifying model. In *Manipulation of Flowering* (ed. J. G. Atherton). pp. 203–223. Proceedings of the 45th Easter School, Faculty of Agricultural Science, University of Nottingham. London: Butterworths.

THORNE, G. N. & WOOD, D. W. (1987). Effects of radiation and temperature on the tiller survival, grain number and grain yield in winter wheat. *Ann. Bot.* **59**, 413–426.

VINCE-PRUE, D. (1975). *Photoperiodism in Plants.* Maidenhead: McGraw-Hill.

VINCE-PRUE, D. (1981). Daylight and photoperiodism. In *Plants and the Daylight Spectrum* (ed. H. Smith), pp. 223–242. London: Academic Press.

VOS, J. (1981). *Effects of Temperature and Nitrogen Supply on Post-floral Growth of Wheat: Measurements and Simulations.* Agricultural Research Report 911. Wageningen, The Netherlands: Pudoc.

WARRINGTON, I. J. & KANEMASU, E. T. (1983). Corn growth response to temperature and photoperiod. II. Leaf initiation and leaf appearance rates. *Agron. J.* **75**, 755–761.

WEIR, A. E., BRAGG, P. L., PORTER, J. R. & RAYNER, J. H. (1984). A winter wheat crop simulation model without water or nutrient limitations. *J. agric. Sci., Camb.* **102**, 371–382.

WRIGHT, D. & HUGHES, LL. G. (1987). Relationships between time, temperature, daylength and development in spring barley. *J. agric. Sci., Camb.* **109**, 365–373.

LOW TEMPERATURE AND THE GROWTH OF PLANTS

C. J. POLLOCK and C. F. EAGLES

Plant and Cell Biology Department, Welsh Plant Breeding Station, Aberystwyth SY23 3EB, Wales, UK

Summary

The growth responses of plants at low temperature are discussed, with particular reference to the reversible effects of low positive temperatures upon the growth of obligately chilling-resistant species. Techniques for the measurement of growth and the control of meristem temperature are described, and evidence concerning the sites of temperature perception and transduction in the developing monocot leaf is reviewed. The rapidity and reversibility of such responses, and their apparent independence of changes in carbon supply or meristem cell turgor are held to support the hypothesis that chilling alters the physical properties of the walls of extending cells within the meristematic zone. The existence of marked genetic variability in these responses facilitates agricultural exploitation of low-temperature growth.

The effects of longer-term chilling upon growth, cell extension and cell division are described, with particular reference to interactions with other environmental variables and with developmental factors which can alter the nature of the temperature response. The relationship between growth reduction and the development of freezing tolerance during chilling is discussed in relation to programmes of crop improvement which will maximize productivity at low temperatures without adversely affecting survival.

Introduction

The responses of plants to temperature change are considered to be governed by interactions between the direct effects of temperature on chemical reactions occurring in the plant and indirect effects upon the catalytic machinery (Hochachka & Somero, 1973; Sutcliffe, 1977). The indirect effects are influenced by both genetic and environmental factors and have been studied extensively (Berry & Raison, 1982 and references therein). Although the ecological aspects of these responses have been investigated (Björkman *et al.* 1974; Larcher & Bauer, 1982), much of the work in this area has been concerned with the agricultural consequences of temperature limitations and the development of ways of overcoming them. Unfortunately, the constraints imposed upon agricultural systems by non-optimal temperatures are of several different kinds, and there is no *a priori* reason for supposing that strategies for overcoming one type of limitation will be effective in overcoming another.

When considering the effects of sub-optimal temperatures on plant growth, it is important to distinguish between limitations attributable to a reversible reduction in the rates of certain key processes and those imposed by injury, where injury can be defined as an inability to restore normal function following removal of the causal constraint. Cold-induced injury can occur over a wide range of temperatures, dependent upon both genotype and environment. Plant species have been classified into three groups, depending upon the temperature ranges within which injury can be observed. Chilling-sensitive plants suffer injury at positive temperatures, usually below 10–15 °C; freezing-sensitive plants are damaged by exposure to any temperature below 0 °C and freezing-resistant plants are able to survive subzero temperatures down to a limit which is characteristic of the genotype (Larcher, 1981). Like most attempts to classify plants by their physiological responses, these distinctions are not absolute, and individual plants can show the attributes of more than one group in the course of a normal growth cycle.

Strategies for agricultural improvement involving altered low-temperature responses have concentrated upon the amelioration of injury by means of breeding techniques and cultivation practice. Reduction of chilling injury can permit extension of the climatic limits for cultivation of crops such as maize (Derieux, 1984; Stamp, 1984) whilst increases in the freezing tolerance of autumn-sown cereals has been a major factor in their successful cultivation in high latitudes (Grafius, 1981). However, improvement of the reversible responses of plants to sub-optimal temperatures is an equally attractive agricultural objective in view of the importance of early-season growth as a determinant of final crop yield. A number of studies have established a relationship between total intercepted radiation and crop yield (Biscoe & Gallagher, 1977; Monteith, 1977) and have emphasized the significance of sub-optimal temperature in affecting leaf growth and the attainment of high leaf area indices (Woolhouse, 1981; Kemp & Blacklow, 1982; Kemp, 1984; Ong & Baker, 1985). The lag between the increases of irradiance and temperature in the spring, particularly in temperate climates (Fig. 1), reinforces the agricultural desirability of maximizing early-season crop growth rates and thus of studying the growth responses of plants in the temperature range below the optimum for growth and above the temperature where injury occurs. Such a range would be encountered during the normal annual cycle.

This review will, therefore, consider:

1. The site of temperature perception by growing plants.
2. How this perception is transduced into altered rates of growth.
3. The extent of variability of these characters observed in natural genotypes and in mutants and their interactions with environmental and developmental factors.
4. The possible consequences of utilization of such variability to improve growth at low temperatures.

Although the majority of examples cited will be drawn from northern temperate crops, where sub-optimal temperatures may occur at any stage of the annual cycle, some reference will be made to tropical species where seedling establishment may be affected by low temperatures (Peacock, 1982).

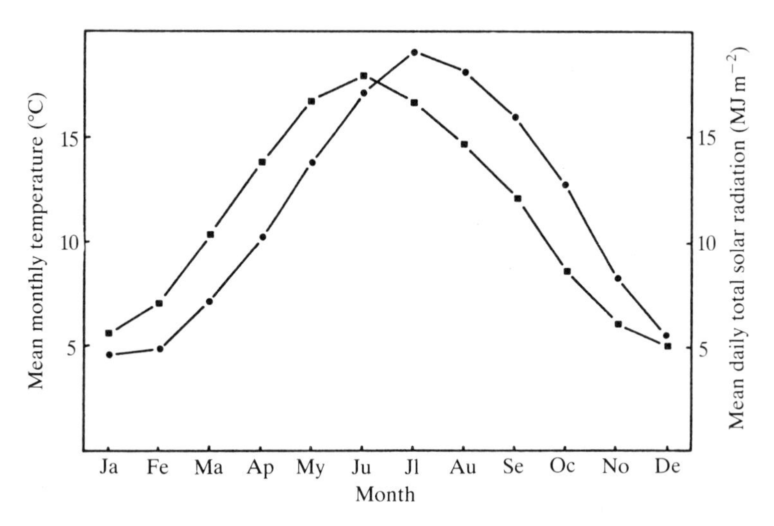

Fig. 1. Mean daily total of solar radiation (■——■) and mean monthly temperatures (●——●) for mainland UK. (Data from Chandler & Gregory, 1976.)

Short-term responses of plant growth to temperature

Measurement and interpretation

The measurement of plant growth is a complex subject which has been discussed extensively (e.g. Causton & Venus, 1981). In this review it is appropriate to outline only those approaches which relate specifically to the effects of temperature change upon growth. The major technical problems which must be overcome in such studies are those associated with the sensitivity of the measuring equipment and the specificity of the site of application of temperature change. Controlled environment studies, field and glasshouse treatments, allied to standard methods of growth analysis, permit the gross or longer-term effects of temperature to be quantified but make it difficult to study the site of temperature perception or the speed of the response. More sensitive auxanometric techniques are available and localized chilling can be imposed but such techniques are not universally applicable. The leaves of most dicotyledonous plants are unsuitable for study using linear auxanometers because of the two-dimensional nature of growth, and disseminated meristems make it difficult to perturb the temperature of the growing zone differentially. In contrast, many monocotyledonous plants have growth patterns which are well suited to high resolution auxanometry and localized temperature regulation (Stoddart *et al.* 1986). Petiolar, stem and root growth in dicotyledonous plants can also be studied using similar techniques. It should be noted, however, that the effects of temperature upon increase in size, total dry weight or structural dry weight may differ, particularly in view of the tendency of chilled plants to accumulate reserve carbohydrates (ap Rees *et al.* 1982), so the choice of a suitable index of growth is important.

Subject to the limitations outlined above, use of sensitive auxanometers has permitted the rapid effects of localized temperature change upon increase in leaf size to be measured under controlled conditions (Watts, 1971; Kleinendorst & Brouwer, 1972; Sharp *et al.* 1979; Kemp & Blacklow, 1982; Stoddart *et al.* 1986). In our current studies on leaf growth a displacement transducer (Penny and Giles, UK) is used to produce a linear voltage change of 10 V for a displacement of 10 mm. Sensitivity is further increased by amplifying the output, which is then directed to a microcomputer. Adequate resolution requires the use of 12- or 16-bit A/D converters which permit growth rate measurements to be made over periods as short as 20 s. Temperature control is achieved by circulating coolant through a grooved brass collar which is normally placed around the basal leaf meristematic zone. Coolant temperature can be regulated using a range of commercially available equipment. The use of a temperature programmer such as the TP 16 (Techne, Cambridge, UK) allows reproducible linear temperature profiles to be generated with a range of cooling and warming rates. Direct measurement of leaf temperature is possible using a thermocouple placed between the growing leaf and the base of the oldest mature leaf. Amplified thermocouple outputs can be interfaced to a microcomputer to allow simultaneous calculation of temperature and leaf growth rate.

As with all auxanometric techniques, care has to be taken to minimize the effects of enclosing and attaching the leaf upon the rate of growth. Use of uniform plant material grown under controlled conditions permits comparison of rates obtained using auxanometers with those derived from less sensitive but non-intrusive techniques (Thomas, 1983). Such preliminary experiments would be necessary whenever a new species is studied and may also be required if material from a different environment or at a different developmental stage is to be measured. Subsequent studies can performed as a routine operation with little attention being required following the installation of the plant in the auxanometer and initiation of data collection and temperature programming. Computer-based data-gathering also facilitates subsequent data manipulation. Straightforward displays of growth rate and temperature or an equivalent graphical representation can be produced easily; subsequently curve-fitting routines can be employed to derive numerical parameters which can be used to rank genotypes, treatments etc. in larger comparative studies.

The nature of the response

The meristematic regions of plants react to temperature change in a variety of ways. Temperature change may alter the rate of leaf initiation and the rate or duration of leaf growth; differential changes in these parameters are associated with anatomical and morphological change (Terry *et al.* 1983). The effects upon leaf growth can be shown to depend more upon meristem temperature than upon whole plant temperature. Watts (1974) showed that extension of maize leaves was unaffected by air temperature if the meristem was maintained at a constant temperature around 30 °C. Control plants with untreated meristems showed a threefold reduction in extension rate between 25 °C and 5 °C (Fig. 2). Peacock (1975) demonstrated that the rate of growth of leaves of *Lolium perenne* was

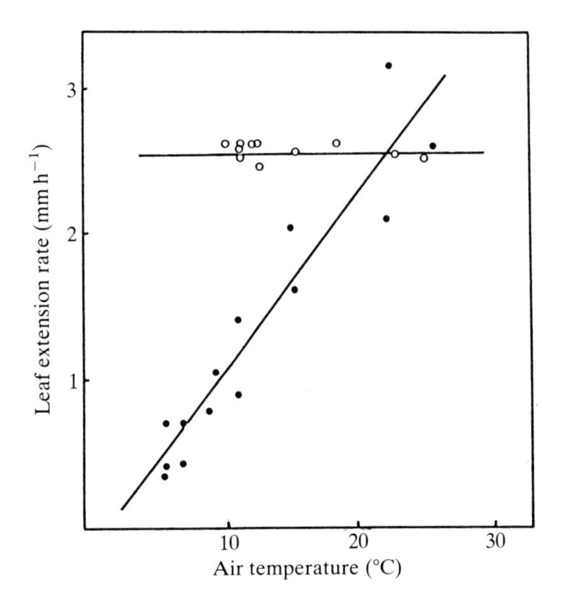

Fig. 2. The effect of air temperature on leaf extension rates in maize (from Watts, 1974). ●——●, leaf meristem at air temperature; ○——○, leaf meristem maintained at 30–34°C.

Table 1. *Temperature sensitivity of growth and mature leaf photosynthesis in* Lolium temulentum *(data from Thomas, 1983; Pollock* et al. *1983)*

	Rate at			Q_{10}	
Parameter	20°C	5°C	2°C	2–20	2–5
Relative leaf extension rate (d^{-1})	0·70	0·11	0·02	7·2	293·7
Relative growth rate (d^{-1})	0·064	0·025	0·008	3·2	44·6
Relative root growth rate (d^{-1})	0·068	0·030	0·015	2·3	10·1
Photosynthetic capacity ($\mu l\ O_2\ h^{-1}\ mg chl^{-1}$)	220	40	24	3·4	5·3
Whole plant relative growth rate (d^{-1})					
control plants	0·09	–	–	–	–
following 41 days at 2°C	0·08	–	–	–	–

largely determined by the temperature of the stem apex and that there was an effective cessation of growth below about 5°C. Similar abrupt changes in growth rate over narrow temperature ranges have been observed with other gramineae (Smillie, 1976; Pollock *et al.* 1983; Thomas, 1983). In the last two studies, relative growth rates were calculated on a dry weight and leaf extension basis for *L. temulentum* grown at a range of temperatures, and the temperature sensitivity of the various growth parameters expressed as Q_{10} values (Table 1). The extreme sensitivity of growth to temperature in the range 2–5°C can be seen, as can the reversible nature of the constraints imposed by these treatments, regardless of duration. Increases in Q_{10} values with decreasing temperature were observed by

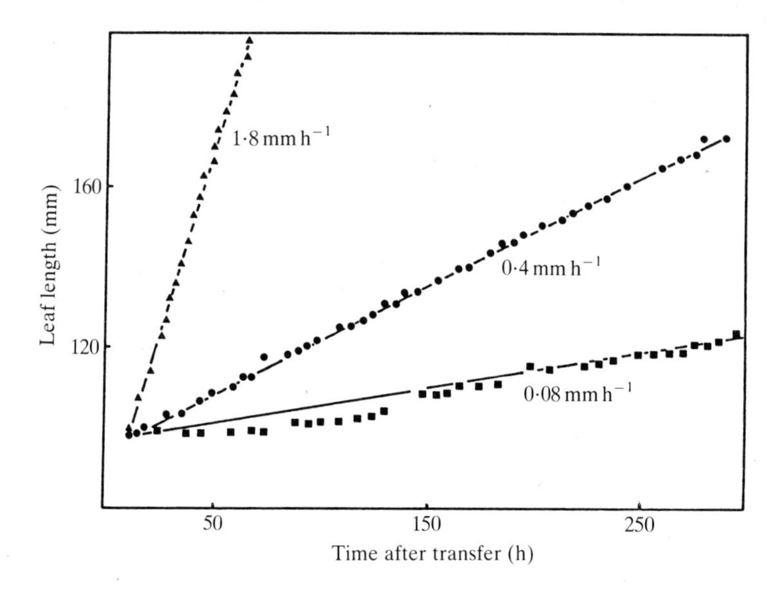

Fig. 3. The effects of growth temperatures on leaf extension in *Lolium temulentum*. Plants were transferred from 20 °C to either 5 °C or 2 °C mid-way through the expansion of the fourth leaf. ▲——▲, 20 °C controls; ●——●, 5 °C; ■——■, 2 °C.) (Data from Thomas & Stoddart, 1984.)

Peacock (1975) and by Kemp & Blacklow (1980), but not by Williams & Biddiscombe (1965) in field studies using *Phalaris tuberosus*. In this case, however, interpretation of the data was complicated by the fact that leaves of differing physiological state were measured at different temperatures (see p. 169).

The speed of the response

Rates of leaf growth are known to change rapidly under the influence of a range of environmental variables (Christ, 1978; Terry *et al.* 1983), but few studies have attempted to characterize the kinetics of growth rate changes under such conditions. Auxanometric measurements of leaf extension rates in seedlings of *L. temulentum*, following transfer of plants from 20 °C to either 5 or 2 °C mid-way through the expansion of the fourth leaf, showed that stable, reduced rates of extension were detectable soon after transfer (Fig. 3). Diurnal fluctuations in extension rate occurred in all treatments, but the mean rates remained remarkably constant until leaf development was almost complete (Thomas & Stoddart, 1984). These observations were supported by the detection of changes in the patterns of assimilate partitioning in cooled plants over a similar time period (Pollock, 1984). The response time of the optical auxanometers used in these studies did not permit the resolution of growth rate changes occurring within the first hour of transfer. Use of the LVDT system and specific meristem chilling, described above, allowed more rapid changes to be measured (Stoddart *et al.* 1986). When stepped temperature changes were imposed on the meristematic region, growth rate responded rapidly, particularly at leaf temperatures below 15 °C (Fig. 4). Cooling

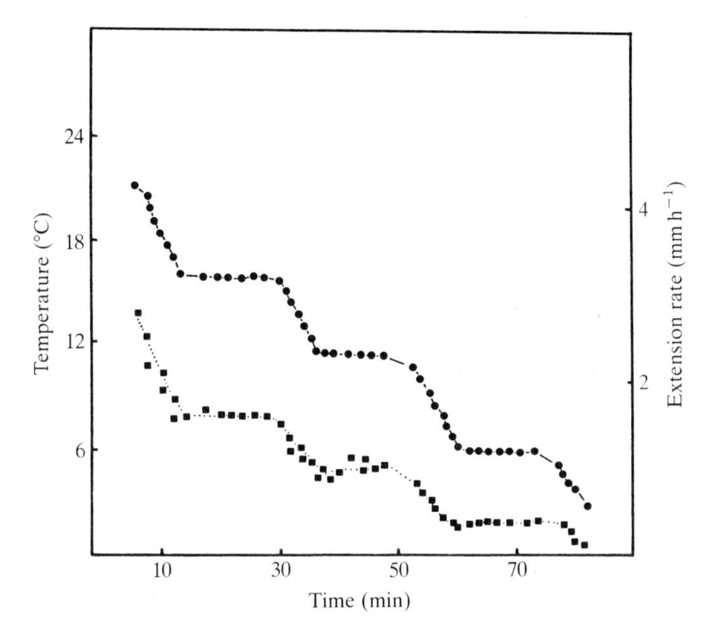

Fig. 4. The effects of rapid meristem temperature change upon leaf extension rate in *Lolium temulentum*. ●——●, temperature; ■——■, extension rate. (Data from Stoddart *et al.* 1986.)

the leaf outside the meristematic zone had no effect upon leaf extension. Under optimal conditions, the LVDT auxanometer can detect length increases equivalent to about 20 s of growth at 20 °C, suggesting that any lag must be shorter than this. Even if lag times in the order of seconds could be measured, it would be extremely difficult to distinguish between delays attributable to the thermal characteristics of the equipment and those caused by a delay between perception of temperature change and transduction into altered leaf extension. The rapidity of the response indicates that the sites of perception and transduction are closely linked. Using seedlings of *L. temulentum* grown at 20 °C it was found that the effects of meristem chilling were wholly reversible when minimum temperatures were above 0 °C and that the rate of response to warming was at least as rapid as that to cooling (Stoddart *et al.* 1986).

Exposure of plants to continuous rather than stepped temperature gradients in the above range causes a smooth reduction in growth rate followed by a resumption of growth and full recovery (Fig. 5). Although temperature sensitivity increases markedly at low positive temperatures, there is no clear evidence of any discontinuity in the response to temperature as has been claimed for other processes (Berry & Raison, 1982). The absence of discontinuities confirms the observations of Bagnall & Wolfe (1978) and Kemp & Blacklow (1980). The characteristic features of Fig. 5 are the resumption of growth during rewarming at a slightly lower temperature (Px) than that at which it ceased during cooling (Pe), together with attainment of a higher rate of growth at intermediate temperatures

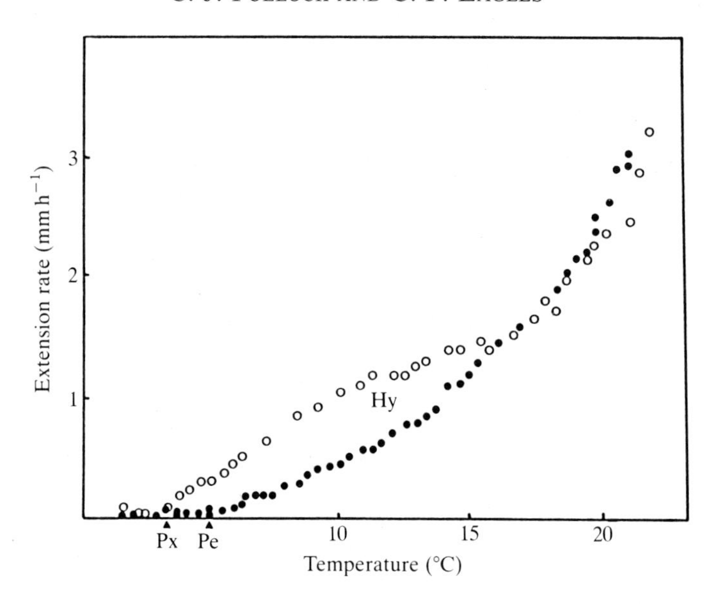

Fig. 5. The effects of cooling and rewarming upon leaf extension rates in *Lolium temulentum*. ●, rate during cooling; ○, rate during rewarming; Pe, temperature at which growth ceases during cooling; Px, temperature at which growth resumes during rewarming; Hy, area of hysteretic growth behaviour.

Table 2. *Temperature characteristics of extension growth in various species*

	Growth temperature (°C)	Temperature (°C) at which growth: Stops during cooling (Pe)	Commences on rewarming (Px)
Lolium temulentum	20	1·80	1·45
	5	0·20	−0·18
Avena sativa			
cv. Margam (spring oats)	20	0·90	0·85
cv. Pennal (winter oats)	20	3·72	3·51
Trifolium repens	20	8·60	8·50
Sorghum bicolor	35	7·60	7·60

during rewarming than that observed during cooling. The area of this hysteretic component can be quantified (Hy). For a range of temperate and tropical species, Pe and Px values are characteristic of both genotype and growth conditions (Table 2), are independent of cooling rates in the range 0·2–2°C min^{-1} and appear to be a good comparative measure of the sensitivity of growth to cold. Hy values can be reduced by chilling unacclimated plants well below Pe and these treatments are associated with an inability to recover fully following rewarming (Table 3). Other factors affecting Hy have not been identified although similar transient increases in growth rate have been observed following alleviation of water stress in

Table 3. *Effect of final meristem temperature on recovery following rewarming of seedlings of* Lolium temulentum *(data from Stoddart* et al. *1986)*

Final meristem temperature (°C)	Relative area of hysteretic component (Hy)	% Recovery of initial growth rate at 20°C
1·5	13·3	95
−0·5	1·9	76
−2·8	−12·3	52
−4·2	−11·9	43
−5·3	−19·6	0

maize (Kleinendorst, 1975). For unacclimated seedlings of *L. temulentum*, Hy is unaffected by duration of exposure to temperatures below Pe or by rates of cooling or rewarming in the range 0·2–2°C min^{-1} (C. J. Pollock, unpublished results) so it is difficult to draw conclusions concerning the physiological significance of this parameter. It is tempting to regard it as the expression of some kind of 'stored growth', but further study will be required to determine its physiological or agricultural relevance. Data of the type outlined above are available for a restricted range of monocotyledonous crop species but the problems of performing similar studies on leaves of dicotyledonous plants remain to be overcome. Studies of clover petiole extension with the LVDT system suggest, however, that tissue responses may be broadly similar (Table 2).

The nature of the limitation

Growth is dependent upon cell division in order to provide a supply of unextended cells, upon those cells having plastic cell walls which will accommodate volume change and upon the development and maintenance of turgor within such cells sufficient to exceed the yield threshold of the walls (Tomos, 1985). The rapidity of the responses described above (Fig. 5) is such that the number of divided, unextended cells is unlikely to change significantly during the course of such an experiment. It can be assumed, therefore, that the effects of cold upon cell division are not the primary site of short-term temperature responses, although they may have considerable significance in the longer term (see p. 171). The development and maintenance of turgor and the modification of wall structure during extension are processes which require energy and carbon skeletons so it is necessary to distinguish between the direct effects of temperature on the metabolism of extending cells and indirect effects mediated through changes in the supply of necessary materials.

Indirect effects of photosynthesis and translocation

The effects of temperature upon the fixation and transport of carbon in higher plants has been studied in a range of systems and the results obtained have often been equivocal. Once again, however, it is important to distinguish between the effects of injurious temperatures outside the normal seasonal range and those which cause a temporary reduction in rate. The overall process of photosynthesis shows temperature dependence, with responses which differ between species and can exhibit acclimation (Berry & Björkman, 1980). The nature of such acclimation has been studied in detail (Baker *et al.* 1988) and these effects may be of considerable importance in determining performance under prolonged exposure to low temperatures such as in the early spring (Parsons & Robson, 1981). However, there are indications that reduced carbon fixation does not pose an immediate limitation to growth following exposure to cold (Terry *et al.* 1983; Körner & Larcher, 1988). Forde, Whitehead & Rowley (1975) showed that leaf growth in *Paspalum dilatatum* was more sensitive to low temperature than was photosynthesis or starch accumulation, and Q_{10} measurements of light-dependent oxygen evolution in leaves of *L. temulentum* were lower in the range 5–2°C than those for leaf growth (Table 1). The widespread occurrence of carbohydrate accumulation in plants exposed to low temperatures also suggests that photosynthetic capacity often exceeds the requirements for growth under these conditions (ap Rees *et al.* 1982; Pollock, 1986; Hendry, 1987), although it has been suggested that reduced carbon supply can limit root growth at low temperature (Crawford & Huxter, 1977). The translocation of fixed carbon at low temperatures by chilling-resistant plants does not appear to be abnormally reduced (Berry & Raison, 1982), having Q_{10} values similar to the viscosity of phloem exudate (Giaquinta & Geiger, 1973). This insensitivity may be associated with substantial accumulation of carbohydrate in the meristematic zone (Pollock, 1984; Schnyder, 1986). In the former study, meristematic carbohydrate contents increased steadily from 6·5 to 24·0 $mg\,g^{-1}$ f.wt in the 4 days immediately following transfer of *L. temulentum* seedlings from 20 to 5°C. During this period no change in growth rate was detected apart from the rapid (<1 h) decline immediately following cooling (Thomas & Stoddart, 1984). With some reservations, associated with the small number of species which have been studied in detail, it appears therefore that the production and movement of fixed carbon is unlikely to represent the primary cause of low-temperature growth inhibition in obligately chilling-resistant species.

Direct effects on turgor and cell wall extensibility

The maintenance of turgor during cell extension requires the uptake of water together with osmotic readjustments in order to maintain a gradient of water potential (Tomos, 1985). Inhibition of water uptake at low temperature has been correlated with reduced growth rate (Kleinendorst & Brouwer, 1970; Watts, 1971; Barlow *et al.* 1977) and low-temperature treatments can reduce water flow in roots (Markhart *et al.* 1979). However, osmoregulation can compensate for increased water deficits, thus maintaining turgor pressure above the yield threshold for meristematic cells (Kleinendorst & Brouwer, 1972; Meyer & Boyer, 1972; Terry *et al.* 1983). Use of the pressure probe method permits direct measurements of turgor pressure to be made in individual plant cells (Tomos & Zimmerman, 1982), and modification of the LVDT auxanometer and cooling collar has allowed simultaneous determinations of meristem cell turgor, leaf temperature and extension rate. A small portion of the encircling mature leaf sheath can be removed without affecting extension rate. This allows insertion of the pressure probe directly into the extending cells of the developing leaf. Unacclimated leaves of *L. temulentum* show no significant alterations in turgor pressure (*ca* 0·5 MPa) in the temperature range 20–2°C, over which growth rate fell from 2·2 to 0·05 mm h^{-1} (A. D. Tomos, personal communication). At least under these circumstances, therefore, it appears that inhibition of growth by cold occurs independently of any changes in meristem cell turgor.

Although the evidence presented above might suggest that the properties of the extending cell wall are directly affected by cold in such a way as progressively to decrease wall plasticity with declining temperature, direct experimental confirmation of this hypothesis is lacking. Indirect evidence of the importance of wall rheology in determining short-term temperature response has come from studies on the 'slender' mutant of barley (Stoddart & Lloyd, 1986). This mutation, inherited as a single recessive gene, has a number of morphological and physiological consequences, among the most striking of which is a depression of Pe in unacclimated seedlings. Normal seedlings stop growing at around 2·5°C whereas mutant material continues to extend until −7°C, at which point the LVDT trace, although still indicating the occurrence of leaf extension, shows abrupt displacements apparently associated with ice formation and tissue disruption. Rewarming from this temperature produces minimal recovery of growth. As with *L. temulentum* seedlings, carbohydrate contents of the meristematic zones remain high and meristem cell turgor is unaffected by temperature change in either mutant or normal seedlings (A. D. Tomos, personal communication). There are marked differences, however, in the physical properties of the cell walls in the two forms, and these differences are restricted to young, expanding leaves. Plots of leaf displacement against applied load were obtained for mature and growing leaves (Fig. 6) using the techniques described by Van Volkenburgh *et al.* (1983). Mature leaves of both normal and mutant plants behaved as rigid solids, with little deformation even under relatively high loading and a small degree of hysteresis indicating predominantly elastic behaviour during stretching and recovery. In contrast, young leaves showed greater deformation under light loading, together with marked hysteresis suggesting substantial plastic deformation. This deformation was much more marked in expanding leaves from the mutant, but the differences were not observed if the seedlings were killed in ethanol prior to measurement, suggesting that this increased plasticity was a function of the living meristematic zone (C. J. Pollock & E. J. Lloyd, unpublished observations). Chemical analysis of the wall carbohydrate composition of leaf meristematic zones from plants grown at both 20° and 5°C showed no significant differences in gross chemical composition (Table 4), suggesting that expression of the mutation was associated with the dynamics of wall turnover during extension. The connection between altered wall rheology and abnormal temperature response in the mutant is not proven, but it represents a working hypothesis which explains the rapidity and localization of the response, its apparent independence

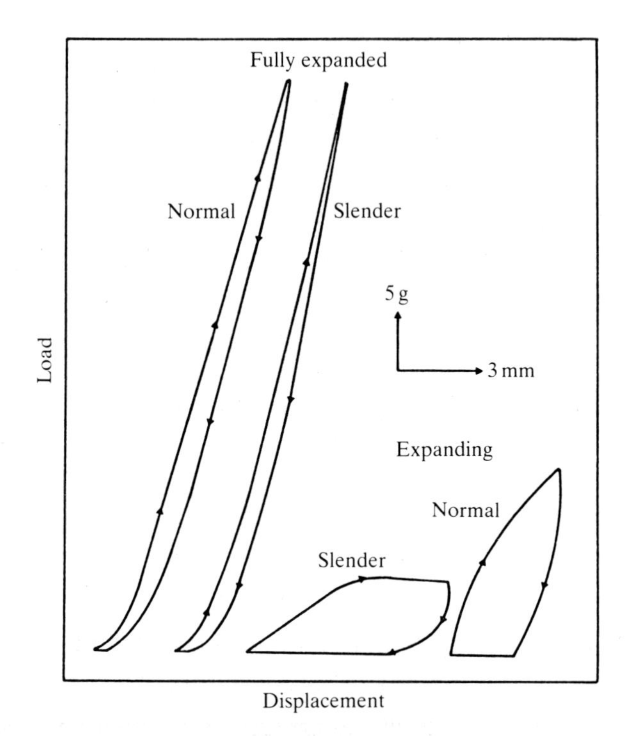

Fig. 6. Load/displacement curves for attached leaves of normal and 'slender' barley genotypes. First-leaf seedlings were clamped into the extensiometer. The applied load was measured by a strain gauge attached to the leaf tip and displacement by a transducer attached to the moving crosshead. Output from both devices was displayed on an X–Y recorder.

Table 4. *Mole percent composition of non-cellulosic cell wall carbohydrates extracted from meristematic zones of 'slender' and normal barley genotypes (C. J. Smith, unpublished observations)*

	Growth temperature			
	5°C		20°C	
	Normal	Slender	Normal	Slender
Rhamnose	Trace present in all			
Arabinose	19·8	20·8	20·9	20·1
Xylose	20·4	20·1	20·9	21·2
Galactose	7·3	7·4	7·3	7·2
Glucose	38·3	38·6	38·9	39·5
Galacturonic acid	8·3	8·2	7·3	7·3
Glucuronic acid	6·0	5·0	4·9	5·0

Table 5. *Structural differences in fourth leaves of* L. temulentum *developed under two contrasting temperature regimes (data from Pollock* et al. *1984)*

	Growth temperature	
	20°C	5°C
Leaf area (cm^2)	11·8	5·4
Leaf fresh weight (mg)	223	111
Chlorophyll (mg $leaf^{-1}$)	0·34	0·15
Mesophyll cell content (cells per leaf $\times 10^{-6}$)	8·2	5·8
Chloroplasts per cell	25	21

of solute and solvent effects and the wide range of biological variation which exists. Elucidation of the nature of the reactions within the wall which exhibit temperature sensitivity, and how they are modified by both genotype and environment remain challenging problems requiring detailed understanding of the processes of cell wall deposition and modification under different conditions.

Long-term responses to low temperature

Morphological, physiological and developmental changes in leaf growth

Prolonged alterations in the patterns of cell division and extension which occur during growth might be expected to produce modifications in plant form, and such changes have been described during overwintering and during prolonged exposure of plants to low temperatures in controlled environment facilities (Dale, 1965; Eagles, 1967; Haycock, 1981; Huner *et al.* 1981; Thomas, 1983; Krol *et al.* 1984; Pollock *et al.* 1984). In general, leaves which develop at low temperatures tend to be smaller (Table 5) and such differences in size have been correlated with reduced epidermal cell number (Auld *et al.* 1978; Paul, 1984). Marked genotypic differences in the degree of such responses have been observed. In studies on ecotypes of *Dactylis glomerata*, Eagles (1967) showed that the morphological changes induced by growth at 5 °C were much more extreme in northern European ecotypes than those observed in plants of Mediterranean origin. Such genotypic differences have been described for a range of temperate species (Cooper, 1964; Robson & Jewiss, 1968; Eagles & Othman, 1981).

Interactions between low temperature and other developmental or environmental factors may complicate the patterns of plant response. Temperate grasses show higher growth rates in the spring than in the autumn (Thomas & Norris, 1977) and such differences are not wholly attributable to temperature change. Under such conditions, growth rates reflect a complex interaction between current temperature and developmental state. Parsons & Robson (1980) transferred field-grown plants of *Lolium perenne* at various stages during the year into constant temperature conditions of 15 °C and measured the subsequent rates of leaf growth. These values were held to represent the potential for leaf growth of plants under the field conditions current at the time of transfer. Potential leaf growth rates fell

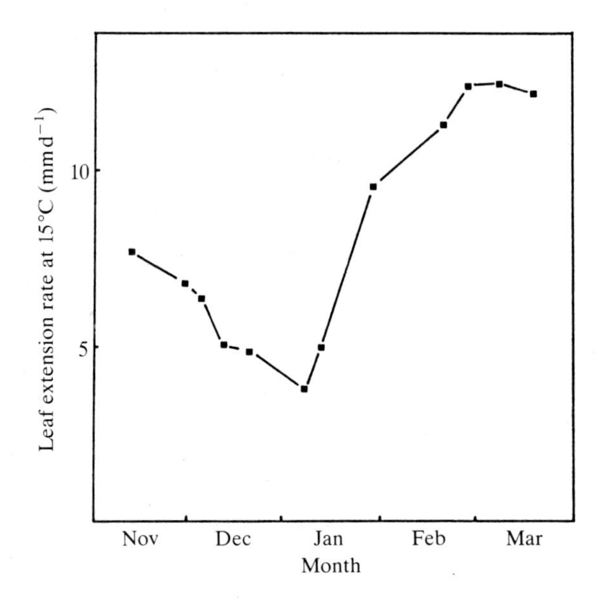

Fig. 7. Potential rates of leaf growth for plants of *Lolium perenne* cv. S.24 measured at different times during the year. (Redrawn from Parsons & Robson, 1980.)

during the autumn, reaching a minimum around midwinter, and then rose sharply during the spring (Fig. 7). Specific leaf photosynthesis showed parallel changes (Parsons & Robson, 1981) and, in related studies, levels of high molecular weight carbohydrate reserves were shown to be inversely related to growth potential (Pollock & Jones, 1979; Pollock, 1981), again suggesting that substrate supply was not a major limitation to growth during this period. Parsons & Robson (1980) correlated the midwinter change in potential leaf growth with the earliest stages of the transition of the stem apex from a vegetative to a reproductive state, and in other studies early flowering has been connected with early onset of spring growth (Kemp, 1983).

Such evidence suggests that temperature sensitivity may be markedly affected by developmental changes such as flowering and it has been shown that other external modulators of growth can alter temperature response. External application of the plant growth regulator GA_3 to lettuce hypocotyls increased the temperature sensitivity of extension growth (Stoddart *et al.* 1978) and administration of GA_3 to unacclimated wheat seedlings lowered the Pe temperature from 2·7° to −2·1°C (Stoddart & Lloyd, 1986). There is no direct evidence to suggest that developmentally or environmentally induced changes in the temperature responses of growth are obligately manifested *via* changes in the metabolism of plant growth regulators, but observations of this kind indicate that such regulation may occur under specific conditions. Prolonged exposure to low temperature may have effects upon other plant processes such as ion uptake or water movement (Sutton, 1969) and limitations in such processes may ultimately restrict growth or alter the overall temperature response of the tissue. Such complex interactions have not, in general, been characterized in detail under controlled conditions. However, the existence of genetic variability in long-term responses does provide useful experimental systems which should permit progress to be made in understanding the mechanisms that determine temperature sensitivity.

Low temperature and cell division

The effects described above, because of their rapidity, must act principally on the processes of cell extension. However, longer-term exposure to cold must affect both cell extension and cell division. There is little evidence concerning the specific effects of low positive temperatures on the mechanism of cell division in obligately chilling-resistant plants but it has been suggested that differential effects of temperature on division and extension account for the effects of chilling upon leaf growth in *Phaseolus vulgaris* (Wilson & Ludlow, 1968). Direct measurements of the effects of temperature upon the components of cell division and extension in roots of *Helianthus* (Burholt & Van't Hof, 1971) showed that, during chilling, the extension rate fell, the duration of the mitotic cycle increased but the mitotic index remained constant. Reduction in temperature also altered the proportion of the total mitotic cycle spent in S phase. Insensitivity of the mitotic index to low temperature has been observed in other species (Van't Hof & Ying, 1964; Lopez-Saez *et al.* 1966; Francis & Barlow, 1988), suggesting that, whilst straightforward to measure, it may not always be a suitable technique for determining the temperature sensitivity of cell division. Studies on cell cultures have also demonstrated differential temperature sensitivity for the various phases of cell division, with the G_1 phase showing little dependence upon temperature and the S phase being more sensitive (Gould, 1977). Low temperatures may also perturb the cell cycle by a direct effect upon the stability of microtubule structure (Rikin *et al.* 1983).

Although studies of this kind indicate the potential significance of cold inhibition of cell division, there is little direct evidence of its significance in determining growth rates *in vivo*. Imposition of abrupt temperature changes has been shown to be followed by a rapid reduction in growth rates which are then stable for some days (Thomas, 1983; Thomas & Stoddart, 1984; p. 162), suggesting a degree of synchronous adjustment of rates of division and extension. However, consideration of such changes is complicated by the potential contribution to growth of cells which had divided but not extended prior to the application of temperature change. Temporal separation of division and extension is seen in extreme forms during dormancy (Perry, 1971) and it has been suggested that growth in the early spring in a number of overwintering species occurs predominantly by extension of previously divided cells, and is associated with large cell volumes and high values for nuclear DNA content (Grime & Mowforth, 1982). Such a response would minimize any disruption to growth at low temperatures caused by reduced rates of cell division. Plants of *L. perenne* transferred from the field to controlled environments at 20°C during winter showed patterns of leaf production which reflected the previous growth conditions. The length of the first

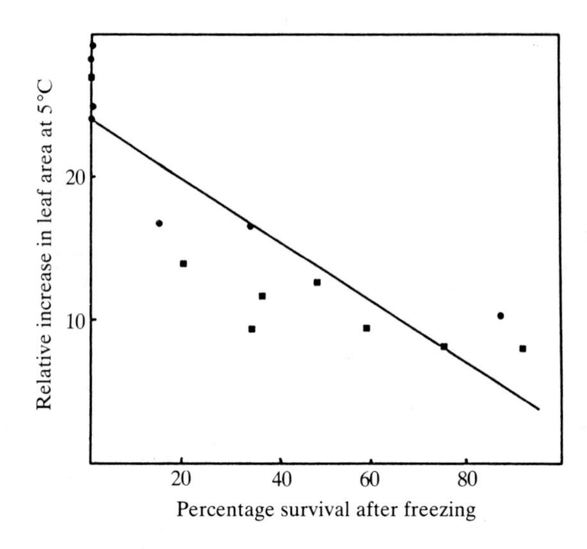

Fig. 8. The relationships between low temperature growth and frost tolerance for ecotypes of *Lolium perenne* (■) and *Dactylis glomerata* (●). Following growth measurements at 5°C, whole plants were exposed to −5°C for 3 days and survival assessed. (Plotted using data of Cooper, 1962.)

two leaves which emerged after transfer was similar to those measured in the field, and was substantially smaller than either those which grew on plants maintained at 20°C throughout or those which emerged subsequently, suggesting that there was some temporal separation of early leaf development and leaf extension (C. F. Eagles, unpublished observations). This separation is not complete, however, since mitotic division can be detected in the basal portion of leaves of *L. temulentum* throughout most of the phase of leaf expansion (Ougham *et al.* 1987).

The paucity of information on the integration of cell division and expansion during plant growth at low temperatures is striking, and a full understanding of temperature responses cannot be achieved without progress in this area. In particular, the study of acclimatory responses, such as increases in meristem cell number, which might help overcome limitations imposed by reduced rates of cell division, would be of considerable assistance in attempts to improve crop performance under such conditions.

Low temperature growth and the development of freezing tolerance

Often the exposure of temperate plants to declining temperatures in the autumn is not only associated with reductions in growth and altered plant form, but also with the development of enhanced ability to resist exposure to sub-zero temperatures. This process is often called 'hardening', and the resultant acclimation as 'hardiness' or 'frost-tolerance'. The successful agricultural exploitation of enhanced low-temperature growth in perennial or overwintering crops cannot be achieved if survival is adversely affected, and it is unfortunate that, in a number of cases, a strong negative correlation has been described between growth at low temperatures and subsequent freezing tolerance (Fig. 8). In more recent studies using quantitative laboratory estimates of freezing resistance (see p. 174), a strong negative correlation was observed between this character and winter growth rate for 86 lines of *L. perenne* (M. O. Humphreys, personal communication). It is not known whether or not this relationship is causal, i.e. whether reduction in growth rate is an obligatory part of the development of hardiness, although the reduction in growth rate on chilling is much more rapid than the detectable development of hardiness. Hardening is an active, energy-requiring process (Levitt, 1980; Li, 1984) requiring sustained exposure to hardening conditions for it to be maximized. In those herbaceous species where true dormancy is absent, marked changes in hardiness can be detected throughout the winter as growth is stimulated or checked by prevailing temperature conditions (Eagles & Fuller, 1982) or by management practice (Hides, 1978). Adequate development of hardiness in the autumn, rapid but non-injurious dehardening during early spring growth and the ability to exploit short periods of favourable weather during the winter are all legitimate agricultural objectives which require optimization of the balance between survival and growth under a range of environments, and which would benefit from a more detailed understanding of the interrelationship between hardening and growth potential.

Improvement of low temperature growth in crop plants

Selection for improved growth at low temperature

Different agricultural systems will require different strategies to optimize performance at low temperature. Glasshouse cultivation of plants with improved growth at low temperatures could lead to reduced energy costs; under these circumstances any adverse effects of a breeding programme on freezing-tolerance would be unimportant. Spring-sown crops which are exposed to occasional frosts must be bred with some degree of hardiness as well as with rapid growth at low positive temperatures, whereas in perennial or winter-sown crops adequate freezing-tolerance must be maintained to ensure survival, even at the expense of some reduction in low-temperature growth. Exploitation of winter growth in herbage species may be complicated by problems of animal access, leaving maximization of carbohydrate storage for subsequent use in spring growth as a more attractive strategy. In all of these cases, however, selection or crossing to produce material with enhanced low-temperature growth would be important, although the emphasis placed upon the maintenance of freezing tolerance would differ.

Selection for growth rates can be made in the field, glasshouse or under controlled environment conditions. The variability of field conditions, particularly year-to-year, is offset by easy accessibility and initial selections are often made there using visual assessment or restricted estimation of biomass increase. Controlled environment studies can give more accurate estimates of temperature relationships, and the use of temperature gradient systems to generate a range of environments between 2° and 10°C (Mason *et al.* 1976) allows growth to be monitored by both conventional growth analysis and by computerized auxanometry (Thomas *et al.* 1984). Meristem cooling collars can be used in more detailed studies to identify optimal temperature responses in individual plants. The wide range of genetic variability identified within groups such as temperate gramineae confirms the applicability of such selection techniques to improving performance within the normal survival range of the species. The detection and utilization of variation outside the normal temperature range for growth or survival may be more difficult, since this may involve alleviation of specific injuries rather than alteration of the balance between reversible responses.

Assessment of winter survival

The necessity of combining good growth in the winter or spring with high levels of winter survival invalidates selection or breeding based solely on enhanced ability to grow at low temperatures. Assessment of winter survival in general and freezing tolerance in particular must form part of crop improvement programmes for such situations (Wilson, 1981). As with selections for enhanced growth, climatic variability makes accurate and reproducible field measurement of freezing tolerance difficult. A number of laboratory-based techniques have been devised in which plants are hardened, preferably under controlled conditions (Marshall *et al.* 1981), and the responses of hardened plants are compared with unhardened controls. Direct measurements of freezing tolerance can be obtained using controlled freezing of whole plants, crowns, tillers or seedlings followed by measurement of survival. These techniques, though time-consuming, can yield accurate, sensitive and reproducible measures of freezing tolerance (Fuller & Eagles, 1978). Removal of replicate material from a cooling bath at progressively lower temperatures followed by analysis of survival *versus* temperature curves by probit analysis allows calculation of LT_{50} values (temperature for 50% kill). Genotypes can be distinguished by differences in their LT_{50} values (Table 6), and performance in this test agrees well with survival in the field (Fuller & Eagles, 1978).

Other correlated responses which have been employed to screen material for frost tolerance include electrolyte leakage (Jenkins & Roffey, 1974) and chlorophyll fluorescence (Barnes & Wilson, 1984). These techniques are experimentally convenient, but the nature of the relationship is complex and variable between species. They are often employed to rank genotypes within species rather than to estimate absolute levels of frost tolerance.

Current progress and future prospects

Although difficulties have been encountered in producing varieties which combine good low-temperature growth with adequate freezing tolerance, there are indications that the inverse relationship between these characters (p. 172) may not be obligatory. Material has been introduced from specific locations where selection pressures have led to the evolution of genotypes with good winter survival and this has been shown to exhibit early spring growth when transferred to

Table 6. LT_{50} *values for plant kill for seedlings of contrasting cultivars of three species*

Species	LT_{50} value (°C)
Lolium perenne	
cv. Grasslands Ruanui	−8·0
cv. S.23	−10·8
cv. Premo	−12·7
Lolium multiflorum	
cv. Grasslands Paroa	−9·5
cv. Optima	−10·6
cv. RvP	−13·6
Avena sativa	
cv. Bulwark	−8·5
cv. Peniarth	−9·6
Kentucky	−12·3

Growth conditions: 15°C, 16-h photoperiod for 18 days.
Hardening conditions: 2°C, 8-h photoperiod for 14 days.
Recovery conditions: 15°C, 16-h photoperiod for 21 days.

less rigorous environments. Based on such material, advanced breeding lines have been produced in *L. perenne* which combine winter-hardiness and early growth in an acceptable agricultural background (Humphreys, 1984). There is also a report where winter growth and frost-hardiness of clover genotypes did not show an inverse relationship (Ollerenshaw & Haycock, 1984). Exploitation of material such as this is facilitated by the use, at appropriate stages in the breeding programme, of the physiological techniques outlined above.

Future progress will, however, depend increasingly upon the utilization of a wider range of variability by the use of novel methods of genetic manipulation. There appears to be no theoretical limit to the accessibility of genetic material to such manipulations and it therefore behoves agricultural physiologists and biochemists to study the extent and nature of variability in a much wider range of species than those currently used. In the case of low-temperature growth, consideration of the growth processes of obligate psychrophiles and herbaceous species where winter is the period of maximum growth (*Muscari atlanticum* for example) will provide not only valuable background information but also access to extreme expressions of characters which might prove of great use in the improvement of plant performance under suboptimal conditions.

Conclusions

The literature on low-temperature effects on plants is dominated by work on injury caused by chilling or freezing and its amelioration. Consideration of the regulation of growth by plants within their normal survival range, whilst less fashionable, does have direct relevance to the improvement of crop performance. Identification of the sites of temperature perception and transduction, and more detailed studies using mutants and genotypic variants, should enable the genetic basis of this regulation to be described and manipulated. In a number of cases, however, our understanding is rudimentary, in particular concerning the effects of cold upon the integration of cell division and cell extension and the long-term effects of chilling upon the solute balance of growing leaves. It has been the aim of this review to highlight areas of ignorance as much as areas of increased knowledge, and we hope that, as a result, interest in these problems will be stimulated.

We are grateful to John Stoddart, Helen Ougham and Howard Thomas for helpful discussions during the preparation of this manuscript, to Eric Lloyd, John Toler, Janet Williams and Jenny Ashton for technical assistance and to Chris Smith (University College of Wales, Swansea), Alan Thomas and Deri Tomos (University College of North Wales, Bangor) for permission to cite unpublished material. The Welsh Plant Breeding Station is a constituent station of the AFRC Institute for Grassland and Animal Production.

References

ap Rees, T., Dixon, W. L., Pollock, C. J. & Franks, F. (1982). Low temperature sweetening of higher plants. In *Recent Advances in the Biochemistry of Fruit and Vegetables* (ed. J. Friend & M. J. C. Rhodes), pp. 41–61. London: Academic Press.

Auld, B. A., Dennett, M. D. & Elston, J. (1978). The effect of temperature change on the expansion of individual leaves of *Vicia faba* L. *Ann. Bot.* **42**, 877–888.

Bagnall, D. J. & Wolfe, J. A. (1978). Chilling sensitivity in plants. Do the activation energies of growth processes show an abrupt change at a critical temperature? *J. exp. Bot.* **29**, 1231–1242.

Baker, N. R., Long, S. P. & Ort, D. R. (1988). Photosynthesis and temperature, with particular reference to effects on quantum yield. In *Plants and Temperature*, Symp. Soc. Exp. Biol., vol. 42 (ed. S. P. Long & F. I. Woodward), pp. 347–375. Cambridge: Company of Biologists.

Barlow, E. W. R., Boersma, L. & Young, J. C. (1977). Photosynthesis, transpiration and leaf elongation in corn seedlings at suboptimal soil temperatures. *Agron. J.* **69**, 95–100.

Barnes, J. D. & Wilson, J. M. (1984). Assessment of the frost sensitivity of *Trifolium* species by chlorophyll fluorescence analysis. *Ann. appl. Biol.* **105**, 107–116.

Berry, J. A. & Björkman, O. (1980). Photosynthetic response and adaptation to temperature. *A. Rev. Pl. Physiol.* **31**, 491–543.

Berry, J. A. & Raison, J. K. (1982). Responses of macrophytes to temperature. In *Encyclopedia of Plant Physiology*, vol. 12a (ed. O. Lange, C. B. Osmond & P. S. Nobel), pp. 277–328. Berlin: Springer.

Biscoe, P. V. & Gallagher, J. N. (1977). Weather, dry matter production and yield. In *Environmental Effects on Crop Physiology* (ed. J. J. Landsberg & C. V. Cutting), pp. 75–100. London: Academic Press.

Björkman, O., Mahall, B., Nobs, M., Ward, W., Nicholson, F. & Mooney, H. (1974). Growth responses of plants from habitats with contrasting thermal environments. An analysis of the temperature dependence under controlled conditions. *Carnegie Instn Yb.* **73**, 748–757.

Burholt, D. R. & Van't Hof, J. (1971). Quantitative thermal-induced changes in growth and cell population kinetics of *Helianthus* roots. *Am. J. Bot.* **58**, 386–393.

Causton, D. R. & Venus, J. C. (1981). *The Biometry of Plant Growth.* London: Edward Arnold.

CHANDLER, T. J. & GREGORY, S. (1976). *The Climate of the British Isles.* 390 pp. London: Longmans.
CHRIST, R. A. (1978). The elongation rate of wheat leaves. II. Effect of sudden light change on the elongation rate. *J. exp. Bot.* **29**, 611–618.
COOPER, J. P. (1962). Light and temperature responses for vegetative growth. *Ann. Rep. Welsh Pl. Breed. Sta 1961*, pp. 16–17.
COOPER, J. P. (1964). Climatic variation in forage grasses. Leaf development in climatic races of *Lolium* and *Dactylis. J. appl. Ecol.* **1**, 45–62.
CRAWFORD, R. M. M. & HUXTER, T. J. (1977). Root growth and carbohydrate metabolism at low temperatures. *J. exp. Bot.* **28**, 917–925.
DALE, J. E. (1965). Leaf growth in *Phaseolus vulgaris.* 2. Temperature effects and the light factor. *Ann. Bot.* **29**, 293–308.
DERIEUX, C. (1984). The main limiting factors of environment. In *Physiologie du Mais* (ed. A. Gallais), pp. 383–387. Paris: INRA.
EAGLES, C. F. (1967). The effect of temperature on vegetative growth in climatic races of *Dactylis glomerata* in controlled environments. *Ann. Bot.* **31**, 31–39.
EAGLES, C. F. & FULLER, M. P. (1982). Evaluation of cold-hardiness in forage grasses and legumes. In *The Utilisation of Genetic Resources in Fodder Crop Breeding* (ed. M. D. Hayward), pp. 172–184. Aberystwyth: Eucarpia Fodder Crops Section.
EAGLES, C. F. & OTHMAN, O. B. (1981). Growth at low temperature and cold-hardiness in white clover. In *Plant Physiology and Herbage Production* (ed. C. E. Wright), Occasional Symposium No. 13, pp. 109–113. Hurley: British Grassland Society.
FORDE, B. J., WHITEHEAD, H. C. & ROWLEY, J. A. (1975). Effect of light intensity and temperature on photosynthetic rate, leaf starch content and ultrastructure of *Paspalum dilatatum. Aust. J. Pl. Physiol.* **2**, 185–195.
FRANCIS, D. & BARLOW, P. W. (1988). Temperature and the cell cycle. In *Plants and Temperature*, Symp. Soc. Exp. Biol., vol. 42 (ed. S. P. Long & F. I. Woodward), pp. 181–202. Cambridge: Company of Biologists.
FULLER, M. P. & EAGLES, C. F. (1978). A seedling test for cold-hardiness in *Lolium perenne. J. agric. Sci., Camb.* **91**, 217–222.
GIAQUINTA, R. T. & GEIGER, D. R. (1973). Mechanism of inhibition of translocation by localised chilling. *Pl. Physiol.* **51**, 372–377.
GOULD, R. (1977). Temperature response of the cell cycle of *Haplopappus gracilis* in suspension culture and its significance to the G1 transition probability model. *Planta* **137**, 29–36.
GRAFIUS, J. E. (1981). Breeding for winter hardiness. In *Analysis and Improvement of Plant Cold-hardiness* (ed. C. R. Olien & M. N. Smith), pp. 161–175. Florida: CRC Press.
GRIME, J. P. & MOWFORTH, M. A. (1982). Variation in genome size: an ecological interpretation. *Nature, Lond.* **299**, 151–153.
HAYCOCK, R. (1981). Environmental limitations to spring production in white clover. In *Plant Physiology and Herbage Production* (ed. C. E. Wright), Occasional Symposium No. 13, pp. 119–123. Hurley: British Grassland Society.
HENDRY, G. (1987). The ecological significance of fructan in a contemporary flora. *New Phytol.* **106**, 201–216.
HIDES, D. H. (1978). Winter hardiness in *Lolium multiflorum* Lam. 1. The effect of nitrogen fertilizer and autumn cutting managements in the field. *J. Br. Grassld Soc.* **33**, 99–105.
HOCHACHKA, P. W. & SOMERO, G. N. (1973). *Strategies of Biochemical Adaptation*, pp. 179–270. Philadelphia: W. Saunders.
HUMPHREYS, M. O. (1984). Breeding for seasonal consistency in perennial ryegrass for pastures. In *Development, Construction and Multiplication of Fodder Crop Varieties* (ed. U. Simon), pp. 59–70. Weihenstephan, Germany: Eucarpia Fodder Crops Section.
HUNER, N. P. A., PALTA, J. P., LI, P. H. & CARTER, J. V. (1981). Anatomical changes in leaves of Puma rye in response to growth at cold-hardening temperatures. *Bot. Gaz.* **142**, 55–62.
JENKINS, G. & ROFFEY, A. P. (1974). A method of estimating the cold-hardiness of cereals by measuring electrical conductance after freezing. *J. agric. Sci., Camb.* **83**, 87–92.
KEMP, D. R. (1983). The regulation of winter growth of temperate pasture grasses. *Proc. XV Int. Grassld Cong.*, Kyoto, Japan, pp. 383–386.

Kemp, D. R. (1984). Temperate pastures. In *Control of Crop Productivity* (ed. C. J. Pearson), pp. 159–184. Sydney, Australia: Academic Press.
Kemp, D. R. & Blacklow, W. M. (1980). Diurnal extension rates of wheat leaves in relation to temperature and carbohydrate concentration of the extension zone. *J. exp. Bot.* **31**, 821–828.
Kemp, D. R. & Blacklow, W. M. (1982). The responsiveness to temperature of the extension rate of leaves of wheat growing in the field under different levels of nitrogen fertilizer. *J. exp. Bot.* **33**, 29–36.
Kleinendorst, A. (1975). An explosion of leaf growth after stress conditions. *Neth. J. agric. Sci.* **23**, 139–144.
Kleinendorst, A. & Brouwer, R. (1970). The effect of temperature of the root medium and of the growing point of the shoot on growth, water content and sugar content of maize leaves. *Neth J. agric. Sci.* **18**, 140–148.
Kleinendorst, A. & Brouwer, R. (1972). The effect of local cooling on growth and water content of plants. *Neth J. agric. Sci.* **20**, 203–217.
Korner, Ch. & Larcher, W. (1988). Plant life in cold climates. In *Plants and Temperature, Symp. Soc. Exp. Biol.*, vol. 42 (ed. S. P. Long & F. I. Woodword), pp. 25–57. Cambridge: Company of Biologists.
Krol, M., Griffith, M. & Huner, N. P. A. (1984). An appropriate physiological control for environmental temperature studies: comparative growth responses of winter rye. *Can. J. Bot.* **62**, 1062–1068.
Larcher, W. (1981). Effects of low temperature stress and frost injury on productivity. In *Physiological Processes Limiting Plant Productivity* (ed. C. B. Johnson), pp. 253–269. London: Butterworths.
Larcher, W. & Bauer, H. (1982). Ecological significance of resistance to low temperature. In *Encyclopedia of Plant Physiology*, vol. 12a (ed. O. Lange, C. B. Osmond & P. S. Nobel), pp. 253–269. Berlin: Springer.
Levitt, J. (1980). *Responses of Plants to Environmental Stress, vol. 1. Chilling, Freezing and High Temperature Stress*, 497 pp. New York: Academic Press.
Li, P. H. (1984). Subzero temperature stress physiology of herbaceous plants. *Hort. Rev.* **6**, 373–416.
Lopez-Saez, J. F., Giminez-Martin, G. & Gonzalez-Fernandez, A. (1966). Duration of the cell division cycle and its dependence upon temperature. *Z. Zellforsch.* **75**, 591–600.
Markhart, A. H., Fiscus, E. L., Naylor, A. W. & Kramer, P. J. (1979). Effect of temperature on water and ion transport in soybean and broccoli systems. *Pl. Physiol.* **64**, 83–87.
Marshall, H. G., Olien, C. R. & Everson, E. H. (1981). Techniques for selection of cold-hardiness in cereals. In *Analysis and Improvement of Plant Cold Hardiness* (ed. C. R. Olien & M. N. Smith), pp. 139–159. Florida: CRC Press.
Mason, G., Grime, J. P. Lumb, A. H. (1976). The temperature-gradient tunnel. A versatile controlled environment. *Ann. Bot.* **40**, 137–142.
Meyer, R. F. & Boyer, J. S. (1972). Sensitivity of cell division and cell elongation to low water potential in soybean hypocotyls. *Planta* **198**, 77–87.
Monteith, J. L. (1977). Climate and the efficiency of crop production in Britain. *Phil. Trans. R. Soc. Ser.* B **281**, 277–294.
Ollerenshaw, J. H. & Haycock, R. (1984). Variation in the low temperature growth and frost tolerance of natural genotypes of *Trifolium repens* L. from Britain and Norway. *J. Agric. Sci., Camb.* **102**, 11–21.
Ong, C. K. & Baker, C. K. (1985). Temperature and leaf growth. In *Control of Leaf Growth* (ed. N. R. Baker, W. J. Davies & C. K. Ong), pp. 175–200. Society for Experimental Biology Seminar Series, No. 27. Cambridge: Cambridge University Press.
Ougham, H. J., Jones, T. W. A. & Evans, M. Ll. (1987). Leaf development in *Lolium temulentum* L.: progressive changes in soluble polypeptide complement and isoenzymes. *J. exp. Bot.* **38**, 1689–1696.
Parsons, A. J. & Robson, M. J. (1980). Seasonal changes in the physiology of S.24 perennial ryegrass (*Lolium perenne* L.). 1. Response of leaf extension to temperature during the transition from vegetative to reproductive growth. *Ann. Bot.* **46**, 435–445.
Parsons, A. J. & Robson, M. J. (1981). Seasonal changes in the physiology of S.24 perennial ryegrass (*Lolium perenne* L.). 2. Potential leaf and canopy photosynthesis during the

transition from vegetative to reproductive growth. *Ann. Bot.* **47**, 249–258.
PAUL, E. M. M. (1984). The response to temperature of leaf area in tomato genotypes. 1. Cell size and number in relation to the area of a leaf. *Euphytica* **33**, 347–354.
PEACOCK, J. M. (1975). Temperature and leaf growth in *Lolium perenne.* II. The site of temperature perception. *J. appl. Ecol.* **12**, 115–123.
PEACOCK, J. M. (1982). Response and tolerance of sorghum to temperature stress. In *Sorghum in the Eighties* (ed. L. R. House, L. K. Mughogho & J. M. Peacock), pp. 143–159. Pantacheru, India: ICRISAT.
PERRY, T. O. (1971). Dormancy of trees in winter. *Science* **171**, 29–36.
POLLOCK, C. J. (1981). Environmental effects on reserve carbohydrate metabolism in *Phleum pratense.* In *Plant Physiology and Herbage Production* (ed. C. E. Wright), Occasional Symposium No. 13, pp. 115–118. Hurley: British Grassland Society.
POLLOCK, C. J. (1984). Sucrose accumulation and the initiation of fructan biosynthesis in *Lolium temulentum* L. *New Phytol.* **96**, 527–534.
POLLOCK, C. J. (1986). Fructans and the metabolism of sucrose in higher plants. *New Phytol.* **104**, 1–24.
POLLOCK, C. J. & JONES, T. (1979). Seasonal patterns of fructan metabolism in forage grasses. *New Phytol.* **83**, 8–15.
POLLOCK, C. J., LLOYD, E. J., STODDART, J. L. & THOMAS, H. (1983). Growth, photosynthesis and assimilate partitioning in *Lolium temulentum* exposed to chilling temperatures. *Physiologia Pl.* **59**, 257–262.
POLLOCK, C. J., LLOYD, E. J., THOMAS, H. & STODDART, J. L. (1984). Changes in photosynthetic capacity during prolonged growth of *Lolium temulentum* at low temperature. *Photosynthetica* **18**, 478–481.
RIKIN, A., ATSMON, D. & GITLER, C. (1983). Quantitation of chill-induced release of a tubulin-like factor and its prevention by abscisic acid in *Gossypium hirsutum* L. *Pl. Physiol.* **71**, 747–748.
ROBSON, M. J. & JEWISS, O. R. (1968). A comparison of British and North African varieties of tall fescue (*Festuca arundinacea*). III. Effects of light, temperature and daylength on relative growth rate and its components. *J. appl. Ecol.* **5**, 191–204.
SCHNYDER, H. (1986). Carbohydrate metabolism in the growth zone of tall fescue leaf blades. In *Current Topics in Plant Physiology and Biochemistry,* vol. 5 (ed. D. D. Randall), pp. 47–60. Columbia: University of Missouri Press.
SHARP, R. E., OSONUBI, O., WOOD, W. A. & DAVIES, W. J. (1979). A simple instrument for measuring leaf extension in grasses and its application in the study of the effects of water stress in maize and sorghum. *Ann. Bot.* **44**, 35–45.
SMILLIE, R. M. (1976). Temperature control of chloroplast development. In *Genetics and Biogenesis of Chloroplasts and Mitochondria* (ed. T. Bucher), pp. 103–110. North-Holland: Elsevier.
STAMP, P. (1984). Chilling tolerance of young plants demonstrated on the example of maize (*Zea mays* L.). *Fortschr. Acker und Pflanz.* **7**, 1–83.
STODDART, J. L. & LLOYD, E. J. (1986). Modification by gibberellin of the growth-temperature relationship in normal and mutant genotypes of several cereals. *Planta* **167**, 364–368.
STODDART, J. L., TAPSTER, S. M. & JONES, T. W. A. (1978). Temperature dependence of the gibberellin response in lettuce hypocotyls. *Planta* **141**, 283–288.
STODDART, J. L., THOMAS, H., LLOYD, E. J. & POLLOCK, C. J. (1986). The use of a temperature-profiled position transducer for the study of low-temperature growth in gramineae. *Planta* **167**, 359–363.
SUTCLIFFE, J. (1977). *Plants and Temperature,* pp. 25–40. London: E. Arnold.
SUTTON, C. D. (1969). Effect of low soil temperature on phosphate nutrition of plants. A review. *J. Sci. Fd Agric.* **20**, 1–3.
TERRY, N., WALDRON, L. J. & TAYLOR, S. E. (1983). Environmental influences on leaf expansion. In *The Growth and Functioning of Leaves* (ed. J. E. Dale & F. L. Milthorpe), pp. 179–205. Cambridge: Cambridge University Press.
THOMAS, H. (1983). Analysis of the response of leaf extension to chilling temperatures in *Lolium temulentum* seedlings. *Physiologia Pl.* **57**, 509–513.
THOMAS, H. & NORRIS, I. B. (1977). The growth responses of *Lolium perenne* to the weather

during winter and spring at various altitudes in mid-Wales. *J. appl. Ecol.* **14**, 949–964.
THOMAS, H., ROWLAND, J. J. & STODDART, J. L. (1984). A microprocessor-based instrument for measuring plant growth. *J. Microcomp. Appl.* **7**, 217–223.
THOMAS, H. & STODDART, J. L. (1984). Kinetics of leaf growth in *Lolium temulentum* at optimal and chilling temperatures. *Ann. Bot.* **53**, 341–347.
TOMOS, A. D. (1985). The physical limitations of leaf cell expansion. In *Control of Leaf Growth* (ed. N. R. Baker, W. J. Davies & C. K. Ong), pp. 1–33. Society for Experimental Biology Seminar Series No. 27. Cambridge: Cambridge University Press.
TOMOS, A. D. & ZIMMERMAN, V. (1982). Determination of water relations parameters of individual higher plant cells. In *Biophysics of Water* (ed. F. Franks & S. F. Mathias), pp. 256–261. Wiley, Chichester.
VAN'T HOF, J. & YING, H-K. (1964). Relationship between the duration of the mitotic cycle, the rate of cell production and the rate of growth of *Pisum* roots at different temperatures. *Cytologia* **29**, 399–406.
VAN VOLKENBURGH, E., HUNT, S. & DAVIES, W. J. (1983). A simple instrument for measuring cell wall extensibility. *Ann. Bot.* **51**, 669–672.
WATTS, W. R. (1971). Role of temperature in the regulation of leaf extension in *Zea mays*. *Nature, Lond.* **229**, 46–47.
WATTS, W. R. (1974). Leaf extension in *Zea mays*. III. Field measurements of leaf extension in response to temperature and leaf water potential. *J. exp. Bot.* **25**, 1085–1096.
WILLIAMS, C. N. & BIDDISCOMBE, E. F. (1965). Extension growth of grass tillers in the field. *Aust. J. agric. Res.* **16**, 4–22.
WILSON, D. (1981). The role of physiology in breeding herbage cultivars adapted to their environment. In *Plant Physiology and Herbage Production* (ed. C. E. Wright), Occasional Symposium No. 13, pp. 95–108. Hurley: British Grassland Society.
WILSON, G. L. & LUDLOW, M. M. (1968). Bean leaf expansion in relation to temperature. *J. exp. Bot.* **19**, 309–321.
WOOLHOUSE, H. W. (1981). Crop physiology in relation to agricultural production: the genetic link. In *Physiological Processes Limiting Plant Productivity* (ed. C. B. Johnson), pp. 1–21. London: Butterworths.

Printed in Great Britain © Society for Experimental Biology 1988

TEMPERATURE AND THE CELL CYCLE

D. FRANCIS[1] *and P. W. BARLOW*[2]

[1]School of Pure and Applied Biology, University of Wales, P.O. Box 915, Cardiff CF1 3TL, UK

[2]University of Bristol, Department of Agricultural Science, Institute of Arable Crops Research, Long Ashton Research Station, Long Ashton, Bristol BS18 9AF, UK

Summary

During the period between successive divisions, a cell traverses three stages of interphase: G_1 (pre-synthetic interphase), S-phase (DNA synthetic interphase) and G_2 (post-synthetic interphase). The time taken for all cells in a meristem to divide (the cell doubling time (cdt)) decreases in response to an increase in temperature. For example, the cdt in root meristems of *Zea mays* decreases 21-fold as the temperature is increased from 3 to 25 °C. Whether all phases of the cell cycle alter proportionately with temperature has been ascertained by comparing data from the root meristem of five species: *Pisum sativum*, *Helianthus annuus, Tradescantia paludosa, Allium cepa* and *Triticum aestivum*. In three of the five species there is a disproportionate lengthening of the G_1 phase at low temperatures. We suggest that arrest in G_1 with the associated $2C$ amount of DNA, confers maximal protection on the genome of a somatic cell to the stress of low temperature. DNA replication has been studied at different temperatures for *Helianthus annuus*, *Secale cereale* and *Oryza sativa*. The rate of DNA replication, per single replication fork, increases when the temperature is raised, while the distance between initiation points (replicon size) remains constant.

The temperature at which the cell cycle has a minimum duration is close to 30 °C in many species, and it seems that this optimum temperature is always near the upper temperature limit of the cell cycle. The rate of cell division determines the rates of organ and cell growth. Thus, temperature has a major effect on the way in which meristematic cells are deployed in organogenesis. The rate of organogenesis, in turn, determines the response of the plant to the growing season. We predict that species growing in sub-arctic conditions comprise cells with low DNA contents and hence have the potentialities for rapid cell cycles so that maximum advantage can be taken of a short growing season. Data from *Triticum aestivum* show that at 5 °C, nucleoli are large compared with those at 10–25 °C. These observations are consistent with high levels of RNA polymerase and cellular RNA found at low compared with high temperatures. These responses may be important in sustaining growth at 5 °C.

Finally, the effects of temperature on developmental transitions are discussed. The picture that emerges is that more is known about low, as opposed to high, temperature as a morphogenetic switch but virtually nothing is known about cell cycle activity during such transitions. A better understanding of temperature and the cell cycle, particularly in relation to geographical distribution, will be forthcoming when data exist for those species endemic to particular latitudes. Only then, will it be possible to judge the relative contributions of cell elongation and cell division to the growth and reproductive strategies adopted by plants at different latitudes and temperatures.

Introduction

Our intention is to review the effects of temperature on the mitotic cell cycle and its component phases in higher plants. Central to the cell cycle is the duplication of nuclear DNA in the period between cell birth and division. Specifically, we shall explore the influence of temperature on DNA replication and cell division and the strategies that plants adopt to match their cell reproduction rate and rate of overall development to the temperature of their environment. Consideration will also be given to nucleolar activity in higher plants grown at different temperatures. The role of cell division during temperature-induced developmental transitions will also be discussed. Finally, some attention will be given to how responses of the cell cycle may enable plants to withstand temperature stress. Effects of temperature on sub-cellular structures such as the cytoskeleton, microtubule organizing centres and membrane elements have recently been described elsewhere (e.g. Barlow, 1987) and will not be referred to here.

Before considering the cell cycle, it is worthwhile considering how the regions of growth are organized in plants. Major meristematic regions are located at the tips of shoots and roots. The true shoot meristem, the apical dome, is the meristematic region above the youngest leaf primordia and the size of the apical dome increases exponentially through a given plastochrone. At the end of the plastochrone part of the dome is invested in one or more primordia; the remainder reconstitutes a dome of minimal size for the start of the next plastochrone (Lyndon, 1976). In the base of the new leaf meristematic activity is retained for a while before cell expansion predominates; both processes give the species-specific size and shape to the leaf (Dale, 1986). The root meristem, on the other hand, does not usually initiate new organs. On its distal surface are the cap initials whose descendants form the root cap, while more proximal cells are progenitors of the epidermal, cortical and vascular tissues. In effect, both the root and shoot meristems are populations of cells which grow away from the majority of their descendants as the latter expand and differentiate into the various tissue systems. Some descendants of these two meristems persist as meristematic regions such as the procambial and cambial zones, intercalary meristems, as well as the pericycle.

Cell cycle

An intrinsic property of a meristematic cell is its ability to grow and divide. A feature of these processes is that nuclear DNA is replicated in a discrete interval, the DNA synthetic-(S)-phase, normally separated in time from mitosis. G_1 and G_2 (pre- and post-synthetic interphase, respectively) are phases normally interpolated between mitosis and S-phase. Howard & Pelc (1953) introduced these terms to describe the so-called cell cycle during which a cell undertakes a series of regulated, or deterministic, events in order to acquire competence for DNA replication and cell division.

The term 'cell cycle' conveys an image of regularity, but cells rarely, if ever, go through two successive cycles in exactly the same way. Indeed, the lack of constancy of the durations of the component phases during successive cycles of cultured mammalian cells has led to an alternative model. This alternative is based on the concept of a random transition of a cell from an inactive state to an active, division-oriented phase (Smith & Martin, 1973). The validity of this so-called 'transition-probability' model as applied to multicellular plant meristems is questionable (Armstrong & Francis, 1985; Francis, 1988). Although the concept of a deterministic cell cycle can be justifiably criticized, it continues adequately to describe the life history of a dividing cell. Moreover, the term 'cell cycle' will probably continue to survive with the proviso that in meristems the cycles are heterogeneous with respect to the durations of the component phases both within and between tissues.

A standard way of measuring the cell cycle in plant meristems is the percentage labelled mitoses (PLM) method which involves using the radioactive nucleoside, thymidine. Meristems are pulse-labelled with methyl-^{3}H-thymidine (^{3}H-TdR) and then chased with cold TdR before being grown on in non-radioactive solution. By subsequent autoradiography, the relationship between the percentage of labelled mitoses and time since labelling is determined. The PLM method monitors the flow of the labelled cells from S-phase through G_2 and mitosis. The interval between peaks of labelled mitoses gives a measure of cell cycle duration, $G_2+1/2M$ is the interval between the labelling period and the 50% intercept of the initial ascending limb of the first peak. M is from where the initial limb rises to where it plateaus, S-phase is obtained from the interval between the 50% intercept of the ascending and descending limbs of the first peak minus the labelling time, and G_1 is obtained by difference (Quastler & Sherman, 1959). Cautious use of low specific activity ^{3}H-TdR is necessary since highly radioactive ^{3}H-TdR can perturb the cell cycle (De la Torre & Clowes, 1974). Also, uniformity of growth conditions is important since transfer of seedlings from solid to aqueous conditions can result in erroneous measures of cell cycle parameters (Thomas & Davidson, 1983).

Other methods of cell cycle analysis involve exposing cells to colchicine and recording rates of accumulation of metaphase cells (Evans *et al.* 1957), or inducing telophase nuclei to form binucleate cells with caffeine and recording the time of appearance of subsequent bimitoses (Gonzalez-Fernandez *et al.* 1966). However, these techniques give a measure of potential cell doubling time which is only equal to cell cycle times if the proportion of actively dividing cells (the so-called 'Growth Fraction') is 100%. This is occasionally so in the shoot meristem (Ormrod & Francis, 1985) but rarely in root meristems (see, for example, Clowes, 1971). Labelling methods are preferred for measures of the cell cycle and its component

Table 1. *Duration of mitosis (M) and cell doubling times (cdt) in hours, in root meristems of* Pisum sativum *and* Vicia faba, *and cdt in the endosperm and embryo of* Hordeum vulgare *grown at various temperatures*

Temperature (°C)	*P. sativum*[a]		*V. faba*[b]		*H. vulgare*[c] (cdt)	
	M	cdt	M	cdt	Endosperm	Embryo
3	–	–	27	264	–	–
8	–	–	13	90	–	–
10	–	–	–	–	15	48
13	–	–	5	35	–	–
15	3	26	–	–	8	24
20	2	19	–	22	6	16
25	1	16	2	12	3	10
30	1	14	1	13	3	8
35	–	–	1	24	3	7

[a] From Brown (1951).
[b] From Murin (1967, 1968).
[c] From Pope (1943).

phases, the PLM method being the most popular (for a review of the various techniques, see Webster & MacLeod, 1980).

Cell cycle phase durations and temperature

Rates of cell division alter with changes of temperature in a tissue- and species-specific manner. For example, Brown (1951) observed a 1·4-fold lengthening of the cell doubling time when primary root tips of *Pisum sativum*, grown for 3 days at 25 °C, were transferred to 15 °C. Mitosis lengthened by a similar factor (Table 1). Murin (1967, 1968) made similar measurements on root meristems of *Vicia faba* cv. Zborovicky, and showed that as the temperature was increased from 3 to 25 °C by 5° increments, a 21-fold shortening occurred to both the duration of mitosis and the cell doubling time. However, increasing the temperature to 30 or 35 °C resulted in a progressive lengthening of cell doubling time although mitosis continued to shorten. Other data in Table 1 indicate that 5 °C increments from 10 to 35 °C result in a 5- to 6·5-fold shortening of cell doubling time in both the embryo and endosperm of *Hordeum vulgare* (Pope, 1943).

The variations in the percentage of the cell cycle occupied by G_1, S, G_2 and M in roots of five species exposed to various temperatures are shown in Fig. 1. The data for *Pisum sativum* indicate that an increase in temperature results in a progressive

Fig. 1. The relationship between durations (h) of the cell cycle, G_1 (□), S (▦), G_2 (□), and M (■), and temperature (°C), in root meristems of *Pisum sativum*, *Helianthus annuus*, *Tradescantia paludosa*, *Allium cepa* and *Triticum aestivum* a and b, adapted from Gudkov *et al.* (1974), Burholt & Van't Hof (1971), Wimber (1966), Gonzalez-Fernandez *et al.* (1971), Bayliss (1972), Kidd (1986) and D. Francis (unpublished data), respectively. The numbers on the histograms refer to the percentage of the cell cycle that each phase occupies.

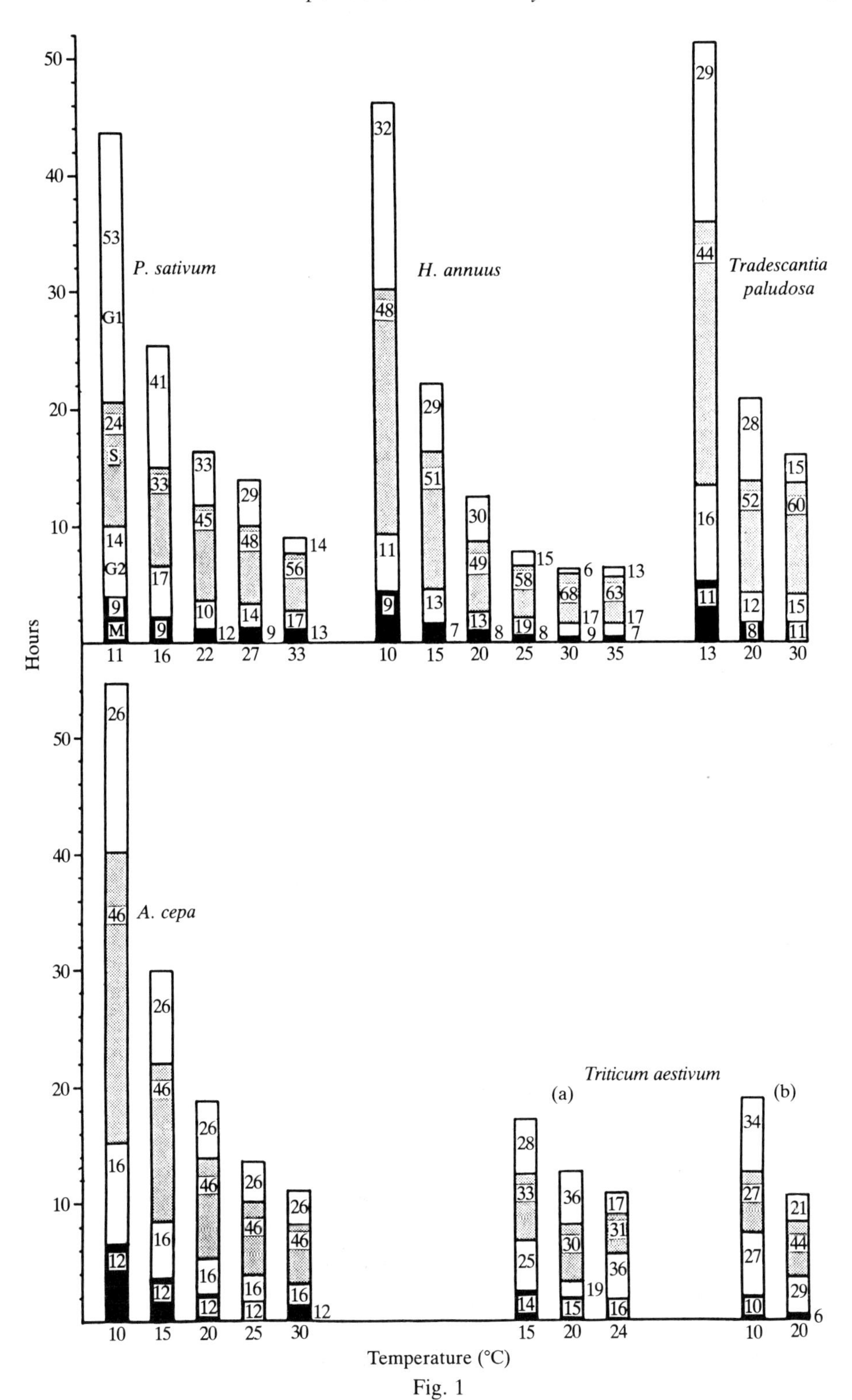

Hours
Temperature (°C)
P. sativum
H. annuus
Tradescantia paludosa
A. cepa
Triticum aestivum
(a)
(b)
G1
S
G2
M

Fig. 1

shortening of the cell cycle. However, although G_1 shortens from about 23 h at 11 °C to 1 h at 33 °C, S-phase shortens from about 11 to 5 h. In other words, while G_1 at 33 °C is one twenty-third of its value at 11 °C, the corresponding S-phase has shortened by about a half. Moreover, G_2 and M shorten by about a quarter. The same trend is apparent in the data for *Helianthus annuus* (Fig. 1). As the temperature is raised in 5 °C increments, G_1 shortens by about one-twentieth whereas S-phase and G_2 shorten by about a fifth and M by a tenth. Both sets of data also indicate that, on a percentage basis, G_1 occupies progressively less, while S-phase occupies progressively more of the shortened cell cycle; G_2 and M remain relatively constant at each temperature. Thus, at low temperatures, dividing cells of each of these species spend a proportionately greater length of time in G_1. Does this confer an advantage to plants exposed to such temperatures?

Pluripotent cells of various animal tissues are characterized by slow mitotic cycles and spend the majority of their life history in G_1 with $2C$ nuclear DNA amounts. These so-called stem cells can be experimentally induced into more rapid rates of cell division (Susumu & Potten, 1985). Cairns (1975) proposed that when immortal stem cells divide in the basal layer of an epithelial tissue, they retain an unblemished copy of the genome by the normal mechanism of semi-conservative DNA replication. Because the $2C$ amount is the smallest amount of nuclear DNA in a somatic cell, it presents the smallest target for mutagens. Thus, under normal circumstances, a slowly cycling cell with a prolonged G_1 will remain a minimum target for the majority of its life history.

Cells of the quiescent centre of angiosperm root meristems are either slow-cycling with long G_1-phases, or are apparently non-cycling with $2C$ nuclear DNA amounts. These cells can replenish the root meristem when cells are removed by decapping (Barlow, 1974), or rendered sterile after exposure to X-rays (Clowes, 1959) or low temperature (Clowes & Stewart, 1967). In the latter case, exposing primary roots of *Zea mays* to 5 °C for 7 to 14 days resulted in extensive damage to the meristem and a cessation of root growth. Upon transfer to 21 °C, quiescent centre cells regenerated the meristem enabling the root to elongate again. Following from these observations, Barlow & Rathfelder (1985) and Barlow & Adam (1989) found that the amount of cell division in the quiescent centre is related to the duration of the cold treatment. This, in turn, seems to be inversely related to the number of cell cycles that occur within proximal regions of the meristem. There is an obvious analogy between these responses and the reaction of mammalian stem cells to stress (Barlow, 1978). However, the effect of cold on the meristematic cell population is not simply that it brings about changes in the relative duration of the phases of the cell cycle; it seems also to perturb a higher level of meristem organization that depends upon interrelationships between its various zones.

Cells in a meristem subjected to low temperatures may traverse the cell cycle slowly so that the longest phase is G_1. According to our hypothesis, the $2C$ amount of DNA confers maximal protection to the genome of a somatic cell against the stress of low temperature. An important aspect of this protection is that phosphorylation of histone H1 begins at the beginning of S-phase and is an essential event for chromatin condensation and, hence, cell division (Bradbury & Matthews, 1982). Temperatures below 5°C may irreversibly damage this cell division-associated machinery (e.g. the uncondensed regions of chromosomes (Darlington & La Cour, 1940) and their susceptibility to misdivision at anaphase (Shaw, 1958)), but a cell resting in G_1 would remain unaffected and be able to resume the cell cycle upon the return of elevated temperatures. This idea is a continuation of the themes developed by Cairns (1975) for mammalian stem cells and by Clowes (1967*a*) for the response of quiescent centre cells to X-rays. Of further interest are the observations that mammalian cell lines which are temperature-sensitive, arrest in G_1 at non-permissive temperatures (Burstin *et al.* 1974; Basilico, 1978).

The hypothesis is supported by observations on *Pisum*, *Helianthus*, *Tradescantia paludosa* and, to a lesser extent, for *Triticum aestivum* (Fig. 1). Data clearly not in accord with this idea are those for *Allium cepa* in which 5°C increments from 10 to 30°C resulted in a remarkably equal shortening of the component phases. However, it must be appreciated that these inter-specific comparisons are with data obtained from different experiments performed at different times. Ideally, comparisons should be made from a single study of corresponding tissues from plants of corresponding developmental age. With this in mind, the data in Fig. 1 are supplemented with unpublished data from primary and seminal roots of *Triticum aestivum* cv. Chinese Spring at 10°C and compared with published data on four-day-old seedlings of the same cultivar grown at 20°C (Kidd *et al.* 1987). The four-day interval at 20°C was chosen so that roots were about 1 cm long and cell division in the meristem was in steady-state. (When an embryo germinates, emergence of the radicle is by cell elongation (Davidson, 1966).) At 10°C, 6 days elapsed before the roots were 1 cm long and, thus, the seedlings were of identical developmental age to those at 20°C yet were chronologically 48 h older. In this case, G_1 is the longest phase at 10°C, occupying 36 % of the cell cycle, whereas at 20°C S-phase predominates. This is another example of a disproportionate lengthening of the component phases as the temperature is lowered. However, as mentioned, not all data are consistent on this point and clearly more investigations are required.

Rates of DNA replication

In eukaryotes, DNA is duplicated during S-phase and its replication commences at multiple initiation sites spaced along the DNA molecule. From each initiation site two replication forks replicate the DNA semi-conservatively until they converge with replication forks from adjacent initiation sites. The initiation site and the twin replication forks spanning a particular stretch of parental DNA constitute a replicon. Higher plant nuclei are characterized by vast numbers of replicons: in *Secale cereale* there are approximately 250 000 replicons per nucleus (D. Francis, unpublished data).

Table 2. *Rate of DNA replication per single replicon fork (fork rate) in µm h⁻¹ and replicon size in µm, measured from DNA fibre autoradiographs of root tips of 4- to 5-day-old seedlings of* Helianthus annuus, Secale cereale *and* Oryza sativa *grown at various temperatures*

	H. annuus[a]		*S. cereale*		*O. sativa*	
Temperature (°C)	Fork rate ($\mu m\,h^{-1}$)	Replicon size (μm)	Fork rate ($\mu m\,h^{-1}$)	Replicon size (μm)	Fork rate ($\mu m\,h^{-1}$)	Replicon size (μm)
10	6	27	–	–	–	–
15	6	24	–	–	–	–
20	8	20	8	22[b]	6	18[b]
23	10	21	12	20–25[c]	–	–
27	–	–	–	–	10	15–20[d]
30	11	23	–	–	–	–
35	12	23	–	–	–	–
38	12	19	–	–	–	–

[a] From Van't Hof *et al.* (1978).
[b] From Kidd *et al.* (1987).
[c] From Francis & Bennett (1982).
[d] D. Francis unpublished.

Temperature could affect the rate of DNA replication in at least three ways. First, the rate of DNA replication per single replication fork could alter. Second, extra initiation sites could be deployed in between primary origins of replication, or third, some primary sites could be inactivated. In the latter two instances, S-phase would alter while the rate of fork movement would remain constant.

To our knowledge, only three studies have measured the kinetics of DNA replication at different temperatures (Table 2). Van't Hof *et al.* (1978) germinated and grew primary roots of *Helianthus annuus* at 21 °C before exposing them to temperatures between 10 and 35 °C. Rates of replication fork movement and replicon size were measured by DNA fibre autoradiography in the apical meristem of 3- to 4-cm-long roots of seedlings at each temperature. Although the rate of replication fork movement doubled between 10 and 35 °C, replicon size remained fairly constant (20–25 µm). Less extensive data for root meristem cells of 4-day-old seedlings of *Secale cereale* and *Oryza sativa* also indicate that changes in temperature affect fork movement but not replicon size (Francis & Bennett, 1982; Kidd, 1986; D. Francis unpublished data).

Minimum cell cycle times

Cell division in root meristems occurs over a wide range of temperatures, yet the precise limits of this range are unknown for many species. The upper limit for the cell cycle is about 35 °C in *Helianthus annuus* and *Vicia faba* (Burholt & Van't Hof, 1971; Murin, 1967), but the lower limit in these species is not known. In roots of *Secale cereale*, however, the cycle still operates near 0 °C (Grif & Valovich, 1973*a*).

Table 3. *The coefficient, Q_5, by which the cell cycle in root meristems is lengthened on either side of the optimum temperature at which the cycle has minimum duration (adapted from Grif, 1981)*

Interval above or below optimum temperature (°C)		Q_5 — Species and optimum temperature (°C): *Allium cepa* (32)	*Helianthus annuus* (32)	*Vicia faba* (28)
Q_{-5}[a]	20–25	2·4	–	–
Q_{-4}	15–20	1·7	1·9	2·4
Q_{-3}	10–15	1·5	1·8	1·6
Q_{-2}	5–10	1·3	1·4	1·5
Q_{-1}	0–5	1·1	1·1	1·3
Q_{+1}	5–10	1·2	–	1·6

[a] Notation by which the coefficient in question is designated. Plus and minus numerals indicate the number of 5-degree increments or decrements the coefficient is away from the optimum.

Nothing is known about the relationship between temperature and the cell cycle in shoot and leaf meristems. It might be anticipated that the relationship here would be the same as for roots, even though cells in shoot meristems reproduce at about two-thirds of the rate in root meristems.

The temperature at which the mitotic cycle has a minimum duration (the optimal temperature) varies in different species of angiosperms. For example, the optima are 28 °C in *Vicia faba* and 32 °C in *Allium cepa*. The cycle is prolonged at temperatures both above and below the optimum. Grif (1981) has recommended a coefficient, Q_5, as a convenient way of expressing the duration of the cell cycle at 5 °C shifts of temperature on either side of the optimum temperature (Table 3). Thus, in roots of *V. faba*, Q_5 for the first 5 °C increment above the optimum is 1·6, while for the 5 °C decrement below the optimum it is 1·3. This asymmetry of response of the cell cycle to temperature and the absence of values at more than one increment above the optimum indicates the closeness of the optimum temperature to the upper temperature limit for cell division. This seems to be a general property of the cell cycle/temperature relationship in all the species of angiosperms examined so far.

The various types of habitat which plants occupy is probably related to the different temperatures at which their cell cycles have a minimum duration. For example, in comparison with terrestrial angiosperms, two lower plants whose cell cycles have been examined have lower optimal temperatures. The cycle in the protonemal apical cell of the fern, *Dryopteris psuedo-mas*, grown in light of low intensity, is fastest at about 20 °C (at higher light intensity the optimum is 22·5 °C) (Dyer & King, 1979). Similarly, in the green alga *Oedogonium cardiacum* the cell cycle is faster at 20 °C than at either 25 or 17·5 °C (Horsley & Fucikovsky, 1962). In the case of the fern cells, as in angiosperms, a small increase in temperature above the optimum causes a greater reduction in cycle duration than a decrease of similar magnitude, perhaps indicating a correspondingly lower 'upper' temperature limit for division.

Cell and organ growth

Nuclear events of the cell cycle are coordinated with the growth of the cytoplasm and of the cell wall over a wide range of temperatures. This is shown by the constancy of meristematic cell length in roots of *Secale cereale* growing at either 1 or 23 °C (Grif & Valovich, 1973*b*) and also by the proportionality between the rates of elongation and cell division in roots of *Zea mays* (Erickson, 1959). The length that cells attain after they leave the meristem is also constant at temperatures ranging from 5 to 25 °C in *Allium cepa* (Lopez-Saez *et al.* 1969), from 15 to 30 °C in *H. annuus* (Burholt & Van't Hof, 1971) and from 15 to 35 °C in *Zea mays* (Erickson, 1959). However, beyond these upper and lower limits the mature cells are shorter. The length of the root meristem of *H. annuus* is also shorter at temperatures above 25 °C. Together, these results suggest that the tight coupling which usually exists between nuclear reproduction and cell growth can be dissociated at extreme temperatures, and that, within the meristem itself, the coordination may be less well regulated in older, more proximal regions than in younger, more distal regions.

Because of the complex interactions between the rates of division, numbers of dividing cells, rates of cell elongation and mature cell size, it does not necessarily follow that an effect of temperature on cell division will inevitably cause a corresponding change in organ growth rate. For example, although the cell cycle in root meristems of *Zea mays* is fastest at 30–35 °C (Verma, 1980), the fastest growth rate of the root system as a whole occurs at 26 °C (Walker, 1969). A similar difference between the optimal temperatures for the cell cycle and root growth has been demonstrated in a number of other species, e.g. pea (Gudkov *et al.* 1974), onion (Lopez-Saez *et al.* 1969) and broad bean (Murin, 1972). In the shoot apex of *Nicotiana tabacum* the plastochrone lengthens from 0·99 days at 15 °C to 1·15 days at 31 °C (Raper *et al.* 1975). It is unlikely that the prolongation of the plastochrone at higher temperatures is due to slower rates of division in the apical meristem, particularly since the rate of leaf growth, which like primordium initiation also depends on cell division, is increased by a rise in temperature. Similar differential effects of temperature on primordium initiation and its subsequent growth have also been reported for various parts of the flower of *Silene coeli-rosa* (Lyndon, 1979). The observations indicating a disparity between rates of division and rates of leaf and flower primordium production suggest that the spatial organization of mitoses in the apex as a whole is sensitive to temperature. We conclude, therefore, that temperature, in addition to influencing division rates in meristematic cells, also has profound effects on the way in which the cells produced by the meristem are deployed in organogenesis.

Nucleotype, cell cycle and temperature

The total amount, or mass, of nuclear DNA is the sum of genic and non-genic DNA and is expressed in *C* values where 1*C* is the amount in the unreplicated nuclear genome of a gamete. Bennett (1971) introduced the term nucleotype to describe the effects of DNA mass on the phenotype independently of the informational content of the nuclear DNA. Logically, the more DNA present in a nucleus the longer the time expected for it to replicate. Indeed, a positive nucleotypic effect of DNA *C* value on the duration of S-phase has been demonstrated in root meristems of diploid dicots (Van't Hof, 1965) and diploid monocots (Evans & Rees, 1971; Kidd *et al.* 1987). However, this relationship is altered when allopolyploids and their parental genotypes are compared. For example, despite a 1·5- to 2·6-fold increase in DNA *C* value, cells in the root meristem of hexaploid triticale complete S-phase and the cell cycle over intervals of about 5 and 12 h, respectively, which are very similar to those in the parental genotypes, diploid *Secale cereale* and allotetraploid *Triticum turgidum* (Kaltsikes, 1972; Kidd, 1986).

As well as having consequences at the cellular level, the nucleotype also has effects on the whole plant, particularly with respect to the minimum generation time (i.e. the time taken from germination to the production of the first set of seeds (Bennett, 1972)) and geographical distribution. High DNA *C* value species tend to be cultivated at temperate latitudes while species of increasingly lower DNA *C* values have been chosen at successively lower latitudes (Bennett, 1976). However, as discussed by Bennett (1987), this latitudinal cline is reversed in summer at very high latitudes.

Available data suggest that species of low DNA *C* value (<5 pg) are widespread, whereas those of high *C* value (>20 pg) are progressively excluded from higher latitudes and hence, harsher climates (see Bennett, 1987, for specific details). The cause of any links between DNA amounts and geographical distribution are unknown but they have been discussed in terms of the positive relationship between *C* value and characters such as cell size and chromosome volume (Bennett, 1973; Cavalier-Smith, 1985). These ideas will not be restated here. Instead, we will consider the so-called DNA *C* value latitude cline in terms of the possible effects of the prevailing temperature at differing latitudes on the rate of DNA replication and cell division.

Is there some property of the cell cycle that favours the establishment and growth of species with low DNA *C* values at northerly latitudes at the expense of species with high DNA *C* values? In Fig. 2, monthly mean temperatures are shown from two continental weather stations in North America, one at Churchill, Manitoba, Canada (58° 47′ N) and the other at Lawton, Oklahoma, USA (34° 37′ N) where winter climates are known to be harsh and mild, respectively (Walter & Lieth, 1960). The relevant difference between these monthly means is that temperatures which would be expected to sustain plant growth (>5 °C) exist only for about one-third of the year at 58° 47′ N but for four-fifths of the year at 34° 37′ N. Most species found in sub-arctic regions of North America are long-

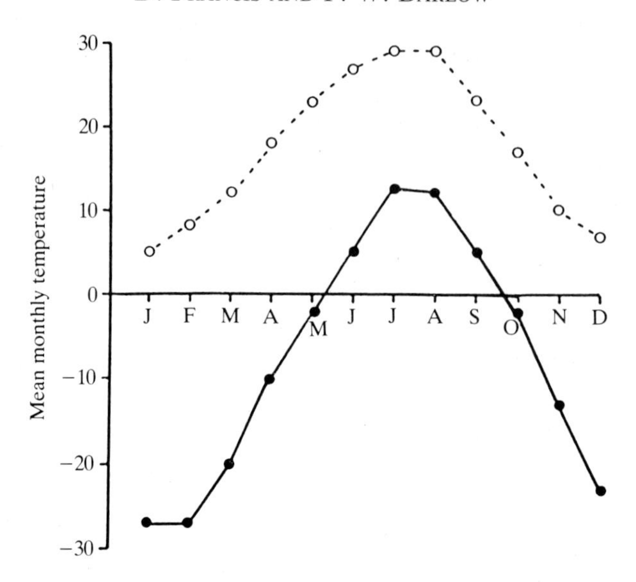

Fig. 2. Monthly mean temperatures (°C) at Churchill, Manitoba, Canada, 58°47′N (●), and Lawton, Oklahoma, USA, 34°37′N (○), adapted from Walter & Lieth (1960).

lived perennials (Chapin & Shaver, 1985; Bliss, 1985) and therefore capable of long periods of quiescence whilst overwintering. However, during the period June to September they must, presumably, be able to grow quickly. At the level of nuclear DNA replication, the kinetics of enzymes involved in DNA unwinding, replication, and stabilization are strictly temperature-regulated. The positive nucleotypic effect of *C* values on cell cycle and S-phase times (see earlier) would be consistent with Bennett's (1987) observation that at high latitudes, selection against species of high DNA *C* values increases. It may be predicted, therefore, that species growing in sub-arctic conditions would be of low DNA *C* value and comprise meristematic cells with short S-phases and rapid cell cycles so that maximum advantage is taken of the very short growing season.

These ideas must carry a number of important reservations. First, a mean temperature at a given latitude ignores any fluctuations in wind, cloud-cover, rainfall and altitude, all of which have profound effects on monthly temperatures. Second, no mention is made here of the type of strategies that may be adopted by species endemic to particular latitudes. Whether the habitat is temporary or more permanent can impose the so-called ruderal and competitive strategies for ephemerals and perennials, respectively (Grime *et al.* 1986). Clearly, the interaction between climate and strategy is complex and has been discussed with greater authority elsewhere (see, for example, Grace, 1987). Nevertheless, it would be of interest to determine whether or not species such as *Eriophorum angustifolium, Abies lasiocarpa* and *Carex bigelowii*, which are typically found in these arctic-alpine domains, have low DNA *C* values and short cell cycles. So far, work on the relationship between temperature and the cell cycle has been performed on species living in temperate environments. A comparison of the cell cycle properties of species from more extreme environments (e.g. between arctic and tropical, alpine and desert) would also be instructive. Moreover, determining ploidy status in such species would be worthwhile particularly since Stebbins (1971) noted that extreme environments do not necessarily favour polyploids compared with diploids.

Nucleolar activity and temperature

Changes in temperature that alter the duration of component phases of the cell cycle will also affect synthetic and organellar activity. Plants growing at low temperatures have large nucleoli (Morcillo *et al.* 1978) which is consistent, in some cases, with higher amounts of cellular RNA and increased activity of RNA polymerase (Siminovitch, 1963).

Recently, a detailed study was made of the effect of temperature on nucleoli in relation to seedling growth in *Triticum aestivum* cv. Chinese Spring. Seedlings were germinated and grown in darkness at 5, 10, 12, 20 or 25°C; the maximum growth rates of the roots ($mm\,h^{-1}$) were 0·6, 1·25, 2·0, 3·1 and 3·1, respectively. Over the range 10–25°C, mean nucleolar volume began to peak 2 days after the start of imbibition, whereas at 5°C a peak was not observed until after 10 days. Interestingly, the largest nucleolar volumes were recorded at 5°C, a finding consistent with increased levels of cellular RNA and RNA polymerase at low temperatures (Siminovitch, 1963). Clearly, for *T. aestivum*, the cell cycle would be expected to be longer and growth rates are lower at 5 compared with 20°C (Jordan *et al.* 1985). The increased size of nucleoli and probable associated increase in the level of rRNA at 5°C (see Fakan & Deltour, 1981; de Barsy *et al.* 1974) are presumably brought about by high levels of RNA polymerase activity. These changes must be necessary to sustain growth at a temperature, which, on a kinetic basis, would be expected to reduce all rates of reaction. In other words, although metabolic rates are much lower at 5 compared with 20°C, cells compensate by boosting RNA levels, and presumably protein, at least to sustain a minimal amount of growth. This may be why some species can grow at near-freezing temperatures (Kimball & Salisbury, 1974).

There are no data which indicate that low temperature blocks the cell cycle, although low-temperature-induced quiescent cells may be deficient in a cell cycle-related protein(s) which prevents them from entering S-phase. This notion has some support from temperature-sensitive mutants of *Chlamydomonas rheinhardi* which have been shown to be deficient in regulatory proteins following transfer to restrictive temperatures (Howell, 1974). Admittedly, most of the data point to the denaturation of cell-cycle-specific proteins and hence the imposition of a block on the cell cycle at high temperatures. However, if low temperature slows or inactivates protein synthesis this would have the same effect of taking out specific proteins necessary for the transition from G_1 to S-phase. Increased nucleolar size, and the suggestion of increased protein levels in plants grown at low temperatures would seem to be at odds with this idea. However, if we assume that perennial plants remain quiescent at temperatures below 5°C, then plants may accumulate proteins during the preceding summer and autumn months (e.g. Siminovitch, 1963). The enlarged nucleoli observed in cells of plants grown at low temperature may be a relic of this sequence of events.

Cell division, developmental transitions and temperature

Two major developmental transitions which can be regulated by temperature are stratification and vernalization. Stratification is a cold treatment of moist seeds which stimulates, or promotes, germination (see, for example, Bewley & Black, 1982). Vernalization is a cold treatment of moist seeds or young seedlings which makes them competent to respond to day lengths and temperatures which can induce flowering. Both responses have been documented at the physiological and whole plant levels (Bewley & Black, 1982; Bernier *et al.* 1981). However, data and ideas on the role, if any, of cell division in these responses are fragmentary.

The initial morphological expression of germination is the emergence of the radicle through the testa by cell elongation. The onset of cell division is very much a secondary event of germination. It is known that, in the root apices of dormant embryos, cells are arrested in either G_1 or G_2, or both, in a species-specific manner (Van't Hof, 1974), though usually cells in G_1 predominate. It could be reasoned that cells in the radicle of seedlings which germinate in response to low temperatures accumulate RNA as a prerequisite for nuclear DNA replication. Indeed, in cultured root tips of *Pisum sativum*, selective inhibition of RNA synthesis prevents G_1 cells from entering S-phase (Webster & Van't Hof, 1970). This in itself is not particularly remarkable although such a requirement for RNA may add credence to the idea that the appearance of large nucleoli in cells of plants grown at low temperature may be of functional significance in relation to the onset of DNA replication and cell division during germination, following a period of cold.

Plants and their meristems (particularly shoot meristems) may experience both rapid fluctuations of mean hourly temperature during the course of a day and slower fluctuations of mean daily or weekly temperature during the changing of the seasons. Heat shock proteins may be important in stabilizing meristematic cellular changes in the face of sudden local increases in temperature and consequently permit the continuation of the cell cycle when cooler temperatures return. On the other hand, seasonal temperature changes may be accompanied by alterations in hormonal balances within the meristems of perennial species. In shoot apices, for example, this may lead to controlled and sequential cessation of internodal cell enlargement and cell cycling during the summer and winter dormancy of buds, respectively.

In many species, a period of winter cold is essential to break winter dormancy of shoots and hence permit the resumption of cell division. However, when division does resume, it is not necessarily stimulated to equal degrees throughout the meristem. In buds of *Stachys sieboldi*, for instance, the stimulation of cell division is most intense in the apical initials and least in the axial zone which, whatever the growing conditions of the meristem, maintains a constant, albeit low, level of cell cycle activity (Tort, 1971). On the other hand, in shoot apices of *Arabidopsis thaliana*, the cell cycles of cells of the axial zone are stimulated most markedly following a period of cold (Besnard-Wibaut, 1977). This effect, however, is associated with the transition of the meristem from the vegetative to the floral state.

The latter example indicates how important interactions of temperature with cycling cells may be in initiating new morphogenetic programmes. Another spectacular example is found in the lily, *Lilium candidum*. Here the absence of a cold period (2°C) causes most of the shoot meristem of the bulb to disappear and be replaced by a new group of meristematic cells that eventually construct a new bulb (Riviere, 1978). Under normal circumstances, a cold period is the stimulus necessary for the meristem to start constructing a flower and the tissue that responds to cold temperature is the shoot apex. This was shown directly by locally cooling (5°C) just the crown of *Apium graveolans*; all other parts remained at 21°C. The plant flowered when returned to a warm house at 21°C (Curtis & Chang, 1930).

A vernalized plant is competent to respond to inductive day lengths. In general, a plant flowers one to several months after vernalization and, as a consequence, often comprises meristems which were not formed at the time of vernalization. Hence, transmission of the cold-induced stimulus received by embryos or juvenile plants could be propagated by DNA replication and cell division. Wellensiek (1964) proposed that not only is the shoot apex the receptive tissue for vernalization but also it must comprise dividing cells.

To our knowledge, there have been very few studies of the effects of vernalizing temperatures on the cell cycle in the shoot apex. Griffiths *et al.* (1985) showed that in *Triticum aestivum* cv. Chinese Spring, cell doubling times were similar in various isogenic lines, some of which required vernalization and others which did not. They concluded that vernalization must involve transient molecular changes in cells of the apex. Such transient changes which do not result in any stable changes in the genome, and are not passed on to the progeny, are consistent with an epigenetic effect (Nanney, 1958). However, in *Pisum sativum*, the progeny of vernalized plants responded to inductive treatments although they had not received a cold treatment themselves (Reid, 1979). In this case a stable change in the genome of the vernalized parents must have occurred, suggesting that competence to flower is achieved by species-specific responses to cold treatments.

Very little, then, is known about the effects of vernalization on the cell cycle in the shoot apex. Perhaps the general effect will be similar to that for root meristems (Fig. 1) and cells will accumulate in G_1 during vernalization. It may also be that during exposure to cold, the proportion of dividing cells in the meristem declines. The growth fraction has been shown to alter at the time of floral transition. For example, in the long day plant *Sinapis alba*, the growth fraction increases from

30–40 % to 50–60 % within 30 h of the start of a single inductive long day (Gonthier *et al.* 1987). This increase occurs at the same time as a shortening of the cell cycle from 86 to 32 h, and coincides with a synchronization of cell divison which is an important marker of floral competence (Bernier, 1971). It would be of interest to discern whether alterations in the growth fraction can be used as markers of changes in the shoot apex during vernalization.

These are a few of many cases illustrating the essentiality of low temperature as a developmental switch. This may help explain the reason for the optimum temperature being close to the upper temperature limit (see pp. 188–189): it could be that there are specific amounts of 'thermal time' which encode specific information that can be translated into particular patterns of development. Further, a slowly dividing, rather than rapidly dividing, population of cells may somehow be better able to coordinate the complex changes of cell polarity and the re-adjustment of positionally specified division rates within the meristem that are necessary for implementing new morphogenetic programmes.

Conclusions

One way for increasing the climatic range for agriculture is to grow crops at higher or lower latitudes than normal. If a plant is to exploit a new habitat it will either have to withstand, or grow at, temperatures beyond its usual range. The data in Table 2 on the effects of temperature on DNA replication suggest that it is kinetically regulated. Thus, if a plant is to grow at low or high temperatures one requirement could be for DNA replicative enzymes to be modified so that they work efficiently at physiologically unusual temperatures. To engineer such plant cells may not necessarily rely on insertion of 'new genes' expressing 'new' replication enzymes, but rather may involve post-translational protein modifications. Such conformational changes could lead to the required efficient catalytic activity at high or low temperatures.

Low and high temperatures have predictable effects in slowing or hastening the cell cycle (Fig. 1). Some of the published data on temperature effects on component phase durations indicate a bias towards a longer G_1-phase in response to low temperature. In other words, as the temperature is lowered G_1 tends to occupy proportionately more of the cell cycle. Given the idea that a 2*C* nuclear DNA amount affords maximal protection of the genome, the G_1 state would seem to offer shelter from stresses linked to prolonged exposure to low temperatures. Clearly, however, more data are required to substantiate this idea. Conversely, at high temperatures, G_1 is the phase that shortens maximally. The variability of G_1 as cell cycle duration alters is well known in both plants (Clowes, 1967*b*) and animals (Cooper, 1982), but whether there are G_1-specific processes which commit a cell to replicate its DNA is unknown. Indeed, even the notion of the necessity of G_1 in a cell cycle has been questioned (Cooper, 1982). Regarding the data in Fig. 1 it seems that G_1 duration is positively correlated with cell cycle duration, and when

G_1 is absent cell division and cell growth may well lack the coordination necessary to achieve normal development.

Clearly, advantages may well be conferred on a plant with the ability to remain quiescent during long periods of drought or cold, and then rapidly to exploit transient periods of elevated temperatures or rainfall. Much also would depend on whether a species adopts a ruderal or competitive strategy (see Grime *et al.* 1986). Nevertheless, a prime requirement in this case would be for rapid rates of division in order to maximize growth during short growing seasons. Grime *et al.* (1986) have argued that tolerance to harsh environments would be best served by perennials consisting of large cells since, although the cells may elongate at relatively slow rates, by virtue of their size they will achieve increases in organ or plant size more rapidly than species comprising smaller cells subjected to the same environmental conditions. We suggest that the prime requirement to achieve rapid growth upon the return of favourable temperatures is a short cell cycle, coupled with rapid rates of DNA replication. Thus, the known positive nucleotypic effect of DNA *C* value on cell cycle times would make low *C* value species best fitted to withstand extreme environments. However, since a positive nucleotypic effect on cell volume exists for at least some species (e.g. Van't Hof & Sparrow, 1963; Lawrence, 1985), our proposal would seem to be at odds with that of Grime *et al.* (1986). It is obvious that primary data obtained from the type of work suggested earlier (see p. 192) are necessary. It may be that in harsh climates a plant grows by cell elongation, followed by rapid rates of cell division, at the start and at the height of the growing season, respectively.

Finally, it is tempting to suggest that by specifically tampering with a cell cycle event, cells could be adapted to replicate and divide at lower temperatures conferring on crop species the ability to grow in new territories. The theme developed in this paper is that there are constraints on the kinetics of the cell cycle beyond which a cell will not grow and divide. Even if an economically important cereal could be engineered to grow in an extreme environment, its metabolism may become perturbed. Such an imbalance could lead to over-production of secondary products with alarming consequences on normal development. A more realistic approach is to select for factors which give a plant the ability to remain viable during extremes of temperature but which would enable the meristematic cells to divide rapidly upon the return of favourable environmental conditions.

Work providing unpublished data was supported by a grant from the Agricultural and Food Research Council (AG 72/48) to D.F.

References

Armstrong, S. W. & Francis, D. (1985). Differences in cell cycle duration of sister cells in secondary roots of *Cocos nucifera* L. *Ann. Bot.* **56**, 803–813.

Barlow, P. W. (1974). Regeneration of the cap of primary roots of *Zea mays*. *New Phytol.* **73**, 937–954.

Barlow, P. W. (1978). The concept of the stem cell in the context of plant growth and development. In *Stem Cells and Tissue Homeostasis, 2nd Symposium of the British Society for Cell Biology* (ed. B. I. Lord, C. S. Potten & R. J. Cole), pp. 87–113. Cambridge: Cambridge University Press.

BARLOW, P. W. (1987). The cellular organisation of roots and its response to the physical environment. In *Root Development and Function* (ed. P. J. Gregory, J. V. Lake & D. A. Rose), pp. 1–26. Cambridge: Cambridge University Press.

BARLOW, P. W. & ADAM, J. S. (1989). The response of the primary root meristem of *Zea mays* L. to various periods of cold. *J. exp. Bot.* (in press).

BARLOW, P. W. & RATHFELDER, E. L. (1985). Cell division and regeneration in primary root meristems of *Zea mays* recovering from cold treatment. *Exp. Env. Bot.* **25**, 303–314.

BASILICO, C. (1978). Selective production of cell cycle specific *ts* mutants. *J. cell. Physiol.* **95**, 179–205.

BAYLISS, M. W. (1972). An analysis of meiosis in *Triticum aestivum*. *Ph.D Thesis, University of Cambridge.*

BENNETT, M. D. (1971). The duration of meiosis. *Proc. R. Soc. Ser.* B **178**, 259–275.

BENNETT, M. D. (1972). Nuclear DNA content and minimum generation time. *Proc. R. Soc. Ser.* B **181**, 109–135.

BENNETT, M. D. (1973). Nuclear characters in plants. *Brookhaven Symp. Biol.* **25**, 344–366.

BENNETT, M. D. (1976). DNA amount, latitude and crop plant distribution. *Env. Exp. Bot.* **16**, 93–108.

BENNETT, M. D. (1987). Variation in genomic form in plants and its ecological implications. *New Phytol. Suppl.* **106**, 177–200.

BERNIER, G. (1971). Structural and metabolic changes in the shoot apex in transition to flowering. *Can. J. Bot.* **49**, 803–819.

BERNIER, G., KINET, J.-M. & SACHS, R. M. (1981). *The Physiology of Flowering*, vol. 2. Boca Raton, Florida: C.R.C. Press.

BESNARD-WIBAUT, C. (1977). Histoautoradiographic processes in the shoot apex of *Arabidopsis thaliana* L. Heynh, vernalised at different stages of development. *Pl. Cell Physiol.* **18**, 949–962.

BEWLEY, J. D. & BLACK, M. (1982). *Physiology and Biochemistry of Seeds in Relation to Germination*, vol. 2, p. 173. Springer-Verlag: Berlin, Heidelberg, New York.

BLISS, L. C. (1985). Alpine. In *Physiological Ecology of North American Plant Communities* (ed. B. F. Chabot & H. Mooney), pp. 41–65. New York, London: Chapman & Hall.

BRADBURY, E. M. & MATTHEWS, H. R. (1982). Chromatin structure, histone modification, and the cell cycle. In *Cell Growth* (ed. C. Nicolini), pp. 411–454. New York: Plenum.

BROWN, R. (1951). The effects of temperature on the durations of different stages of cell division in the root tip. *J. exp. Bot.* **2**, 96–110.

BURHOLT, D. R. & VAN'T HOF, J. (1971). Quantitative thermal induced changes in growth and cell population kinetics of *Helianthus* roots. *Am. J. Bot.* **58**, 386–393.

BURSTIN, S. J., MEISS, H. K. & BASILICO, C. (1974). A temperature-sensitive cell cycle mutant of the BHK cell line. *J. cell. Physiol.* **84**, 397–408.

CAIRNS, J. (1975). Mutation selection and the natural history of cancer. *Nature, Lond.* **255**, 197–200.

CAVALIER-SMITH, T. (1985). Cell volume and the evolution of the eukaryotic genome size. In *The Evolution of Genome Size* (ed. T. Cavalier-Smith), pp. 105–184. London: Wiley.

CHAPIN, F. S. & SHAVER, G. R. (1985). Arctic. In *Physiological Ecology of North American Plant Communities* (ed. B. F. Chabot & H. Mooney), pp. 16–40. London, New York: Chapman & Hall.

CLOWES, F. A. L. (1959). Reorganisation of root apices after irradiation. *Ann. Bot.* **23**, 205–210.

CLOWES, F. A. L. (1967*a*). The quiescent centre. *Phytomorphology* **17**, 132–140.

CLOWES, F. A. L. (1967*b*). Synthesis of DNA during mitosis. *J. exp. Bot.* **18**, 740–745.

CLOWES, F. A. L. (1971). The proportion of cells that divide in root meristems of *Zea mays* L. *Ann. Bot.* **35**, 249–261.

CLOWES, F. A. L. & STEWART, H. E. (1967). Recovery from dormancy in roots. *New Phytol.* **66**, 115–123.

COOPER, S. (1982). The continuum model: application to G1 arrest and G0. In *Cell Growth* (ed. C. Nicolini), pp. 315–336. New York: Plenum Press.

CURTIS, O. F. & CHANG, H. T. (1930). The relative effectiveness of the temperature of the crown with that of the rest of the plant upon the flowering of the celery plant. *Am. J. Bot. Suppl.* **17**, 1047–1048.

DALE, J. E. (1986). Plastic responses of leaves. In *Plasticity in Plants. Symp. Soc. Exp. Biol.* vol. 40 (ed. D. H. Jennings & A. J. Trewavas), pp. 287–306. Cambridge: Company of Biologists.

DARLINGTON, C. D. & LA COUR, L. F. (1940). Nucleic acid starvation of chromosomes in *Trillium*. *J. Genet.* **40**, 185–213.

DAVIDSON, D. (1966). The onset of mitosis and DNA synthesis in roots of germinating beans. *Am. J. Bot.* **53**, 491–495.

DE BARSY, T., DELTOUR, R. & BRONCHART, R. (1974). Study of nucleolar vacuolation and RNA synthesis in embryonic root cells of *Zea mays*. *J. Cell Sci.* **16**, 95–112.

DE LA TORRE, C. & CLOWES, F. A. L. (1974). Thymidine and the measurement of rates of mitosis in meristems. *New Phytol.* **73**, 919–925.

DYER, A. F. & KING, M. A. L. (1979). Cell division in fern protonema. In *The Experimental Biology of Ferns* (ed. A. F. Dyer), pp. 307–354. London: Academic Press.

ERICKSON, R. O. (1959). Integration of plant growth processes. *Am. Nat.* **43**, 225–235.

EVANS, G. H. & REES, H. (1971). Mitotic cycles in dicotyledons and monocotyledons. *Nature* **233**, 350–351.

EVANS, H. J., NEARY, G. J. & TONKINSON, S. M. (1957). The use of colchicine as an indicator of mitotic rate in broad bean root meristems. *J. Genet.* **55**, 487–502.

FAKAN, S. & DELTOUR, R. (1981). Ultrastructural visualisation of nucleolar organiser activity during early germination of *Zea mays* L. *Expl Cell Res.* **135**, 277–288.

FRANCIS, D. (1988). Control of DNA replication. In *DNA Replication in Plants* (ed. J. A. Bryant & V. L. Dunham). Boca Raton, Florida: C.R.C. Press.

FRANCIS, D. & BENNETT, M. D. (1982). Replicon size and mean rate of DNA synthesis in rye (*Secale cereale* L. c.v. Petkus Spring). *Chromosoma* **86**, 115–122.

GONTHIER, R., JACQMARD, A. & BERNIER, G. (1987). Changes in cell-cycle duration and growth fraction in the shoot meristem of *Sinapis alba* during floral transition. *Planta* **170**, 55–59.

GONZALEZ-FERNANDEZ, A., GIMENEZ-MARTIN, G. & DE LA TORRE, C. (1971). The duration of the interphase periods at different temperatures in root tip cells. *Cytobiologie* **3**, 367–371.

GONZALEZ-FERNANDEZ, A., LOPEZ-SAEZ, J. F. & GIMENEZ-MARTIN, G. (1966). Duration of the division cycle in binucleate and mononucleate cells. *Expl Cell Res.* **43**, 255–267.

GRACE, J. (1987). Climatic tolerance and the distribution of plants. *New Phytol. Suppl.* **106**, 113–130.

GRIF, V. G. (1981). The use of temperature coefficients for studying the mitotic cycle in plants. *Tsitologiya* **23**, 166–173 (In Russian).

GRIF, V. G. & VALOVICH, E. M. (1973*a*). The mitotic cycle of plant cells at the minimal temperature for mitosis. *Tsitologiya* **15**, 1510–1514 (In Russian).

GRIF, V. G. & VALOVICH, E. M. (1973*b*). Effect of positive low temperatures on cell growth and division in seed germination. *Tsitologiya* **15**, 1362–1369 (In Russian).

GRIFFITHS, F. E. W., LYNDON, R. F. & BENNETT, M. D. (1985). The effects of vernalisation on the growth of the wheat shoot apex. *Ann. Bot.* **56**, 501–511.

GRIME, J. P., CRICK, J. C. & RINCON, J. E. (1986). The ecological significance of plasticity. In *Plasticity in Plants. Symp. Soc. Exp. Biol.* vol. 40 (ed. D. H. Jennings & A. J. Trewavas), pp. 5–29. Cambridge: Company of Biologists.

GUDKOV, I. N., PETROVA, S. A. & ZEZINA, N. V. (1974). The effect of temperature on the duration of mitotic cycle stages in cells of the pea root meristem and its radioresistance. *Fiziologiya i biokhimiya kul'turnykh rastenii* **6**, 257–262 (in Russian).

HORSLEY, R. J. & FUCIKOVSKY, L. A. (1962). Further growth and radiation studies with filamentous algae. *Int. J. Rad. Biol.* **4**, 409–428.

HOWARD, A. & PELC, S. R. (1953). Synthesis of desoxyribonucleic acid in normal and irradiated cells and its relation to chromosome breakage. *Heredity Suppl.* **6**, 261–273.

HOWELL, S. H. (1974). An analysis of the cell cycle in temperature sensitive *Chlamydomonas reinhardi*. In *Cell Cycle Controls* (ed. G. M. Padilla, I. L. Cameron & A. Zimmerman), pp. 235–249. New York: Academic Press.

JORDAN, E. G., COOPER, P. J., MARTINI, G., BENNETT, M. D. & FLAVELL, R. B. (1985). The

effect of temperature and ageing on root apical meristems during seedling growth of *Triticum aestivum* L.: a specific effect on nucleoli. *Pl. Cell Env.* **8**, 325–331.
KALTSIKES, P. J. (1972). The mitotic cycle in an amphidiploid (Triticale) and its parental species. *Can. J. Genet. Cytol.* **13**, 656–662.
KIDD, A. D. (1986). Studies on DNA replication and the cell cycle in the root meristem of fifteen monocotyledonous angiosperms of heterogenous DNA *C* values. *Ph.D. Thesis, University of Wales.*
KIDD, A. D., FRANCIS, D. & BENNETT, M. D. (1987). Replicon size, mean rate of DNA replication and the duration of the cell cycle and its component phases in eight monocotyledonous species of contrasting DNA *C* values. *Ann. Bot.* **59**, 603–609.
KIMBALL, S. L. & SALISBURY, F. B. (1974). Plant development under snow. *Bot. Gaz.* **135**, 147–149.
LAWRENCE, M. E. (1985). *Senecio* L. (Asteraceae) in Australia: nuclear DNA amounts. *Aust. J. Bot.* **33**, 221–232.
LOPEZ-SAEZ, J. F., GONZALEZ-BERNALDEZ, F., GONZALEZ-FERNANDEZ, A. & GARCIA-FERRERO, G. (1969). Effect of temperature and oxygen tension on root growth, cell cycle and cell elongation. *Protoplasma* **67**, 213–221.
LYNDON, R. F. (1976). The shoot apex. In *Cell Division in Higher Plants* (ed. M. M. Yeoman), pp. 285–314. London: Academic Press.
LYNDON, R. F. (1979). Rates of growth and primordial initiation during flower development in *Silene* at different temperatures. *Ann. Bot.* **43**, 539–551.
MORCILLO, G., KRIMER, D. B. & DE LA TORRE, C. (1978). Modifications of nucleolar components by growth temperature in meristems. *Expl Cell Res.* **115**, 95–102.
MURIN, A. (1967). The effect of temperature on the mitotic cycle and its time parameters in root tips of *Vicia faba. Naturwissenschaften* **53**, 312–313.
MURIN, A. (1968). Einfluss der Temperatur auf die Mitose und den Mitosezyklus. *Acta Fac. Rec. Nat. Univ. Comen., Bot.* **14**, 83–120.
MURIN, A. (1972). The effect of temperature on cell production and cell elongation in roots of *Vicia faba. Biologia (Bratislava)* **27**, 73–75.
NANNEY, D. L. (1958). Epigenetic control systems. *Proc. natn. Acad. Sci. U.S.A.* **44**, 712–717.
ORMROD, J. C. & FRANCIS, D. (1985). Effects of light on the cell cycle during the first day of floral induction in *Silene coeli-rosa L. Protoplasma* **124**, 96–105.
POPE, M. N. (1943). The temperature factor in fertilization and growth of the barley ovule. *J. agric. Res.* **66**, 389–402.
QUASTLER, H. & SHERMAN, F. G. (1959). Cell population kinetics in the intestinal epithelium of the mouse. *Expl Cell Res.* **17**, 420–438.
RAPER, C. D., JR, THOMAS, J. F., WANN, M. & YORK, E. M. (1975). Temperatures in early post-transplant growth: influence on leaf and floral initiation in tobacco. *Crop Sci.* **15**, 732–733.
REID, J. B. (1979). Flowering in *Pisum*: effect of the parental environment. *Am. J. Bot.* **44**, 461–467.
RIVIERE, S. (1978). Influence de la temperature sur la floraison du Lis blanc. *Can. J. Bot.* **56**, 110–120.
SHAW, G. W. (1958). Adhesion loci in the differentiated heterochromatin of *Trillium* species. *Chromosoma* **9**, 292–304.
SIMINOVITCH, D. (1963). Evidence for increase in ribonucleic acid and protein synthesis in autumn for increase in protoplasm during the frost hardening of black locust bark cells. *Can. J. Bot.* **41**, 1301–1308.
SMITH, J. A. & MARTIN, L. (1973). Do cells cycle? *Proc. natn. Acad. Sci. U.S.A.* **70**, 1263–1267.
STEBBINS, G. L. (1971). *Chromosomal Evolution in Higher Plants*. London: Edward Arnold.
SUSUMU, T. & POTTEN, C. S. (1985). Recruitment of cells in the small intestine into rapid cell cycle by small doses of external or β-radiation. *Int. J. Radiat. Biol.* **48**, 361–370.
THOMAS, J. E. & DAVIDSON, D. (1983). Cell and nuclear sizes in *Vicia faba* roots: changes during germination and in response to levels of ambient water. *Ann. Bot.* **51**, 353–361.
TORT, M. (1971). Etude histoautoradiographique de la synthese d'ADN dans les bourgeons de Crosne du Japon dormant ou non dormant. *Botaniste* **54**, 351–361.
VAN'T HOF, J. (1965). Relationships between mitotic cycle duration, S-period duration and the

average rate of DNA synthesis in the root meristem cells of several plants. *Expl Cell Res.* **39**, 48–54.
Van't Hof, J. (1974). Control of the cell cycle in higher plants. In *Cell Cycle Controls* (ed. G. M. Padilla, I. L. Cameron & A. Zimmerman), pp. 77–85. New York: Academic Press.
Van't Hof, J., Bjerknes, C. A. & Clinton, J. H. (1978). Replicon properties of chromosomal DNA fibres and the duration of DNA synthesis of sunflower root tip meristematic cells at different temperatures. *Chromosoma* **66**, 161–171.
Van't Hof, J. & Sparrow, A. H. (1963). A relationship between DNA content, nuclear volume and minimal cycle time. *Proc. natn. Acad. Sci. U.S.A.* **49**, 897–902.
Verma, R. S. (1980). The duration of G1, S, G2 and mitosis at four different temperatures in *Zea mays* L. as measured with ^{3}H-thymidine. *Cytologia* **45**, 327–333.
Walker, J. M. (1969). One-degree increments in soil temperatures affect maize seedling behaviour. *Proc. Soil Sci. Soc. Amer.* **33**, 729–736.
Walter, H. & Lieth, H. (1960). *Klimadiagramm-Weltatlas*. Jena: Fischer.
Webster, P. L. & MacLeod, R. D. (1980). Characteristics of root apical meristem cell population kinetics: a review of analyses and concepts. *Env. Exp. Bot.* **20**, 335–358.
Webster, P. L. & Van't Hof, J. (1970). DNA synthesis and mitosis in meristems: requirements for RNA and protein synthesis. *Am. J. Bot.* **57**, 130–139.
Wellensiek, S. J. (1964). Dividing cells as the prerequisite for vernalisation. *Pl. Physiol.* **39**, 832–835.
Wimber, D. E. (1966). Duration of the nuclear cycle in *Tradescantia* root tips at three temperatures as measured with H^3-thymidine. *Am. J. Bot.* **53**, 21–24.

TEMPERATURE AND THE PARTITIONING AND TRANSLOCATION OF CARBON

J. F. FARRAR

School of Plant Biology, University College of North Wales, Bangor, Gwynedd LL57 2UW, UK

Summary

The partitioning of carbon within sources and sinks, and its transport between them, is considered in relation to temperature. The characteristic accumulation of non-structural carbohydrates in both sources and sinks at low temperature is due partly to growth being more sensitive than photosynthesis to reductions in temperature, and partly to the differential sensitivity that enzymes of carbohydrate metabolism show to temperature. Translocation in the phloem is reduced by low temperature, due partly to viscosity and partly, possibly, to displacement of the contents of sieve elements; cooling slowly has much less effect than cooling rapidly. Partitioning in the whole plant has two partial processes: acquisition, the rate of import into a sink region, and allocation, the proportional distribution of assimilate between two or more competing sinks. Each of these can be affected by temperature treatment of the sink, of the source, or of the transport path. Allocation between the two halves of a barley root system held at different temperatures could not be explained by effects of temperature on metabolism, sucrose uptake or viscosity of transport in phloem.

Introduction

Carbon accounts for about 40 %, and compounds of carbon for about 90 %, of plant dry weight; a plant is, to a first approximation, little more than the result of the fixation, partitioning and loss of carbon. Temperature is but one of the many exogenous and endogenous variables that affects the partitioning and loss of carbon. I will first consider partitioning (product partitioning) within the mature source leaves that provide the substrate for between-organ partitioning and then within the sinks that consume it. The properties of the transport pathway connecting source and sink will be considered before dealing with the interactions between sink, source and pathway which determine partitioning in the whole plant.

Some examples of the effects of temperature on the growth of whole plants will be given by way of introduction: it is these that this chapter must seek to explain. An increase in shoot/root ratio is commonly seen as the temperature of the whole plant rises, at least over those temperatures an increase in which increases growth rate (van Dobben, 1962; Friend, 1965; Rajan *et al.* 1971; Pearcy, 1976; Went, 1944). Jones (1971) could best explain the response of shoot/root ratio in

Phaseolus vulgaris to temperature by simply that root and shoot were in constant allometric ratio to each other, independent of temperature, and temperature was simply varying the mass of the whole plant; it is not clear how widespread this situation is. With *Atriplex lentiformis*, grown with night temperatures lower than day, a rise in mean temperature causes a rise in shoot/root ratio, and within the shoot the proportion of leaf to stem rises (Pearcy, 1976). When the temperature of whole tomato plants is lower during the night than the day, overall growth and shoot/root ratios are lower than when the same temperature is maintained day and night (Went, 1944). In soybean, increased night temperatures increases net partitioning to seeds (Seddigh & Jolliff, 1986).

When different temperatures are applied simultaneously to different parts of the plant (the normal condition in the real world) a variety of effects are seen. Kleinendorst & Veen (1983) grew cucumber plants at all combinations of three root and three shoot temperatures (12, 18, 24°C). Whilst shoot/root ratio increased at increasing shoot and root temperature, shoot temperature had a greater effect on root growth than did root temperature; the stem/leaf ratio was increased by high shoot temperature but unaffected by root temperature. Although fruit production varied in a complex manner with temperature, in general it was promoted by high shoot temperature and little affected by root temperature; depending on treatment, this was due to either fruit number or individual fruit weight being increased. These authors concluded that at the low photon fluence rates (PFR) at which their plants were grown, shoot temperature had a major effect on plant form, but at higher PFR root temperature became relatively more important. Such interactions with other environmental factors probably explain the varied responses to differential shoot and root temperatures reported by Cooper (1973). Following a change in temperature, the rates of growth of different parts of the plant may change relative to each other with time, as found in *Lolium perenne* (Clarkson *et al.* 1986). A simplified and highly controlled experimental system is necessary for more analytical information to emerge. Brouwer (Fig. 1; 1981) provides an elegant example from maize. When the temperature of the whole root system was increased from 10 to 25°C, both shoot and root mass increased with constant shoot/root ratio; from 25–40°C shoot/root ratio rose as root mass fell faster than that of the shoot. Shoot/root ratio and total plant mass could be kept constant if just half of a split root system was held at 25°C whilst the other was at any temperature between 10 and 40°C; now the distribution of mass between the two halves of the root system was highly temperature dependent. Below 25 and above 35°C, less went to the half with varying temperature, whereas from 25–35°C less went to the half at 25°C. In every case, total root growth (the sum of the halves of the split root system) was unaltered by temperature.

Temperature can clearly affect the net partitioning of assimilates in whole plants, in spite of the dependence of the effects of temperature on interactions with other environmental variables and even though some effects may be on whole plant growth giving, via allometry, an apparent effect on partitioning. These whole

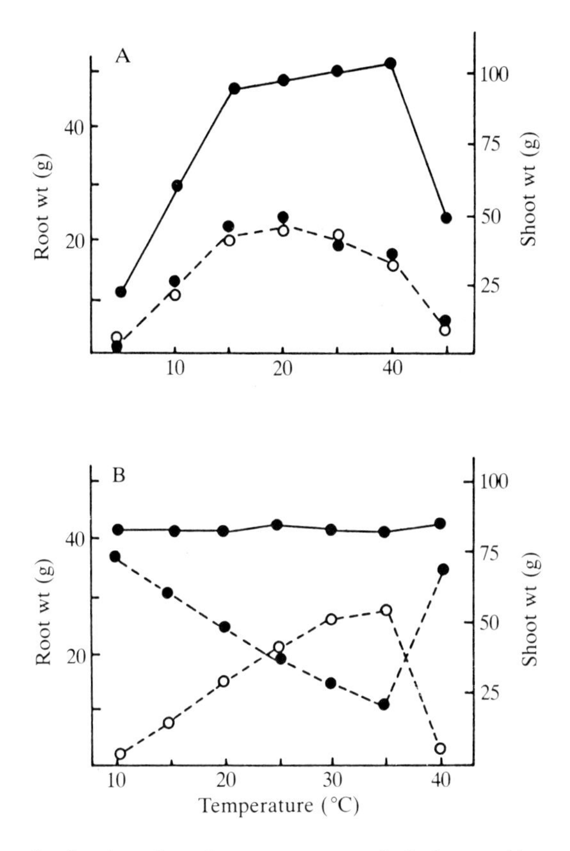

Fig. 1. The growth of maize when the temperature of whole root (A, top) or half of the split root system (B, bottom) was varied from 10 to 40°C. The temperature of the control half of the split root system was 25°C. Shoot weights, solid lines; root weights, dotted lines; symbols for roots: solid, control; open, treated. Redrawn from Brouwer (1981).

plant data are purely descriptive, and nothing in them suggests how such effects are mediated: for this, it is necessary to consider in detail those processes that contribute to the final form of the plant. To this I now turn.

Sources

It is generally held that a source leaf determines the rate of production and export of assimilate, but not the way in which it is partitioned between sinks. As we shall see, this is an oversimplification, as the rate of export from a source is at least partly under the control of the rest of the plant. A diagram of a generalized source is shown in Fig. 2. The soluble carbohydrate that accumulates is commonly either sucrose or starch, but fructans, other oligosaccharides including the raffinose series, and alditols are not infrequent. It is convenient to consider sucrose as the end-point of photosynthesis, and to consider the production and fates of sucrose as central to carbon partitioning; there is both spatial compart-

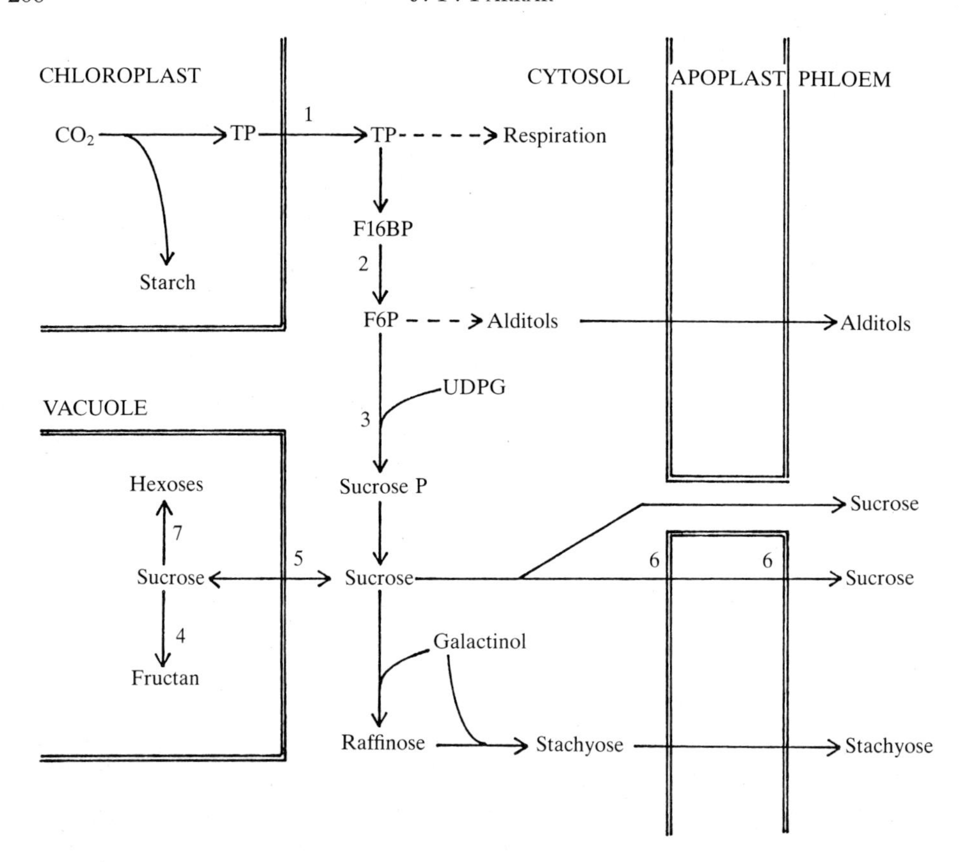

Fig. 2. Some major pathways in a generalized source. Dotted lines represent multi-reaction steps. Key: 1, phosphate translocator; 2, fructose bisphosphatase; 3, sucrose phosphate synthases; 4, sucrose–sucrose fructosyl transferase and others; 5, sucrose carrier (tonoplast); 6, sucrose carrier (plasmalemma); 7, acid invertase. The synthetic route for alditols is uncertain.

mentation, particularly between apoplast, cytosol and vacuole, and product partitioning of carbon. Rather than produce sucrose, four things could happen: synthesis of starch; synthesis of alditols, respiration, or a reduction in the rate of photosynthesis. The unravelling of the role of fructose 2,6-bisphosphate has led to a great clarification of the fine controls that operate in some species (Huber, 1986). Product partitioning between starch and sucrose (or, on a wider canvas, between starch and alditols, sucrose, or oligosaccharides produced from sucrose) is partly modulated by F2,6bP (Sicher, 1986); the effects of temperature on this system are not known. The other main control point of sucrose synthesis is also critical in partitioning between starch and sucrose; this is the activity of sucrose phosphate synthase (Sicher, 1986; Huber, 1986). Four major fates await sucrose synthesized in the cytosol: translocation (which may or may not involve movement through the apoplast), temporary storage in vacuoles in the mesophyll, or synthesis of oligosaccharides such as fructan or the raffinose series.

Temperature could have an immediate effect on such a system with no change in the amount of enzymatic machinery (fine control) by changes in, for example, the relative rates of individual enzyme-catalysed reactions. Alternatively and more slowly it could alter the amounts of certain enzymes and transport systems involved (coarse control). Direct effects on the source itself will be considered first.

Commonly, carbohydrates accumulate in the leaves of plants that are cooled (Labhart *et al.* 1983; Pollock, 1986*a*; ap Rees *et al.* 1983). Whilst it can often be detected in a single photoperiod, net accumulation may continue for several weeks after transfer to a lower temperature (Pollock *et al.* 1983). The simplest explanation for increased carbohydrate content is that growth falls more than the rate of photosynthesis. Effects on particular enzymes, mentioned below, show that more specific mechanisms also operate. Much of the sucrose and fructan will accumulate in vacuoles, but how compartmentation is affected by temperature is unknown. The accumulation of soluble carbohydrates can impose a considerable osmotic burden on a leaf with resulting problems of turgor regulation. For example, sucrose in a barley leaf at 20°C can account for 15 % of total solutes (Farrar & Farrar, 1986). Therefore, the characteristic accumulation of fructans in many species (Pollock, 1986*a*) may serve to reduce the osmotic burden that would result from disaccharide accumulation. The types of fructan that accumulate include those of intermediate chain length in leaves of *Phleum pratense*, a species capable of producing high d.p. (degree of polymerization) fructans (Smith, 1968). In barley, which usually has low d.p. fructans, those of d.p. above 3 accumulate at low temperature (Wagner *et al.* 1983). Leaves of *Festuca pratensis* produce fructans of d.p. 45–50 in the cold, whilst those of lower d.p. accumulate elsewhere in the plant (Labhart *et al.* 1983). The type of fructan accumulating is clearly not a simple function of the genetics of the species concerned.

Carbohydrates accumulate in leaves in the light (accounting for 20–60 % of photosynthesis, depending on species) and are mobilized for export during darkness. At least in two species of *Poa*, it appears that the increased carbohydrate consequent on cooling is seen as an increased basal amount of carbohydrate present throughout the diel cycle 6 d after transferring plants from 19 to 7 °C (Table 1). The accumulation of carbohydrate is lower during the day at 7°C in *P. annua*, but higher in the arctic-alpine *P.* × *jemtlandica*, than at 19°C. These data show that fructans show the largest response to a fall in temperature, a three- to fourfold rise in background concentration as compared with 1·5–2 for sucrose and 1·5 for starch (Borland & Farrar, 1985, 1987). It is striking that the rate of carbohydrate accumulation at the start of the light period in *P.* × *jemtlandica* is greater at 7 than at 19°C, in spite of lower photosynthesis, and the rates of export in the dark are better maintained in *P.* × *jemtlandica* than in *P. annua* at 7 °C (Fig. 3; Borland & Farrar, 1987).

Such changes in carbohydrate status must result from the properties of the enzymes of carbohydate metabolism. The cytosolic fructose 1,6-bisphosphatase (F1,6bPase) of spinach and pea is cold-labile *in vitro* (Weeden & Buchanan, 1983),

Table 1. *The content of carbohydrates* ($mg\,g^{-1}$) *in leaves of two species of* Poa *grown at two temperatures. Content at the end of 8-h dark period; the figures in parentheses indicate increase over the subsequent 16-h light period. The activities of acid invertase and sucrose phosphate synthase* (μmol *sucrose* $mg\,protein^{-1}\,h^{-1}$) *in leaves sampled 6 h into the photoperiod are also shown. (Data from Borland & Farrar, 1985, 1987)*

	Poa annua		*Poa × jemtlandica*	
Temperature (°C)	7	19	7	19
Sucrose, hexose	128 (63)	67 (88)	114 (78)	64 (63)
Fructan	58 (12)	14 (7)	79 (40)	20 (8)
Starch	81 (21)	45 (24)	60 (33)	40 (29)
Acid invertase	0·50	0·85	0·53	0·90
SPS	0·59	0·86	0·22	0·20

but the importance of this *in vivo* is unknown: it would be hard to reconcile with an increased accumulation of sucrose and fructan, both dependent on sucrose synthesis, but would provide a mechanism for increased accumulation of starch. A little more is known about sucrose phosphate synthase (SPS) which has a Q_{10} of 1·7 (2–10°C) in *Lolium temulentum* (Pollock, 1986*a*). This enzyme, which can rapidly change activity (Housley & Pollock, 1986), shows increases in activity when soybean leaves are warmed, and decreases when they are cooled, which last for several days (Rufty *et al.* 1985). In two species of *Poa* from the thermally contrasting habitats, a drop in temperature from 19 to 7°C causes a fall in activity of SPS in *P. annua* (warm habitat) but not in *P.* × *jemtlandica* (cold habitat); conversely acid invertase shows a fall in activity in both species (Table 1).

The fructan-synthesizing enzymes sucrose-sucrose fructosyl transferase (SST) and fructan-fructan fructosyl transferase have Q_{10}s of 1·4, 1·6, and 1·5 respectively (2–10°C); the reason that fructans accumulate in the cold may thus be partly that these enzymes continue to operate at low temperature (Pollock, 1986*b*; Wagner *et al.* 1983). It is also due to an increase in the activities of these enzymes: certainly the activity of SST increases greatly in excised, illuminated leaves of barley and *Lolium temulentum* accumulating sucrose and fructan (Wagner *et al.* 1983; Housley & Pollock, 1985). Invertase from *Lolium temulentum* has a Q_{10} of 2·0, and that from barley is similar (Wagner *et al.* 1983). Sucrose hydrolysis will thus be more temperature-sensitive than fructan synthesis, again favouring fructan synthesis at low temperature in those species capable of it.

Starch commonly accumulates at low temperatures. In the chilling-resistant species *Lolium temulentum*, however, not only is starch accumulation in the light lower at low temperatures (as it is in a number of other temperate species) but *in vitro* enzymes of sucrose synthesis were less temperature-sensitive than those of starch synthesis (Pollock & Lloyd, 1987). Barley leaves also accumulate starch less rapidly in the light at 10 than at 20°C (Sicher & Kremer, 1986). Although ADPG pyrophosphorylase is less temperature-sensitive (Q_{10} of 2) than ADPG starch

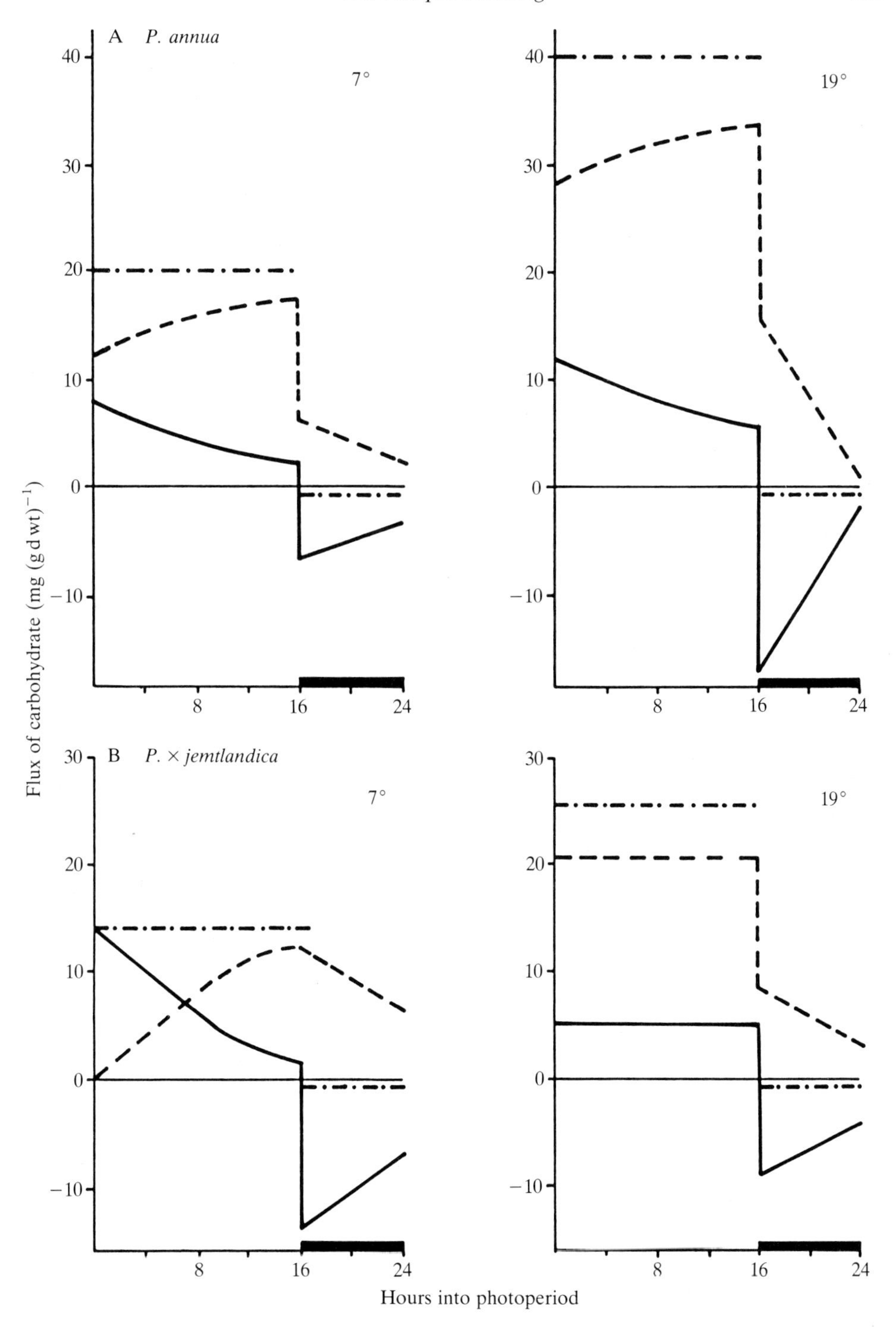

Fig. 3. Rates of CO_2 exchange (—·—), translocation (– – –), and carbohydrate accumulation (——) at two temperatures (7 °C, left; 19 °C, right) in mature leaf blades of *Poa annua* (A) and *P.* × *jemtlandica* (B). Plants were sampled over a single diel cycle (dark period indicated by solid bar on *x*-axis) 5 d after transferring plants to 7 °C from 19 °C, at which temperature they had been grown. From data given in Borland & Farrar (1985, 1987).

synthase (Q_{10} of 3·9) (from leaves of *Lolium temulentum*; Pollock, 1986*a*), the synthetic pathway is clearly temperature-sensitive. That accumulation proceeds in spite of this may be due to much slower remobilization of starch at night at low temperature (Smith & Struckmeyer, 1974; Chatterton *et al.* 1972). The C_4 species *Panicum maximum* shows a greater delay in starch remobilization below 12 °C, a threshold which could not be lowered by hardening (Lush & Evans, 1974). *Digitaria decumbens* grown at 30 °C shows great accumulation of starch in leaves following transfer to 10°C for 1 or 2 nights, and this starch was remobilized on transfer back to 30° (West, 1973). The degree of starch accumulation by *Eragrostis curvula* was greater in a variety showing poor growth at low temperatures than in one that grew well at such temperatures (West, 1973). Although the pathway of starch degradation *in vivo* is not certain, it may be amylase that is responsible for the reduced degradation of starch at low temperatures. In *Digitaria decumbens* grown at 10°C amylase activity is much lower than in plants grown at 30 °C; the more cold-tolerant species *Dactylis glomerata* shows no such effect (Kabasi *et al.* 1972).

The ultimate fate of most sucrose in leaves is to be exported in the phloem: the rate at which sucrose is made available for export is the sum of its production in photosynthesis, plus remobilization of sucrose stored in vacuoles, of starch, and of fructans, less storage. All of these processes probably occur simultaneously in illuminated leaves (Farrar & Farrar, 1985; Stitt & Steup, 1985) and all but photosynthesis and possibly starch production in darkened ones. Sucrose available for export must then be loaded into the phloem. The temperature dependence of phloem loading is little explored. The active uptake of sucrose from the apoplast, thought to be the critical step in phloem loading in most species, is not surprisingly slower at 3 than at 20 °C (Giaquinta, 1980). Sucrose transport into phloem is also greater when the sucrose content of the phloem is low, and experimentally this lowering can be achieved by washing sugar-beet leaf discs in buffer at 20°C; washing at 0°C prevents a subsequent increase in sucrose uptake (Giaquinta, 1980). In intact plants, therefore, temperature may affect phloem loading either directly, via proton-linked sucrose transport, or indirectly via the sucrose content of phloem in source leaves responding to changes in temperature elsewhere in the plant. Effects of turgor also need investigation.

The approach of Moorby & Jarman (1975) is of considerable use in describing export of assimilate from leaves in a rigorous fashion. They monitored continuously the translocatory efflux of C-14 from leaves briefly fed $^{14}CO_2$; typically, a two-phase exponential curve is found, and Moorby & Jarman (1975) showed that this could be modelled as a two-compartment system with just four parameters: the slope and percentage of efflux for each exponential phase. Each phase corresponds to discrete pools of sucrose, at least in barley (Farrar & Farrar, 1986). Using this approach, it has been shown that whole plants of *Pennisetum typhoides* show a rise in the initial rate of efflux of isotope as temperature rises, but this initial rapid loss is sustained for a shorter time the higher the temperature (Pearson & Derrick, 1977). Presumably the turnover rate of readily translocated sucrose is

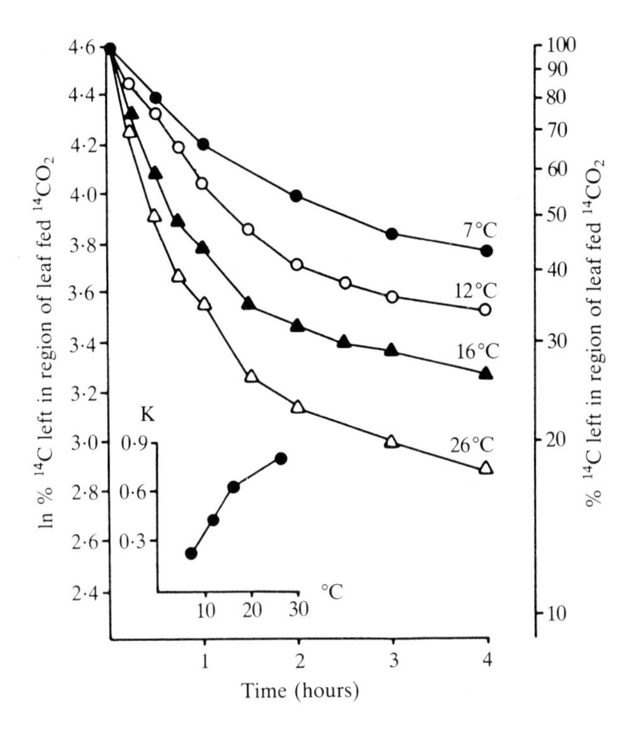

Fig. 4. Efflux of ^{14}C from leaves of maize at four temperatures. Bands of maize leaves were fed $^{14}CO_2$ for 1 min on plants held at the temperatures shown, and ^{14}C efflux was monitored by a Geiger–Muller tube attached to the leaf at the fed band. Inset: K, the initial slope of the curve, as a function of temperature (calculated from the authors' data). Redrawn from Hofstra & Nelson (1969).

higher at high temperature. Maize grown at 25°C, and then held at a range of temperatures from 7 to 26°C, shows a similar pattern, with both a larger proportion and a more rapid loss of ^{14}C from the warmer plants (Fig. 4; Hofstra & Nelson, 1969); this work suggests that turnover of vacuolar sucrose is less temperature sensitive than turnover of the readily translocated pool.

A variety of effects of temperature on parts of the plant remote from leaves have an influence on leaves themselves, and at least some of these may operate via the concentration of sucrose in the phloem. A small increase in export from a bean leaf was seen about 2h after warming an adjacent pod, but at 10°C increase in temperature of a beet root was not followed by increased export from a source leaf, even 8h later (Geiger & Fondy, 1985). Cooling roots of *Saccharum officinarum* resulted in a reduced export of ^{14}C from leaves 6 and 24h later (Nelson, 1963; Crafts & Crisp, 1971). Such divergence of response is not surprising: the complexity of carbon partitioning within leaves will ensure that no simple, universal response to manipulation of remote parts of the plant will be seen. Leaves of tomato plants on which individual fruits were heated to 30°C showed a greater translocatory loss of ^{14}C in the first, rapid, phase of efflux than controls at 20°C, but no great change in slope of either exponential phase (Moorby

& Jarman, 1975). The readily translocated pool of sucrose is probably depleted relatively more under extra demand from a sink. Work of this type, coupled with analyses of carbohydrates within leaves, is needed on a range of species subjected to a variety of treatments.

The sucrose which is readily available for translocation is distributed between the cytosol, apoplast and phloem, and the effect of temperature on each of these pools needs to be understood. Ntsika & Delrot (1986) have shown that on cooling the petiole of *Vicia faba* to block export from the leaf, phloem loading was reduced within 10 min, starch accumulated within 30 min, and apoplastic sucrose (but not hexose) rose within 1 h; neither photosynthesis nor intracellular sugars were affected. Since the change in apoplastic sugars was slower than other variables, it was considered not to be part of the events controlling a change in carbon partitioning following reduced export. It was also shown that, on rewarming the petiole, the concentration of sucrose in the apoplast fell. It will be the readily translocated sucrose that responds first to temperature treatments elsewhere in the plant to affect the rate of export from a source leaf, but as this pool is small effects will not be confined to it for long.

There is much to be said, in the face of a normal plant containing numerous sinks and sources, for working on a simplified system. Single-rooted soybean leaves are one such source–sink system, and Sawada *et al.* (1987) have shown that cooling the roots alone results in an increase in sucrose and starch (and smaller increases in hexoses) in the leaf over a 4 d-period of cooling progressively to 6°C; only fructose did not fall in concentration when the roots were returned to 20°C.

In summary, sources can perceive temperature changes either directly or, when those changes occur in remote parts of the plant, by feedback mechanisms probably similar to those that normally control carbon partitioning (such mechanisms are scarcely understood). They way in which sources perceive temperature is at least partly by the differential temperature sensitivity of key enzymes; it is not yet certain whether any more sophisticated mechanisms need be invoked, for example involving altered amounts of enzyme protein, seen on prolonged exposure to a new temperature. To simplify, temperature affects the fate of assimilate more than its rate of production.

Sinks

Sinks, defined as regions of the plant showing a net import of carbohydrate, are of two basic types: meristematic and storage. Storage tissue nearing maturity behaves differently from indeterminate meristems in several ways. Thus, the generalized sink shown in Fig. 5 would show unloading of the phloem (reviewed by Thorne, 1986) into the apoplast were it a storage tissue such as sugar beet taproot or sugar cane internode, and via plasmodesmatal connections to the symplast were it a meristem such as a maize or barley root tip. The predominant flux of carbon compounds is also different – typically to vacuole or amyloplast in

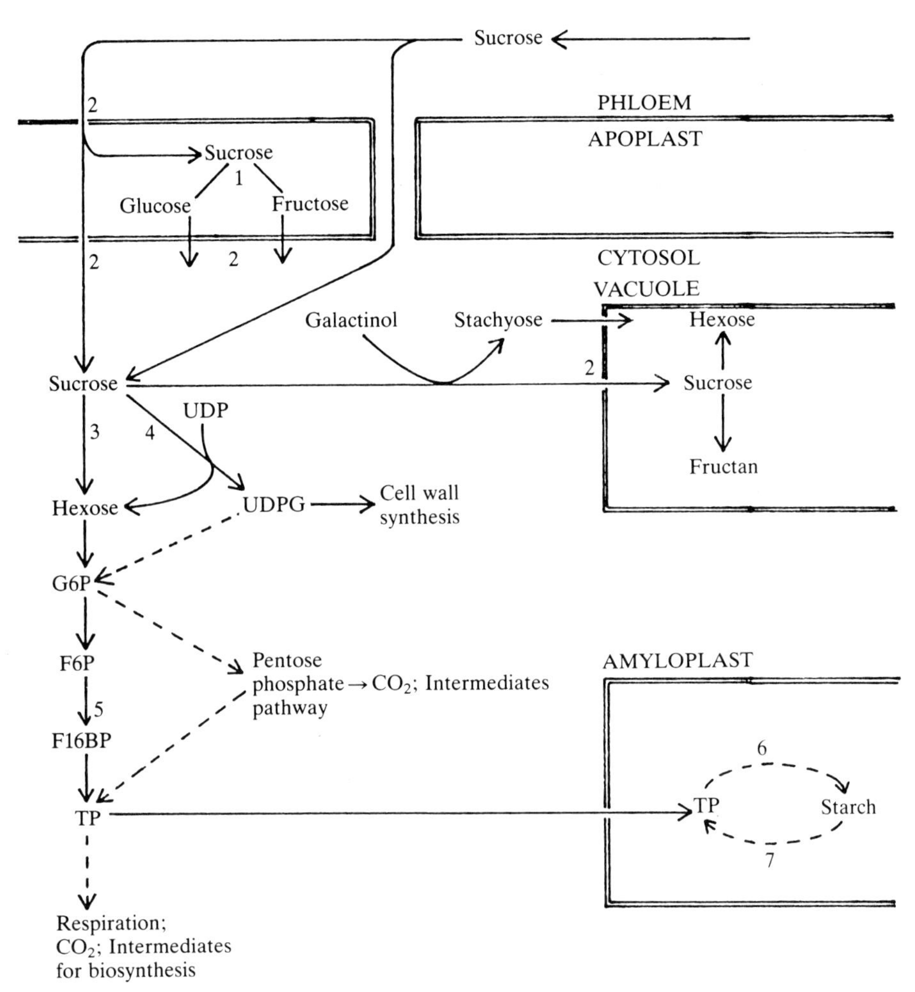

Fig. 5. Some major pathways in a generalized sink. Dotted lines represent multi-reaction steps. Key: 1, acid invertase; 2, sucrose and hexose carriers; 3, neutral invertase; 4, sucrose synthase; 5, phosphofructokinase; 6, ADPG pyrophosphorylase and starch synthetase; 7, starch phosphorylase and amylase.

storage sinks, and to respiration and structural compounds via intermediary metabolism in meristematic sinks.

Temperature can alter sink metabolism simply by speeding or slowing individual reactions; by altering rates of diffusion, particularly where diffusion is a necessary and possibly rate-limiting part of sink growth; by changing rates of active transport across membranes; and by more specific effects which include coarse control of enzyme complement. Since, as will be seen below, the temperature of a sink will affect the rate of transport into it, temperature may have both direct effects and others attendant on an altered flux of assimilate.

An increase in content of soluble carbohydrate commonly results from the cooling of sinks: the potato tuber is a classic example (ap Rees *et al.* 1983). Such an increase could be due to continued import into a sink with reduced metabolism, or alternatively to mobilization of polymeric reserves there. Whilst roots of *Festuca pratensis* show an increase in carbohydrate (Labhart *et al.* 1983) root tips of pea and maize show a complex behaviour following transfer of the plants bearing them from 20°C to a range of temperatures down to 2°C (Crawford & Huxter, 1977): here the only consistent finding was a transitory fall in sugar content immediately after cooling.

Not surprisingly, different carbohydrates show distinct responses to temperature changes. The ratio of glucose to sucrose falls in pea and maize root tips following cooling below 20°C (Crawford & Huxter, 1977). Roots of *Festuca pratensis* show a rise in fructan with d.p. ≃ 30, although less marked than in other tissues on the plant, when grown at low temperatures (Labhart *et al.* 1983). Whilst leaf sheaths of *Phleum pratensis* accumulate long chain fructans when chilled, roots accumulate those of intermediate d.p. (Smith, 1968). Starch concentrations in roots and nodules of soybean are not affected by a 10°C-temperature change (Schweitzer & Harper, 1980). Just 10 days of raised temperature (33/25°C day/night compared with 21/16°C in controls) was sufficient to reduce final weight and starch content of wheat grains (Bhullar & Jenner, 1985). The importance of distinguishing between types of sink is illustrated by the way in which sugar contents of cooled roots of maize correlate with rates of respiration and growth. Cooled roots with low sugar contents can be induced to grow and respire faster by exogenous glucose (Crawford & Huxter, 1977). No such relationship would exist in a storage sink, and it seems that in maize roots the ability to utilize sugar was less affected by temperature than the ability to supply it.

The changes in content of sugars are a reflection of metabolism within the sink. Little is known of diel changes, or of turnover of major carbohydrates at various temperatures; more is known of the dynamics of carbon partitioning in sinks (particularly in tomato fruits and potato tubers). Fruits cooled from 25°C (control) to 5°C on intact tomato plants had reduced rates of starch accumulation, whilst their sucrose concentration rose and hexose concentration fell (Walker & Ho, 1977). Heating fruits to 35°C resulted in falling concentrations of sucrose and starch, but a rise in hexoses. Since these changes were coupled to reduced import of carbon at low temperature (and even net export in some fruits) it is not possible to tell if they are due to the reduced flux of carbon or to temperature *per se*. Walker & Ho (1977) considered that import was inversely related to mean sucrose concentration within the fruit, and so the rate of utilization of sucrose may be the critical temperature-dependent step. This was tested by injecting ^{14}C-sucrose into young tomato fruits held at 5 or 25°C on intact plants at 25°C. Forty-eight hours after injection, only 2% of the label was in sucrose in the fruits at 25°C, but 32% at 5°C. When labelled glucose, fructose, malate or citrate were injected, less marked temperature effects were obtained, suggesting that an early step in sucrose metabolism might be particularly temperature-sensitive, although other processes in intermediary metabolism are also affected (Walker *et al.* 1978). In both mature and young tomato fruits, only at 5°C was there appreciable production of sucrose from injected hexoses (Walker *et al.* 1978). In short, both the supply and the metabolism of sucrose respond to temperature in tomato fruits.

A classic example of sugar accumulation is low-temperature sweetening of potato tubers. When potato plants grown at 22/18 °C day/night had a proportion of their tubers cooled to 8°C, these tubers showed reduced rates of growth. No differences in the sucrose, starch, or reducing-sugar content of these young growing tubers was found after 5 d cooling, but 20 h after supplying $^{14}CO_2$ to the shoot the ratio of ^{14}C in starch/(ethanol soluble compounds) was far greater in control than cooled tubers, implying that a change in carbon partitioning towards sweetening had begun (Engels & Marschner, 1986*b*). At least some of this change is due to altered activities of enzymes. The *in vitro* activities of ADPG-pyrophosphorylase, starch synthase and starch phosphorylase from potato tubers, measured at 30 °C, showed no effect of cooling tubers to 8 °C for 5 d from control temperatures of 22 °C. However, the Q_{10}s of these three enzymes were markedly different *in vitro*: ADPG-pyrophosphorylase was almost inactive below 12 °C, and as temperature increased above 12 °C it showed far more response to temperature than the other two enzymes. Starch synthase and phosphorylase both showed Q_{10}s less than 2 (5–20 °C). These differences are likely to be of significance *in vivo* as there was a positive relationship between import of ^{14}C into a tuber and its *in vitro* activity of both ADPG-pyrophosphorylase, and to a lesser extent starch synthase (Engels & Marschner, 1986*b*).

A further and detailed account of how cooling alters partitioning in potato tubers is the work of ap Rees *et al.* (1983). They show, using assays of enzyme activity and ^{14}C-glucose feeding, that the rises in sucrose and hexoses that occur when tubers are stored at 2 °C is mainly due to the relatively greater sensitivity to low temperature of some of the enzymes of respiratory metabolism compared with those of sucrose synthesis. Starch, rather than recently imported sucrose, must be the source of the sugars that accumulate on sweetening, since detached tubers show sugar accumulation. Hexose phosphate is produced from starch by slow degradation at low temperature (and this process may be different from that in source leaves showing retarded degradation of starch in the cold). The hexose phosphate would be respired at normal temperatures, but is instead metabolized to sucrose, some of which is hydrolysed to hexoses. A main cause of this behaviour is the dissociation at low temperature of the oligomeric phosphofructokinase (PFK). However, if the product of starch degradation in the amyloplast is triose phosphate, the PFK may be less important (Hawker, 1985). A secondary cause is the increased activity of acid invertase following storage at low temperature. Most of the enzymes assayed, however, showed no change in activity with time, at chilling temperatures (ap Rees *et al.* 1983).

A quite different type of response has been reported from root tips of pea where the K_m of acid invertase from plants treated at low temperatures (2 °C) was, at $1{\cdot}6\,mol\,m^{-3}$, much lower than that of $5{\cdot}3\,mol\,m^{-3}$ for the enzyme from plants at 20 °C; no such effect was found for maize root tips (Crawford & Huxter, 1977). This could represent the increased relative abundance of a different isozyme of acid invertase, as temperature treatment lasted for 120 h before assay and this enzyme is turned over rapidly.

The development of, and starch deposition in, grains of cereals is temperature-sensitive. Below optimal temperatures, ^{14}C allocation to grains is low compared with that to the rest of the plant (Wardlow, 1970) and starch deposition is substantially reduced whilst the sucrose concentration is higher (Jenner, 1982). Even though the Q_{10} for respiration of wheat kernels is only 1·5–2, the allocation of ^{14}C to the grain from the flag leaf is only dependent on the temperature of the ear (Chowdhury & Wardlaw, 1978).

Less work has concerned the effects of elevated temperatures. Kraus & Marschner (1984) have shown that potato tubers exposed to 30°C for 6 d cease growing; non-treated tubers on the same plant continue to grow. At this temperature, ^{14}C is still readily incorporated into soluble sugars within the tuber, but much less enters starch and, significantly, the activity of ADPG pyrophosphorylase is halved after 6 d at 30°C.

Transport of sugars across membranes is a crucial part of sink metabolism, especially in the case of apoplastic unloading or of storage in vacuoles. Saftner *et al.* (1983) show that the rate of active uptake of sucrose by sugar beet discs is temperature dependent, with a Q_{10} of 3; this uptake was thought to be largely but not exclusively into into vacuoles. Barley roots show temperature-dependent uptake of sucrose over the range 5–30 °C, with a Q_{10} of 1·6, regardless of whether the uptake is from 5 or 50 mol m^{-3} sucrose (that is, on the saturable or non-saturable part respectively of the biphasic uptake curve linking sucrose concentration to rate of uptake). Uptake at 5 °C, 20 °C and 30 °C is equally sensitive to the protonophore FCCP (p-trifluoromethoxy(carbonylcyanide)phenylhydrazone) and the sugar analogue phloridzin (Fig. 9; J. F. Farrar and C. L. Jones, unpublished). Thus the rate, but possibly not the mechanism, of sucrose uptake is altered by temperature in this cold-tolerant species. A full description is needed of how temperature affects the proton-motive force which drives active transport. Temperature-dependence of loss of carbon from roots of wheat, barley and maize (Whipps, 1984; Kolek *et al.* 1981) may be partly due to altered membrane permeabilities to the carbon compounds concerned, but more precise evidence is needed. Wolswinkel (1978) has shown that stem segments of *Vicia faba* lose ^{14}C more readily in a washout experiment at 0°C than at 25 °C, especially when parasitized by *Cuscuta*. Barley roots show no change in the ease with which ^{14}C sugars are washed out of them over the range 5–35°C (J. F. Farrar and C. L. Jones, unpublished).

Within sinks compounds must move from vascular tissue to site of utilization by diffusion. Since diffusion will be temperature-sensitive with a Q_{10} of $\simeq$2 (Johnson & Thornley, 1985), there is a little explored temperature-sensitive component of the transport of assimilate into a sink. A model of, and experiments on, transport within the spikelet of wheat have indicated the possible importance of such a system (Thornley *et al.* 1981; Jenner, 1985).

The rates of synthetic processes and of respiration in sinks will themselves be temperature-dependent. It is necessary to make allowance for respiration when calculating gross import into a sink. This may, in some cases, be relatively straightforward. Szaniawski (1981) has shown that for sunflower roots at 10°C, 20°C and 30°C, the proportion of photosynthetically produced assimilate respired by the roots is independent of temperature at 14 %. Although it might be expected that maintenance respiration would be greater at higher temperatures, as protein turnover would be faster, respiration associated with growth and perhaps that associated with net uptake of ions should remain at a constant proportion of the assimilate imported into the sink (Farrar, 1985): this is what the data of Szaniawski (1981) suggest. There is some evidence that, at low temperature, a higher proportion of mitochondrial electron transport is via the alternative oxidase (Lambers, 1985). As this pathway is non-phosphorylating, an increased proportion of the the carbon entering a sink may need to be respired at low temperature to give the same yield of ATP. Unfortunately there seem to be no sets of data where the carbon balance of a sink is described at different temperatures. A far larger proportion of ^{14}C from labelled sucrose injected into tomato fruits was used for respiration at the control than at lowered temperatures (Walker *et al.* 1978). When maize seedlings were cooled to 2°C, sugar content, respiration rate, and growth rate of the root tips all fell, implying that the cause of temperature-reduced growth was inadequate supply of carbohydrate to the root tip; growth and respiration could be restored by exogenous application of sucrose (Crawford & Huxter, 1977).

The sensing of temperature by sinks seems singularly imprecise. It is likely that the potential of sinks to grow may be temperature-dependent in a developmental sense. For example, high temperature over the period of meiosis during wheat grain development reduces grains set (Zeng *et al.* 1985). Beyond this, the response of enzymes, membrane transport systems and limitations of diffusion outlined earlier may represent the commonest way in which sinks respond to temperature. Effects of temperature treatments to remote parts of the plant will probably be mainly reflected in altered supply of assimilate to sinks.

The transport pathway

There are many experimental observations of the effect of temperature – especially low temperature – on translocation. These experiments will first be interpreted here as if Münch pressure flow operates, this currently being the most widely accepted hypothesis (Grusak & Lucas, 1984; Wright & Fisher, 1980). On this hypothesis, temperature treatments applied at any point on the phloem pathway can affect either the physical state of the contents of the lumen of the phloem, or the integrity and activity of the plasmalemma bounding the phloem either directly or via the supply of energy to it. The state of this bounding membrane will determine the rate of leakage from the phloem and the reloading of it. It is important to remember that loose phrases such as 'the rate of translocation' conceal three fundamental characteristics of movement in the phloem: volume flux of solution, mass transfer of sucrose and velocity of an average molecule of

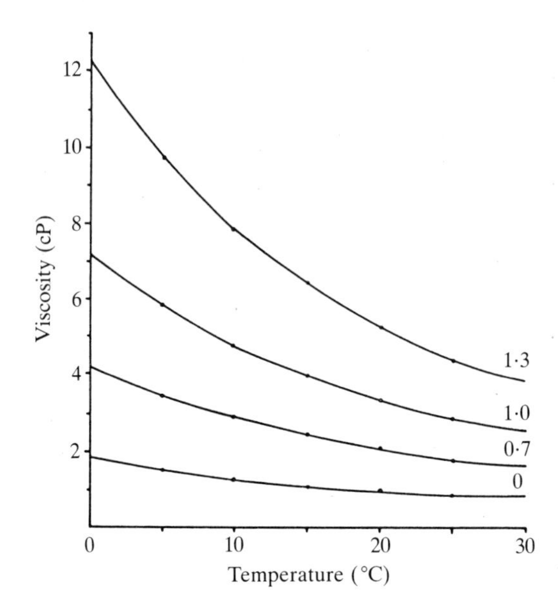

Fig. 6. Effect of temperature on the viscosity (η, absolute or dynamic viscosity, centipoise) of sucrose solutions of various concentrations (in mol dm^{-3}). Drawn from data in Bates (1942) and Dawson *et al.* (1986).

sucrose. It is also pertinent that the process of loading within source leaves may have a large effect on transport monitored remotely from it.

According to the Münch pressure flow hypothesis, a difference in hydrostatic pressure is the driving force for translocation, as defined by the Poiseuille equation: $J = \Delta P . r^2/(8\eta l)$ (where J is volume flux through unit cross sectional area of tubes of radius r and length l, ΔP is the pressure difference between the ends of the system and η is dynamic viscosity). Crafts & Crisp (1971) consider that creeping rather than laminar flow may occur through sieve pores which are wider than they are long; this is defined by $J = \Delta P . r/(3\pi . \eta)$. In either case, flux is inversely proportional to viscosity, and viscosity is highly dependent upon both concentration of sucrose and temperature (Fig. 6). Other things being equal, a fall in temperature should lead to an immediate fall in the rate of translocation. This has been shown by numerous workers applying cooling treatments to as little as a 2 cm-length of petiole or stem (Chamberlain & Spanner, 1978; Geiger & Sovonick, 1975; Giaquinta & Geiger, 1973; Lang, 1978; Minchin *et al.* 1983; Webb, 1971). This fall has been linked quantitatively to changes in viscosity of phloem sap in yucca (Tammes *et al.* 1969) and in *Nymphoides geminata* (Lang, 1978). Giaquinta & Geiger (1973) collected sieve-tube exudate from castor bean and showed that its fluidity ($1/\eta$) had a Q_{10} of 1·32 over the temperature range 0–30 °C, compared with 1·34 for 10 % sucrose; unfortunately, they did not measure the sucrose content of the sap, the viscosity of which may have been modified by its high protein content. They then measured the dependence of translocation velocity on temperature in *Beta vulgaris* and *Phaseolus vulgaris*. Both species showed, above 1 °C and 10 °C respectively, a Q_{10} of 1·1–1·35, arguing strongly for a purely physical effect of temperature on translocation via viscosity – especially since respiration rate in petioles of the bean has a Q_{10} of 3·85.

The initial, large drop in translocation rate caused by local application of low temperature would be expected to be transitory since ΔP across the treated zone will increase due to a pressure build up on the source side. As P increases so translocation rate should rise again whilst cooling is maintained and this effect should be less marked the longer the cooled zone (as $\Delta P/l$ will be smaller). Whether this recovery is seen in practice is species dependent, but chilling resistant species such as *Beta vulgaris* and *Lolium temulentum* do show such recovery (Geiger & Sovonick, 1975; Lang & Minchin, 1987). The net result is that, at equilibrium, translocation rate is relatively insensitive to temperature (Minchin & Thorpe, 1983).

Changing viscosity alone seems to explain the initial effects of low temperature (but above a threshold) in both chilling-sensitive and resistant species. This and the recovery of translocation are quite compatible with the Münch pressure flow hypothesis being a sufficient explanation of transport in phloem. Unfortunately, experiments using temperature as a modifier of phloem transport have mainly been designed, and their results used, to argue for or against a particular mechanism of phloem transport. This promises more polemic than progress, and regrettably this promise is fulfilled in the literature to which I now turn.

To appreciate the complexity of this literature, one only need consider that some plants are chill sensitive, some not; some have P-protein in their phloem, some do not; some have phloem that is very leaky to sugars, others are less leaky. Further, some experimenters cool small sections of plants, others much more; some cool to near zero, others only to a smaller degree, and often without knowing the curve relating temperature to translocation for that species. (Excellent examples of this relationship are given by Webb & Gorham (1965), Giaquinta & Geiger (1973), and Marowitch *et al.* (1986); even these were the product of widely differing experimental protocols.) Some studies have measured the rate of cooling achieved, most have not; most measured flux of isotope, few flux of sugar or solution. Most important, of course, is the possibility that Münch pressure flow is not the sole or dominant mechanism of transport in phloem: arguments for this view have been well summarized by Fensom (1981).

The first finding seemingly inconsistent with pressure flow is that in some cases cooling has no or very little effect on translocation. In many monocots such as *Lolium temulentum* (Wardlaw, 1972), *Zea mays* and *Triticum aestivum* (Faucher *et al.* 1982; Wardlaw, 1974) but not in dicots (Lang & Minchin, 1987) temperatures near zero have very little effect. The phloem of at least *Z. mays* and *T. aestivum* lacks P-protein and it is possible that this is correlated with the lack of effect of low temperature (Chamberlain & Spanner, 1978; Faucher *et al.* 1982). Reconciliation of this finding with any mechanism of phloem transport is difficult, especially as cooling in these experiments was probably too rapid to allow a build-up of pressure gradients over the cooled region, sufficient to overcome the effects of viscosity.

Translocation in stems of *Salix viminalis* is also resistant to cooling to 0°C (Watson, 1975) but as cooling took at least 20 min (Weatherley & Watson, 1969) and the stems have a large thermal capacity, this may be an example of slow cooling, which is discussed later. The most striking feature of the findings of Watson (1975) is that translocation was hardly affected by cooling as much as a 65 cm length of stem. The question here must be how the inevitable effects of viscosity were overcome; although increasing the concentration of phloem sap will uncouple fluxes of sucrose and solution, the viscosity of more concentrated solutions shows even greater temperature dependence (Fig. 6). One possibile explanation, which avoids the maintenance of flow through the same conduits, is raised later.

The second finding that is inconsistent with pressure flow is that in some species cooling has too much effect to be explained solely by viscosity. Examples of this include both chilling resistant (*Beta vulgaris*) and chilling sensitive (*Sorghum*; *Phaseolus vulgaris*) species (Giaquinta & Geiger, 1973; Wardlaw & Bagnall, 1981). There seems to be a threshold temperature, higher in sensitive than in resistant species, below which translocation is greatly reduced. Physical blockage of sieve pores by displaced sieve tube contents, or by deposition of callose, might contribute to this effect (Giaquinta & Geiger, 1973; Majumdar & Leopold, 1967; Wardlaw & Bagnall, 1981), but there is no full explanation of how it might work, and the mechanism by which a physical blockage may be cleared during recovery at low temperature is unexplained (Fensom, 1981; Minchin & Thorpe, 1983). Small reductions of 2–5°C in temperature (Pickard *et al.* 1978; Minchin & Thorpe, 1983) also reduce translocation. These reductions may not be sufficiently drastic to displace tube contents, but could have significant effects on viscosity (Lang, 1978). Respiration rate seems to drop less with temperature than does translocation (Coulson *et al.* 1972), and so reduced energy supply for an active mechanism of phloem transport need not be invoked as a cause of reduced translocation.

It is now clear that the rate at which stems are cooled can affect dramatically the effects of cooling. In stems of *Ipomoea purpurea* cooled by 10°C transient inhibition of translocation occurred if the $t_{\frac{1}{2}}$ of cooling was 3 s (fast cooling, cold shock), but not 40 s. Cooling by as little as 2·5°C, applied as a cold shock, gave interruption of flow (Minchin & Thorpe, 1983), whereas an equally rapid increase of 5°C had no effect (Pickard *et al.* 1978). Rapid cooling of sugar beet from 25°C to 1°C ($t_{\frac{1}{2}} \approx 100$ s) resulted in a halving of net translocation, whereas slow cooling (in steps of 4°C every 16 min) had no effect (Grusak & Lucas, 1985). Although the time course was longer than used on *Ipomoea*, the conclusion is that the rate of cooling matters. Grusak & Lucas (1985) showed that rewarming the beet petiole and then cooling it again resulted in a 10 % fall in translocation. They localized this response to the upstream part of the cooled region, whereas the response to rapid cooling was distributed throughout the cooled region.

Recovery of translocation during cooling does not always accord with predictions from Münch pressure flow. Firstly, chill-sensitive species such as *Phaseolus vulgaris* do not show rapid (2 h) recovery at low temperature (Geiger & Sovonick,

1975). This is probably due to rather general effects of cooling on membranes and metabolism in chill-sensitive species. Secondly, predictions of a model of recovery from cooling, based on assumptions of Münch pressure flow, were not confirmed by experiment (Grusak & Lucas, 1984). Neither length nor position of a cold block on sugar beet petioles, nor rate of unloading at the sink, affected recovery. Grusak & Lucas (1984) suggest that that recovery from cooling involves an increase in the number of sieve tubes that are translocating through the cooled region, by transfer of assimilate through anastomoses between sieve tubes and vascular bundles under the increased pressure gradient. Recovery of translocation also occurs when the cooled region is rewarmed: even chill-sensitive species can recover on rewarming (Geiger & Sovonic, 1975; Webb, 1971) whilst chill-resistant plants can show translocation more rapid than before treatment (Grusak & Lucas, 1984). The latter finding could also be explained by the cooperation of an increased number of sieve tubes. The graminaceous species that retain high rates of translocation on cooling have numerous transverse veins connecting the main veins, and also intermediate veins that are not normally concerned with net translocation (Altus & Canny, 1982); it would be worth seeking an anatomical explanation for their cooling-resistance.

Leakage (or, more precisely, apoplastic unloading along the phloem path), presumably usually followed by reloading, is normal in phloem. In stems of *Phaseolus vulgaris* loss of ^{11}C to a solution bathing the apoplast of the stem was reduced by transfer from 30 to 15 °C (Minchin & Thorpe, 1984; Fig. 7). A similar investigation at chilling temperatures would help decide if membrane leakage of chill-sensitive species contributes to reduced translocation. Leakage will also change turgor by altering the balance of solutes across the phloem bounding membrane, and indeed measurements of the water relations of phloem, at a range of temperatures, are badly needed.

This rapid survey of temperature effects on translocation suggests very strongly that too much has been deduced from tracer profiles, kinetic or temporal. The challenge must now be an integrated exploration of anatomy (which bundles and sieve tubes operate? does this change with treatment?), water relations, the concentrations of sugars in phloem sap and the apoplast, and exchange between them, and tracer work. Until this has been done, the basic effects of temperature on translocation will not be understood, and a major contribution to the question of the mechanism of phloem transport is unlikely.

Whole plants treated with low temperatures would be expected to show effects due to the phloem pathway and to sources and sinks all acting in concert. Whole-plant experiments have shown an optimum for translocation at 20–30 °C (Geiger & Sovonick, 1975). Using the chilling-sensitive plants *Phaseolus vulgaris* and soybean, Marowitch *et al.* (1986) have reinvestigated translocation when the temperature of whole plants is varied. Since photosynthesis also varies with temperature, it was convenient to express data as the ratio, translocation/photosynthesis. As temperatures fell from 35 to 4 °C plants translocated a smaller proportion of the carbon fixed in photosynthesis, good evidence to support the

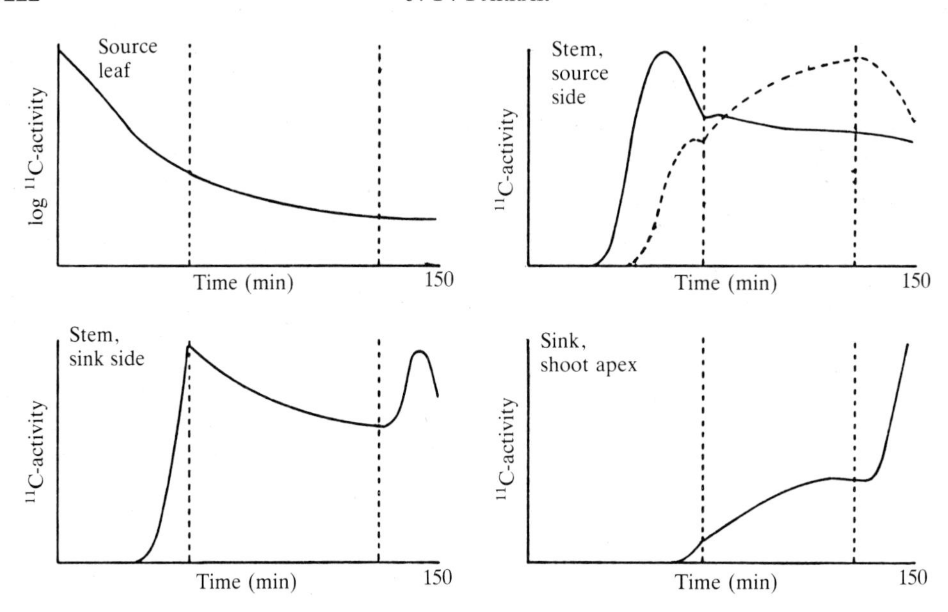

Fig. 7. Temporal profiles of ^{11}C (corrected for decay) in various parts of *Phaseolus vulgaris* following feeding of $^{11}CO_2$ to a source leaf, and cooling a 5-cm portion of the stem to 6°C over the period shown. The stem was steam-girdled below the single source leaf, and the sink was the shoot apex beyond the cold zone. Values on the *y*-axes cannot be compared between graphs. Two Geiger–Muller tubes were attached to the source side of the stem: the dashed line is for that closer to the cold zone. Redrawn from Minchin *et al.* (1983).

classical generalization that temperature affects translocation more than photosynthesis. For translocation, uncorrected for changing photosynthesis, in the two soybean varieties used, a Q_{10} of about 2·1 was found; in *P. vulgaris* Q_{10} rose steadily from 1·16 at 35 °C to 4·47 at 10 °C. This measure of translocation, including as it does loading and unloading of the phloem and activity of source and sink (see, for example, Fig. 4), is likely to have rather complex causes underlying the response to temperature: but it provides a useful link between transport itself and the problems of partitioning in the whole plant.

Partitioning in the whole plant

This section concerns the effects of general or localized temperature treatments on partitioning between organs of the whole plant. It is generally held that control of partitioning between sinks resides in competition between the sinks themselves, but just how is unknown. To make progress, it is first necessary to identify the various types of partitioning that occur. The most basic classification is between (a) partitioning between a single source and a single sink, where direction is not in question but flux is, and (b) partitioning of carbon from a single source between two (or more) sinks, where the relative flux to the sinks is at issue. All partitioning between organs can be reduced to a combination of these, the first of which I will term *acquisition* and the second *allocation*, keeping partitioning as a generic term. It may further prove useful to classify allocation according to the pathway of unloading in the sinks, since these pathways are presumably subject to different types of control. This will yield three types of allocation: between sinks unloading apoplastically (e.g., grains on a cereal ear), between sinks unloading symplastically (e.g., apical meristems of root and shoot), and between sinks unloading apoplastically and those unloading symplastically (tap and fibrous roots). Much of the diffiulty in approaching partitioning in a rigorous manner may have been due to insufficient appreciation of the diversity of its partial processes.

Acquisition: partitioning between a source and a sink

The independence of acquisition from allocation is elegantly demonstrated by the data of Brouwer (1981) for maize (Fig. 1) where total net import into the root system is unaltered, both absolutely and relative to the shoot, by temperature treatment of one half of a split root system. However, allocation between the halves of the root system is highly temperature-dependent. Acquisition of carbon by the root system was changed greatly by subjecting the whole of it to different temperatures. Activity of the sink, and of the source, and the capacity of the transport pathway joining them, may each alter acquisition and short- and long-term applications of temperature treatment may have different effects.

Moorby *et al.* (1974) cooled tomato fruits from 26 °C (the temperature of the rest of the plant) to 10 °C. The velocity of phloem transport through the petiole of the subtending leaf fell from 1·02 to 0·86 cm min^{-1}. A return to control rates followed rewarming and removal of the fruit caused a fall to 0·67 cm min^{-1}. Thus, that part of translocation concerned with the manipulated fruit showed substantial response to temperature. Mass transfer was not explored. Similarly Moorby & Jarman (1975) warmed tomato fruits to 30 °C and caused a large increase in export from a leaf which was confined entirely to the initial (1·5 h) phase of export. This rate was both more rapid, and a higher proportion of total export, than in controls. Walker & Ho (1977) showed that import into tomato fruit was enhanced by warming to 35 °C (from 25 °C) and reduced by cooling to 5 °C. Cooling of large fruits to 5 °C caused a net export of carbon. Import was positively correlated with the accumulation of starch and insoluble residue: these fractions account for nearly 30 % of carbon imported, independent of temperature treatment. Although there was greater hydrolysis of sucrose at 35 °C, there was no suggestion that the rate of production of storage and structural material within the sink was controlling import (see section above, Sinks). Since the localization of sucrose – phloem, apoplast or vacuole – is unknown, it is premature to support the suggestion of Walker & Ho (1977) that concentration gradients of sucrose between source and sink are controlling the rate of acquisition.

Further evidence that acquisition by a sink is largely determined by the sink itself comes from work with pods of pea (Thorne, 1982). The import of ^{14}C into cotyledons in the developing seeds in the pod shows strong temperature and oxygen dependence. Whilst sucrose uptake by cotyledons at a range of concentrations shows little dependence on O_2, phloem unloading does, suggesting that apoplastic unloading may be one critical step in mediating the effect of temperature in at least some sinks. When both temperature and O_2 are favourable for pod growth, cotyledonary metabolism and sucrose uptake is rapid: so is import of ^{14}C. Before it can be concluded that metabolic demand within the pod is a further factor controlling import, it is necessary to determine if acquisition satisfies this demand by comparing the potential and actual growth rates of sinks. Egli & Wardlaw (1980) showed that excised soybean cotyledons grew three times as fast *in vitro* as on the plant at high temperatures (33/28 °C) and at about the same speed at lower temperatures (24/19 °C); here the ability of the plant to supply assimilate was limiting at the higher temperature.

The imposition of changed temperature on a sink such as a tomato fruit results in changes in metabolism of the source leaves complicating interpretation of this type of experiment. Hurewitz & Janes (1983) showed that raising the temperature of tomato roots resulted in increased net photosynthesis of the leaves, the starch content of which however fell, and downward translocation rose slightly. Such feedback cannot always overcome direct effects of temperature on the source. When leaves of *Digitaria decumbens* were held at 10 °C, their increased content of starch was not remobilized even when sink demand was raised by warming the rest of the plant or shading other leaves (Garrard & West, 1972). Whatever the nature of the signal that alters the metabolism of the source leaf, understanding what controls acquisition by a sink must allow for alteration in source metabolism.

There is much to be said for studying acquisition by using a simplified system, and single leaves that have been allowed to root may be a valuable experimental tool. Sawada *et al.* (1987) used rooted soybean leaves with the roots placed in low (6 °C rather than 20 °C) temperatures for a 4-day period. Root growth only slowed towards the end of this period, and after sucrose and starch – but not hexose – contents of the roots had risen. The carbohydrate content of the leaves also increased during the cooling treatment. Clearly, ability to grow rather than substrate supply limited root growth at low temperatures, and acquisition was not entirely growth-led as carbohydrate accumulation could occur in the sink.

Allocation: partitioning between two or more sinks supplied from a single source

The relative flux between two sinks fed from a single source can be determined by the pathways connecting then, or the sinks themselves. The effects of temperature on competing pathways alone have not been investigated experimentally, but can be assessed theoretically by assuming that phloem transport can be described by the Poiseuille equation. Since $J = \Delta P \pi r^2 / 8 \eta l$, fluxes J_1 and J_2 to two competing sinks will be in the ratio $J_1/J_2 = \eta_2 l_2 \Delta P_1 / \eta_1 l_1 \Delta P_2$. Viscosity, η, will change with temperature, and the question becomes the proportion of temperature-dependent change in partitioning that it can account for. The predicted effect of viscosity is shown in Fig. 8, with the data of Brouwer (1981) superimposed on it. If we assume that these data reflect gross partitioning (they are of dry weight), it is clear that viscosity cannot account for more than about one-third of the change

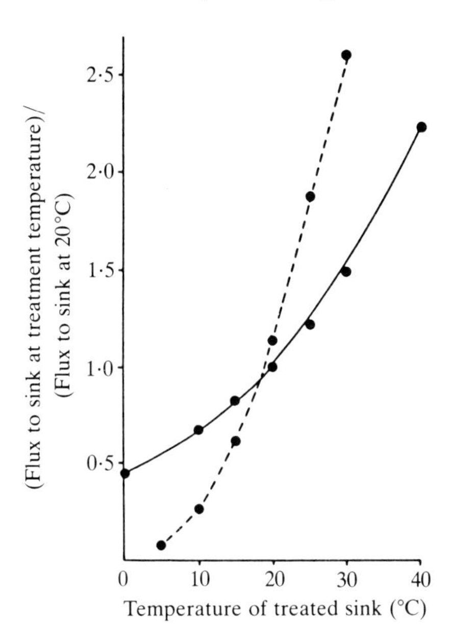

Fig. 8. The predicted effect of viscosity on allocation between two sinks held at different temperatures (solid line) compared with the data of Brouwer (1981) (broken line) for growth of maize with split root systems. Data are expressed as the ratio (value at treatment temperature)/(value at 20 °C control temperature). Viscosity is calculated for 1 mol dm^{-3} sucrose.

seen. Viscosity may be a significant factor in allocation, but is probably not an overriding one, in accord with the evidence for major temperature-mediated effects on sink metabolism given earlier. Unfortunately 'sink strength', the ability of a sink to accumulate assimilate in competition with other sinks, cannot yet be measured.

A fruiting plant provides an example of allocation between symplastically and apoplastically unloading sinks: cereals are particularly convenient as a tiller, being monocarpic, finally has just the grain (apoplastic unloading) at one end and root meristems (symplastic) at the other. Wardlaw (1970) has shown that 2–3 d after anthesis, temperature has little effect on allocation between these two. However, by 9–10 d, wheat plants at 27/22 °C allocate four times as much to the ear relative to the root as those at 15/10 °C. It is tempting to suggest that the common observation – of which this is an example – that reproductive sinks are strong ones is simply because they use apoplastic unloading. Certainly, both in wheat (Chowdhury & Wardlaw, 1978; Chen-Au *et al.* 1963; Wardlaw *et al.* 1980) and pea (Williams & Marinos, 1977) allocation to the apoplastically unloading fruit is highly sensitive to the temperature of the fruit itself. It is possible that apoplastic unloading is more responsive than symplastic to temperature.

There are few reports of allocation between sinks unloading apoplastically. The cereal ear itself contains a series of separate, apoplastically unloading sinks, allocation between which is little affected by temperature (Thornley *et al.* 1981). If pods of *Phaseolus vulgaris* are warmed (following 24-h cooling), assimilate is allocated to them at the expense of pods that had been left untreated (Geiger & Fondy, 1985).

Allocation between sinks unloading symplastically will now be considered. Hurewitz & Janes (1983) showed that as they increased the root temperature of tomato seedlings from 15 to 35 °C, these roots obtained an increasingly large share of leaf-fed ^{14}C relative to the growing shoot system held at 22 °C. Fujiwara & Suzuki (1961) obtained qualitatively similar results for barley. More detailed experiments on non-fruiting tomato by Hori & Shishido (1977) showed that, as night temperature of the whole plant rose from 9 to 24 °C, a decreasing proportion of ^{14}C was translocated to the regions below the fed leaf, and more above. If the plants were kept at these night temperatures for a week before applying ^{14}C, the main difference was that a higher proportion of ^{14}C moving up the plant from the fed leaf remained in the stem rather than accumulating at the apex. Fruiting plants showed essentially similar behaviour, except that the proportion of ^{14}C entering the fruit was unchanged by temperature on the first night of temperature treatment, but after two weeks of pretreatment at the various night temperatures, fruits of plants in a 24 °C night imported about 15 times as much ^{14}C as those in a 12 °C night, although fruits were only twice as big (Hori & Shishido, 1977). When cucumber plants were similarly exposed to night temperatures varying from 5 to 30 °C, appreciable effects on allocation were only found at the 7- rather than 20-leaf stage. At 5 °C much less material moved upwards from the fed leaf, the stem apex appearing to be a very poor sink at this temperature (Kanahama & Hori, 1980). Cucumber given temperatures between 15 and 30 °C for the complete diel cycle showed maximal allocation upwards from the fed leaf at 25 °C, at each of two irradiances (Shishido *et al.* 1987).

Sucrose unloading in potato tubers is predominantly symplastic (Oparka & Prior, 1987). Elegant work by Engels & Marschner (1986*a*) showed that when an individual tuber was cooled to 8 °C, its growth rate fell, whilst non-cooled (22/18 °C day/night) tubers on the same plant grew faster than previously. This, like the data of Brouwer (1981) for maize roots, shows that in control plants sinks were growing below the rate of which they were capable at that temperature, indicating that their growth rate may be constrained by substrate supply reduced by competition from other sinks. The total rate of growth of all tubers on the plant was unaffected when some, but not all, were cooled. Although cooled tubers were capable of recovering their pre-cooling growth rate they did not always do so in competition with uncooled tubers (Engels & Marschner, 1986*a*). Little change was seen in contents of reducing sugars, sucrose and starch in cooled and non-cooled tubers. A much smaller proportion of ^{14}C fed to the shoot as $^{14}CO_2$ entered starch from ethanol-soluble pools in cooled tubers (Engels & Marschner, 1986*b*). The behaviour of enzymes in this system was discussed earlier; a correlation was found between import of ^{14}C and activities of ADPG-pyrophosphorylase and starch synthase in control tubers of differing growth rate, but this was not reflected in enzyme changes on cooling. In particular, tubers that had been cooled and rewarmed had activities of these enzymes similar to the controls, but lower growth rates. Clearly other factors are involved in determining allocation beyond activities of major enzymes in sinks. Competitive sink strength and potential growth rate do not seem to be related, at least in potato tubers. Engels & Marschner (1986*b*) conclude that the cross-sectional area of phloem in the stolon subtending a tuber may have a major role in determining allocation between tubers.

The importance of the phloem pathway has also been emphasized by Grusak & Lucas (1986). Sugar beet plants reduced to one source and two sink leaves had cooling treatments applied to the petiole of the source leaf. The authors concluded that when cooling was applied slowly, from 25 °C to 1 °C over 90 min, a change in pathway connections could occur such that allocation between the two sink leaves and unmonitored sinks changed: the site effecting these changes was not the cooled region itself. A rapid, physical, transmission of a signal such as a pressure change was postulated to account for this effect. The effect could be mimicked by dropping the turgor in the source leaf petiole with osmotica. Change in turgor as a mediator of partitioning is currently enjoying extensive interest (Wolswinkel, 1985). Unfortunately, there is no information on how temperature might affect turgor relations in such a way as to modify partitioning: it could be via direct changes to membranes or via solute potential differences across membranes.

We have begun to use allocation between the two halves of a barley root system, and still attached to the plant, as an example of allocation between symplastically unloading sinks. This system has the advantage that vascular supply to each seminal root is equivalent, and both temperature and the chemical environment of the split roots are readily and separately variable. In control roots at 20 °C, respiration does not respond to added sugars, and since growth and respiration are stoichiometrically related it is unlikely that growth is substrate-limited in the short term (Bingham & Farrar, 1988). Although the roots can take up extra sugars, this results in storage and not growth; the rate of sucrose uptake can substantially exceed its rate of utilization. We have examined the effects of temperature on short-term allocation of photosynthetically fixed ^{14}C between the halves of a split root system (Fig. 9). Whilst temperatures below 10 °C markedly reduce allocation, it is striking that from 10 °C–30 °C the allocation between the control (20 °C) and treated halves varies less than two-fold. The flux of ^{14}C to the control half was unaffected by the temperature of the treated half. By contrast, uptake of sucrose at 5 or 50 mol m^{-3} shows a $Q_{10} \approx 1{\cdot}5$ (Fig. 9), and respiration of roots on an intact plant responds to temperature greatly and with a Q_{10} that falls from 2·8 to 1·2 over the range 4–34 °C (J. F. Farrar and C. L. Jones, unpublished). The temperature dependence of the patterns of allocation, of respiration and sucrose uptake, and of viscosity of phloem sap, can be emphasized by plotting the ratios of values or rates of each at treatment and control temperature (Fig. 10). This plot predicts allocation as determined by individual variables; it can be seen that the variables predict widely divergent allocation patterns in response to temperature. I suggest two alternative explanations. Perhaps the linkage between allocation and metab-

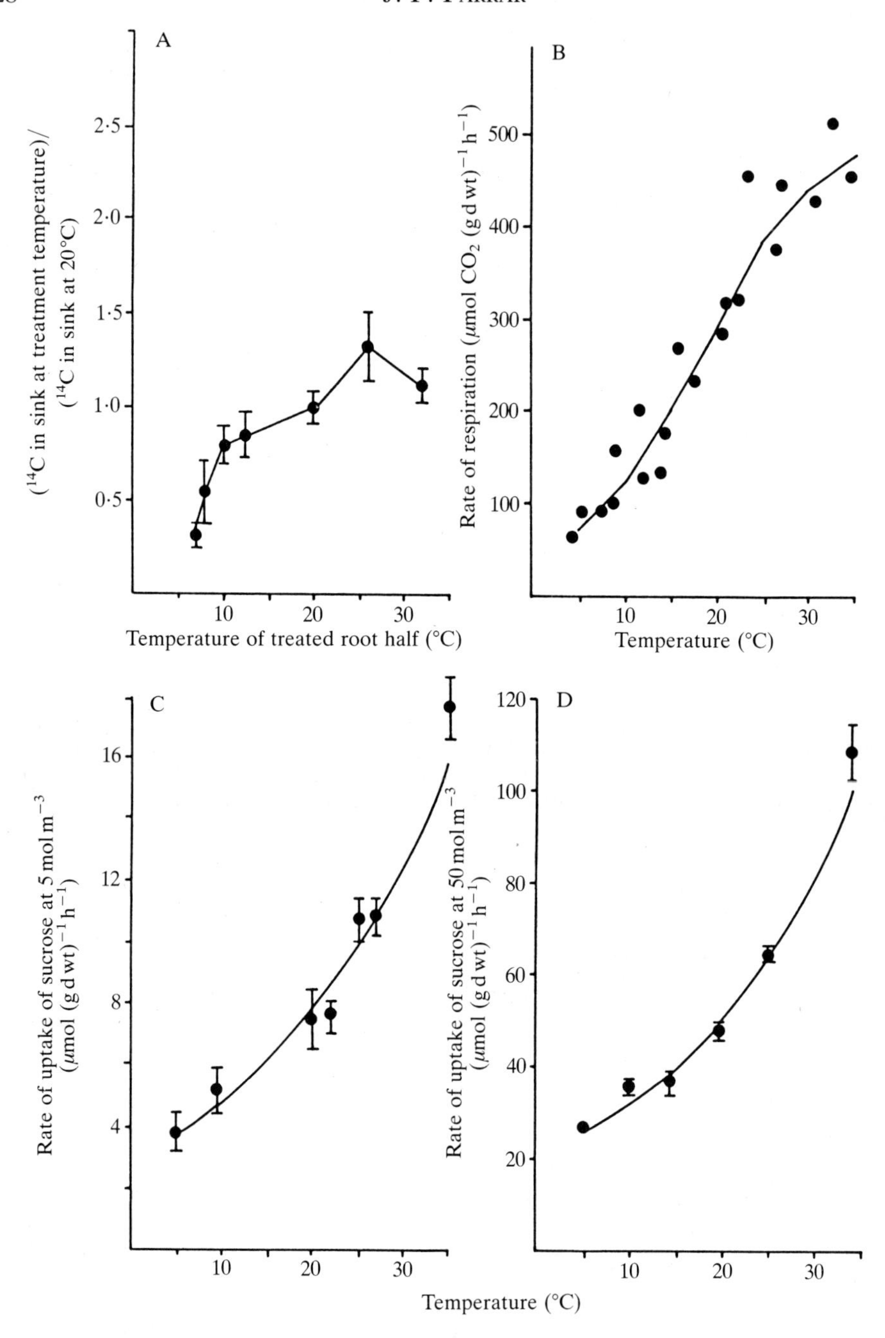

Fig. 9. The effect of temperature on barley roots. A. Allocation over 3 h of ^{14}C, fed as $^{14}CO_2$ to the second leaf blade for 20 min, to the two halves of a root system split between 20°C (control) and treatment temperatures. Absolute counts going to the control half were not affected by temperature. B. Rate of respiratory CO_2 efflux from roots on intact plants, measured by IRGA. Plants were grown at 20°C and the respiration measured within 2 h of applying the temperatures indicated. C,D. Uptake of 5 and 50 mol m^{-3} sucrose respectively by freshly excised barley seminal roots, measured by incorporation of ^{14}C from ^{14}C-sucrose. Data of Farrar and Jones (unpublished).

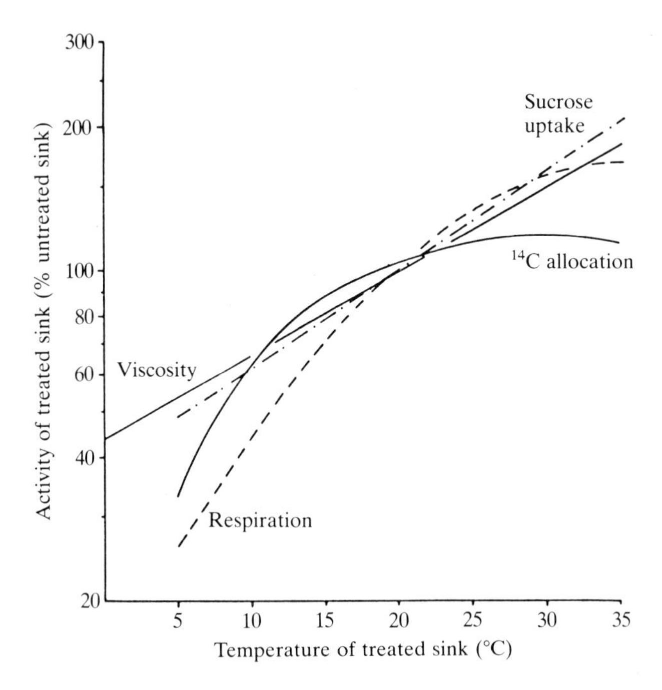

Fig. 10. A comparison of the measured short-term allocation of ^{14}C between two halves of a barley root system at different temperatures, and allocation predicted from the temperature dependence of viscosity, sucrose uptake and respiration. Control temperature is 20°C. Note the semilogarithmic scale.

olism in a sink is poor because events within the phloem (e.g. turgor) are poorly coupled to events in receiver cells. Alternatively, above control temperatures substrate limitation of allocation is seen, perhaps due to transport resistances between phloem and receiver cells in the root, which are not measured by sucrose uptake (direct access to receiver cell apoplast) or respiration (direct access to cytosolic sugars). At temperatures below that of the control, the ability to metabolize sucrose, although buffered by spare storage capacity, may limit allocation.

Although, in the short term, root growth is not substrate-limited, at control temperatures these roots are below their potential rates of growth (Farrar & Jones, 1986) and respiration (Bingham & Farrar, 1988). It may be that they are, at 20°C, growing at a rate adjusted to a substrate supply determined by competition between individual seminal roots. Certainly when root halves are cooled for 48 h before assessing allocation, less ^{14}C goes to the cooled half than before treatment, whilst 48-h pre-warming results in warmed halves receiving much less ^{14}C than control halves (J. F. Farrar and C. L. Jones, unpublished). This indicates the independence of short- and long-term responses to temperature and, more fundamentally, the way in which the growth rate of a sink will partly be a function of its previous supply of assimilate. Its potential ability to grow may not only be substantially greater than its actual rate, but also it may take some time (days) in favourable conditions for that ability to be expressed. The concept of sink strength needs to be extended to allow for regulatory changes at the gene level to occur over periods of 12–48 h, and needs to be coupled to those features of phloem physiology, such as turgor (Wolswinkel, 1985), that control acquisition from a source.

Conclusions

This survey of the effects of temperature on partitioning has been unpleasantly similar to a mere catalogue of experimental findings. To catalogue is usually a sign of a lack of real understanding, and I fear that in this case it is just so. The problem is two-fold: we do not understand the control of partitioning, and the effects of temperature upon it are far from simple or clear cut. However, no theories which seek to explain partitioning can be complete unless they embrace the effects of temperature and, similarly, it may be that temperature is a variable useful in experiments aimed at understanding partitioning. Temperature primarily affects (one could say, is sensed by) two types of process: physical, such as diffusion and, via viscosity, translocation in the phloem; and enzymatic. The key point about enzymatic effects is the differential sensitivity of various enzymes, with obvious implications for the relative fluxes through the pathways they serve. Within sources and within sinks, enzymatic (and to a lesser extent diffusive) processes dominate the response to altered temperature. More detailed information on the temperature sensitivity of individual enzymes should lead to a clear understanding of the behaviour of isolated sources and sinks, as long as carrier systems are also investigated. It matters not whether sucrose is removed from the cytosol in a sink by hydrolysis prior to metabolism, or by loading into a vacuole: the end result is less sucrose in the cytosol and a steeper gradient favouring the entry of more from outside the cell. The temperature-dependence of the process of loading and unloading is urgently in need of study.

It is when acquisition by a source from a sink, or allocation between competing sinks, is considered that complications really start. Temperature treatment of one part of a plant will soon affect the behaviour of a remote part, via the altered fluxes of sugars to or from it. It is this integration of the plant that provides the greatest problem and the greatest challenge. It is likely that control of plant growth is distributed between source and sink, and that mutual regulation of one by the other – greatly complicated by competition between sinks – occurs continuously. Two time-scales may operate. Over hours, fine control by the existing machinery will respond to changes in metabolite pools and the activities of individual enzymes; over days, coarse control by changing the amounts of enzymes and transport systems will occur. We are currently investigating whether sucrose itself can bring about coarse control.

In the introduction to this chapter some whole-plant responses to temperature were described. As these were of plants growing over periods of days or weeks, they will involve coarse control. Nothing I have said brings us much nearer to being able to explain them. Indeed, the real explanation of, for example, the relatively greater allocation to roots at low temperature might not be primarily a function of carbon partitioning *per se*. The activity of the sinks, whether determined by effects of temperature on cell division or on energy required to take up nutrients or maintain membranes and solute status, is probably paramount, and this will then indirectly dictate partitioning. The mechanisms by which this is brought about are worthy of close attention.

References

Altus, D. P. & Canny, M. J. (1982). Loading of assimilates in wheat leaves. I. The specialisation of vein types for different activities. *Aust. J. Pl. Physiol.* **9**, 571–581.

ap Rees, T., Dixon, W. L., Pollock, C. J. & Franks, F. (1983). Low temperature sweetening of higher plants. In *Recent Advances in the Biochemistry of Fruits and Legumes* (ed. J. Friend & M. J. C. Rhodes), pp. 41–61. London: Academic Press.

Bates, F. Y. (1942). *Polarimetry, Saccharimetry and the Sugars*. Washington: U.S. Government.

Bhullar, S. S. & Jenner, C. F. (1985). Differential responses to high temperatures of starch and nitrogen accumulation in the grain of four cultivars of wheat. *Aust. J. Pl. Physiol.* **12**, 363–375.

Bingham, I. & Farrar, J. F. (1988). Regulation of respiration rate in roots of barley. *Physiologia Pl.* **73**, 278–285.

Borland, A. M. & Farrar, J. F. (1985). Diel patterns of carbohydrate metabolism in leaf blades and leaf sheaths of *Poa annua* L. and *Poa × jemtlandica* (Almq.) Richt. *New Phytol.* **100**, 516–531.

Borland, A. M. & Farrar, J. F. (1987). The influence of low temperature on diel patterns of carbohydrate metabolism in leaves of *Poa annua* L. and *Poa × jemtlandica* (Almq.) Richt. *New Phytol.* **105**, 255–263.

Brouwer, R. (1981). Co-ordination of growth phenomena within a root system of intact maize plants. In *Structure and Function of Plant Roots* (ed. R. Brouwer), pp. 269–276. The Hague: Junk.

Chamberlain, I. S. & Spanner, D. C. (1978). The effect of low temperatures on the phloem transport of radioactive assimilates in the stolon of *Saxifraga saramentosa* L. *Plant Cell Environ.* **1**, 285–290.

Chatterton, N. J., Carlson, G. E., Hungerford, W. E. & Lee, D. R. (1972). Effect of tillering and cool nights on photosynthesis and chloroplast starch in pangola. *Crop Sci.* **12**, 206–208.

Chen-Au, H., Shin-San, W. & Fu-Te, W. (1983). The effect of temperature on the physiological changes of wheat during grain development. *Acta Bot. Sinica,* **11**, 338–339.

Chowdhury, S. I. & Wardlaw, I. F. (1978). The effect of temperature on kernel development in cereals. *Aust. J. Agric. Res.* **29**, 205–223.

Clarkson, D. T., Hopper, M. J. & Jones, L. H. P. (1986). The effect of root temperature on the uptake of nitrogen and the relative size of the root system in *Lolium perenne*. *Pl. Cell Environ.* **9**, 535–546.

Cooper, A. J. (1973). *Root Temperature and Plant Growth.* Farnham: Commonwealth Agricultural Bureaux.

Coulson, C. L., Christy, A. L., Cataldo, D. A. & Swanson, C. A. (1972). Carbohydrate translocation in sugar beet petioles in relation to petiolar respiration and adenosine 5′-triphosphate. *Pl. Physiol.* **49**, 919–923.

Crafts. A. S. & Crisp, C. E. (1971). *Phloem Transport in Plants.* San Francisco: Freeman.

Crawford, R. M. M. & Huxter, T. J. (1977). Root growth and carbohydrate metabolism at low temperatures. *J. Exp. Bot.* **28**, 917–925.

Dawson, R. M. C., Elliott, D. C., Elliott, W. H. & Jones, K. M. (1986). *Data for Biochemical Research.* Oxford: Clarendon.

EGLI, D. B. & WARDLAW, I. F. (1980). Temperature response of seed growth characteristics of soybean. *Agron. J.* **72**, 560–564.
ENGELS, CH. & MARSCHNER, H. (1986*a*). Allocation of photosynthate to individual tubers of *Solanum tuberosum* L. I. Relationships between tuber growth rate and enzyme activities of the starch metabolism. *J. Exp. Bot.* **37**, 1795–1803.
ENGELS, CH. & MARSCHNER, H. (1986*b*). Allocation of photosynthate to individual tubers of *Solanum tuberosum* L. II. Relationship between growth rate, carbohydrate concentration and ^{14}C-partitioning within tubers. *J. Exp. Bot.* **37**, 1804–1812.
FARRAR, J. F. (1981). Respiration rate of barley roots: its relation to growth, substrate supply and the illumination of the shoot. *Ann. Bot.* **48**, 53–63.
FARRAR, J. F. (1985). The respiratory source of CO_2. *Pl. Cell Environ.* **8**, 427–438.
FARRAR, J. F. & JONES, C. L. (1986). Modification of respiration and carbohydrate status of barley roots by selective pruning. *New Phytol.* **102**, 513–521.
FARRAR, S. C. & FARRAR, J. F. (1985). Carbon fluxes in leaf blades of barley. *New Phytol.* **100**, 271–283.
FARRAR, S. C. & FARRAR, J. F. (1986). Compartmentation and fluxes of sucrose in intact leaf blades of barley. *New Phytol.* **103**, 645–657.
FAUCHER, M., BONNEMAIN, J.-L. & DOFFIN, M. (1982). Effets de refroidissements localises sur la circulation liberienne chez quelques especes avec ou sans proteines-P et influence du mode de refroidissement. *Physiol. Veg.* **30**, 395–405.
FENSOM, D. S. (1981). Problems arising from a Münch-type pressure flow mechanism of sugar transport in phloem. *Can. J. Bot.* **59**, 425–432.
FRIEND, D. J. C. (1965). The effect of light and temperature on the growth of cereals. In *The Growth of Cereals and Grasses* (ed. F. L. Milthorpe & J. D. Ivins), pp. 181–199. London: Butterworths.
FRIEND, D. J. C., HELSON, V. A. C. & FISHER, J. E. (1965). Changes in the leaf area ratio during growth of Marquis wheat, as affected by temperature and light intensity. *Can. J. Bot.* **43**, 15–28.
FUJIWARA, A. & SUZUKI, M. (1961). Effects of temperature and light on the translocation of photosynthetic products. *Tohoku J. Agric. Res.* **12**, 363–367.
GARRARD, L. A. & WEST, S. H. (1972). Suboptimal temperature and assimilate accumulation in leaves of pangola digitgrass (*Digitaria decumbens* Stent.). *Crop Sci.* **12**, 621–623.
GEIGER, D. R. & FONDY, B. R. (1985). Responses of export and partitioning to internal and environmental factors. In *Regulation of Sources and Sinks in Crop Plants*, Monograph 12 (ed. B. Jeffcoat, A. F. Hawkins & A. D. Stead), pp. 177–194. Bristol: British Plant Growth Regulator Group.
GEIGER, D. R. & SOVONICK, S. A. (1975). Effects of temperature, anoxia and other metabolic inhibitors on translocation. In *Transport in Plants I. Phloem Transport*. Encycl. Plant Physiol. 1 (ed. M. H. Zimmermann & J. A. Milburn), pp. 256–286. Berlin: Springer.
GIAQUINTA, R. (1980). Mechanism and control of phloem loading of sucrose. *Ber. Deutsch Bot. Ges.* **93**, 187–201.
GIAQUINTA, R. T. & GEIGER, D. R. (1973). Mechanism of inhibition of translocation by localized chilling. *Pl. Physiol.* **51**, 372–377.
GRUSAK, M. A. & LUCAS, W. J. (1984). Recovery of cold-inhibited translocation in sugar beet. I. Experimental analysis of an existing mathematical recovery model. *J. Exp. Bot.* **35**, 389–402.
GRUSAK, M. A. & LUCAS, W. J. (1985). Cold-inhibited phloem translocation in sugar beet. II. Characterization and localization of the slow-cooling response. *J. Exp. Bot.* **36**, 745–755.
GRUSAK, M. A. & LUCAS, W. J. (1986). Cold-inhibited phloem translocation in sugar beet. III. The involvement of the phloem pathway in source-sink partitioning. *J. Exp. Bot.* **37**, 277–288.
HAWKER, J. S. (1985). Sucrose. In *Biochemistry of Storage Carbohydrates in Green Plants* (ed. P. M. Day & R. A. Dixon), pp. 1–51 New York: Academic Press.
HOFSTRA, G. & NELSON, C. D. (1969). The translocation of photosynthetically assimilated ^{14}C in corn. *Can. J. Bot.* **47**, 1435–1442.
HORI, Y. & SHISHIDO, Y. (1977). Studies on translocation and distribution of photosynthetic assimilates in tomato plants. *Tohoku J. Agric. Res.* **28**, 26–40.
HOUSLEY, T. L. & POLLOCK, C. J. (1985). Photosynthesis and carbohydrate metabolism in detached leaves of *Lolium temulentum* L. *New Phytol.* **99**, 499–502.

HUBER, S. C. (1986). Fructose 2,6-bisphosphate as a regulatory metabolite in plants. *A. Rev. Pl. Physiol.* **37**, 233–246.

HUREWITZ, J. & JANES, H. W. (1983). Effect of altering the root-zone temperature on growth, translocation, carbon exchange rate, and leaf starch accumulation in the tomato. *Pl. Physiol.* **73**, 46–50.

JENNER, C. F. (1982). Storage of starch. In *Plant Carbohydrates I. Intracellular Carbohydrates.* Encycl. Plant Physiol. 13A (ed. F. W. Loewus & W. Tanner), pp. 700–747. Berlin: Springer.

JENNER, C. F. (1985). Transport of tritiated water and ^{14}C-labelled assimilate into grains of wheat. I. Entry of THO through and in association with the stalk of the grain. *Aust. J. Pl. Physiol.* **12**, 573–586.

JOHNSON, I. R. & THORNLEY, J. H. M. (1985). Temperature dependence of plant and crop processes. *Ann. Bot.* **55**, 1–24.

KABASSI, P., WEST, S. H. & GARRARD, L. A. (1972). Amylolytic activity in leaves of a tropical and a temperate grass. *Crop. Sci.* **12**, 58–60.

KANAHEMA, K. & HORI, Y. (1980). Time course of export of ^{14}C-assimilates and their distribution patterns as affected by feeding time and night temperature in cucumber plants. *Tohoku J. Agric. Res.* **30**, 142–152.

KLEINENDORST, A. & VEEN, B. W. (1983). Responses of young cucumber plants to root and shoot temperatures. *Neth. J. Agric. Sci.* **31**, 47–61.

KOLEK, J., MISTRIK, I. & HOLOBRADA, M. (1981). The response of *Zea mays* roots to chilling. In *Structure and Function of Plant Roots* (ed. R. Brouwer), pp. 327–335. The Hague: Junk.

KRAUS, A. & MARSCHNER, H. (1984). Growth rate and carbohydrate metabolism of potato tubers exposed to high temperature. *Potato Res.* **27**, 297–303.

LABHART, C., NOSBERGER, J. & NELSON, C. J. (1983). Photosynthesis and degree of polymerisation of fructan during reproductive growth of meadow fescue at two temperatures and two photon flux densities. *J. Exp. Bot.* **34**, 1037–1046.

LAMBERS, H. (1985). Respiration in intact plants and tissues: its regulation and dependence on environmental factors, metabolism and invaded organisms. In *Respiration of Higher Plant Cells.* Encycl. Plant Physiol. 18 (ed. R. Douce & D. Day), pp. 418–473. Berlin: Springer.

LANG, A. (1978). Interactions between source, path and sink in determining phloem translocation rate. *Aust. J. Pl. Physiol.* **5**, 665–674.

LANG, A. & MINCHIN, P. E. M. (1986). Phylogenetic distribution and mechanism of translocation inhibition by chilling. *J. Exp. Bot.* **37**, 389–398.

LUSH, W. M. & EVANS, L. T. (1974). Translocation of photosynthetic assimilate from grass leaves, as influenced by environment and species. *Aust. J. Pl. Physiol.* **1**, 417–431.

MAJUMDAR, S. K. & LEOPOLD, A. C. (1967). Callose formation in response to low temperature. *Pl. Cell Physiol.* **8**, 775–778.

MAROWITCH, J., RICHTER, C. & HODDINOTT, J. (1986). The influence of plant temperature on photosynthesis and translocation rates in bean and soybean. *Can. J. Bot.* **64**, 2337–2342.

MCDUFF, J. H., WILD, A., HOPPER, M. A. & DHANSA, A. (1986). Effects of temperature on parameters of root growth relative to nutrient uptake. *Pl. Soil* **94**, 321–332.

MINCHIN, P. E. H., LANG, A. & THORPE, M. R. (1983). Dynamics of cold induced inhibition of phloem transport. *J. Exp. Bot.* **34**, 156–162.

MINCHIN, P. E. H. & THORPE, M. R. (1983). A rate of cooling response in phloem translocation. *J. Exp. Bot.* **34**, 529–536.

MINCHIN, P. E. H. & THORPE, M. R. (1984). Apoplastic phloem unloading in the stem of bean. *J. Exp. Bot.* **35**, 438–450.

MOORBY, J. & JARMAN, P. D. (1975). The use of compartmental analysis in the study of the movement of carbon through leaves. *Planta* **122**, 155–168.

MOORBY, J., TROUGHTON, J. H. & CURRY, B. G. (1974). Investigations of carbon transport in plants. II. The effects of light and darkness and sink activity on translocation. *J. Exp. Bot.* **25**, 937–944.

NELSON, C. D. (1963). Effect of climate on the distribution and translocation of assimilates. In *Environmental Control of Plant Growth* (ed. L. T. Evans), pp. 149–173. New York: Academic Press.

Ntsika, G. & Delrot, S. (1986). Changes in apoplastic and intracellular leaf sugars induced by the blocking of export in *Vicia faba. Physiol. Pl.* **68**, 145–153.

Oparka, K. J. & Prior, D. A. M. (1987). ^{14}C sucrose efflux from the perimedulla of growing potato tubers. *Pl. Cell Environ.* **10**, 667–675.

Pearcy, R. W. (1976). Temperature effects on growth and CO_2 exchange rates in control and desert races of *Atriplex lentiformis. Oecologia* **26**, 245–255.

Pearson, C. J. & Derrick, G. A. (1977). Thermal adaptation of *Pennisetum*: leaf photosynthesis and photosynthate translocation. *Aust. J. Pl. Physiol.* **4**, 763–769.

Pickard, W. F., Minchin, P. E. H. & Troughton, J. H. (1978). Transient inhibition of translocation in *Ipomoea alba* L. by small temperature reductions. *Aust. J. Pl. Physiol.* **5**, 127–130.

Pollock, C. J. (1986*a*). Environmental effects on sucrose and fructan metabolism. *Curr. Topics Pl. Biochem. Physiol.* **5**, 32–46.

Pollock, C. J. (1986*b*). Fructans and the metabolism of sucrose in vascular plants. *New Phytol.* **104**, 1–24.

Pollock, C. J., Lloyd, E. J., Stoddart, J. L. & Thomas, H. (1983). Growth, photosynthesis and assimilate partitioning in *Lolium temulentum* exposed to chilling temperatures. *Physiologia Pl.* **59**, 257–262.

Pollock, C. J. & Lloyd, E. J. (1987). The effect of low temperature upon starch, sucrose and fructan synthesis in leaves. *Ann. Bot.* **60**, 231–235.

Rajan, A. K., Betheridge, B. & Blackman, G. E. (1971). Interrelationships between the nature of the light source, ambient air temperature, and the vegetative growth of different species within growth cabinents. *Ann. Bot.* **35**, 923–943.

Rufty, T. W., Huber, S. C. & Kerr, P. S. (1985). Association between sucrose-phosphate synthase activity in leaves and plant growth rate in response to altered aerial temperature. *Pl. Sci.* **39**, 7–12.

Saftner, R. A., Daie, J. & Wyse, R. E. (1983). Sucrose uptake and compartmentation in sugar beet taproot tissue. *Pl. Physiol.* **72**, 1–6.

Sawada, S., Kawamura, H., Hayakawa, T. & Kasai, M. (1987). Regulation of photosynthetic metabolism by low-temperature treatment of roots of single-rooted soybean plants. *Pl. Cell Physiol.* **28**, 235–241.

Schweitzer, L. E. & Harper, J. E. (1980). Effect of light, dark and temperature on root nodule activity of soybeans. *Pl. Physiol.* **65**, 51–56.

Seddigh, M. & Joliffe, G. D. (1986). Remobilisation patterns of C and N in soybeans with different sink-source ratios induced by different night temperatures. *Pl. Physiol.* **81**, 136–141.

Shishido, Y., Challa, H. & Krupa, J. (1987). Effects of temperature and light on the carbon budget of young cucumber plants studied by steady-state feeding with $^{14}CO_2$. *J. Exp. Bot.* **38**, 1044–1054.

Sicher, R. C. (1986). Sucrose biosynthesis in photosynthetic tissue: rate-controlling factors and metabolic pathway. *Physiologia Pl.* **67**, 118–121.

Sicher, R. C. & Kremer, D. F. (1986). Effects of temperature and irradiance on non-structural carbohydrate accumulation in barley primary leaves. *Physiologia Pl.* **66**, 365–369.

Smith, D. (1968). Carbohydrates in grasses. IV. Influence of temperature on the sugar and fructosan complement of timothy plant parts at anthesis. *Crop. Sci.* **8**, 331–334.

Smith, D. & Struckmer, B. E. (1974). Gross morphology and starch accumulation in leaves of alfalfa plants grown at high and low temperatures. *Crop. Sci.* **14**, 433–436.

Stitt, M. & Steup, M. (1985). Starch and sucrose degradation. In *Respiration of Higher Plant Cells*. Encycl. Plant Physiol. 18 (ed R. Douce & D. Day), pp. 347–390. Berlin: Springer.

Szaniawski, R. K. (1981). Shoot: root functional equilibria. Thermodynamic stability of the plant system. In *Structure and Function of Plant Roots* (ed. R. Brouwer), pp. 357–360. The Hague: Junk.

Tammes, P. M. L., Vonk, G. R. & Van Die, J. (1969). Studies on phloem exudation from *Yucca flaccida*. VII. The effect of cooling on exudation. *Acta. Botan. Neerl.* **18**, 224–229.

Thorne, J. H. (1982). Temperature and oxygen effects on ^{14}C-photosynthate unloading and accumulation in developing soybean seeds. *Pl. Physiol.* **69**, 48–53.

THORNE, J. H. (1986). Sieve tube unloading. In *Phloem Transport* (ed. J. Cronshaw, W. Lucas & R. T. Giaquinta), pp. 211–224. New York: Liss.

THORNLEY, J. H. M., GIFFORD, R. M. & BREMNER, P. M. (1981). The wheat spikelet – growth response to light and temperature – experiment and hypothesis. *Ann. Bot.* **47**, 713–725.

VAN DOBBEN, W. H. (1962). Influence of temperature and light conditions on dry-matter distribution, development rate and yield in arable crops. *Neth. J. Agric. Sci.* **10**, 377–389.

EFFECTS OF TEMPERATURE ON CELL MEMBRANES

PETER J. QUINN

Department of Biochemistry, King's College London, Campden Hill, London W8 7AH, UK

Summary

The plasma membrane and cytoplasmic membranes of plants, like those of animal cells, are composed of lipids and proteins that are often glycosylated. Likewise, the composition from one membrane type to another is highly heterogeneous. There is some evidence to suggest that the composition, particularly of the lipid component, may change in response to environmental conditions such as temperature, water stress, etc. as well as during growth, development and ultimately senescence of the cell. It is believed that these changes are required to adjust the physical characteristics of membrane structures so that they may perform their necessary physiological tasks when environmental factors change. If the environmental conditions are altered beyond the normal limits within which the plant survives, the cell membranes are often found to undergo gross structural changes. These structural perturbations include phase separation of the membrane constituents and are associated with characteristic disturbances of function such as loss of selective permeability and transport processes. In most instances, the observed phase separations appear to be driven by phase changes in the membrane lipids. Some lipids extracted from algae and plant membranes are known to exist in a bilayer gel phase when dispersed in aqueous systems at the growth temperature while other lipid fractions are in a liquid-crystalline state. Nearly all membranes contain varying proportions of their lipid complement that do not form bilayer structures under such conditions and most commonly adopt an hexagonal-II arrangement. In this review the importance of lipid phase behaviour is discussed in the context of membrane stability at different temperatures.

Introduction

Analysis of the plasma and cytoplasmic membranes of a wide variety of plants and algae indicates that their composition is highly heterogeneous. The amount and type of protein present is unique to each morphologically distinct membrane and the ratio of protein to lipid also differs over wide limits from one membrane type to another. Furthermore, the polar lipid composition is extremely complex and although several distinct classes of polar lipid can be recognized each of these classes is comprised of a whole range of molecular species which differ in the length, extent of unsaturation and positional distribution of the substituent fatty acids. It can be argued that the variation in protein composition is necessary to allow different membranes to perform specific tasks. It is not clear, however, why the lipid composition is so complex. This is particularly so in view of the fact that many membrane-bound enzymes, especially from animal cell membranes, can be isolated from native membranes and their functions reconstituted in single molecular species of lipid.

Plants, unlike higher animals, are heterothermic organisms and are able to grow and survive at temperatures and under other environmental conditions that vary over relatively wide limits. It is commonly observed that when the environment is altered within physiological limits, the lipid molecular species may change, particularly with respect to the hydrocarbon substituents, in a manner which suggests an adjustment to suit the new conditions.

The observed changes in membrane composition that follow shifts in temperature and other environmental conditions suggests that some mechanism(s) is able to recognize the change and translate this into biochemical activity culminating in a modification of the structure. These biochemical events may be operational at the level of the genome whereby environmental changes result in induction of the synthesis of new enzymes or the enhanced production of existing enzymes capable of acting on membrane constituents themselves or of augmenting synthesis of new membrane molecules. In other cases, the device may consist of a change in the rate of turnover of membrane molecules mediated by a change in the physical state of the substrate upon which enzymes act. Where a number of enzymes act in a concerted fashion the change in environmental conditions may affect the cooperativity between such enzymes. All of these factors require that the environmental changes are sufficiently gradual to enable appropriate biochemical responses to take place. If not, then the limits of membrane stability are prescribed by the rate at which the biochemical machinery responsible for preserving homeostasis of membrane composition within the normal growth limits of the organism can operate.

Temperature is one of the most widely investigated influences on plant growth and development. Changes in growth temperature, for example, are known to be associated with changes in membrane lipid composition. These changes often involve alteration in the proportions of the various lipid classes present in the membrane and the nature of their associated fatty acyl chains. These changes may involve alteration in the length, degree of unsaturation and the position at which they are acylated to the membrane lipids. The physical significance of such changes has been the subject of considerable debate (Quinn, 1981). It is commonly believed, however, that the changes in membrane lipids are required to maintain the fluidity of plant membranes within narrow limits required for the efficient function of the particular membrane. Such changes, therefore, can be regarded as an adaptive change which results in the preservation of a constant membrane fluidity under conditions of varying environmental temperature. The loss of membrane stability and irreversible changes in structure associated with exposure of plant membranes to temperatures beyond the normal range of temperature for growth is probably the major limiting factor in plant growth conditions.

In this review the composition of plant cell membranes will be examined. Examples of how environmental temperature can alter membrane composition in selected membranes will be given. The effect of temperature on the thermotropic

Table 1. *Major polar lipids of the higher plant chloroplast and a cyanobacterium*

	Relative proportion (mole %)	
Lipid class	Spinach	*Anacystis nidulans* (38°C)
Monogalactosyldiacylglycerol	38	54
Digalactosyldiacylglycerol	29	14
Sulphoquinovosyldiacylglycerol	18	11
Phosphatidylglycerol	12	21
Phosphatidylcholine	3	—

behaviour of membrane lipids and how membranes respond to heat stress and chilling will be described.

Membrane composition

Any determination of the composition of subcellular membranes relies on a subfractionation of plant tissues and the isolation of pure membrane fractions. Methods of fractionating animal tissues have been devised for a wide variety of cell membrane types but plant tissues pose a number of particular problems which are by and large avoided in animal tissues. The presence of a cell wall, for example, means that homogenization of plant tissues is difficult to achieve without damaging the constituent cell membranes. In other cases such as chloroplasts, starch granules may be present which can also inflict damage during the preparation of functional organelles. Furthermore, differential centrifugation of the type used for fractionating organelles from animal tissues does not separate plant cell organelles efficiently and cross-contamination, particularly from damaged organelles, is often a serious problem. Although lysosomes in animal cells contain lytic enzymes, in general methods of membrane fractionation have been devised to preserve lysosomes intact. Because of the difficulties in homogenization of plant tissues and the variety of lytic enzymes in subcellular fractions, such as vacuoles and other membrane-bound organelles, the release of these enzymes can modify membrane composition drastically and bring about a significant alteration in properties.

Notwithstanding these difficulties, methods for the isolation of functional chloroplasts and other plant subcellular organelles have been successfully developed and analyses of their constituent membranes have been published. Table 1 shows the principal polar lipid composition of the photosynthetic membrane of higher plant chloroplasts and, for comparison, the composition of the blue-green alga, *Anacystis nidulans*. Unlike animal cell membranes and other plant cell membranes, the major polar lipid classes of photosynthetic membranes are not phospholipids but galactolipids. The dominant galactolipid is monogalactosyldiacylglycerol and there are lesser amounts of digalactosyldiacylglycerol. A sulphonated glycolipid, sulphoquinovosyldiacylglycerol, and the phospholipids, phosphatidylcholine and phosphatidylglycerol are represented as relatively minor

Table 2. *The major fatty acyl residues of polar lipids extracted from different organs of soyabean* (Glycine max) *plants. Data derived from Tattrie & Veliky (1973)*

	Fatty acid (mole %)					
Organ	16:0	18:0	18:1	18:2	18:3	Other
Roots	51	8	7	11	4	19
Stems	40	4	2	28	17	9
Leaves	27	4	2	13	36	21
Seeds	21	8	5	47	5	4

components of photosynthetic membranes. The major difference between the photosynthetic membranes of different species is the type of fatty acyl residues associated with each of the polar lipids. Despite these differences, it is remarkable that higher plants and blue-green algae, which differ markedly in terms of their evolution and habitat, contain similar proportions of the major membrane lipid classes. The differences are due almost entirely to the nature of the fatty acyl substituents. In the higher plant, the photosynthetic membrane contains mostly trienoic fatty acids of which linolenate predominates, whilst in blue-green algae the major fatty acyl residues are palmitic and palmitoleic. The phosphatidylglycerol component of the chloroplast is characterized by the presence of a *trans*Δ-3-hexadecanoic acid, which differs from the phospholipid class present in the endoplasmic reticulum and mitochondria which contains fatty acids with conventional *cis*-unsaturated double bonds. Comparing the composition of lipids in chloroplasts of higher plants with those of the cyanobacteria provides support for the idea that chloroplasts originated from a symbiotic association of eukaryotic cells and cyanobacteria.

The lipid composition of other plant tissues and organelles is remarkably similar in plants grown under identical environmental conditions but significant differences are often associated with development and ageing of particular organs or their constituent organelles as well as differences in environmental factors. This is illustrated in Table 2 which presents the fatty acid composition of roots, stems, leaves and seeds of soyabean plants. It can be seen that the predominant fatty acids associated with the membrane polar lipids of roots, stems and seeds are palmitic and linoleic acids while linolenic acid dominates the composition of leaf lipids. In general, the proportion of lipid present in vegetative plant tissues is often less than and quite different from, the lipid content of seeds of the same plant. This may reflect the storage and nutritive function of the seed. The presence of certain exotic plant lipids, such as cholesterol, brassicasterol, petroselinic and erucic acids, cyclopentanyl and epoxy fatty acids, which are found abundantly in some seeds but only in trace amounts in other plant tissues, is more difficult to explain. Other changes in lipid composition associated with organ development have been reported (Ichihara & Noda, 1977). The total weights of lipids and their associated fatty acids of safflower seedlings during germination and development, for example, are found to decrease during the first seven days of growth but at the same time there is a marked increase in linolenic and a corresponding decrease in linoleic acid, while palmitic and oleic acids remain relatively constant. Many of these changes are believed to be controlled by the production of gibberellins in the embryo which is responsible for the initiation of synthesis and secretion of hydrolytic enzymes from the aleurone layer into the endosperm (Yomo & Varner, 1971). The relationship between these hormonal effects in other plants is not, however, as well defined. It has been suggested, for example, that the hormones from the embryonic axis exert a direct control over the level of glyoxylate-cycle enzymes like isocitrate lyase in the cotyledons of some plants but no such control appears to exist in cotyledons or endosperm of other species (Marriott & Northcote, 1975; Tester, 1976). In cucumber seedlings, lipid mobilization during germination does not appear to be associated with the level of isocitrate lyase and enzymes concerned with β-oxidation of fatty acids of the cotyledons but the synthesis of these enzymes appears to be inhibited by the presence of the testa, the removal of which, during normal development of the plant, was suggested as the main controlling factor in lipid mobilization (Slack *et al.* 1977). Changes in polar lipid composition associated with leaf development have been reported in bean (Fong & Heath, 1977) and cucumber (Ferguson & Simon, 1973) and show that the fresh weight and the chlorophyll content of primary leaves of these plants increase during development. These changes may reflect a synthesis and degradation of chloroplast membranes and other ultrastructural changes associated with leaf development.

Effect of temperature on membrane lipid composition

Changes in the lipid composition in response to temperature have been studied most extensively in the cyanobacteria. The relationship between growth temperature and fatty acid composition of total lipids in *Anacystis nidulans* was first reported by Holton *et al.* (1964). It was found that a reduction of the growth temperature caused a decrease in 16:0 and an increase in 16:1 fatty acyl residues of the polar lipid fraction; the average chain length of fatty acids was also found to decrease. The mode of these changes in fatty acid composition of each of the major lipid classes with growth temperature was found to be similar to that of the total cell lipids (Sato *et al.* 1979; Murata *et al.* 1979).

Multicellular species of cyanobacteria such as *Anabena variabilis*, in contrast to the unicellular *Anacystis*, contain polyunsaturated fatty acids and respond to changes in growth temperature in a different way. In this organism, when the growth temperature is lowered there is a decrease in 18:1 and 18:2 and a corresponding increase in α-18:3 at the carbon-1 position of all the major lipid classes. The proportion of carbon-16 acids remains fairly constant except for a slight decrease in 16:1 and an increase in 16:2 at the carbon-2 position of monogalactosyldiacylglycerol and digalactosyldiacylglycerol (Sato *et al.* 1979; Sato & Murata, 1980*a*). A summary of the temperature dependence of the molecular

Table 3. *Effect of growth temperature on molecular species composition of the polar lipids on* Anabaena variabilis. *Data from Sato & Murata (1982)*

Lipid	Growth Temperature (°C)		Molecular species							
		C1	18:0	18:1	18:2	18:3	18:1	18:2	18:3	18:3
		C2	18:0	16:0	16:0	16:0	16:1	16:1	16:1	16:1
MGluDG	38		24	60	10	0	—	—	—	—
	22		22	40	24	6	—	—	—	—
MGalDG	38		1	25	23	1	11	35	—	—
	22		2	2	12	34	—	3	32	12
DGDG	38		1	16	24	0	16	38	—	—
	22		1	4	20	19	1	9	37	4
SQDG	38		10	48	26	—	—	—	—	—
	22		2	10	16	58	—	—	—	—
PG	38		1	56	41	—	—	—	—	—
	22		—	10	26	61	—	—	—	—

species composition in *Anabena variabilis* is presented in Table 3. It can be seen that a decrease in growth temperature from 38°C to 22°C provokes a decrease in 18:1/16:0 and 18:2/16:0 molecular species and an increase in α-18:3/16:0 in all the major lipid classes and a corresponding decrease in 18:1/16:1 and 18:2/16:0 and increases in α-18:3/16:1 and α-18:3/16:2 in the monogalactosyldiacylglycerol and digalactosyldiacylglycerol lipids. These changes in fatty-acid composition with growth temperature can be regarded as an adaptive response of the organism to a change in environmental temperature (Ono & Murata, 1982; Murata *et al.* 1984). The nature of these changes is likely to result in a maintenance of the fluid properties of the membranes at lower growth temperatures (Quinn, 1981).

Many studies have related changes in growth temperature to changes in the membrane lipids in the tissues of higher plant cells (Hitchcock & Nichols, 1971; Harwood, 1975; Quinn & Williams, 1978; Wintermans & Kuiper, 1982). The changes observed are usually most conspicuous in membranes of non-photosynthetic organelles and there has been some confusion as to the relevance of these changes in respect of plant growth and development. Nevertheless, growth at low temperature usually results in an increase in the proportion of lipid to protein, a small increase in the proportion of phospholipids to galactolipids and a decrease in overall lipid saturation. The relative significance of these changes tends, however, to vary over relatively wide limits. Increases in unsaturated fatty acid content during acclimation, for example, are reported in numerous studies (Kuiper, 1970; De la Roche *et al.* 1972; Wilson, 1978; Ketchie & Kuiper, 1979), but are not seen in other situations (De la Roche *et al.* 1975; Wilson & Crawford, 1974). More recent studies of Chapman *et al.* (1983*a*,*b*) on membrane lipid changes in peas grown under different temperature regimes have suggested that changes in the lipid/protein ratio is the most significant factor in the photosynthetic membrane rather than changes within molecular species of lipids. The variation in amounts of

Table 4. *Differences in the proportion of saturated fatty acids, average number of double bonds per lipid* (n) *and lipid: chlorophyll ratio in spinach leaves grown under winter* (W) *and summer* (S) *conditions. Data from Chapman* et al. *(1983*a*)*

	Season	Thylakoid Saturated fatty acids (%)	Thylakoid n	Total leaf Saturated fatty acids (%)	Total leaf n
Total lipid					
	W	15·5	4·71	24·3	3·99
	S	13·9	4·76	29·2	3·69
Mole lipid/mole chlorophyll					
Galactolipid	W	2·9		3·1	
	S	2·1		2·7	
Phospholipid	W	1·2		2·0	
	S	0·8		1·2	

saturated fatty acids and the average number of double bonds per lipid molecule in photosynthetic and non-photosynthetic membranes of spinach leaves grown under winter and summer conditions are presented in Table 4. It can be seen that changes in lipid class distribution and lipid unsaturation are small compared with changes in lipid/chlorophyll ratios in thylakoid membranes with growth under these two conditions. The observed changes in lipid/chlorophyll ratio are interpreted as reflections of altered lipid/protein ratios, which is suggested by Chapman *et al.* (1983*a*) to be the dominant factor in the maintenance of membrane fluidity in the thylakoids at lower temperatures.

Effects of temperature shift on photosynthetic algae

The biochemical processes involved in lipid modification of plant membranes have been examined by observing changes in fatty acid and membrane polar lipid composition on changing environmental temperature. An organism that has been particularly useful in studies of this type is the unicellular green alga *Dunaliella salina*. This organism has many unique features which recommend its use as an experimental system. Since it has no cell wall it can be disrupted readily and fractionated conveniently. The lipid composition is typical of green algae and similar to that of most higher plants. Most importantly, it is able to tolerate a wide range of temperatures and salinities (Brown & Borowitzka, 1979) so that studies of the response to changing environmental conditions can be conveniently undertaken.

Such studies have been reported by Lynch & Thompson (1982). They subjected a homogeneous population of cells of *Dunaliella* growing logarithmically at 30 °C to chilling at 12 °C and performed a lipid analysis on membrane fractions at intervals after the temperature shift. They found that in both chloroplast and microsomal fractions there was an increase in the number of double bonds in the fatty acyl residues of phospholipids during acclimation to lower temperature. In general, microsomal phospholipids responded more quickly and to a greater extent than the chloroplast phospholipids. Little change in the relative proportions of phospholipid classes was observed in chloroplasts and microsomes despite the alterations in the fatty acyl residues. In contrast to the pattern of phospholipid changes, the chloroplast glycolipids altered dramatically on exposure to low temperature and, in particular, there was a significant increase in the proportion of digalactosyldiacylglycerol relative to monogalactosyldiacylglycerol. At the same time, there were only slight changes in fatty acid distribution within either of these glycolipid classes. The differences between phospholipid and glycolipid responses to the temperature shift appear to result from the fact that phospholipids are synthesized predominantly on the microsomal membranes by a so-called eukaryotic pathway while the glycolipids are synthesized in the chloroplast by what is referred to as a prokaryotic pathway (Quinn, 1988).

Similar studies of temperature shifts have been reported in *Anabena variabilis* (Sato & Murata, 1980*a*,*b*). It was found that 10 h after a shift in temperature from 38° to 28°C there was a decrease in 16:0 and a concomitant increase in 16:1 acylated at the carbon-2 position of monogalactosyldiacylglycerol, but the total amount of lipid remained constant. The ratio of 16:0 to 16:1 is eventually restored to that present in cells grown at 38°C. This type of transient desaturation of 16:0 to 16:1 is not observed in the other major lipid classes such as digalactosyldiacylglycerol. The introduction of the *cis* double bonds in the 16-carbon chain is regarded as an acute acclimation response to restore membrane fluidity until other, longer-term processes, can be brought into operation. Using radioisotopic techniques Sato & Murata (1981) showed that the desaturation of the 16:0 hydrocarbon chain occurs in the lipid-bound form and does not require *de novo* synthesis (Sato *et al.* 1986). Slower changes in the 18-carbon fatty acids are observed, in particular, an increase in α-18:3 in monogalactosyldiacylglycerol, sulphoquinovosyldiacylglycerol and phosphatidylglycerol at the expense of 18:1 and 18:2 fatty acids. Studies with inhibitors of protein synthesis suggest that *de novo* synthesis is required to desaturate the 18:1 and 18:2 fatty acids. By contrast, an increase in growth temperature by shifting the organisms from a low to a high temperature causes an increase in *de novo* synthesis of fatty acids and a suppression of desaturation of existing fatty acids. The net result is an increase in 16:0 and a decrease in 16:1 in monogalactosyldiacylglycerol without changes in the fatty acid composition of other major lipid classes.

Effect of temperature shift on higher plants

Thermal adaptation in higher plants to enable them to grow under different environmental conditions is partly genetic, involving the natural selection of different genotypes that can thrive in particular environments, and partly phenotypic, involving adjustments of functional properties of the plant to local environmental conditions. These can be regarded as adaptation on the one hand and acclimation on the other. The ability of higher plants to acclimate to high and low temperatures respectively is, in general, related to their normal growth conditions. Plants that grow normally in cold environments, for example, appear to have fairly limited potential to acclimate to high temperatures, whilst those that grow in warm climates tend to acclimate more readily to high temperatures but have relatively limited scope for growth at low temperatures. The ability of plants to grow at high temperatures is therefore largely a property of the individual species of plant.

Exposure of higher plant tissues to elevated temperatures is known to result in changes in the pattern of proteins that are synthesized (Barnett *et al.* 1980). This has been investigated in soyabean hypocotyls grown at 30° and exposed to a temperature of 40° (Key *et al.* 1981). A change in protein synthesis under these conditions is manifest in the appearance of a cohort of 'heat shock' proteins which disappear and are replaced by normal proteins within 3 to 4 h of returning the plant to its original growth temperature. Studies of maize seedlings subjected to elevated temperatures also suggest evidence of heat shock proteins (Cooper & Ho, 1983).

Lipid composition and saturation of the leaf or chloroplast membrane lipids of plants adapted to growth at high temperatures have not yet been subjected to systematic investigation. Pearcy (1978) has reported that an increase in growth temperature leads to an elevation in saturated leaf lipids of the desert plant, *Atriplex lentiformis*. Similarly, a detailed analysis of the changes in lipid composition of chloroplasts isolated from *Nerium oleander* adapted to growth at different temperatures has also been observed by Raison *et al.* (1982). They observed a significant change in linolenic acid in the total lipid extract, which decreased significantly on acclimation to higher temperatures. The change could be largely attributed to decreases in fatty acid of the galactolipid fraction in which linolenic acid predominates; this was accompanied by a corresponding increase in 16:0, 18:1 and 18:2 fatty acids of these lipids. The changes in fatty acids associated with the phospholipid fraction were considerably less than those in the glycolipids. As in *Dunaliella*, the differences between acclimation observed in the phospholipid and glycolipid fractions was likely to be due to their different modes of biosynthesis. In all cases, however, the variations in fatty acids observed were relatively rapid in onset after a temperature shift and significant changes were observed during the first day of acclimation but the whole process was essentially complete within 14 days.

The effect of low temperature on membrane stability

The changes in lipid composition associated with acclimation and adaptation of plant cell membranes to different environmental temperatures has repercussions on the physical stability of the membrane and its ability to function under different conditions. Phase transitions in membrane lipids, for example, have long been proposed as a primary event in such physiological phenomena as chilling injury of higher plants of tropical origin (Lyons, 1973; Raison, 1973). Nevertheless, the existence of phase transitions is difficult to demonstrate unequivocally because of the highly complex nature of the membrane systems that are present in higher plant cells. Lipid phase separations in membranes of unicellular and other photosynthetic microorganisms have been clearly demonstrated, however. Freeze-fracture electron microscopy, for example, has been used to study the phase behaviour of the blue-green alga *Anacystis nidulans* subjected to exposure at chilling temperatures (Furtado *et al.* 1979). When cells cultured at a temperature of 38 °C were thermally quenched for freeze-fracture at 35 °C a random distribution of membrane associated particles was observed in both the plasma membrane and the photosynthetic membrane of the alga. If the same cells were cooled to 15 °C before thermal quenching a lateral phase separation of membrane-associated particles was observed and large domains in both the plasma membrane and the photosynthetic membrane, which were devoid of particles, was created. These smooth regions were interpreted as gel-phase lipid domains from which the intrinsic membrane proteins had been excluded. In control experiments where cells were grown at low temperatures and thermally quenched from 15°C a random distribution of membrane-associated particles was observed and this correlates with changes in lipid composition of the membrane. Thus the membrane lipids possess fatty acyl residues which are considerably less saturated than when the organism is grown at high temperature.

The question as to which lipids are located in the phase separated domains and which lipids separate together with the proteins has been investigated by differential scanning calorimetry, low-angle X-ray diffraction and freeze-fracture electron microscopy (Mannock *et al.* 1985). The results of the calorimetric study are illustrated in Fig. 1. This shows that the organisms cultured at 28°C exhibit a broad endotherm in their heating scans and a mid-point temperature of 14°C, which presumably corresponds with the transition of gel phase lipid domains into a liquid-crystalline configuration. A similar endotherm is observed in cells cultured at 38 °C although, judging from the enthalpy of these transitions, the proportion of lipid in gel phase in cells cultured at 38° is greater than that in the corresponding membranes of cells cultured at 28°C. There is, however, a dramatic change in the mid-point temperatures of the endotherms observed in heating thermograms of total polar lipid extracts of organisms cultured at 28°C and 38°C which occur at 2°C and 4°C respectively. This observation clearly shows that it is the high-melting-point lipids which phase separate from the membrane proteins in the intact biological membrane. The question remains as to which class of lipids have higher melting points and which lipids have low melting points? This has important implications with regard to the ability of membranes to restore a normal distribution of the components after thermally induced phase separations. It is well known, for example, that phase separations of the type observed in *Anacystis nidulans* result in irreversible changes and loss in viability of cells which are unable to repair their leaky membranes (Brand *et al.* 1979).

Changes in the physical properties of *Anacystis nidulans* subjected to chilling have been compared with those of *Anabaena variabilis* by Murata (1987). A collation of this information is presented in Table 5 together with the effects on photosynthetic and other membrane-associated functions. It can be seen in *Anacystis nidulans* that the temperature of onset of phase separation, judged by freeze-fracture electron microscopy, carotenoid absorption and spin-label methods provides almost identical values for plasma membrane in cells grown at 28 °C and 38 °C. In addition, chlorophyll fluorescence of the photosynthetic membrane gives an accurate indication of the onset of phase separation in the thylakoids from cells grown at the two temperatures. It may be concluded from this table that, at the particular growth temperature, both the plasma membrane and the thylakoid membrane are in a liquid-crystalline state. On cooling, the thylakoid membranes show the first evidence of phase separation. At about 10° below this phase separation an onset of phase separation in the plasma membrane is first detected. It is also clear that the onset of phase separations in both types of membrane is reflected in the growth temperature of the algae.

In view of the very different physical behaviour of the plasma membrane and the thylakoid membranes it is rather surprising that there was no discernible difference in their polar lipid composition or substituent fatty acids (Omata & Murata, 1983). The explanation for the differences in onset temperature of phase separations may simply reflect the differences in the ratio of membrane lipids to

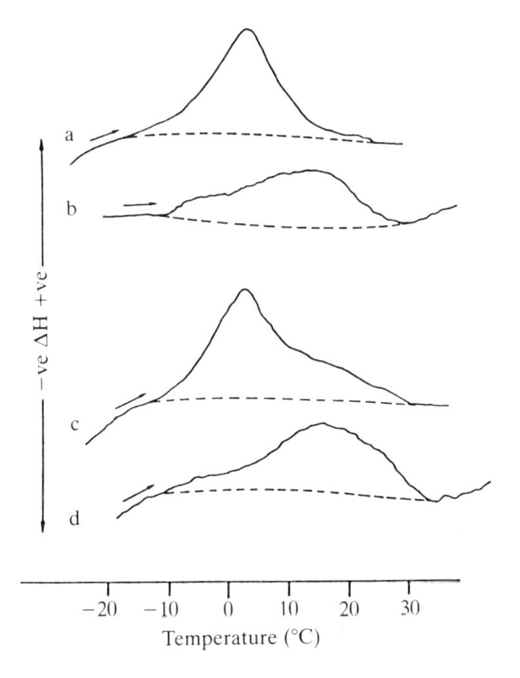

Fig. 1. Differential scanning calorimetric heating curves of aqueous dispersions of total polar lipid extracts of *Anacystis nidulans* cultured at 28 °C (a) and 38 °C (c). Thermograms of whole cell suspensions cultured at 28 °C (b) and 38 °C (d) are also shown. Data redrawn from Mannock *et al.* (1985).

Table 5. *Physiological and biophysical changes associated with cooling of two species of cyanobacteria grown a two different temperatures. Data collated by Murata (1987)*

	A. nidulans		*A. variabilis*	
Parameter	28°	38°	22°	38°
Temperature of irreversible change				
Photosynthesis	5	16	<0	<0
Hill reaction	5	15		
Leakage of electrolytes	7	16	<0	<0
Onset of lipid phase separation				
Freeze-fracture electron microscopy	5	16	<0	<0
Carotenoid absorption	5	15		
Chlorophyll fluorescence from thylakoid membrane	13	25		
Spin-label				
Plasma membrane	5	13		
Thylakoid membrane	14	23	7	17
X-ray diffraction				
Thylakoid membrane	16	26		
Differential scanning calorimetry	13	25		

protein in these two membranes. During cooling the high-melting-point lipids separate into gel phase domains and exclude the membrane proteins together with the lower melting-point lipids. Thus the higher the protein to lipid ratio the larger will be the proportion of membrane lipids that segregate together with the protein and the domains of gel phase lipid will be correspondingly smaller. The rate of cooling influences the ultimate size of phase separated domains because a coalescence of these small domains is required to observe extensive lipid phase separation in the membrane.

It can be seen from the data presented in Table 5 that critical phase separations occur in both thylakoid and plasma membranes of *Anacystis nidulans* grown either at 28° or 38°C at temperatures above zero. In contrast, the chilling-insensitive cyanobacterium *Anabaena variabilis* does not appear to undergo phase separations at temperatures in the region where chilling damage occurs to *Anacystis nidulans*. In a careful analysis of the physiological and structural changes to the membranes of these cyanobacteria, Murata (1987) concluded that phase transitions that take place in thylakoid membranes are not responsible for irreversible damage to viable cells but phase transitions in the plasma membrane are. In *Anacystis nidulans*, these transitions are observed at chilling temperatures and they are directly related to low temperature damage and loss of viability.

The relationship between lipid phase behaviour and chilling injury in higher plants is less clear than in the case of cyanobacteria. The involvement of thylakoid membranes in chilling-induced injury is, so far, not convincing. Garber (1977), for example, has reported very similar responses for thylakoids isolated from chilling-resistant plants (spinach) compared with chilling-sensitive plants (cucumber). The extent of chilling-induced injury has also been found to depend on whether the chloroplasts are illuminated during the chilling stress (Powles *et al.* 1980). The deleterious effects of chilling stress, in general, can be accounted for, in part, by cell-mediated damage and, in part, by photodynamic damage associated with an inability of the photosynthetic apparatus to function efficiently at low temperatures. Whether the membrane lipids or the fluidity of membrane are concerned in these processes remains unresolved.

The general consensus is that bulk membrane lipids of chilling-sensitive plants do not undergo gel-to-liquid crystalline phase transitions of the conventional type but there remains a distinct possibility that some minor lipid fraction does undergo such a transition. One interesting factor in this regard is the apparent differences in acidic lipid composition of the chloroplasts of chilling-sensitive and chilling-resistant species. Chloroplasts of chilling-sensitive species appear to contain higher proportions of the dipalmitoyl and 1-palmitoyl,2-(trans)-Δ^3-hexadecenoyl species of phosphatidylglycerol (Murata, 1982, 1983; Murata *et al.* 1982). The presence of a small fraction (about 1·5 %) of dipalmitoylphosphatidylglycerol in polar lipid extracts of leaves of mung bean, a chilling-sensitive species, has been reported by Raison & Wright (1983). Analysis of the corresponding fractions of the chilling-resistant species of pea and wheat, however, revealed only trace amounts of this lipid. Examination of the physical properties of the different extracts by differential scanning calorimetry indicated that whilst the mung bean samples showed a small endotherm spanning the range 3–20 °C and an exotherm between 12 °C and −5 °C, the pea extract showed only an endotherm between 2 °C and −10 °C. It was suggested that these thermal events reflect lamellar phase transitions induced by relatively small amounts of high-melting-point lipids forming monotectic mixtures with structurally dissimilar, lower-melting-point lipids. Other studies reported by Pike & Berry (1980) have shown that the fluorescence yield of *trans*-parinaric which preferentially distributes into gel phase lipid (Sklar *et al.* 1979), shows a marked increase below 10 °C if added to lipid extracts of chilling-sensitive species such as corn. No equivalent increases were observed for chilling-resistant species such as barley. Pike (1982) extended these studies to phospholipids of a series of warm- and cool-season desert annuals and showed that the temperature at which these fluorescence increases occur is markedly higher for the warm-season annuals even when the two sets of plants are grown at the same temperature. Measurements using the *cis* isomer of parinaric acid, which shows little discrimination between gel and liquid-crystalline domains showed no such increases. These observations may suggest the formation of small areas of gel-phase lipids by the solidification of domains of higher-melting-point lipids present in the chilling-sensitive species and which are sensed by the *trans* isomer but not by the *cis* isomer of parinaric acid.

Effect of high temperatures on membrane stability

The biophysical and functional properties of plant membranes exposed to high temperature reveal characteristic damage to membranes. It appears that the upper limit of temperatures at which particular plant species survive is a function of the stability of the chloroplast membrane. Leaves and isolated chloroplasts, for example, show marked reductions in their photosynthetic activity following exposures to temperatures above about 40–45 °C (Quinn & Williams, 1985). Measurements of changes in chlorophyll *a* fluorescence emission under such conditions (Krause & Santarius, 1975; Schreiber *et al.* 1976; Schreiber & Berry, 1977) indicate that the light-harvesting apparatus of photosystem-II is particularly susceptible to thermal damage. There is now convincing evidence that heating briefly to relatively high temperatures results in irreversible functional reorganization of the photosynthetic apparatus and particularly photosystem-II (Schreiber & Armond, 1978). Freeze-fracture studies (Armond *et al.* 1980; Gounaris *et al.* 1984) have shown that incubation of isolated chloroplasts in suspension at elevated temperatures leads to a progressive dissociation of the supramolecular complex corresponding to the photosynthetic light-harvesting unit of photosystem-II and the consequent destacking of the thylakoid membrane. Differential scanning calorimetric studies (Cramer *et al.* 1981) have indicated a series of endothermic transitions corresponding to order-disorder transitions of different structural domains within the chloroplast membrane. The lowest temperature transition, with a transition maximum at 42–44 °C, correlates with the release of manganese from the thylakoid membrane, loss of oxygen evolution ability by the chloroplasts and a decrease in the redox potential of high potential cytochrome *b*-559. It appears to correspond to the thermal disruption of a protein component on the donor side of photosystem-II and as such probably reflects part of the dissociation process observed in freeze-fracture studies. A breakdown of the photosynthetic apparatus is also interpreted from changes in fluorescence yield of chlorophyll *a* associated with photosystem-II. The effect of growth temperature on the threshold temperatures of fluorescence yield of chlorophyll *a* measured as changes in the ratio f_0/f_{max} (where f_0 is the initial, and f_{max} the final fluorescence yield for heated leaves), on heating leaves of *Atriplex lentifornis* is illustrated in Fig. 2. Acclimation to growth at 45 °C day/30 °C night, as opposed to 20 °C day/15 °C night, results in a shift of the threshold temperature, as reflected by these fluorescence measurements, from about 40 °C to 48 °C. Similar shifts in fluorescence yield are seen for other desert plants, such as *Larrea divearicarta* (Armond *et al.* 1978) and *Nerium oleander* (Raison *et al.* 1982). Acclimation of those species normally growing at low temperatures is probably much less marked. Santarius & Muller (1979), for example, noted only a 3° increase in the thermostability of the chloroplasts of spinach exposed to temperatures of 35 °C for three days.

The structural changes in thylakoid membranes of higher plant chloroplasts subjected to thermal stress have been examined by freeze-fracture electron microscopic techniques (Gounaris *et al.* 1983). It was reported that a normal morphology was preserved during brief exposure (5 min) of the chloroplast suspension to temperatures of up to 35 °C. Incubation at between 35° and 45 °C caused complete destacking of the grana and higher temperatures caused a phase

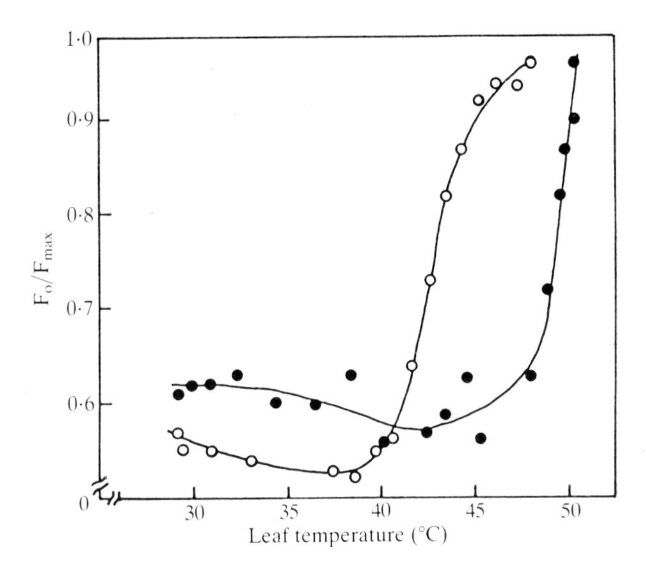

Fig. 2. The effect of leaf temperature on the ratio of initial (F_0) to maximum (F_{max}) fluorescence of *Atriplex lentiformis* grown under different temperature regimes. ●, 45°/30°C; ○, 20°/15°C. The temperature was increased at a constant rate of 1°/min from 30°C. Data from Pearcy *et al.* (1977).

separation of non-bilayer lipids into stable aggregates of cylindrical inverted micelles. The interpretation of the structural changes observed at temperatures greater than 45°C is based on phase conditions that result in a release of the constraints imposed by interaction of the non-bilayer lipid, monogalactosyl-diacylglycerol, with other membrane components and its segregation into domains of inverted micelles of lipid. Gross phase separations of this type require that the shift in thermal stability of the stacked membrane is relatively large because three-dimensional aggregates of lipids are not observed if the chloroplast membrane is destacked by manipulation of the ionic environment before heat stress.

The factors which are responsible for destacking of the granal membranes have been postulated in a model proposed by Barber (1980). According to this model, thylakoid stacking and related phenomena are explained in terms of the effect of cations on electrostatic charges on the surface of the membrane. A difference in surface charges of light-harvesting chlorophyll *a*/*b*-protein complexes associated with photosystem-II and P-700–chlorophyll *a* protein complexes associated with photosystem-I is said to result in a randomization of the complexes laterally in the membrane in conditions where the surface charges on the membrane are unscreened. It is argued that addition of counter ions to screen these charges reduces electrostatic repulsion between photosystem-I complexes, allowing a reorganization of the membrane system in which the two photosystems become spatially segregated. The formation of the grana stacks is explained by a reduction in coulombic repulsion forces between opposing membrane surfaces (Mullet & Arntzen, 1980; Ryrie *et al.* 1980).

Table 6. *The density and size of membrane-associated particles observed on the exoplasmic and protoplasmic fracture faces of heat-treated (5 min, 45°C) broad bean* (Vicia faba) *chloroplasts. Data from Gounaris* et al. *(1984)*

Fracture face (EF, Exoplasmic; PF, protoplasmic; s, stacked; u, unstacked)	Mean particle diameter (nm)	Particle density (μm^{-2})
Unheated		
EF_s	$11{\cdot}2 \pm 2{\cdot}4$	1238 ± 163
EF_u	$9{\cdot}3 \pm 2{\cdot}1$	348 ± 59
PF_s	$7{\cdot}8 \pm 2{\cdot}0$	4583 ± 322
PF_u	$9{\cdot}2 \pm 2{\cdot}1$	3580 ± 263
Heated		
EF_s	—	—
EF_u	$8{\cdot}9 \pm 3{\cdot}0$	818 ± 38
PF_s	$6{\cdot}6 \pm 1{\cdot}7$	7290 ± 258
PF_u	$9{\cdot}3 \pm 2{\cdot}4$	4346 ± 478

In studies reported by Gounaris *et al.* (1984) freeze-fracture electron microscopy showed that, on heating, normal granal stacks are progressively replaced by modified thylakoid attachment sites in which contact between opposing membranes becomes restricted to regions of focal contact. Changes in the size distribution of particles observed in the exoplasmic face of stacked and unstacked regions and corresponding regions of the protoplasmic face due to heat treatment are summarized in Table 6. The most notable difference is a disappearance of intramembranous particles in the exoplasmic profiles of the attachment sites formed during exposure to high temperature. The reverse occurs outside the stacked regions of the exoplasmic face, where the density of particles increases significantly. Changes in particle size in regions of contact between the membranes of the exposed exoplasmic and protoplasmic faces are shown in the form of difference histograms in Fig. 3. These data are interpreted as a dissociation of light-harvesting units of photosystem-II in which the antennae complexes cluster together, maintaining regions of membrane adhesion, whilst excluding the core complexes of photosystem-II and light-harvesting units of photosystem-I. A model to illustrate this process is shown in Fig. 4. In this model the electrostatic interactions maintaining contact between the membrane are not perturbed by the brief exposure to high temperature but a shift in the phase behaviour of the membrane lipid leads to dissociation of the oligomeric complexes of the photosystems. This is consistent with functional changes (Schreiber & Armond, 1978) in heat-stressed chloroplasts. The underlying reason for the change in the observed photosynthetic functions is a phase change of monogalactosyldiacylglycerol which is said to package the light-harvesting chlorophyll *a*/*b*-protein complexes together with photosystem-II core protein complex into efficient functional units localized within the region of the grana stacks (Quinn & Williams, 1983). A similar

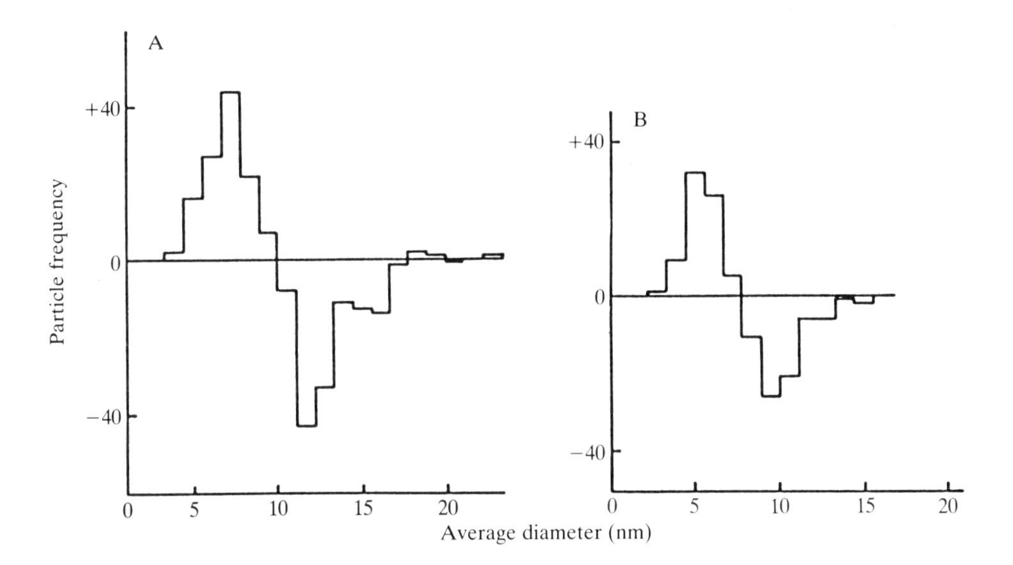

Fig. 3. Particle size difference histograms (heated minus control) of the intramembranous particles found in the exoplasmic (A) and protoplasmic (B) fracture faces of the stacked regions of the chloroplast thylakoid membrane. Data derived from Table 6.

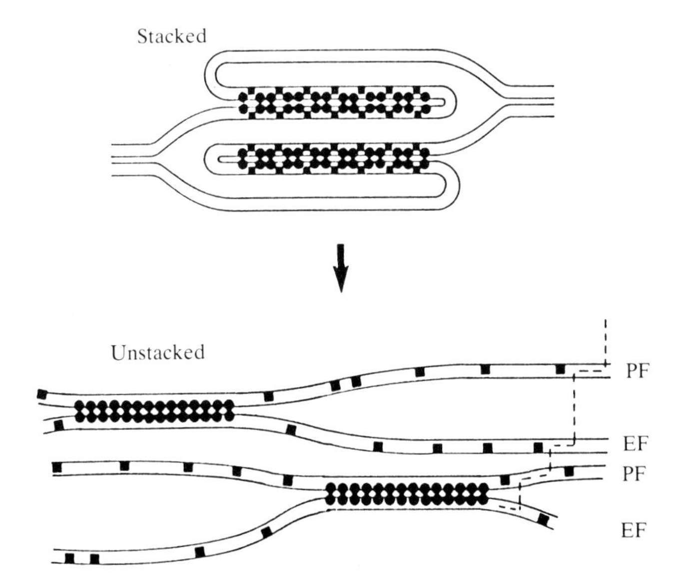

Fig. 4. Schematic model showing the effect of the dissociation of chlorophyll *a*/*b* light-harvesting proteins, ●, from photosystem-II core proteins, ■, on the organization of the photosynthetic membrane. PF, protoplasmic fracture face; EF, exoplasmic fracture face.

functional role of monogalactosyldiacylglycerol and other non-bilayer-forming lipids such as phosphatidylethanolamines has been proposed on the basis of reconstitution studies of calcium transport protein of sarcoplasmic reticulum in which the efficiency of the pumping is markedly improved by the presence of non-bilayer-forming lipids in the reconstituted membrane (Navarro *et al.* 1984).

Thylakoid membranes can also be stabilized at high temperatures by altering the chemical composition of the membrane lipids. One way to achieve this is by the use of homogeneous water-soluble hydrogenation catalysts which are able to saturate membrane lipids *in situ* when chloroplasts are incubated under an atmosphere of hydrogen gas in the dark (Vigh & Joo, 1983; Vigh *et al.* 1983, 1985*a*,*b*). The effect of hydrogenation of the membrane lipids is to create particle-free patches corresponding to phase-separated gel-phase lipids in the membrane. Examination of the freeze-fracture profiles of thylakoid membranes of chloroplasts subjected to heat treatment indicates that saturation of the membrane lipids prevents the tendency of thylakoid membrane to destack and to vesiculate at high temperatures (Thomas *et al.* 1986). Measurements of the chlorophyll *a* fluorescence emission and the thermal properties of the membranes using differential scanning calorimetry suggests that the membrane is stabilized at high temperatures by saturation of the lipids and that the interaction between the pigment-protein complexes of the photosystem-II light-harvesting apparatus is preserved by altering the phase behaviour of the membrane lipids. These observations are consistent with a number of studies (Schreiber & Berry, 1977; Pearcy *et al.* 1977; Raison *et al.* 1983) that suggested acclimation of plants, particularly those native to hot desert climates, to growth at high temperatures involves stabilization of the photosystem-II complex and is associated with changes in the degree of saturation of the membrane lipids (Pearcy, 1978; Raison *et al.* 1983) and/or changes in thylakoid lipid/protein ratios (Chapman *et al.* 1983*a*,*b*).

Conclusion

The polar lipids of plant membranes exhibit complex polymorphic behaviour in response to temperature. In general, all membranes contain molecular species of lipid that, when dispersed alone in aqueous systems, will form liquid-crystalline or gel-phase bilayers or one of a number of non-bilayer phases. It is generally believed that, at the growth temperature, the lipids are all arranged in a fluid bilayer configuration which acts as a matrix for the intrinsic and extrinsic membrane proteins. When the temperature is altered, so that it is either above or below the limits of growth of the plant, the membranes become unstable. The instability is manifest as a phase separation of the membrane components, which is driven by the creation of either gel-phase bilayer domains at low temperatures or the phase separation on non-bilayer lipid domains at high temperatures. In the latter case, gross phase separation into tubular micelle arrangements is required for identification by freeze-fracture electron microscopy, but it is likely that where there is not sufficient lipid for bulk phase-separation of this type, inverted micelles sandwiched within the lipid bilayer may form. Such structures would be indistinguishable from other membrane-associated particles believed to be associated with the intrinsic membrane proteins.

One way that the limiting temperature for plant growth may be extended could be by manipulation of the molecular species of membrane lipid. This would require the presence of increased proportions of polyunsaturated fatty acyl residues to withstand exposure to low temperatures or to increase saturation of membrane lipids to adapt plants to grow at high temperatures. The possibility of extending the range of growth temperatures and hence productivity of crop plants offers a considerable challenge to plant breeders and molecular biologists.

References

ARMOND, P. A., BJORKMAN, O. & STAEHELIN, L. A. (1980). Dissociation of supramolecular complexes in chloroplast membranes. A manifestation of heat damage to the photosynthetic apparatus. *Biochim. Biophys. Acta* **601**, 433–441.

ARMOND, P. A., SCHREIBER, U. & BJORKMAN, O. (1978). Photosynthetic acclimation to temperature in the desert shrub *Larrea divaricata. Pl. Physiol.* **61**, 411–415.

BARBER, J. (1980). An explanation for the relationship between salt-induced thylakoid stacking and the chlorophyll fluorescence changes associated with changes in spillover of energy from photosystem II to photosystem I. *FEBS Lett.* **118**, 1–10.

BARNETT, T., ALTSCHULER, M., MCDANIEL, C. N. & MASCARENHAS, J. P. (1980). Heat shock induced proteins in plant cells. *Dev. Gen.* **1**, 331–340.

BRAND, J. J., KIRCHANSKI, S. J. & RAMIREZ-MITCHELL, R. (1979). Chill-induced morphological alterations in *Anacystis nidulans* as a function of growth temperature. *Planta* **145**, 63–68.

BROWN, A. D. & BOROWITZKA, L. J. (1979). Halotolerance of *Dunaliella*. In *Biochemistry and Physiology of Protozoa*, vol. 1 (ed. M. Levandowsky & S. H. Hunter), pp. 139–190. New York: Academic Press.

CHAPMAN, D. J., DE-FELICE, J. C. & BARBER, J. (1983*a*). Influence of winter and summer growth conditions on leaf membrane lipids of *Pisum. Planta* **157**, 218–223.

CHAPMAN, D. J., DE-FELICE, J. G. & BARBER, J. (1983*b*). Growth temperature effects on thylakoid membrane lipid on protein content of pea [*Pisum sativum* cultivar Feltham First] chloroplasts. *Pl. Physiol.* **72**, 255–258.

COOPER, P. & HO, T. D. (1983). Heat shock proteins in maize. *Pl. Physiol.* **71**, 215–222.

CRAMER, W. A., WHITMARSH, J. & LOW, P. S. (1981). Differential scanning calorimetry of chloroplast membranes: identification of an endothermic transition associated with the water-splitting complex of photosystem II. *Biochemistry* **20**, 157–162.

DE LA ROCHE, I. A., ANDREWS, C. J., POMEROY, M. K., WEINBERGER, P. & KATES, M. (1972). Lipid changes in winter wheat seedlings (*Triticum aestivum*) at temperatures inducing cold hardiness. *Can. J. Bot.* **50**, 2401–2409.

DE LA ROCHE, I. A., POMEROY, M. K. & ANDREWS, C. J. (1975). Changes in fatty acid composition in wheat cultivars of contrasting hardiness. *Cryobiology* **12**, 506–512.

FERGUSON, C. H. R. & SIMON, E. W. (1973). Membrane lipids in senescing green tissues. *J. exp. Bot.* **24**, 307–316.

FONG, F. & HEATH, R. L. (1977). Age-dependent changes in phospholipids and galactolipids in primary bean leaves (*Phaseolus vulgaris*). *Phytochemistry* **16**, 215–217.

FURTADO, D., WILLIAMS, W. P., BRAIN, A. P. R. & QUINN, P. J. (1979). Phase separations in membranes of *Anacystis nidulans* grown at different temperatures. *Biochim. Biophys. Acta* **555**, 352–357.

GARBER, M. P. (1977). The effect of light and chilling temperatures on chilling-sensitive and chilling-resistant plants. *Pl. Physiol.* **59**, 981–985.

GOUNARIS, K., BRAIN, A. P. R., QUINN, P. J. & WILLIAMS, W. P. (1983). Structural and functional changes associated with heat-induced phase-separations of non-bilayer lipids in chloroplast thylakoid membranes. *FEBS Lett.* **153**, 47–51.

GOUNARIS, K., BRAIN, A. P. R., QUINN, P. J. & WILLIAMS, W. P. (1984). Structural reorganisation of chloroplast thylakoid membranes in response to heat stress. *Biochim. Biophys. Acta* **766**, 198–208.

HARWOOD, J. L. (1975). Effect of the environment on the acy lipids of algae and higher plants. In *Recent Advances in the Chemistry and Biochemistry of Plant Lipids* (ed. T. Galliard & F. J. Mercer), pp. 43–93. New York: Academic Press.

HITCHCOCK, C. & NICHOLS, B. B. (1971). *Plant Lipid Biochemistry*. London: Academic Press.

HOLTON, R. W., BLECKER, H. H. & ONORE, M. (1964). Effect of growth temperature on the fatty acid composition of a blue-green alga. *Phytochemistry* **3**, 595–602.

ICHIHARA, K.-I. & NODA, M. (1977). Distribution and metabolism of polyacetylenes in safflower. *Biochim. Biophys. Acta* **487**, 249–260.

KETCHIE, D. O. & KUIPER, P. J. C. (1979). Fatty acid levels in apple leaves of different age as affected by temperature. *Physiologia Pl.* **46**, 93–96.

KEY, J. L., LIN, C. Y. & CHEN, Y. M. (1981). Heat shock proteins in higher plants. *Proc. natn. Acad. Sci. U.S.A.* **78**, 3526–3530.

KRAUSE, G. H. & SANTARIUS, K. A. (1975). Relative thermostability of the chloroplast envelope. *Planta* **127**, 285–299.

KUIPER, P. J. C. (1970). Lipids in alfalfa leaves in relation to cold hardiness. *Pl. Physiol.* **45**, 684–686.

LYNCH, D. V. & THOMPSON, G. A. (1982). Low temperature-induced alterations in the chloroplast and microsomal membranes of *Dunaliella salina*. *Pl. Physiol.* **69**, 1369–1375.

LYONS, J. M. (1973). Chilling injury in plants. *A. Rev. Pl. Physiol.* **24**, 445–466.

MANNOCK, D. A., BRAIN, A. P. R. & WILLIAMS, W. P. (1985). Phase behaviour of the membrane lipids of the thermophilic blue-green alga *Anacystis nidulans*. *Biochim. Biophys. Acta* **821**, 153–164.

MULLET, J. E. & ARNTZEN, C. J. (1980). Simulation of grana stacking in model membrane system. Mediation by a purified light-harvesting pigment-protein complex from chloroplasts. *Biochim. Biophys. Acta* **589**, 100 117.

MURATA, N. (1982). Fatty acid compositions of phosphatidylglycerols from plastids in chilling-sensitive and chilling-resistant plants. In *Effects of Stress on Photosynthesis* (ed. R. Marcelle, H. Clijsters & M. Van Pouke), pp. 285–293. Dordrecht: Martinus Nijhoff Dr W. Junk Publishers.

MURATA, N. (1983). Molecular species composition of phosphatidylglycerols from chilling-sensitive and chilling-resistant plants. *Pl. Cell Physiol.* **24**, 81–86.

MURATA, N. (1987). Unique characteristics of cyanobacterial glycerolipids. In *The Metabolism, Structure and Function of Plant Lipids* (ed. P. K. Stumpf, J. B. Mudd & W. D. Ness), pp. 603–612. New York: Plenum.

MURATA, N., ONO, T. & SATO, N. (1979). Lipid phase of membrane and chilling injury in the blue-green alga *Anacystis nidulans*. In *Low Temperature Stress in Crop Plants: The Role of the Membrane* (ed. J. M. Lyons, D. Graham & J. K. Raison), pp. 337–345. New York: Academic Press.

MURATA, N., SATO, N., TAKAHASHI, N. & HAMAZAKI, Y. (1982). Molecular species composition of phosphatidylglycerols from chilling-sensitive and chilling-resistant plants. *Pl. Cell Physiol.* **23**, 1071–1079.

MURATA, N., WADA, H. & HIRASAWA, R. (1984). Reversible and irreversible inactivation of photosynthesis in relation to the lipid phases of membranes in the blue-green alga (cyanobacteria) *Anacystis nidulans* and *Anabena variabilis*. *Pl. Cell Phsyiol.* **25**, 1027–1032.

NAVARRO, J., TOVIO-KINNUNCAN, M. & RACKER, E. (1984). Effect of lipid composition on the calcium/adenosine 5′-triphosphate coupling ratio of the Ca^{2+}-ATPase of sarcoplasmic reticulum. *Biochemistry* **23**, 130–135.

OMATA, T. & MURATA, N. (1983). Isolation and characterization of the cytoplasmic membranes from the blue-green alga (cyanobacterium) *Anacystis nidulans*. *Pl. Cell Physiol.* **24**, 1101–1112.

ONO, T. & MURATA, N. (1982). Chilling-susceptibility of the blue-green alga *Anacystis nidulans*. III. Lipid phase of cytoplasmic membrane. *Pl. Physiol.* **69**, 125–129.

PEARCY, R. W. (1978). Effect of growth temperature on the fatty acid composition of leaf lipids in *Atriplex lentiformis* (Torr.) Wats. *Pl. Physiol.* **61**, 484–486.

PEARCY, R. W., BERRY, J. A. & FORK, D. C. (1977). Effects of growth temperature on the thermal stability of the photosynthetic apparatus of *Atriplex lentiformis* (Torr.) Wats. *Pl. Physiol.* **59**, 873–878.
PIKE, C. S. (1982). Membrane lipid physical properties in annuals grown under contrasting thermal regimes. *Pl. Physiol.* **70**, 1764–1766.
PIKE, C. S. & BERRY, J. A. (1980). Membrane phospholipid phase separations in plants adapted to or acclimated to different thermal regimes. *Pl. Physiol.* **66**, 238–241.
POWLES, S. B., BERRY, J. A., BJORKMAN, O. (1980). Photoinhibition of intact attached leaves of C_4 plants: dependence on carbon dioxide and oxygen partial pressures. *Carnegie Instn YB.* **79**, 157–160.
QUINN, P. J. (1981). The fluidity of cell membranes and its regulation. *Prog. Biophys. molec. Biol.* **38**, 1–104.
QUINN, P. J. (1988). Regulation of membrane fluidity in plants. In *Advances in Membrane Fluidity*, vol. 3, pp. 293–321 (ed. R. C. Aloa, L. M. Gordon & C. C. Curtain). New York: Alan R. Liss.
QUINN, P. J. & WILLIAMS, W. P. (1978). Plant lipids and their role in membrane function. *Progr. Biophys. molec. Biol.* **34**, 109–173.
QUINN, P. J. & WILLIAMS, W. P. (1985). The structural role of lipids in photosynthetic membranes. *Biochim. Biophys. Acta* **737**, 223–266.
QUINN, P. J. & WILLIAMS, W. P. (1987). The phase behaviour of lipids in photosynthetic membranes. *J. Bioenerg. Biomembr.* **19**, 605–624.
RAISON, J. K. (1973). The effect of temperature-induced phase changes on the kinetics of respiratory and other membrane-associated enzyme systems. *J. Bioenerg.* **4**, 285–309.
RAISON, J. K., ROBERTS, J. K. M. & BERRY, J. A. (1983). Acclimation of the higher plant, *Nerium oleander* to growth temperature: correlations between the thermal stability of chloroplast (thylakoid) membranes and the composition and fluidity of their polar lipids. *Biochim. Biophys. Acta* **688**, 218–228.
RAISON, J. K. & WRIGHT, L. C. (1983). Thermal phase transitions in the polar lipids of plant membranes. Their induction by disaturated phospholipids and their possible relation to chilling injury. *Biochim. Biophys. Acta* **731**, 69–78.
RYRIE, I. J., ANDERSON, J. M. & GOODCHILD, D. J. (1980). The role of the light-harvesting chlorophyll *a*/*b*-protein complex in chloroplast membrane stacking. Cation-induced aggregation of reconstituted proteoliposomes. *Eur. J. Biochem.* **107**, 345–354.
SANTARIUS, K. A. & MULLER, M. (1979). Investigations on heat resistance of spinach leaves. *Planta* **146**, 529–538.
SATO, N., MURATA, N., MIURA, Y. & UETA, N. (1979). Effect of growth temperature on lipid and fatty acid compositions in the blue-green algae *Anabena variabilis* and *Anacystis nidulans*. *Biochim. Biophys. Acta* **572**, 19–28.
SATO, N. & MURATA, N. (1980*a*). Temperature shift-induced responses in lipids in the blue-green alga, *Anabena variabilis*. The central role of diacylmonogalactosylglycerol in thermo-adaptation. *Biochim. Biophys. Acta* **619**, 353–366.
SATO, N. & MURATA, N. (1980*b*). Desaturation of fatty acids in lipids in response to the growth temperature in the blue-green alga *Anabena variabilis*. In *Biogenesis and Function of Plant Lipids* (ed. P. Mazliak, P. Benvenisk, C. Costes & R. Douce), pp. 207–210. Amsterdam: Elsevier/North Holland Biomedical Press.
SATO, N. & MURATA, N. (1981). Studies on the temperature shift-induced desaturation of fatty acids in monogalactosyldiacylglycerol in the blue-green alga (cyanobacterium) *Anabena variabilis*. *Pl. Cell Physiol.* **22**, 1043–1050.
SATO, N. & MURATA, N. (1982). Lipid biosynthesis in the blue-green alga, *Anabena variabilis*. II. Fatty acids and lipid molecular species. *Biochim. Biophys. Acta* **710**, 279–289.
SATO, N., SEYAMA, Y. & MURATA, N. (1986). Lipid-linked desaturation of palmitic acid in monogalactosyldiacylglycerol in the blue-green alga (cyanobacterium) *Anabena variabilis* studied *in vivo*. *Pl. Cell Physiol.* **27**, 819–835.
SCHREIBER, U. & ARMOND, P. A. (1978). Heat-induced changes of chlorophyll fluorescence in isolated chloroplasts and related heat-damage at the pigment level. *Biochim. Biophys. Acta* **502**, 138–151.
SCHREIBER, U., COLBOW, K. & VIDAVER, W. (1976). Analysis of temperature-jump chlorophyll

fluorescence induction in plants. *Biochim. Biophys. Acta* **423**, 249–263.
SCHREIBER, U. & BERRY, J. (1977). Heat-induced changes of chlorophyll fluorescence in intact leaves correlated with damage of the photosynthetic apparatus. *Planta* **136**, 233–238.
SKLAR, L. A., MILJANICH, G. P. & DRATZ, E. A. (1979). Phospholipid lateral phase separation and the partition of *cis*-parinaric acid and *trans*-parinaric acid among aqueous, solid lipid and fluid lipid phase. *Biochemistry* **18**, 1707–1716.
SLACK, P. T., BLACK, M. & CHAPMAN, J. M. (1977). The control of lipid mobilization in *Cucumis* cotyledons. *J. exp. Bot.* **28**, 569–577.
TATTRIE, N. H. & VELIKY, I. A. (1973). Fatty acid composition of lipids in various plant cell cultures. *Can. J. Bot.* **51**, 513–516.
THOMAS, P. G., DOMINY, P. J., VIGH, L., MANSOURIAN, A. R., QUINN, P. J. & WILLIAMS, W. P. (1986). Increased thermal stability of pigment-protein complexes of pea thylakoids following catalytic hydrogenation of membrane lipids. *Biochim. Biophys. Acta* **849**, 131–140.
VIGH, L. & JOO, F. (1983). Modulation of membrane fluidity by catalytic hydrogenation affects the chilling susceptibility of the blue-green alga, *Anacystis nidulans*. *FEBS Lett.* **162**, 423–427.
VIGH, L., JOO, F., VAN HASSELF, P. R. & KUIPER, P. J. C. (1983). Hydrogenation of model and biomembranes using a water-soluble ruthenium phosphine catalyst. *J. molec. Catal.* **22**, 15–22.
VIGH, L., JOO, F. & CSEPLO, A. (1985*a*). Modulation of membrane fluidity in living protoplasts of *Nicotiana plumbaginifolia* by catalytic hydrogenation. *Eur. J. Biochem.* **146**, 241–244.
VIGH, L., JOO, F., DROPPA, M., HORVATH, L. I. & HORVATH, G. (1985*b*). Modulation of chloroplast membrane lipids by homogeneous catalytic hydrogenation. *Eur. J. Biochem.* **147**, 477–481.
WILSON, J. M. (1978). Leaf respiration and ATP levels at chilling temperatures. *New Phytol.* **80**, 325–334.
WILSON, J. M. & CRAWFORD, R. M. M. (1974). Leaf fatty-acid content in relation to hardening and chilling injury. *J. exp. Bot.* **25**, 121–131.
WINTERMANS, J. F. G. M. & KUIPER, P. J. C. (eds) (1982). *Biochemistry and Metabolism of Plant Lipids*. Amsterdam: Elsevier.
YOMO, H. & VARNER, J. E. (1971). In *Current Topics in Development Biology* (ed. A. A. Moscona & A. Monroy), no. 6, pp. 111–114. London: Academic Press.

TEMPERATURE SHOCK PROTEINS IN PLANTS

HELEN J. OUGHAM and CATHERINE J. HOWARTH

Plant and Cell Biology Department, Welsh Plant Breeding Station, Aberystwyth SY23 3EB, Wales, UK

Summary

Plant tissue generally responds rapidly to sudden increases in temperature by curtailing or abolishing normal protein synthesis and producing new polypeptides known as heat shock proteins (HSP). Some of the methods used for monitoring the expression of heat shock genes are described, and the characteristics of the heat shock response in higher plants are discussed with special reference to tropical cereals. The possible role for heat shock proteins in conferring thermotolerance upon plant tissue is considered. The behaviour of plant tissue subjected to temperature decreases has been much less intensively studied, and varies greatly according to species and the nature of the cold treatment. No homology has yet been detected between the heat shock and the cold shock response in any plant system. Cold-induced changes in gene expression observed in a wide range of plant species are discussed with particular reference to parallel changes in cold-hardiness. Marked contrasts have been observed between the response of temperate grasses and that of tropical cereals to cold treatment, and these are discussed in relation to growth and survival at suboptimal temperatures.

Introduction

The biosynthesis of proteins is a very thermosensitive metabolic process. Changes in temperature can affect both the amounts and types of polypeptides produced. The response of living tissue subjected to a sudden temperature rise is almost always characterized by a cessation or reduction in synthesis of the normal complement of polypeptides and rapid production of large amounts of a group of proteins known collectively as heat shock proteins (HSP). It is not the purpose of this paper to describe the heat shock response in detail, as many excellent reviews already cover its biochemistry and molecular biology. Books edited by Schlesinger, Ashburner & Tissières (1982) and Nover (1984) and reviews by Neidhardt *et al.* (1984) and Lindquist (1986) provide comprehensive introductions to the study of heat shock response in a wide range of eukaryotic and prokaryotic systems. Kimpel & Key (1985) give a very readable and concise overview of heat shock in higher plants, and the organization and regulation of plant heat shock genes are described in detail elsewhere (Key *et al.* 1985*a*; Key *et al.* 1985*b*; and Schöffl *et al.* 1986). Papers by Matters & Scandalios (1986) and Sachs & Ho (1986) discuss changes in plant gene expression caused by a range of environmental stresses including heat shock. The first part of this chapter will outline some of the methodology which has been applied to the study of temperature-induced changes

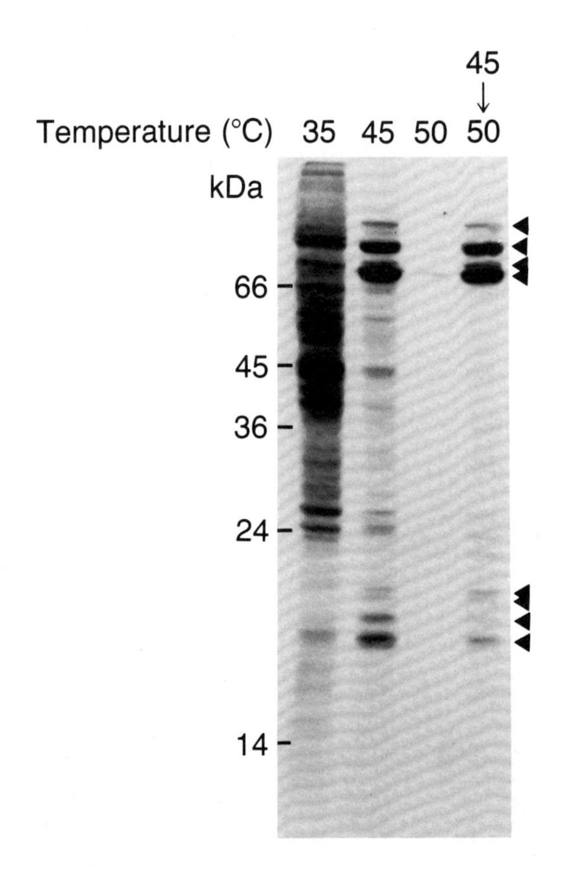

Fig. 1. Effect of temperature on protein synthesis by 40-h old millet seedlings. Seedlings were incubated in the presence of [^{35}S]-methionine for 2 h at 35, 45 or 50 °C, or at 50 °C following a 30-min pretreatment at 45 °C. Radioactive proteins were separated by SDS–polyacrylamide gel electrophoresis and visualized by fluorography. Arrows indicate heat shock proteins. Bars show the positions of molecular weight markers.

in gene expression, and consider those features of the heat shock syndrome which are of particular relevance to the survival and functioning of higher plants at supra-optimal temperatures. The second part will concentrate upon processes occurring as a result of temperature decreases.

Heat shock proteins

Methods for investigating heat shock protein synthesis

Fig. 1 illustrates some typical data from an experiment on the heat shock response. Plant tissue is incubated for a period in the presence of a radioactive amino acid (most commonly [^{35}S-methionine], but tritiated and [^{14}C]-labelled amino acids are also used), at a control or heat shock temperature. Proteins are then extracted from the tissue and separated according to molecular weight by polyacrylamide gel electrophoresis (PAGE), most commonly using a denaturing SDS–PAGE system based on that of Laemmli (1970). After electrophoresis the gel may be stained (for example, using Coomassie Brilliant Blue or silver stain) to visualize the total polypeptide complement of the tissue concerned. When studying heat shock proteins, however, it is generally the *changing* pattern of polypeptide synthesis which is of interest rather than the total protein complement. This is because although HSP are synthesized rapidly in response to heat shock, they do not always accumulate to levels permitting their detection by staining techniques. Using the sensitive silver staining method, however, several groups have been able to show accumulation of HSP in cotton (Burke, Hatfield, Klein & Mullet, 1983), tomato (Nover & Scharf, 1984), soybean (Key *et al.* 1985*a*; Mansfield & Key, 1987), desert succulents (Kee & Nobel, 1985) and several other monocot and dicot species (Mansfield & Key, 1987). Polypeptides synthesized during the period of incubation with radioactive amino acid will usually incorporate the amino acid. If the gel is dried and exposed to an X-ray film, these newly-synthesized polypeptides will therefore appear as black bands upon the autoradiograph. A common variant is fluorography, in which the gel is impregnated with a fluor before drying. The fluor intercepts radioactive emissions (β-particles) and converts them into light emissions; it is this light, rather than the β-particles themselves, which darkens the X-ray film. Fluorography provides greater sensitivity than autoradiography, so is particularly useful where proteins are not very strongly labelled. Fig. 1 is a fluorograph showing the polypeptides synthesized by millet seedlings grown at 35 °C and subsequently incubated at either 35 °C (control) or 45 °C (heat shock). The heat shock proteins (marked) can be seen to fall into two main categories: high molecular weight (over 50 kDa) and low molecular weight (14–30 kDa). This distribution is typical and characteristic of that seen in most higher plants. The high molecular weight HSP are known to have close homologies with those produced in other eukaryotes, and certain of them, e.g. HSP70 (i.e. the protein with a molecular weight of approximately 70 kDa) are even related to bacterial HSP (Craig *et al.* 1982). It is the low molecular weight group which is the special feature of higher plants. Mansfield & Key (1987) surveyed low molecular weight HSP synthesis in a number of plant species, finding that dicot plants produce a higher proportion of these proteins than do monocots, and that the properties of the proteins vary considerably from one species to another, though in all cases they accumulate to stainable levels. It is believed that proteins in this group may have a thermoprotective role. This aspect will be discussed later.

While one-dimensional SDS–PAGE is a powerful technique for separating and visualizing proteins, it sometimes lacks sufficient resolution. Two-dimensional electrophoresis can provide this. In this technique, proteins are separated on the basis of two different physical or chemical properties in mutually perpendicular directions. Usually isoelectric focussing is followed by SDS–PAGE. Fig. 2 illustrates the fluorograms obtained from 2D-gel separations of proteins made in sorghum seedlings at normal or heat shock temperatures. The 70 kDa molecular

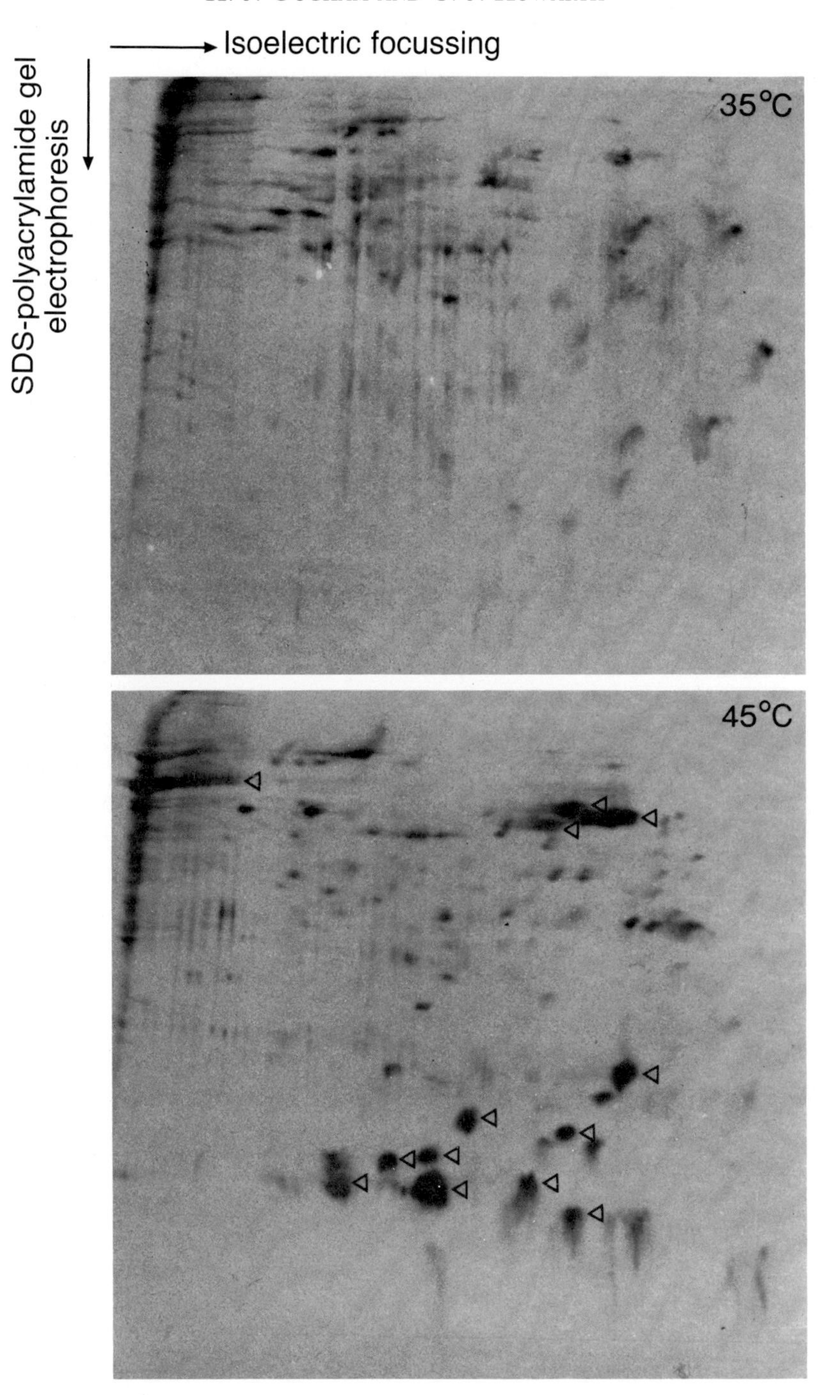

Fig. 2. Two-dimensional separations of proteins synthesized at normal and heat shock temperatures. Sorghum seedlings were exposed to [^{35}S]-methionine for 2 h at 35 or 45°C, and the extracted proteins separated by isoelectric focussing in the first dimension followed by SDS–polyacrylamide gel electrophoresis in the second dimension. Fluorography was used to visualize the radioactive proteins. Arrows show positions of HSP.

weight range can clearly be seen to contain a number of distinct polypeptides which on a 1D-separation would have appeared as a single intense band. The complexity of the low molecular weight HSP is also shown. The major limitation of 2D-electrophoresis is the difficulty of obtaining quantitative data and accurate comparisons, since variations in running conditions are compounded when two separation techniques are used consecutively. Photographic presentation of results alone is not always sufficient to illustrate subtle changes occurring in a very complex polypeptide pattern. A number of commercial companies are, therefore, now producing analytical systems based on image analysis or densitometry which can collect and process data, correct for distortions, compare two or more separations and calculate the relative amounts of labelled material in spots.

The temperature threshold for heat shock protein synthesis

In almost all eukaryotic tissues, the synthesis of heat shock proteins is induced rapidly (within minutes) when the temperature is raised above the requisite threshold. This threshold varies from species to species; for example, in maize it is approximately 35 °C (Cooper & Ho, 1983); in sorghum it is between 37 and 40 °C according to genotype (Ougham & Stoddart, 1986 and unpublished data); in temperate grasses it is between 30 and 35 °C (Ougham, 1987 and unpublished data) and in soybean it lies between 35 and 37·5 °C (Key *et al.* 1985*b*). It is absolute temperature rather than the magnitude of the temperature increase which appears to be significant in inducing the HS response, and this probably reflects the optimum temperature range for the species concerned. Above the threshold, the exact nature of the response depends on both temperature and duration of the heat treatment. For example, in millet seedlings grown at 35 °C, heat shock proteins are first detectable at about 40 °C. At 45 °C, HSP synthesis is maximal and normal protein synthesis is greatly curtailed (Fig. 1). At 50 °C, there is a small amount of residual HSP synthesis but normal proteins are no longer made. A 55 °C treatment is sufficient to abolish protein synthesis completely. For temperate plants lower temperatures are sufficient both to induce maximal HSP synthesis and to abolish protein synthesis.

Heat shock proteins and thermotolerance

An important aspect of HSP synthesis is its effect upon thermotolerance in the tissue concerned. For example, if soybean seedlings grown at 30 °C are exposed to 45 °C for 2 h, little protein synthesis of any kind occurs and the seedlings do not survive when they are returned to 30 °C. But if the 45 °C treatment is preceded by 2 h at 40 °C, during which HSP synthesis occurs, at 45 °C the seedlings produce large amounts of HSP and they resume normal growth when returned to 30 °C (Key *et al.* 1985*b*). A similar effect can be produced if the 2-h pretreatment at 40 °C is replaced by 10 min at 45 °C (a period insufficient to reduce viability) and 1 h at 30 °C. Similar behaviour in other species including barley (Marmiroli *et al.* 1986*a*), sorghum (Ougham & Stoddart, 1986) and millet (Howarth, in press), has added weight to the inference that inducing the synthesis of heat shock proteins leads to an increase in thermotolerance. Fig. 1 provides an illustration of heat-shock-induced increase in the heat tolerance of HSP synthesis. Increased thermotolerance is manifested both as an ability to continue protein synthesis at a temperature which would otherwise prevent it, and as an increase in subsequent viability of material subjected to a heat treatment which would kill most tissue not so pretreated. Heat tolerance induced by exposure to high but non-lethal temperatures is clearly analogous to cold-hardening (acquired resistance to otherwise damaging low temperatures by acclimation at suboptimal but non-injurious temperatures), a phenomenon which will be discussed in more detail later.

The time course of heat shock protein synthesis

For plants growing in a natural environment rather than under laboratory conditions, gradual temperature increases (during the course of the day) are likely to be at least as common as sudden rises of the 'heat shock' type, though the latter may occur due to, for example, alternation of sun and clouds. Gradual temperature rises have been shown to induce HSP synthesis in a number of species, including soybean (Key *et al.* 1985*b*). Moreover, during prolonged high temperature treatment (whether the temperature increase was attained rapidly or gradually) there are alterations in the pattern of proteins being synthesized. Necchi, Pogna & Mapelli (1987) designated two classes of HSP – 'early' and 'late' – observed in five cereal species. Early HSP were synthesized within 4 h of a temperature rise (from 20 to 40 °C) and late HSP appeared only after 7 h or more at 40 °C. Similarly, Cooper & Ho (1983) observed progressive changes in the pattern of HSP over a 10-h exposure of maize root tissue to 40 °C. Kee & Nobel (1986) showed a correlation between HSP synthesis and acquired thermotolerance during prolonged exposure (3–10 days) of desert succulents to elevated growing temperatures. Such findings raise a question concerning nomenclature. The term 'heat shock proteins' has connotations which make it strictly applicable only to those molecules produced rapidly as a result of abrupt temperature rises, and it may be that in the light of currently available information researchers will be compelled to replace the concise 'HSP' with alternative definitions, for example, 'proteins whose synthesis is induced or greatly enhanced by temperature increases'!

Exceptions to the heat shock response

A very small number of eukaryotic tissues studied do not exhibit the normal heat shock response. Synthesis of HSP normally requires transcription of heat shock genes, a process which can be demonstrated by the appearance of heat shock messenger RNA within a few minutes of a temperature rise (Key *et al.* 1985*b*). These heat shock mRNAs, like the HSP themselves, are absent, or present at low levels, in control tissue. However, in *Xenopus* oocytes, the response does not depend upon *de novo* synthesis; instead, mRNAs synthesized during oogenesis and normally stored in an inactive form are translated as a consequence of heat shock (Bienz & Gurdon, 1982). In higher plants, the most intensively studied exception to the standard response is that shown by pollen.

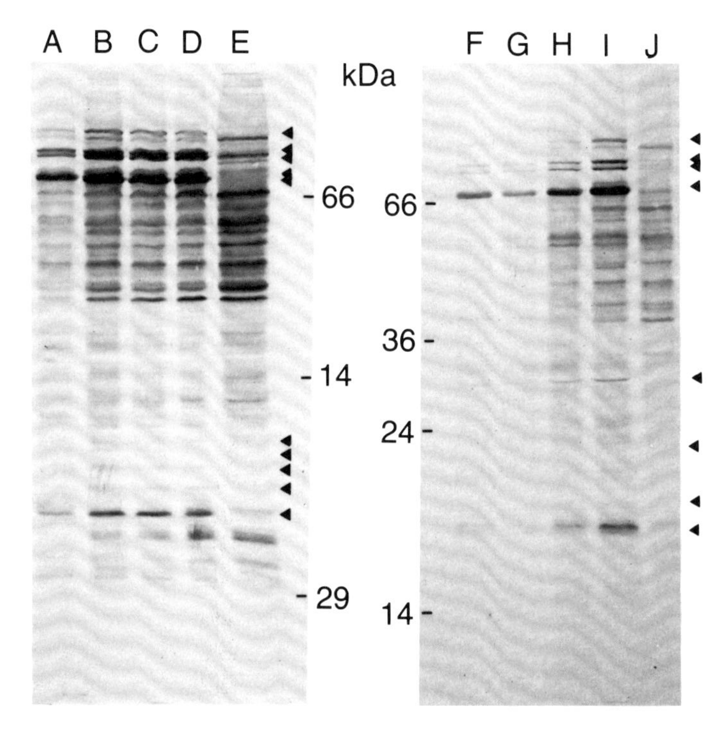

Fig. 3. Increasing capacity for HSP synthesis during the first 8 h of imbibition in two sorghum varieties (A–E, variety IS18530; F–J, variety IS4845). Embryos were dissected from dry seeds and imbibed at 35 °C. During the last 2 h of imbibition, [^{35}S]-methionine was added and the temperature either raised to 45 °C or maintained at 35 °C. Radioactive proteins were separated by SDS–polyacrylamide gel electrophoresis and visualized by fluorography. A and F, 2 h at 45 °C; B and G, 2 h at 35 °C, 2 h at 45 °C; C and H, 4 h at 35 °C, 2 h at 45 °C; D and I, 6 h at 35 °C, 2 h at 45 °C; E and J, 8 h at 35 °C. Arrows indicate HSPs, bars show positions of molecular weight markers.

Germinating maize pollen fails to synthesize any of the normal HSP, in contrast to all other maize tissues studied (Cooper *et al.* 1984), though two novel polypeptides are produced in response to a heat shock. In growing pollen tubes of *Tradescantia*, despite the fact that a prior exposure to gradually increasing temperature protects against otherwise injurious exposure to 41 °C, there is no apparent concomitant synthesis of HSP (Xiao & Mascarenhas, 1985). Since this study also failed to demonstrate the presence of preformed HSP in pollen grains, the mechanism for induction of thermotolerance in pollen remains obscure but appears not to depend upon the heat shock response. Restricted ability to produce HSP is also a characteristic of young seedlings of the tropical cereals sorghum and millet. Fig. 3 shows the increasing ability of two sorghum varieties to synthesize HSP during the first 8 h of imbibition by the seed. After 2 h of imbibition the capacity to synthesize the low molecular weight HSP in particular is very limited; this capacity increases progressively during the next 6 h. Genotypes differ with respect to the time at which they are first able to make HSP. As seen in Fig. 4, once acquired, the ability

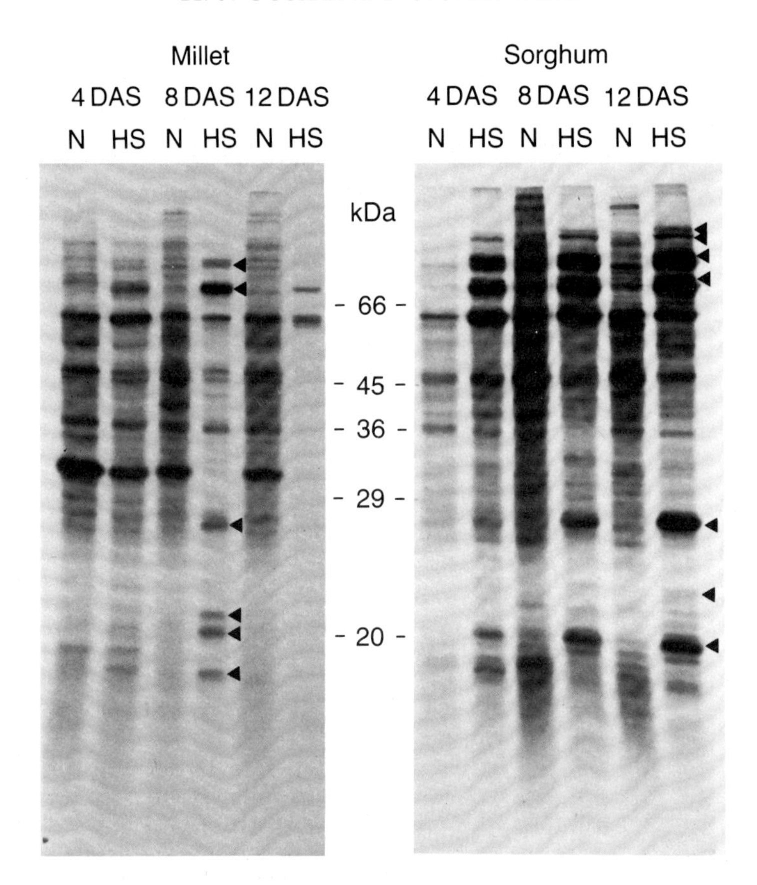

Fig. 4. Capacity for HSP synthesis by millet and sorghum seedlings of different ages. Seedlings were germinated for 4–12 days at 35 °C, then labelled with [^{35}S]-methionine for 2 h at 35 or 45 °C. Radioactive proteins were separated and visualized as described for Fig. 3. DAS, days after sowing. N, normal temperature (35 °C). HS, heat shock (45 °C). Arrows indicate HSP, bars show positions of molecular weight standards.

to make HSP is retained by sorghum seedlings studied at 4, 8 and 12 days after sowing. In contrast, millet seedlings, which are able to synthesize HSP early on during imbibition, progressively lose this ability between 4 and 12 days after sowing. Though at this stage they constitute only circumstantial evidence, these observations are consistent with the experience of plant breeders and agronomists that sorghum seedlings are most susceptible to high soil temperatures during early germination (Peacock, 1982), whereas millet germinates well but fails during the seedling establishment phase if the temperature is too high. Characterization of the biochemical response of plant tissue to high temperatures may thus assist in the development of more thermotolerant varieties.

Other factors which induce the heat shock response

Increased temperature is not the only factor which induces the synthesis of HSP. Many stresses, including water stress (Bewley *et al.* 1983; Heikkila *et al.* 1984) and wounding (Heikkila *et al.* 1984; Theillet *et al.* 1984), bring about changes in gene expression in higher plants, and in some cases these include the synthesis of a subset of the HSP and heat shock mRNAs, though usually at levels much lower than those induced by heat shock. The treatment which elicits a response most closely resembling the heat shock syndrome is exposure to sodium arsenite (Key *et al.* 1985*b*); the reason for this is unclear, since other compounds known to affect respiration or phosphorylation are not particularly effective in inducing HSP synthesis (Czarnecka *et al.* 1984). HSP synthesis caused by stresses other than high temperature does in some cases appear to confer thermotolerance upon plant tissue (Czarnecka *et al.* 1984; Chen *et al.* 1986), as is also the case in yeast (Plesset *et al.* 1982) and mammalian tissue (Li, 1983). It has also been observed that water and heavy metal stress treatments can confer thermotolerance upon maize tissue without a detectable increase in HSP synthesis (Bonham-Smith *et al.* 1987). Results like that of Bonham-Smith *et al.*, and work on acquired thermotolerance in pollen without HSP synthesis (Xiao & Mascarenhas, 1985), strongly imply the existence of other cellular mechanisms, not involving HSP, which are capable of providing heat tolerance. Despite the astonishing degree of conservation of the heat shock response across the spectrum of living organisms, suggesting that it has an essential function, it is clear that the capacity for HSP synthesis is not the sole requirement if tissue is to withstand supra-optimal temperatures, though under many conditions it may be the limiting factor.

Possible functions for heat shock proteins

The heat shock response has been intensively studied, and much is known about the organization and regulation of heat shock genes (Key *et al.* 1985*a*; Schöffl *et al.* 1986), but there is still relatively little information about the function of HSP. Where other methods of characterizing a particular protein fail, it is possible to determine the primary structure by sequencing the cloned gene encoding that polypeptide, and make inferences about its function by comparison with other known protein sequences. This approach is likely to prove valuable in identifying the roles of some of the HSP. Schlesinger (1986) reviews data indicating likely functions for HSP70 (the most highly conserved HSP and in many tissues one of the most abundant to be synthesized following a heat shock). Its properties suggest that it may have a structural role in protecting cell components from high temperatures. In both *Drosophila* (Velasquez & Lindquist, 1984) and maize (Cooper & Ho, 1987), much of the HSP70 synthesized during heat shock is localized in the nucleus, and it has been suggested that it helps to protect ribonuclear proteins (Pelham, 1985). Other HSP have also been shown to have specific subcellular localizations. For example, Vierling *et al.* (1986) showed transport of certain HSPs into chloroplasts of soybean, pea and maize. The molecular weights of these HSPs varied from species to species, but included at least one major polypeptide in the low molecular weight range (21–27 kDa) in each case. Lin *et al.* (1984), Neumann *et al.* (1984), and Cooper & Ho (1987) showed localization of HSP to other subcellular compartments including mitochondria, plasma membrane and ribosomes. Certain plant species have been shown to concentrate a large proportion of low molecular weight HSP into cytoplasmic aggregates known as heat shock granules (Neumann *et al.* 1984; Nover, 1984); these are believed to protect messenger RNA and other cell components. The likelihood that their role is primarily structural is enhanced by the considerable sequence homology between some of the low molecular weight heat shock proteins and the bovine protein α-crystallin, which is a major structural component of the eye lens.

One polypeptide which has been identified as a HSP, at least in animal cells (Bond & Schlesinger, 1985), is the small protein ubiquitin. So called because of its almost universal occurrence in eukaryotic tissues, it functions in the ATP-dependent degradation of (in some cases aberrant) proteins. Vierstra (1987) reviews this role in plants. Based on the finding that ubiquitin is a HSP, Munro & Pelham (1985) have formulated a hypothesis (as yet unproven) for the induction of the heat shock response by accumulation of proteins damaged by heat or other stresses. Ubiquitin has not yet been shown to be a HSP in plants, but in view of the very high homology of its gene sequence, gene organization and expression between barley and mammalian tissue (Gausing & Barkardottir, 1986) it clearly merits detailed study.

Few other HSPs in plants have been characterized. Leland & Hanson (1985) showed induction of a specific N-methyltransferase in expanding barley leaves grown at high temperatures for long periods, but believed the mechanism for its induction to be distinct from the heat shock response; there was no evidence that it conferred thermotolerance upon the tissue. The search for functions for many of the HSP may be hampered by the fact that some at least are gratuitously induced by heat shock because they possess appropriate gene regulatory sequences. One of the characteristic features of heat shock proteins is the presence of a highly conserved DNA sequence (the so-called HS consensus sequence or heat shock box) preceding the coding region of the gene. This sequence appears to be required for transcription of a gene in response to heat shock (Nover, 1984; Schöffl *et al.* 1986; Key *et al.* 1985*a*), but it does not follow that all proteins whose genes possess this feature, and whose synthesis is induced by heat shock, necessarily have a function, whether in conferring thermotolerance or otherwise.

Heat shock proteins as a model genetic system

The limited data available concerning the functions of HSP do not prevent the heat shock response from being a useful model system for workers interested in gene regulation. The potential to induce a rapid and major reprogramming of gene expression by applying a defined environmental stimulus can be exploited in many ways. For example, mature grass leaf tissue, which is incapable of cell division, normally undergoes very restricted gene expression other than that associated with photosynthetic functions (Ougham, 1987; Ougham *et al.* 1987). Nevertheless,

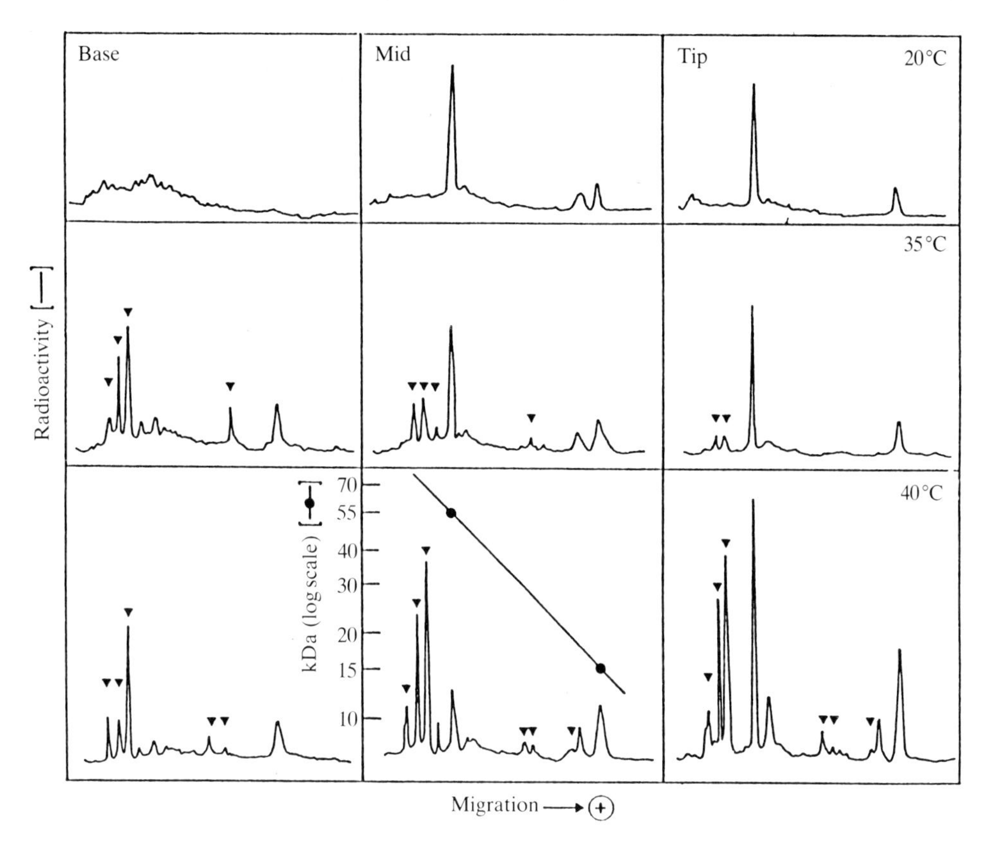

Fig. 5. HSP synthesis in all parts of developing grass leaves. Fourth leaves of *Lolium temulentum* were exposed to [^{35}S]-methionine for 4 h at 20°C or 2 h at 35 or 40°C. Radioactive proteins were extracted separately from base, middle or tip of the leaf, separated by SDS–polyacrylamide gel electrophoresis and visualized by fluorography. Each track on the fluorogram was scanned using a densitometer; peak area is proportional to band intensity and hence to amount of radioactivity in the labelled polypeptide. Closed triangles, HSP; closed circles, positions of large and small subunits of ribulose bisphosphate carboxylase.

when a heat shock is applied, all parts of the leaf from the meristematic basal region to the mature blade are capable of the synthesis of HSP, as shown in Fig. 5. Unless this response is mediated exclusively at the level of translation of pre-existing mRNA this result suggests that access to the heat shock genes is not restricted in this tissue, though it is transcriptionally inactive in comparison with other parts of the plant. The heat shock syndrome is also being used to study regulation of foreign genes in transgenic plants. It has been shown (Schöffl & Baumann, 1985; Schöffl *et al.* 1986; and Baumann *et al.* 1987) that a soybean heat shock gene is accurately expressed in sunflower cell culture and transformed tobacco plants in response to high temperatures. It may, therefore, be possible in the future to use regulatory sequences derived from heat shock genes to direct the expression of foreign genes inserted into plants by genetic engineering techniques.

Responses to low temperature

While much attention has been given to the effects of high temperature upon gene expression, comparatively little work has been directed towards the low temperature range. Plants growing in the temperate regions of the world may experience temperatures of 0°C or below for some part of the year; even in tropical regions they may encounter low night temperatures; and their capacities to survive and grow at suboptimal temperatures vary greatly (Levitt, 1980; Baker *et al.* 1988). Physiologically, plants may be classified into three groups – chilling-sensitive, freezing-sensitive and freezing resistant – according to their abilities to withstand low temperatures (Larcher, 1981). Pollock & Eagles (1988) discuss the concepts of chilling and freezing injury, and cold acclimation, in more detail. They emphasize in particular the importance of distinguishing between reversible alterations in rates of processes with temperature, and irreversible changes due to injury. We have not attempted to relate the studies reviewed below to these classifications, largely because the data are not only scanty but in many cases consist exclusively of biochemical measurements without parallel physiological work. It is to be hoped that increasing collaboration between plant physiologists, biochemists and molecular biologists will lead to the synthesis of a cohesive picture of the way in which plants respond to low temperatures, at all levels from the field crop to the subcellular.

Methods for studying gene expression at low temperatures

In considering the changes in gene expression which may occur in response to decreased temperature, it must be acknowledged that the concept of a 'cold shock' is less clear-cut than that of a heat shock. Large temperature reductions have the primary effect of reducing the rate of enzyme-catalysed reactions, including protein synthesis. Hence, whereas heat-induced alterations in gene expression can be detected within minutes or at most a few hours of application of heat stress, it often takes many hours for new gene expression to become measurable at low temperatures. This is a particular problem in the case of *in vivo* protein labelling studies. Not only the rate of protein synthesis itself, but also the process of radioisotope uptake, is likely to be decreased with decreasing temperature. Labelling of intact grass leaves by feeding [^{35}S]-methionine from the base is hampered by the reduced rate of transpiration at temperatures below 10°C (Ougham, 1987). Some of these problems can be overcome by studying gene expression at the level of the messenger RNA, by extracting it directly from tissues treated in different ways and examining the polypeptides whose synthesis it directs in a cell-free translation system. The disadvantages of this approach are, firstly, that it gives no information about *in vivo* regulation of protein synthesis at the translational level, since it does not distinguish between those mRNAs undergoing active translation and those present in the tissue but not currently being translated. Secondly, differences in polypeptide complement may reflect post-translational modification processes, such as removal of signal peptides or glycosylation, and

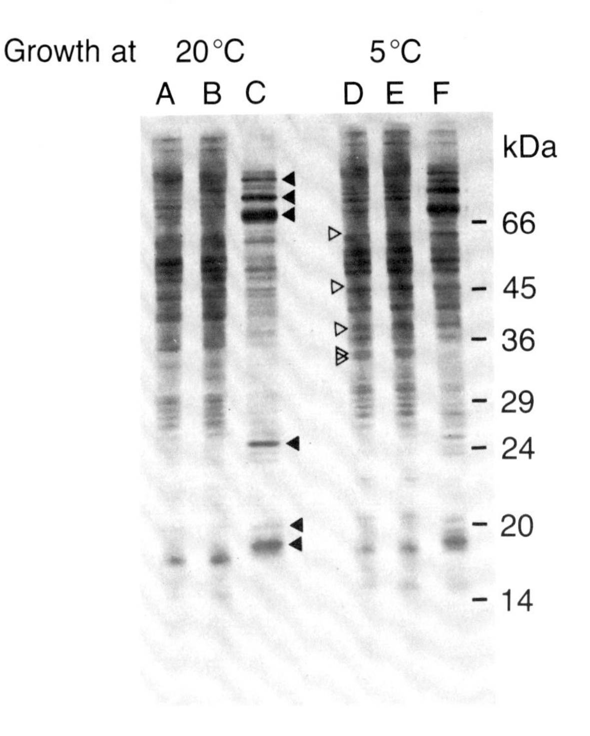

Fig. 6. Proteins synthesized at 5, 20 and 35 °C by leaf base tissue from *Lolium temulentum* grown at 20 or 5 °C. Leaves were incubated with [^{35}S]-methionine for 24 h at 5 °C, for 4 h at 20 °C or for 3 h at 35 °C. Radioactive proteins were extracted and visualized as described for Fig. 3. A–C, plants grown at 20 °C; D–F, plants grown at 5 °C. A and D, labelling at 5 °C; B and E, labelling at 20 °C; C and F, labelling at 35 °C. Closed triangles, HSP; open triangles, proteins whose labelling intensity altered in cold-hardened plants. Bars indicate positions of molecular weight standards.

these changes will not be revealed by an examination of mRNAs. Finally, translation of mRNAs from a given plant tissue in a heterologous cell-free system (for example, legume or brassica RNAs in a wheat germ or rabbit reticulocyte lysate cell-free translation system) cannot be assumed to produce a polypeptide profile perfectly representative of its tissues of origin, though it may be extremely effective in detecting *differences* between treatments.

Cold-hardening and cold-induced changes in protein and RNA complements

Comparatively few investigations have focussed on changes in protein synthesis brought about by rapid temperature decreases. Where studies have been carried out over prolonged periods at low temperatures, it has been observed that in many plant species (see, for example, Fig. 6) progressive alterations in gene expression may result from several days' or weeks' exposure, and these alterations may appear in parallel with an increase in chilling or freezing tolerance. Such acquired low-temperature tolerance is often referred to as 'hardening', and it has long been proposed (Weiser, 1970) that altered gene expression is a necessary precondition for hardening to occur. A number of workers have addressed the question of whether the appearance of specific polypeptides or mRNA species may be causally related to the hardening phenomenon. Studies on the effect of cold acclimation upon plant protein composition have shown that the soluble protein content of several species, including spruce (Kandler *et al.* 1979), alfalfa (Faw *et al.* 1976; Mohapatra *et al.* 1987*a*,*b*), wheat and rye (Cloutier, 1983), increases during periods of exposure to low temperatures. However, the soluble protein content of spinach seedlings does not increase during 14-days' exposure to 5°C (Guy & Haskell, 1987). Certain polypeptide species exhibit specific increases in abundance or rate of synthesis (Cloutier, 1983; Marmiroli *et al.* 1986*b*; Mohapatra *et al.* 1987*a*,*b*; Guy & Haskell, 1987; Meza-Basso *et al.* 1986; Kang & Titus, 1987). Sarhan & Perras (1987) showed that during hardening in wheat the major change was the accumulation of a high molecular weight (200 kDa) polypeptide. In most cases it is necessary to carry out *in vivo* radiolabelling to demonstrate altered patterns of synthesis, because the changes are too subtle to be apparent against the complex background of total polypeptides. Guy *et al.* (1985), Meza-Basso *et al.* (1986) and Mohapatra *et al.* (1987*a*,*b*) have also identified increases in levels of certain RNA species as a result of cold exposure in spinach, rapeseed and alfalfa.

Time-scales and temperature thresholds for cold-induced protein synthesis

The time-scale for alteration of protein synthesis varies from species to species as does the time-scale for development of hardiness. For example, Mohapatra *et al.* (1987*a*,*b*) studied two genotypes of alfalfa and found that the cultivar Saranac attained maximum freezing resistance after 2-days' acclimation at 4°C, while the more hardy cultivar Anik developed freezing resistance gradually over a 17-day period. In both cases, the acquisition of freezing resistance was paralleled by the synthesis of acclimation-specific mRNAs and the polypeptides which they encoded. Similar results were obtained by Guy & Haskell (1987) and Guy *et al.* (1987), who demonstrated that the cold-hardiness of spinach increases for up to 7 days at 5°C, but that the majority of this increase occurs during the first day of acclimation, which is also the period within which the bulk of changes to mRNA and protein synthesis occur. Marmiroli *et al.* (1986*b*) were able to show alterations in protein synthesis within 4 h of exposure of barley seedlings to low temperatures – 'cold shock' proteins appeared and the synthesis of some other proteins was repressed. The pattern of response varied according to tissue (roots or seeds) and genotype (a winter and a spring barley were compared). This study differed from others discussed here in that *in vivo* labelling was carried out at a range of temperatures, including 0°C. At 6°C, changes in rates of synthesis of cold-induced and cold-repressed polypeptides represented comparatively minor changes against a background similar to that observed at 24 or 18°C; but at 0°C, synthesis of most polypeptides other than the 'cold shock' proteins was largely abolished. This syndrome is similar to the full heat shock response. Does tissue so treated survive, and if so does it attain increased cold tolerance? It would be interesting to discover whether the other species studied, in which cold acclimation and analysis of gene expression were carried out at a single low temperature, also have characteristic

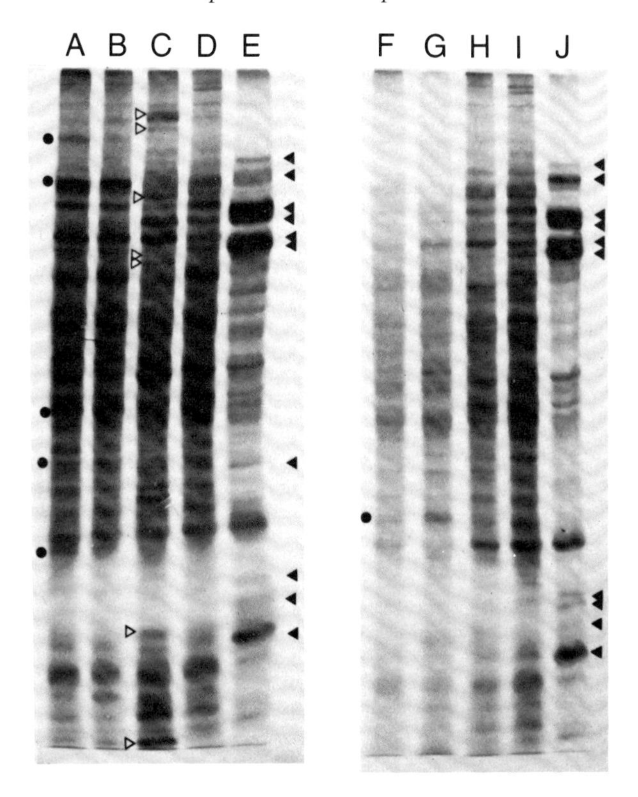

Fig. 7. Effect of low and high temperatures on protein synthesis by two sorghum varieties (A–E, variety IS18530; F–J, variety IS4845). Seedlings were germinated for 40 h at 35 °C, then incubated with [^{35}S]-methionine under the following conditions: A and F, 5 °C, 16 h; B and G, 10 °C, 16 h; C and H, 20 °C, 2 h; D and I, 35 °C, 2 h; E and J, 45 °C, 2 h. Radioactive proteins were separated and visualized as described for Fig. 3. ►, HSP; ▷, '20 °C shock proteins'; ●, cold shock proteins.

ranges in which cold-related polypeptides are first detectable or maximal. As with heat shock proteins, it is likely that the temperature required to elicit a response will depend on the plant species and the ambient temperature range to which it is best adapted. For example, the temperate grass *Lolium temulentum*, which grows down to temperatures of 2 °C (Stoddart *et al.* 1986; Thomas, 1983; Thomas & Stoddart, 1984), does not significantly alter its pattern of polypeptide synthesis (Fig. 6) in response to a sudden temperature drop from 20 to 5 °C (Ougham, 1987), though the rate of radioisotope incorporation is greatly reduced. Subsequent experiments in which the temperature was further reduced to −2 °C similarly failed to reveal altered patterns of gene expression over a 24-h labelling period (Ougham, unpublished data). In contrast, the tropical cereal *Sorghum bicolor*, which grows at ambient temperatures which may frequently exceed 35 °C, responds dramatically to low temperatures. Fig. 7 shows the synthesis of polypeptides by two different sorghum genotypes at 5, 10, 20, 35 and 45 °C, in seedlings grown at 35 °C. Both genotypes exhibit the standard heat shock response at 45 °C (a temperature which for many plant species would be high enough to abolish all

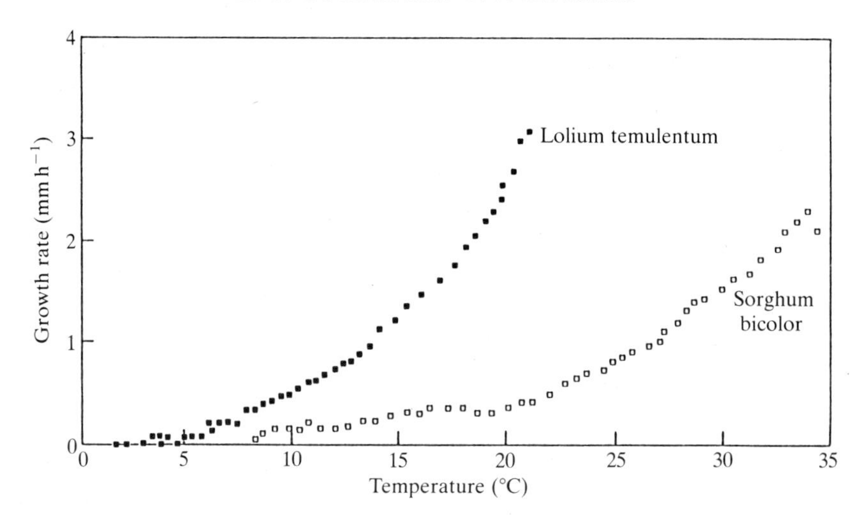

Fig. 8. Effect of cooling on growth rates of *Lolium temulentum* (■) and *Sorghum bicolor* (□) seedlings in the range 0–35 °C. Growth was measured using a linear voltage displacement transducer (Stoddart *et al.* 1986).

protein synthesis), but cold-specific alterations are also apparent. These include, for one genotype, the appearance (or greatly enhanced synthesis) of polypeptides (denoted '20 °C shock proteins') which are not evident at any other temperature. Indeed, each labelling temperature produces a different labelling pattern, and the response to heat shock is completely different from that produced by cold shock. These examples serve to illustrate the diversity of biochemical responses to cold in plants. In seeking explanations for such diversity, it is useful to consider the temperature range to which each plant species is adapted. Fig. 8 contrasts the growth rates of *L. temulentum* and *Sorghum bicolor* seedlings as temperature is reduced from 35 °C to 0 °C. While sorghum (a chilling-sensitive plant) grows poorly below 20 °C and ceases growth altogether at about 8 °C, the temperate grass *Lolium* grows well down to 5 °C. It is possible to expose it for short periods to 0 °C without damage. Hence a 5 °C-treatment may represent a cold shock for *Sorghum* while for *Lolium* it falls within the range of temperatures where growth can occur without the need for altered protein synthesis. Prolonged exposure to 5 °C does cause some reprogramming in gene expression, as shown in Fig. 6; plants grown at 5 °C synthesize a pattern of polypeptides somewhat different from those made by 20 °C-grown plants. These alterations are likely to be related to hardening. It is noteworthy that 5 °C-grown plants do not perceive a temperature rise of 15 °C as a heat shock; they continue to make the same proteins at 20 °C as at 5 °C, and like 20 °C-grown plants only produce HSP above 30 °C. This result confirms the observation that the threshold for induction of HSP synthesis depends on plant species rather than growth conditions: that is, it is genetically rather than environmentally programmed. It will be exciting to discover whether each plant species similarly possesses a 'cold shock' threshold below which the pattern (rather than the level) of gene expression alters.

Other factors inducing cold tolerance

By analogy with the heat shock treatment, can other environmental stimuli which alter protein synthesis enhance cold tolerance? Abscisic acid has been shown to induce freezing resistance in cultured wheat and rye cells (Chen & Gusta, 1983) and other plant tissues, but the mechanism for this induction is not clear, nor is it known whether gene expression is affected by abscisic acid in this system. Experiments on winter wheat and rye (Cloutier, 1983) showed that the application of desiccation stress induces increased tolerance of low temperatures as measured by LT_{50} (the temperature required to kill 50 % of plants). Pollock & Eagles (1988) discuss the use of LT_{50} measurements in more detail. In some varieties, desiccation stress provided a greater degree of cold tolerance than did a prior cold-conditioning treatment. However, though both cold conditioning and desiccation treatments brought about a change in the pattern of soluble polypeptides, there was insufficient homology between the two groups of altered polypeptides to suggest that cold conditioning and desiccation stress induce cold-hardiness by a common mechanism. In this respect it appears that the situation is fundamentally different from the heat shock system, where many different treatments may elicit HSP synthesis and consequent heat tolerance. Guy & Haskell (1987) showed that, although spinach leaf tissue is capable of the synthesis of such heat shock proteins (at 40 °C) and polypeptides termed cold acclimation proteins (at 5 °C), there is no homology between the HSP and the cold acclimation proteins (see also Fig. 7); moreover, heat shock does not induce cold acclimation and cold shock does not induce heat tolerance.

The nature and possible functions of cold-induced proteins

As in the case of heat shock proteins, little is yet known about the identity of proteins whose synthesis is induced or enhanced at low temperatures. Complements of some enzymes – for example, glutathione reductase (Guy & Carter, 1984), invertase (Roberts, 1982) and ribulose bisphosphate carboxylase (Huner & McDowall, 1979) – are modified during exposure to cold. Because both photosynthetic capacity and partitioning of assimilate are altered during prolonged periods at low temperatures (Pollock, 1984; Pollock *et al.* 1983, 1984), components of the photosynthetic apparatus and enzymes of carbohydrate metabolism are among the proteins whose synthesis might be affected. However, an alteration in a metabolic process may not be reflected in the pattern of labelled proteins separated on a gel, or the messenger RNA complement of a tissue; for example, enzymes already present in a tissue may be activated or inactivated at low temperatures, or the subcellular distribution of proteins may alter. Baker *et al.* (1988) report that chilling maize leaves brought about a decrease in photosynthetic capacity and accumulation of a 31 kDa polypeptide in the thylakoids. This has been tentatively identified as a precursor of the smaller chloroplast protein CP29. In this case, therefore, it seems that alteration in post-translational processing of a polypeptide, rather than regulation at the transcriptional or translational level, is responsible for the observed chilling-induced change in polypeptide profile. Sarhan & Chevrier (1985) showed a stimulation of activity of RNA polymerases, and also an increase in the ratio of RNA polymerase I to RNA polymerase II activity, after prolonged exposure of winter wheat to hardening temperatures, but these increases in activity may have reflected regulation at the protein level rather than elevated transcription levels. In some cases a non-protein component may be primarily responsible for an observed change. For example, Krupa *et al.* (1987) found that the lipid composition of light-harvesting complex II in rye altered as a result of exposure to hardening temperatures; while the polypeptide moiety of the complex itself was unchanged, presumably during hardening either the synthesis or the regulation of enzymes of lipid biosynthesis must alter. There is as yet no direct evidence for induction of transcription of any identifiable gene by low temperature treatment, either prolonged or of the rapid 'cold shock' type.

Conclusions

Temperature changes, whether gradual or rapid, cyclical or progressive, increases or decreases, are features of the natural environment in which higher plants must survive and function. In order to analyse the mechanisms which they have evolved to do so, plant scientists are increasingly focussing their attention on subcellular processes such as the heat shock response. Integrating the results obtained from biochemical and molecular biology approaches with the known behaviour of plants and plant parts under temperature stress should greatly enhance our understanding of plant adaptation processes. It is hoped that this review will stimulate interest amongst researchers in all areas of plant science in some of the techniques now available for investigating these processes.

We would like to thank colleagues at the Welsh Plant Breeding Station, especially Howard Thomas, Chris Pollock and John Stoddart, for helpful discussions and critical reading of the manuscript; Teresa Davies, Susan Jones and Simon McAdam for excellent technical assistance; Ian Sant for photographic work; and Margaret Mack for typing. We wish particularly to acknowledge the invaluable contribution of Oscar Fish. Work on sorghum and millet described here was funded by the UK Overseas Development Administration. The Welsh Plant Breeding Station is a constituent station of the AFRC Institute for Grassland and Animal Production.

References

BAKER, N. R., LONG, S. P. & ORT, D. R. (1988). Photosynthesis and temperature stress: the importance of quantum yield. In *Plants and Temperature*, Symp. Soc. Exp. Biol., vol. 42 (ed. S. P. Long & F. I. Woodward), pp. 347–375. Cambridge: The Company of Biologists Ltd.

BAUMANN, G., RASCHKE, E., BEVAN, M. & SCHÖFFL, F. (1987). Functional analysis of sequences required for transcriptional activation of a soybean heat shock gene in transgenic tobacco plants. *EMBO J.* **6**, 1161–1166.

Bewley, J. D., Larsen, K. M. & Papp, J. E. T. (1983). Water stress-induced changes in the pattern of protein synthesis in maize seedling mesocotyls: a comparison with the effects of heat shock. *J. exp. Bot.* **34**, 1126–1133.

Bienz, M. & Gurdon, J. B. (1982). The heat shock response in *Xenopus* oocytes is controlled at the translational level. *Cell* **29**, 811–819.

Bond, U. & Schlesinger, M. J. (1985). Ubiquitin is a heat shock protein in chicken embryo fibroblasts. *Molec. cell. Biol.* **5**, 949–956.

Bonham-Smith, P., Kapoor, M. & Bewley, J. D. (1987). Establishment of thermotolerance in maize by exposure to stresses other than a heat shock does not require heat shock protein synthesis. *Pl. Physiol.* **85**, 575–580.

Brown, G. N. (1978). Protein synthesis mechanisms relative to cold hardiness. In *Plant Cold Hardiness and Freezing Stress: Mechanisms and Crop Implications*, (ed. P. H. Li & A. Sakai), pp. 153–163. London: Academic Press Inc.

Burke, J. J., Hatfield, J. L., Klein, R. R. & Mullet, J. E. (1985). Accumulation of heat shock proteins in field-grown cotton. *Pl. Physiol.* **78**, 394–398.

Chen, T. H. H. & Gusta, L. V. (1983). Abscisic acid-induced freezing resistance in cultured plant cells. *Pl. Physiol.* **73**, 71–75.

Chen, Y.-M., Kamisaka, S. & Masuda, Y. (1986). Enhancing effects of heat shock and gibberellic acid on the thermotolerance in etiolated *Vigna radiata*. I. Physiological aspects on thermotolerance. *Physiologia Pl.* **66**, 595–601.

Cloutier, Y. (1983). Changes in the electrophoretic patterns of the soluble proteins of winter wheat and rye following cold acclimation and desiccation stress. *Pl. Physiol.* **71**, 400–403.

Cooper, P. & Ho, T.-H. D. (1983). Heat shock proteins in maize. *Pl. Physiol.* **71**, 215–222.

Cooper, P. & Ho, T.-H. D. (1987). Intracellular localisation of heat shock proteins in maize. *Pl. Physiol.* **84**, 1197–1203.

Cooper, P., Ho, T.-H. D. & Hauptmann, R. M. (1984). Tissue specificity of the heat shock response in maize. *Pl. Physiol.* **75**, 431–441.

Craig, E., Ingolia, T., Slater, M., Manseau, L. & Bardwell, J. (1982). *Drosophila*, yeast, and *E.coli* genes related to the *Drosophila* heat-shock genes. In *Heat-shock: From Bacteria to Man* (ed. M. J. Schlesinger, M. Ashburner & A. Tissières), pp. 11–18. Cold Spring Harbor, New York: Cold Spring Harbor Laboratory.

Czarnecka, E., Edelman, L., Schöffl, F. & Key, J. L. (1984). Comparative analysis of physical stress responses in soybean seedlings using cloned heat shock cDNAs. *Pl. molec. Biol.* **3**, 45–58.

Faw, W. F., Shih, S. C. & Jung, G. A. (1976). Extractant influence on the relationship between extractable proteins and cold tolerance of alfalfa. *Pl. Physiol.* **57**, 720–723.

Gausing, K. & Barkardottir, R. (1986). Structure and expression of ubiquitin genes in higher plants. *Eur. J. Biochem.* **158**, 57–62.

Guy, C. L. & Carter, J. V. (1984). Characterization of partially purified glutathione reductase from cold-hardened and nonhardened spinach leaf tissue. *Cryobiology* **21**, 454–464.

Guy, C. L. & Haskell, D. (1987). Induction of freezing tolerance in spinach is associated with the synthesis of cold acclimation induced proteins. *Pl. Physiol.* **84**, 872–878.

Guy, C. L., Hummell, R. L. & Haskell, D. (1987). Induction of freezing tolerance in spinach during cold acclimation. *Pl. Physiol.* **84**, 868–871.

Guy, C. L., Niemi, K. J. & Brambl, R. (1985). Altered gene expression during cold acclimation of spinach. *Proc. natn. Acad. Sci. U.S.A.* **82**, 3673–3677.

Heikkila, J. J., Papp, J. E. T., Schultz, G. A. & Bewley, J. D. (1984). Induction of heat shock protein messenger RNA in maize mesocotyls by water stress, abscisic acid, and wounding. *Pl. Physiol.* **76**, 270–274.

Huner, N. P. A. & MacDowall, F. D. H. (1979). Changes in the net charge and subunit properties of ribulose bisphosphate carboxylase-oxygenase during cold hardening of puma rye. *Can. J. Biochem.* **57**, 155–164.

Kandler, O., Dover, C. & Ziegler, P. (1979). Frost hardiness in spruce. I. Control of frost hardiness, carbohydrate metabolism and protein metabolism by photoperiod and temperature. *Ber. Dt. Bot. Ges.* **92**, 225–241.

Kang, S.-M. & Titus, J. S. (1987). Specific proteins may determine maximum cold resistance in apple shoots. *J. hort. Sci.* **62**, 281–285.

KEE, S. C. & NOBEL, P. S. (1986). Concomitant changes in high temperature tolerance and heat-shock proteins in desert succulents. *Pl. Physiol.* **80**, 596–598.
KEY, J. L., GURLEY, W. B., NAGAO, R. T., CZARNECKA, E. & MANSFIELD, M. A. (1985*a*). Multigene families of soybean heat shock proteins. In *Molecular Form and Function of the Plant Genome*, NATO ASI Series A, vol. 2 (ed. L. van Vloten-Doting, G. S. P. Groot & T. Hall), pp. 81–100. New York: Plenum.
KEY, J. L., KIMPEL, J. A., LIN, C. Y., NAGAO, R. T., VIERLING, E., CZARNECKA, E., GURLEY, W. B., ROBERTS, J. K., MANSFIELD, M. A. & EDELMAN, L. (1985*b*). The heat shock response in soybean. In *Cellular and Molecular Biology of Plant Stress* (ed. J. L. Key & T. Kosuge), pp. 161–179. New York: Alan R. Liss, Inc.
KIMPEL, J. A. & KEY, J. L. (1985). Heat shock in plants. *Trends biochem. Sci.* **117**, 353–357.
KRUPA, Z., HUNER, N. P. A., WILLIAMS, J. P., MAISSAN, E. & JAMES, D. R. (1987). Development at cold-hardening temperatures: the structure and composition of purified rye light harvesting complex II. *Pl. Physiol.* **84**, 19–24.
LAEMMLI, U. K. (1970). Cleavage of structural proteins during the assembly of the head of the bacteriophage T4. *Nature, Lond.* **227**, 680–685.
LARCHER, W. (1981). Effects of low temperature stress or frost injury on productivity. In *Physiological Processes Limiting Plant Productivity* (ed. C. B. Johnson), pp. 253–269. London: Butterworths.
LELAND, T. J. & HANSON, A. D. (1985). Induction of a specific N-methyltransferase enzyme by long-term heat stress during barley leaf growth. *Pl. Physiol.* **79**, 451–457.
LEVITT, J. (1980). Responses of plants to environmental stresses. Vol. 1. Chilling, freezing and high temperature stresses (2nd Edition). London: Academic Press, Inc.
LI, G. C. (1983). Induction of thermotolerance and enhanced heat shock protein synthesis in Chinese hamster fibroblasts by sodium arsenite and ethanol. *J. cell. Physiol.* **115**, 116–122.
LIN, C. Y., ROBERTS, J. K. & KEY, J. L. (1984). Acquisition of thermotolerance in soybean seedlings. *Pl. Physiol.* **74**, 152–160.
LINDQUIST, S. (1986). The heat-shock response. *A. Rev. Biochem.* **55**, 1151–1191.
MANSFIELD, M. A. & KEY, J. L. (1987). Synthesis of the low molecular weight heat shock proteins in plants. *Pl. Physiol.* **84**, 1007–1017.
MARMIROLI, N., RESTIVO, F. M., STANCA, M. O., TERZI, V., GIOVANELLI, B., TASSI, F. & LORENZONI, C. (1986*a*). Induction of heat shock proteins and acquisition of thermotolerance in barley seedlings (*Hordeum vulgare* L.). *Genet. Agr.* **40**, 9–25.
MARMIROLI, N., TERZI, V., STANCA, M. O., LORENZONI, C. & STANCA, A. M. (1986*b*). Protein synthesis during cold shock in barley tissues: comparison of two genotypes with winter and spring growth habit. *Theor. appl. Genet.* **7**, 190–196.
MATTERS, G. L. & SCANDALIOS, J. G. (1986). Changes in plant gene expression during stress. *Dev. Genet.* **7**, 167–175.
MEZA-BASSO, L., ALBERDI, M., RAYNAL, M., FERRERO-CADINAOS, M.-L. & DELSENY, M. (1986). Changes in protein synthesis in rapeseed (*Brassica napus*) seedlings during a low temperature treatment. *Pl. Physiol.* **82**, 733–738.
MOHAPATRA, S. S., POOLE, R. J. & DHINDSA, R. S. (1987*a*). Changes in protein patterns and translatable messenger RNA populations during cold acclimation of alfalfa. *Pl. Physiol.* **84**, 1172–1176.
MOHAPATRA, S. S., POOLE, R. J. & DHINDSA, R. S. (1987*b*). Cold acclimation, freezing resistance and protein synthesis in alfalfa (*Medicago sativa* L. cv. Saranac). *J. exp. Bot.* **38**, 1697–1703.
MUNRO, S. & PELHAM, H. (1985). What turns on heat shock genes? *Nature, Lond.* **137**, 477–478.
NECCHI, A., POGNA, N. E. & MAPELLI, S. (1987). Early and late heat shock proteins in wheats and other cereal species. *Pl. Physiol.* **84**, 1378–1384.
NEIDHARDT, F. C., VANBOGELEN, R. A. & VAUGHAN, V. (1984). The genetics and regulation of heat shock proteins. *A. Rev. Genet.* **18**, 295–329.
NEUMANN, D., SCHARF, K.-D. & NOVER, L. (1984). Heat shock induced changes of plant cell ultrastructure and autoradiographic localization of heat shock protein. *Eur. J. cell. Biol.* **34**, 254–264.
NOVER, L. (ed.) (1984). Heat Shock Response Of Eukaryotic Cells. Berlin: Springer-Verlag.
NOVER, L. & SCHARF, K.-D. (1984). Synthesis, modification and structural binding of heat shock proteins in tomato cell cultures. *Eur. J. Biochem.* **139**, 303–313.

OUGHAM, H. J. (1987). Gene expression during leaf development in *Lolium temulentum*: Patterns of protein synthesis in response to heat-shock and cold-shock. *Physiologia Pl.* **70**, 479–484.

OUGHAM, H. J., JONES, T. W. A. & EVANS, M. LL. (1987). Leaf development in *Lolium temulentum* L.: progressive changes in soluble polypeptide complement and isoenzymes. *J. exp. Bot.* **38**, 1689–1696.

OUGHAM, H. J. & STODDART, J. L. (1986). Synthesis of heat-shock protein and acquisition of thermotolerance in high-temperature tolerant and high-temperature susceptible lines of *Sorghum*. *Pl. Sci.* **44**, 163–167.

PEACOCK, J. M. (1982). Response and tolerance of sorghum to temperature stress. In *Sorghum in the Eighties*. Proceedings of the International Symposium on Sorghum, ICRISAT, India, 1981, pp. 143–160. Patancheru, A.P., India: ICRISAT.

PELHAM, H. (1985). Activation of heat-shock genes in eukaryotes. *Trends Genet.* **1**, 31–35.

PLESSET, J., PALM, C. & MCLAUGHLIN, C. S. (1982). Induction of heat shock proteins and thermotolerance by ethanol in *Saccharomyces cerevisiae*. *Biochem. Biophys. Res. Commun.* **108**, 1340–1345.

POLLOCK, C. J. (1984). Sucrose accumulation and the initiation of fructan biosynthesis in *Lolium temulentum* L. *New Phytol.* **96**, 527–534.

POLLOCK, C. J. & EAGLES, C. F. (1988). Low temperature and the growth of plants. In *Plants and Temperature*, Symp. Soc. Exp. Biol., vol. 42 (ed. S. P. Long & F. I. Woodward), pp. 157–179. Cambridge: The Company of Biologists Ltd.

POLLOCK, C. J., LLOYD, E. J., STODDART, J. L. & THOMAS, H. (1983). Growth, photosynthesis and assimilate partitioning in *Lolium temulentum* exposed to chilling temperatures. *Physiologia Pl.* **59**, 257–262.

POLLOCK, C. J., LLOYD, E. J., THOMAS, H. & STODDART, J. L. (1984). Changes in photosynthetic capacity during prolonged growth of *Lolium temulentum* at low temperature. *Photosynthetica* **18**, 478–481.

ROBERTS, D. W. A. (1982). Changes in the forms of invertase during the development of wheat leaves growing under cold-hardening and non-hardening conditions. *Can. J. Biol.* **60**, 1–6.

SACHS, M. M. & HO, T.-H. D. (1986). Alteration of gene expression during environmental stress in plants. *A. Rev. Pl. Physiol.* **37**, 363–376.

SARHAN, F. & CHEVRIER, N. (1985). Regulation of RNA synthesis by DNA-dependent RNA polymerases and RNases during cold acclimation in winter and spring wheat. *Pl. Physiol.* **78**, 250–255.

SARHAN, F. & PERRAS, M. (1987). Accumulation of a high molecular weight protein during cold hardening of wheat (*Triticum aestivum* L.). *Pl. cell Physiol.* **28**, 1173–1179.

SCHLESINGER, M. J. (1986). Heat shock proteins: the search for functions. *J. cell Biol.* **103**, 321–325.

SCHLESINGER, M. J., ASHBURNER, M. & TISSIÈRES, A. (ed.) (1982). *Heat Shock: From Bacteria to Man.* Cold Spring Harbor, New York: Cold Spring Harbor Laboratory. 431 pp.

SCHÖFFL, F. & BAUMANN, G. (1985). Thermoinduced transcripts of a soybean heat shock gene after transfer into sunflower using a Ti plasmid vector. *EMBO J.* **4**, 1119–1124.

SCHÖFFL, F., BAUMANN, G., RASCHKE, E. & BEVAN, M. (1986). The expression of heat-shock genes in higher plants. *Phil. Trans. R. Soc. Lond. B* **314**, 453–468.

STODDART, J. L., THOMAS, H., LLOYD, E. J. & POLLOCK, C. J. (1986). The use of a temperature-profiled position transducer for the study of low-temperature growth in gramineae. *Planta* **167**, 359–363.

THEILLET, C., DELPEYROUX, F., FISZMAN, M., REIGNER, P. & ESNAULT, R. (1982). Influence of the excision shock on the protein metabolism of *Vicia faba* L. meristematic root cells. *Planta* **155**, 478–485.

THOMAS, H. (1983). Analysis of the response of leaf extension to chilling temperatures in *Lolium temulentum* seedlings. *Physiologia Pl.* **57**, 509–513.

THOMAS, H. & STODDART, J. L. (1984). Kinetics of leaf growth in *Lolium temulentum* at optimal and chilling temperatures. *Ann. Bot.* **53**, 341–347.

VELASQUEZ, J. M., DIDOMENICO, B. J. & LINDQUIST, S. (1980). Intracellular localisation of heat shock proteins in *Drosophila*. *Cell* **20**, 679–689.

VELASQUEZ, J. M. & LINDQUIST, S. (1984). Hsp70: nuclear concentration during environmental stress and cytoplasmic storage during recovery. *Cell* **36**, 655–662.
VIERLING, E., MISHKIND, M. L., SCHMIDT, G. W. & KEY, J. L. (1986). Specific heat shock proteins are transported into chloroplasts. *Proc. natn. Acad. Sci.* **83**, 361–365.
VIERSTRA, R. D. (1987). Ubiquitin, a key component in the degradation of plant proteins. *Physiologia Pl.* **70**, 103–106.
WEISER, C. J. (1970). Cold resistance and injury in woody plants. *Science* **169**, 1269–1278.
XIAO, C.-M. & MASCARENHAS, J. P. (1985). High temperature-induced thermotolerance in pollen tubes of *Tradescantia* and heat-shock proteins. *Pl. Physiol.* **78**, 887–890.

TEMPERATURE DEPENDENT FACTORS INFLUENCING NUTRIENT UPTAKE: AN ANALYSIS OF RESPONSES AT DIFFERENT LEVELS OF ORGANIZATION

DAVID T. CLARKSON[1], *MICHAEL J. EARNSHAW*[2], *PHILIP J. WHITE*[2] *and H. DAVID COOPER*[1]

[1] University of Bristol, Long Ashton Research Station, [2] University of Manchester, School of Biological Sciences

Summary

Ion fluxes show a characteristically biochemical dependence on temperature when observed at the membrane level and over short periods after a perturbation of temperature. The primary active transport systems are enzymic and are dependent both on substrate supply and on changes in protein conformation. The hydrophobic parts of the proteins are surrounded by lipid molecules whose physical state may crucially affect conformation changes. These lipids may undergo transitions from a fluid to a gel state at temperatures occurring in the natural environment. It will be noted that the concepts developed in model systems of pure phospholipid/protein interactions cannot be very readily applied to the spatially heterogeneous assemblies of lipid molecules and transport proteins in real cell membranes.

While it is obvious that ion transport rates are responsive to temperature changes in a given cell, it is difficult to explain exactly which components of the transport process become limiting. We will show that, on cooling, the membrane potential can initially be greatly disturbed when temperature is changed and that this may be related to ATP supply to H^+-translocating ATPase. This affects the driving force for all other solutes. When temperature is lowered the permeability coefficients for most ions are reduced and yet it is commonly found that diffusive efflux of ions increases in the cold. We attempt to explain this paradox on the basis of driving forces and metabolic regulation of ion transport.

Acclimatory changes occur on extended exposure of a cell or an organism to a reduced growth temperature. Some of these changes occur at the membrane level and relate to lipid composition and modulation of carrier activity. Others involve changes in the relative size and sometimes the morphology of the root system. We will show that these processes lessen the temperature dependence of ion transport and ensure that the intake of nutrients does not limit growth at low temperatures. These acclimatory changes are seen as part of the general process of regulation of nutrient uptake.

Introduction

After mitosis has been completed, each new cell requires an input of the appropriate quantities of ions to support its expansion and metabolism. Most of the ions utilized in growth will have been recently absorbed from the surrounding soil or water. Dependence on external supplies may be lessened, in young seedlings, by seed reserves or, in older plants, by internal redistribution from senescing organs or specialized storage tissues. It follows that any factor which influences the rate of cell production in an individual will also influence its demand for mineral nutrients. Roots respond to this demand by varying their rate of nutrient absorption. These familiar, but most important, facts bear strongly on any attempt to analyse the effects of temperature on nutrient uptake since the most evident plant response to temperature is a change in growth rate.

Ion transport can be shown to have a characteristically biochemical dependence on temperature if observed at the membrane level in model systems or intact cells. There is, however, an interplay between transport, metabolism and molecular structure in which each feature may have a characteristic response to temperature. A change in temperature may affect the substrate supply to a transport enzyme, its structure or its freedom to operate in a lipid environment. In the case of a primary active transport, such as H^+-efflux across the plasmalemma, all other transport processes will be influenced by such effects since they alter the intensity and, sometimes, the direction of the driving forces on ion movement.

A most interesting aspect of this subject relates to the ability of cells and whole plants to set in motion physiological and morphological changes which appear to ensure that the uptake of ions does not become a factor limiting growth within the temperature range tolerated by a species. These adaptive changes may have the effect, observable in laboratory conditions, of greatly augmenting the capacity for nutrient uptake for a period after the plant has been returned to a more favourable growth temperature. We give examples of this behaviour in species ranging from single cell algae to cereal plants. It seems possible that these responses are manifestations of the normal regulatory processes which match ion uptake to growth demand.

In the first section of this chapter we will review what is known about the interaction of ion-stimulated ATPases, temperature and the state and composition of membrane lipids. The timescale over which we consider the response to temperature changes is measurable in seconds or less. Next we consider the immediate effects, measurable within a few minutes or hours, of temperature changes on ion fluxes across the plasma membrane and point out the difficulties in interpreting apparently straight-forward experiments.

Longer term adjustments to the composition, physiology and morphology of organisms occur over a timescale of days or weeks and may lead to an apparent reduction in temperature sensitivity of transport processes. There are changes in membrane properties and in the relative size of the root system during acclimation; the latter factor can have a very potent effect on rates of ion transport and is of great significance when roots are cooled relative to shoots. We discuss this situation at some length because it is a common occurrence in the field in temperate latitudes. The temperature of the soil below 5 cm tends to fluctuate less than that of the air and there may be major diurnal shifts in the temperature of roots relative to that of leaves. In temperate latitudes the mean temperature experienced by roots during spring days may be much lower than that affecting the shoot; this situation may be reversed in autumn.

Temperature and the activity of membrane-bound systems

Attempts to explain the modulation of ion transport by temperature in intact organisms have frequently taken the reductionist approach and explanations have been sought at the level of isolated membranes. The interpretation of data produced by this approach still rests heavily on the fluid-mosaic model of membrane structure formulated by Singer & Nicholson in 1972. Since its original formulation, the model has been changed to take account of the concept of lateral heterogeneity in both lipids and proteins. A given lipid may occur in discrete domains which differ markedly in composition from others within a single leaflet of the bilayer, for example, boundary lipids associated with specific membrane proteins (Quinn, 1981; Benga & Holmes, 1984). It has been appreciated for some time that the activity of membrane-bound ATPases from animal (e.g. Wheeler *et al.* 1975) and plant (Imbrie & Murphy, 1984) cells is strongly determined by the lipids associated with them. Ideally, then, it is the behaviour of these lipids during a change of temperature which should be measured. Because of the heterogeneity of natural membranes, much work has concentrated on the behaviour of simplified systems containing one or a few lipids of known composition. While this undoubtedly facilitates research it may not produce results which are immediately useful in understanding the response of transport systems to temperature.

Thermal transitions in lipid bilayers

Quinn (1988) has dealt with the details of changes in membranes with temperature. This section deals with specific changes which may influence ion uptake. Lipid bilayers undergo a thermally-induced transition from an ordered crystalline state at low temperature to a disordered fluid state at high temperature. The transition temperature (T_c) is relatively sharp in the case of a pure phospholipid and its value highly dependent upon the lipid species (Melchior & Stein, 1976; Quinn & Chapman, 1980; Quinn, 1981; Benga & Holmes, 1984). For phospholipids with identical saturated fatty acids, pronounced polar group specificity occurs in which T_c decreases in the order: ethanolamine > serine > glycerol = choline. Reduction in fatty acid chain length similarly results in a decrease in T_c as well as a decline in the passive permeability of the lipid bilayer to solutes. A reduction in T_c also occurs with the incorporation of double bonds, with *cis* being more effective isomers than *trans*, but leads to an increase in passive permeability due to increased membrane fluidity (Quinn, 1988). Clearly, bilayers formed from single lipids are far removed from natural membranes but numerous studies have been carried out using lipid mixtures (Melchior & Stein, 1976; Quinn & Chapman, 1980; Lee, 1983). Mixtures of two lipids may produce two independent phase transitions, from which one might conclude that there had been separation of the two lipids into distinct phases, or a single broad transition; the outcome depending on the degree of cooperativity. As might be expected, heterogeneous mixtures of naturally occurring lipids or microsomal membranes show extremely broad phase transitions. It is necessary to exercise caution in relating absolute values of T_c in bilayers derived from naturally occurring phospholipids, to the temperature regime experienced by the organism. Factors in the environment, such as inorganic cation concentrations, pH and other membrane components such as sterols and proteins, can produce marked changes in T_c (Quinn & Williams, 1978; Quinn & Chapman, 1980; Quinn, 1981).

Free sterols in membranes are of particular interest in relation to cold acclimation because they can affect the phase transition of phospholipids. These molecules intercalate with phospholipids where they produce the effect of condensing hydrocarbon chains at high temperature while preventing the formation of the crystalline state at low temperature by hindering the orderly packing of hydrocarbon chains. There is a consequent broadening of T_c and, at high sterol/phospholipid ratio, the elimination of the phase transition. The decrease in membrane fluidity at temperatures higher than the T_c leads to a reduction in passive permeability of ions and uncharged solutes in membranes of high sterol content (Rottem *et al.* 1973; McKersie & Thomson, 1979; Ashcroft *et al.* 1983; Stubbs, 1983).

Viscotropic regulation of transport enzymes

Model membranes have helped interpret the effects of temperature on membrane-bound enzymes whose activity appears to be regulated by membrane fluidity, a process commonly known as 'viscotropic regulation' (Melchior & Stein, 1976; Stubbs, 1983). Membrane ATPases of microorganisms frequently exhibit biphasic Arrhenius plots where the abrupt change in temperature sensitivity, or activation energy (E_a) correlates closely with T_c of the cell membrane measured by a variety of physical techniques (Dufour & Goffeau, 1980; Le Grimellec & Leblanc, 1980; Silvius & McElhaney, 1980; Ahlers, 1981; Dufour & Tsong, 1981). Comparable changes in E_a have also been found in other microbial transport proteins (Thilo *et al.* 1977; Le Grimellec & Leblanc, 1980). Results of this kind have undoubtedly encouraged researchers on higher plants to look for similar interactions. In many of the above cases organisms were used which were adapted to growth at relatively high temperatures and whose membranes contained a major fraction of saturated fatty acids, or had their ATPase reconstituted in bilayers high in saturated fatty acids. This has been overlooked by many in their search for phase transitions in membranes of temperate plants in which most of the phospholipids have poly-unsaturated acyl chains. Correlations have, indeed, been found between changes of activation energy of the plasma membrane K^+-stimulated, Mg^{2+}-dependent ATPase and changes in lipid ordering measured by physical methods; an example of this is shown in Fig. 1. Correlations of this kind in different species may not all have a common cause. In chilling-sensitive plants

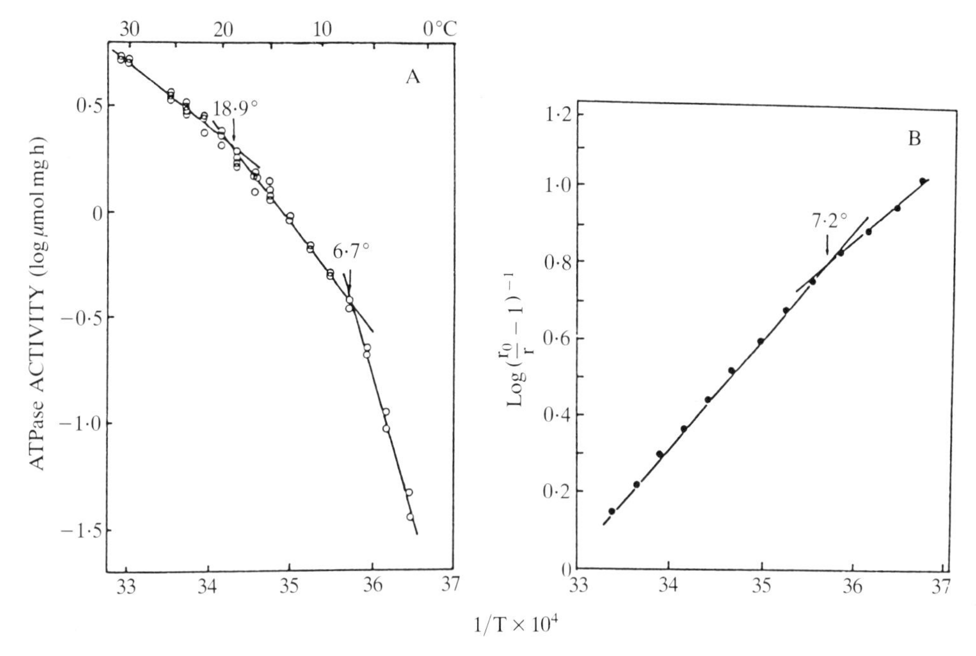

Fig. 1. Arrhenius plots in a plasmalemma fraction from mung bean (*Vigna radiata*) showing marked discontinuities with temperature. A. K^+-stimulated, Mg^{2+}-dependent ATPase activity and, B, fluorescent anisotropy parameters of 1,6-diphenyl-1,3,5-hexatriene. Note that only the 'break' in ATPase activity at the lower temperature is matched by a change in ordering of the fluorescent probe. (Yoshida *et al.* 1986.)

breaks in Arrhenius plots of ATPase activity versus temperature have been correlated with thermal transitions of bulk lipids and there are examples in non-chilling-sensitive species as well (Wright *et al.* 1982; Douglas & Walker, 1984; Yoshida *et al.* 1986). In other instances breaks in Arrhenius plots may be related to changes in the ordering of phospholipid headgroups in circumstances when the bulk lipid is well above the temperature at which thermal transitions have been completed (Caldwell & Haug, 1981).

There are undoubtedly difficulties in understanding the significance of signals from membrane probes used to measure fluidity. The signals from bulky spin labels (electron paramagnetic resonance spectroscopy) or fluorescent probes relate to the lipid environment in the immediate vicinity of the probe. The information is, therefore, highly specific but there is uncertainty about the exact location of the probe and considerable doubt as to its concentration in the crucial boundary layers of lipid around membrane proteins with respect to the bulk phase (Quinn, 1981; Stubbs, 1983). Non-invasive techniques like differential scanning calorimetry are, by contrast, relatively insensitive and reveal only large-scale changes in bulk lipids.

Recent work has made it clear that temperature-dependent changes in the activity of membrane ATPases may have nothing to do with the fluidity of the acyl chains of phospholipids but are related to effects on ordering of headgroups at the membrane surface (Caldwell & Whitman, 1987). Earlier it had been shown that the Km of a barley root plasma membrane ATPase changed with temperature, being at a minimum value for a range of temperature from 15 to 32 °C. The marked increases in the value of the Km on either side of this range coincided with marked discontinuities in the motion of spin-labelled probes reporting on ordering near the membrane surface (Caldwell & Haug, 1981). A further study, using a spin label covalently linked to membrane proteins, revealed a major change in mobility in the membrane surface at around 12 °C, again correlating with the change in the Km of the ATPase (Caldwell & Whitman, 1987). Meanwhile the acyl chains in the membrane interior remained in a liquid-crystalline state throughout; in barley, solid to liquid crystalline transitions would have been completed at −7 °C. In the surface the lateral diffusion of membrane proteins may be hindered by the formation of quasi-crystalline lipid clusters by the bonding of headgroups; this process was greatly intensified at temperatures below 12 °C and by increasing calcium concentration in the range 1–50 mol m^{-3} (Caldwell & Haug, 1981).

In our own work we have studied the response of ion-stimulated ATPases from plasma membrane and tonoplast-enriched membrane fractions separated by continuous sucrose density gradient centrifugation from roots of *Secale cereale* (rye) grown under different temperature regimes. Arrhenius plots (Fig. 2) of the characteristic ATPases activities from these membranes are monophasic, the activity declines exponentially as the temperature of the membranes is lowered and there is no evidence of any changes in Km nor of cold inactivation of the enzymes (cf. Graham & Paterson, 1982; Wright *et al.* 1982). Such results are not unexpected since the bulk lipid phase transitions in plant cell membranes generally occur below 0 °C (Quinn & Williams, 1978; Raison & Wright, 1983). It has been proposed that phase transition in a small proportion of phospholipids with both acyl chains saturated, e.g. dipalmitoylphosphatidylglycerol, can produce the well-marked T_c and evidence of increased ordering seen with fluorescent probes in chilling-sensitive species such as *Vigna radiata* (mung bean); other considerations indicated that only 7 % of the bulk lipid was in a gel state at a temperature as low as −8 °C (Raison & Wright, 1983). This work suggests that it may be species-dependent variations in the abundance of this critical class of disaturated phospholipids which may determine the response of membrane-bound enzymes to temperature, rather than phase changes in the bulk lipid. Raison & Wright (1983) were, however, unable to detect lipids of this kind in *Triticum aestivum* (wheat).

Freeze-fracture electron microscopy of membrane surfaces has provided a further opportunity to observe the behaviour of lipids at or below critical temperatures for growth or for some transport process of interest. Plasma membrane from cells at normal growth temperatures reveals numerous intercalated membrane particles (IMPs) distributed on a featureless background which is comprised of the acyl chains of lipids. With progressive cooling it has been observed in some cells that these IMPs become crowded into discrete areas separated by increasingly large particle-depleted zones. This type of behaviour

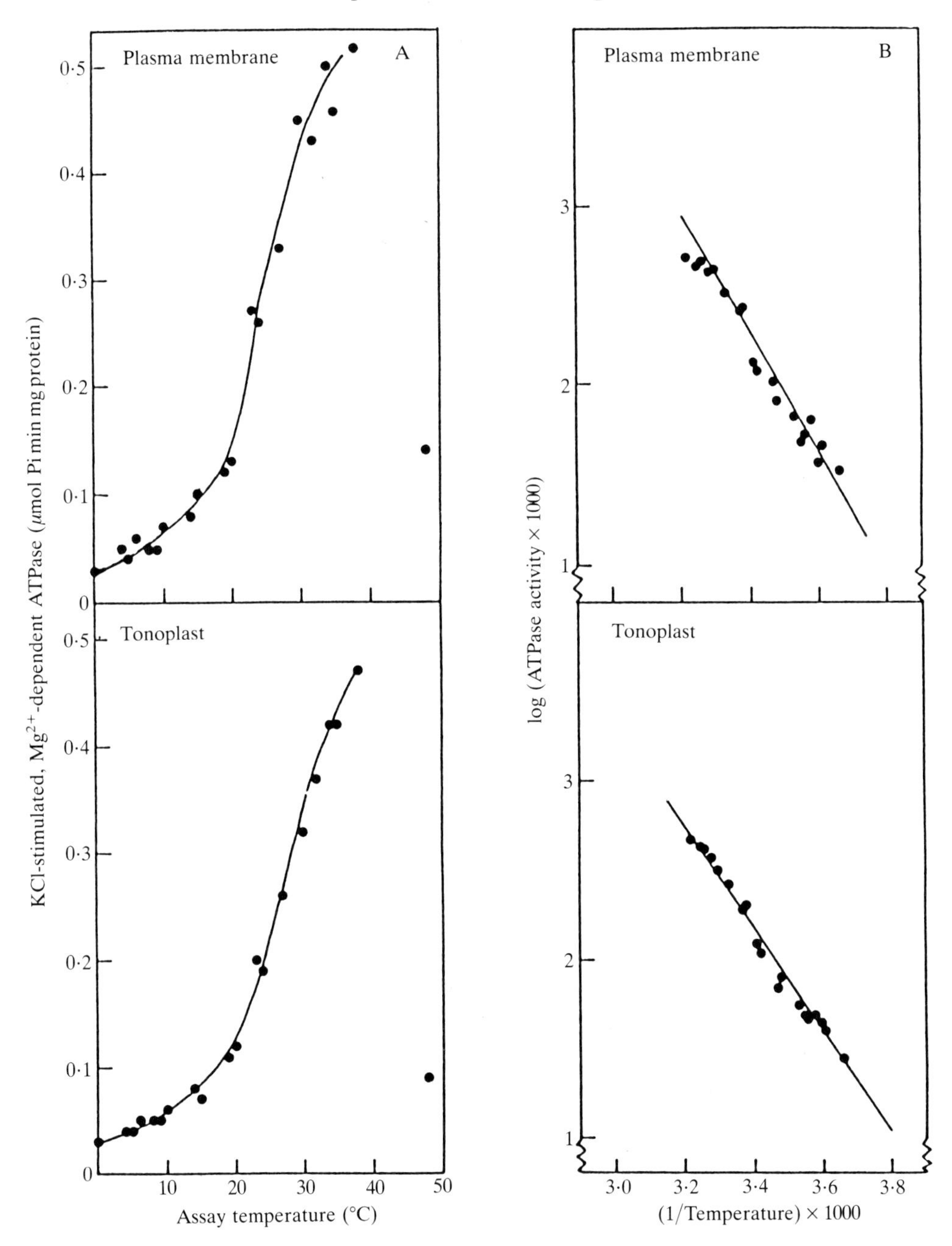

Fig. 2. A. Variation of ATPase activity with temperature in plasmalemma- and tonoplast-enriched fractions from rye roots. B. Arrhenius transformations of the data. (White, 1987.)

correlates well with other evidence of T_c in *Tetrahymena* (Thompson, 1980), *Anacystis nidulans* (Armond & Staehelin, 1979) and for the tonoplast of *Saccharomyces* (Moeller *et al.* 1981) and suggests that phase separation occurs in the membrane, some areas remaining fluid (those in which the IMPs are found) and others freezing. Similar evidence has been seen in the plasma membrane of chilling-injured *Persea americana* (avocado) fruits (Platt-Aloia & Thomson, 1987), but no convincing phase separation could be found in the plasma membrane of roots of *Zea mays* or *Cucurbita pepo* (marrow) at temperatures between 4 and 0°C even though growth and metabolism are grossly disturbed at these temperatures (Robards & Clarkson, 1984).

Temperature-mediated changes in membrane composition

Lowering the growth temperature of *Escherichia coli* results in a shortening of the length, and an increase in the number of double bonds, in the acyl chains of phospholipids. As a consequence it is estimated that the membrane has a similar viscosity at all growth temperatures (Sinensky, 1974). This process, 'homeoviscous adaptation', has profoundly influenced thinking about the way in which higher plants may respond to temperature (Lyons, 1973; Berry & Raison, 1981; Kuiper, 1985). It has been proposed, for example, that reduced temperature sensitivity and increased capacity for ion transport, seen in the roots of plants acclimated to low temperature, might be related to changed membrane composition combined with increased numbers of ion porters (Clarkson, 1976; Markhart *et al.* 1979; Glass, 1983). In this, and in subsequent sections of the paper, we will be exploring this proposition.

There is a great deal more published information about the lipid composition of shoot tissue during cold acclimation than there is for roots. Growth at low temperature produces shoot tissue which is relatively enriched in phospholipid due, in some cases, to a proliferation of membrane material on a per cell basis. There is also an increase in the number of *cis*-unsaturated positions per acyl chain (Willemot, 1979; Kuiper, 1985) which would serve to prevent phase transitions in the membrane. The impetus of these studies has usually been an interest in mechanisms of freezing resistance but over the years it has become evident that there is little correlation between cold hardening and the extent of acyl chain desaturation. Recent studies have suggested a direct relationship between the sterol/phospholipid ratio, which usually decreases during hardening, and the freezing tolerance of a number of species (Horvath *et al.* 1987 and references therein). The ratio may change because of increased rates of phospholipid synthesis in cool conditions (see Kinney, Clarkson & Loughman, 1987*a*) or an actual decrease in abundance of sterols. The lower ratio may be responsible for the lower temperature at which the broad phase transition commences in cold-acclimated wheat plants (Horvath *et al.* 1987; Vigh *et al.* 1987).

Cold acclimation brings about similar changes in lipid composition in root tissue. There is an overall increase in phospholipid content and unsaturation of acyl chains (Willemot, 1979; Clarkson *et al.* 1980; Kinney *et al.* 1987*b*). In the roots of rye there appears to be no change in the abundance of sterol in the total lipid extracted from roots, nor in the distribution of sterol classes (Kinney, 1983).

Most of the above analyses made use of whole tissues or crude microsomal fractions. The plasma membrane is, however, the membrane of particular interest with respect to both ion fluxes and freezing tolerance. As the plasma membrane is usually richer in sterols than other membranes and contains a lower proportion of unsaturated acyl chains in phospholipids (Yoshida & Uemura, 1984), it follows that changes in composition of total tissue lipids may not reflect what is happening to the plasma membrane during cold acclimation. Indeed, it has been shown that composition changes of the plasma membrane are small relative to those of other endomembranes. There was either little change in sterol content, giving a small decrease in sterol/phospholipid ratio (Uemura & Yoshida, 1984; Yoshida & Uemura, 1984) or sterol and phospholipid increase in parallel (Lynch & Steponkus, 1987). Detailed analysis of this kind is in its infancy. This is partly related to difficulties in preparing 'pure' membrane fractions and partly due to uncertainties about such factors as phospholipase activity during the necessarily extended periods between tissue homogenization and the final separation of membrane fractions. Early attempts at characterizing the lipids of the plasma membrane produced some bizarre results (Keenan *et al.* 1973) which suggested extensive degradation by phospholipases.

Before we consider our results with the roots of rye we will briefly preview some facts which will be considered in detail later. When rye plants are grown with their roots at 8°C and shoots at 20°C (DG-plants) or with both organs at 8°C (CG-plants) their capacity for ion uptake per unit root weight at an equivalent physiological age is increased and temperature sensitivity decreased relative to plants grown at 20°C (WG-plants). Such effects are often most pronounced in plants transferred from WG to DG conditions for 3–5 d; these transferred plants are described as DT-plants (further details see White *et al.* 1987).

Membrane fractions from DG-plants had a higher protein content especially in the tonoplast fraction (Table 1). There was also an enrichment of phospholipids in membranes from these plants. Where the whole plant was grown at 8°C these changes were not found, but there was an increase in the abundance of free sterols with a consequent *increase* in the sterol/phospholipid ratio. Changes such as these, in the cold-acclimated roots, might be expected to lead to a decrease in membrane fluidity. This is quite contrary to the expectations raised in our preceding discussion. For instance, it has been suggested that the correlation between sterol enrichment and an increase in the E_a of a plasma-membrane ATPase in *Citrus* genotypes, is due to a decrease in membrane fluidity (Douglas & Walker, 1984), as is the case for the Na^+-K^+-stimulated ATPase of animal cells (Sinensky *et al.* 1979). By contrast, cold acclimation of rye roots results in no change in the E_a of the characteristic ion-stimulated ATPases of either plasma membrane or tonoplast (Table 2) despite general shifts in lipid composition (Table 1). These results indicate that the ATPase activity is not regulated by the 'average' fluidity of the membranes. A wide range of other membrane-bound enzymes is also unaffected by bulk membrane fluidity (Melchior & Stein, 1976) including the ATPase activity of microsomal membranes from wheat (Tánczos *et al.* 1984), the Ca^{2+}-stimulated, Mg^{2+}-dependent ATPase of sarcoplasmic reticulum (East *et al.* 1984) and the

Table 1. *Effect of growth temperature pretreatment upon the protein, free sterol and phospholipid contents of tonoplast- and plasma membrane-enriched fractions isolated from warm-grown (WG, 20°C), differential-grown (DG, 20°C shoot; 8°C root) and cold-grown rye plants (CG, 8°C). Contents are expressed on the basis of 10-g fresh weight of rye roots. (Data from Cooke, White and Burden, unpublished.)*

Growth temperature pretreatment	Membrane fraction	Protein μg	Sterols	Phospholipids nmol	Phospholipid/protein $\mu mol/mg$	Sterol/phospholipid mol/mol
WG	Plasma membrane	267	63	397	1·49	0·16
	Tonoplast	411	58	633	1·54	0·09
DG	Plasma membrane	319	90	560	1·76	0·16
	Tonoplast	552	133	1113	2·02	0·12
CG	Plasma membrane	326	95	456	1·40	0·21
	Tonoplast	387	120	573	1·48	0·21

Table 2. *Specific activities and activation energies of the KCl-stimulated Mg^{2+}-dependent ATPases of plasma membrane and tonoplast fractions isolated from rye roots following pretreatment under a range of temperature regimes. Specific activities of the ATPases are expressed close to the temperature optimum at 34°C and activation energies presented ± S.E. Data from White (1987).*

Growth temperature pretreatment	Plasma membrane		Tonoplast	
	Specific activity $\mu mol\ Pi\ min^{-1}\ mg\ protein^{-1}$	Activation energy $kJ\ mol^{-1}$	Specific activity $\mu mol\ Pi\ min^{-1}\ mg\ protein^{-1}$	Activation energy $kJ\ mol^{-1}$
WG	0·37	60·5 ± 2·8	0·33	47·8 ± 1·4
DT	0·31	54·1 ± 4·5	0·30	52·4 ± 1·9
DG	0·63	59·9 ± 2·4	0·35	53·8 ± 1·9
CG	0·60	57·1 ± 3·6	0·41	53·0 ± 3·2

sugar transporter of the human erythrocyte (Carruthers & Melchior, 1986). Enzyme activity in these cases may be regulated by such factors as membrane surface potential and membrane thickness which are determined by phospholipid structure.

Extended periods of low temperature acclimation in rye plants resulted in an increase in the specific activity of ATPases in the isolated membrane fractions from rye roots, the effect being most marked in the plasma membrane (Table 2). This response is common in many observations of plant cell membrane ATPases (Kuiper, 1972) and probably reflects complex changes in membrane protein synthesis and turnover occurring when plants are exposed to low temperature (Uemura & Yoshida, 1984). It seems unlikely that the greater ATPase activity can explain the enhanced rates of ion transport in cold acclimated roots since there was no effect seen in DT-plants 3 days after transfer to 8°C even though transport was much greater after this period (Clarkson, 1976).

Short-term effects of temperature on ionic relations of cells

The transport of ions across plant membranes is coupled to the electrical potential and/or pH gradient, which is regulated by the primary active transport of H^+ at both plama membrane and tonoplast. Temperature may affect such membrane transport processes in three different ways.

1. Substrate supply to the primary active transport mechanism may change. The substrate is usually thought of as ATP, but pyrophosphate may be utilized at the tonoplast (Rea & Poole, 1986) in some cases, NAD(P)H at the plasma membrane (Serrano, 1985).
2. The conformational change of proteins which occur when ions bind to them, and which are usually thought of as being an intrinsic feature of their transporting properties, may be restricted by low temperature.
3. The physical properties of the lipid environment may alter so as to hinder necessary conformational changes in the membrane protein.

While there is little evidence for restriction of ion transport by cold lability of membrane transport proteins, the modulation of membrane transport processes via temperature effects on the lipid environment has been observed *in vitro*. When the scale of our enquiry is shifted from *in vitro* model systems to living cells, it becomes harder to decide what sort of perturbation has produced an observed effect since direct effects of temperature at the membrane level may be offset by changes due to the regulatory mechanisms leading to cellular homeostasis.

Evidence from electrophysiology

The electrical potential difference between the cell interior and the surroundings provides a sensitive indicator to changes in current (ionic) movement across the plasma membrane. In the giant internodal cells of *Chara corallina* kept in darkness, a temperature switch from 25 to 4°C was accompanied by an instantaneous depolarization of the membrane potential of about 70 mV, followed by an approx. 40 % recovery over a 10-min period (Lucas, 1984). Similar results were obtained from observations on storage root discs from *Beta vulgaris* (red beet) by Poole (1974). In both *Chara* and *Beta* a switch back to the initial temperature promptly restored the potential to its initial value. These results are consistent with an abrupt decrease in the activity of the plasma membrane H^+-pump, on which the membrane potential partly depends (see Clarkson, 1984). All other transport processes will be influenced, to a greater or lesser extent, by the value of the membrane potential. Isolated strips of mesophyll cells from leaves of *Zea mays* and *Avena sativa*, were both depolarized by approx. 30 % when incubated in the dark at 8 °C in comparison with 21 °C (Jennings & Tattar, 1979) but this situation was immediately reversed by transfer of the cells to the warmer temperature. In *Avena*, but not in *Zea*, the membrane potential at 8 °C could be increased slowly up to the value of that in cells at 21 °C if the light was switched on. The result can be interpreted in terms of an improved substrate (ATP) supply to the H^+-translocating ATPase in the plasma membrane when photosynthesis commences. Other work on the dynamics of membrane potential changes in *Neurospora* (Slayman *et al.* 1973) and *Chara* (Keifer & Spanswick, 1979) has shown how rapidly metabolic blockade can influence the proton pump. It seems likely that the immediate effects on membrane potential may result from a shift in the substrate supply for primary active transport.

Permeability measurements

Many Charophytes are species with excitable cells which produce an action potential when stimulated. This property can be usefully exploited to separate 'electrical effects' due to some perturbation from 'metabolic effects' such as we have just discussed. In *Nitella flexilis* the magnitude and the rise-time of the action potential are strongly affected by temperature (Blatt, 1974). The rise-time increased by 2 orders of magnitude between 28 and 3 °C and Arrhenius plots of the rise and decay of the action potential both showed very marked discontinuities of slope near 13·5 °C. The action potential in *Chara corallina* was also found to have components which were highly sensitive to temperature; the major difference with *Nitella* was that the absolute magnitude of the action potential was not affected by temperature and that Arrhenius plots of the rise-time were curved very steeply in the range below 15 °C. Thus, in this species, there was a continuous, rather than an abrupt, change in activation energy (Beilby & Coster, 1976).

The action potential results from a transient, large increase in the permeability of the plasma membrane to Cl^-; gated channels open and a very large efflux of Cl^- occurs down its energy gradient into the external medium. Evidently, this passive, diffusional process can be strongly modulated by temperature and suggests that the channel protein is affected directly or indirectly by the physical state of the membrane lipids. At low temperature the availability of open channels, or the rate at which they open, or passage through them, may be influenced by the fluidity of the lipids in the vicinity of the pore. Much other work shows that the permeability of membranes to ionic diffusion decreases with temperature. Early work on *Chara corallina* and *Griffithsia pulvinata* showed that the permeability coefficients for potassium, P_{K^+}, and, sodium, P_{Na^+}, both decreased with temperature in the range 25–3 °C. In *Chara*, at temperatures up to 13 °C the value of P_{K^+} decreased relatively faster than P_{Na^+} (Hope & Aschberger, 1970).

Efflux and influx measurements

Results of permeability measurements seem paradoxical in the light of the general experience that when tissues, especially if derived from chilling-sensitive plants, are placed in cool solutions, the efflux of both anions and cations *increases*. For instance, the efflux of K^+ from carrot roots and potato tubers increased exponentially as the temperature was lowered from 25 to 5 °C (Murata & Tatsumi, 1979). Similar results were seen in the aquatic plant *Spirodela* where phosphate efflux increased 2·5-fold when the temperature was lowered from 25 to 5°C; by contrast the influx at 5°C was only 17 % of its value at 25°C, and was about the same magnitude as the efflux (Bieleski & Ferguson, 1983). Behaviour of this kind is very hard to relate to the constant ionic composition in cells and tissues which have become acclimatized to different temperatures. As we saw above, the principal processes generating ATP and utilizing it in electrogenic H^+ transport may temporarily collapse when a cell is first placed in cool conditions. The consequences of this might be as follows: depolarization of the membrane potential changes the direction of the thermodynamic driving force on, say, the K^+ ion so that it tends to diffuse out of the cell rather than in; since the ATP-driven H^+ efflux from the cell decreases, coupled K^+/H^+ movement which accounts for a major fraction of the influx will also decrease. For a while, therefore, the cell may have a net efflux of ions (e.g. K^+, Bange, 1979). In the case of an ion which is assimilated into organic compounds in the cell, a reduction of temperature may inhibit the rate at which assimilation occurs. In such circumstances, the pool size of unassimilated ion in the cell may increase rapidly, even though influx is lower. Since the ionic concentrations in the cytoplasm seem to be subject to strict homeostasis, efflux may increase as a result of the controls normally operating. Thus, both thermodynamic and metabolic influences can be at work to produce the results so frequently reported. Again the missing consideration is that of time. Clearly, if a reduction in temperature gives rise initially to an efflux of ions, the situation is bound to change with time as a new steady state is approached which is compatible with the new rate of ATP production and H^+-pumping, or when the regulatory pool has been adjusted to its correct value. These strictures apply equally to measurements of influx and efflux of ions, but may be doubly confounded where net uptake (i.e. influx *minus* efflux) is measured.

Notwithstanding the difficulties mentioned above, there are numerous reports showing how ion uptake varies with temperature. The cells and tissues are usually

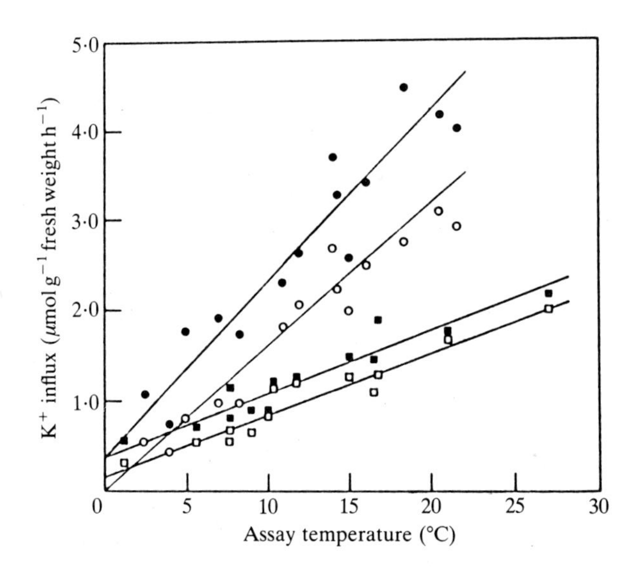

Fig. 3. The apparently linear dependence of (^{86}Rb)K^+ uptake in rye roots on temperature. Two populations were selected from each of the WG and DT treatments, differing markedly in shoot/root ratio.(○, □, 20°C-grown plants, WG; ●, ■, 8°C root, 20°C shoot, DT. Mean shoot/root ratios were 2·52 (○) and 1·76 (□) for WG plants and 2·80 (●) and 2·25 (■) for DT plants.) (White *et al.* 1987.)

grown (and acclimated) at one temperature and observed at a range of others. The shape of the temperature response curve, plotted on arithmetic scales, has been found to vary. It can be linear, at least over a defined range of 'physiological' temperatures (e.g. K^+, Fig. 3, White *et al.* 1987; Zsoldos & Karvarly, 1979), sigmoidal/exponential (e.g. K^+, Bravo-F & Uribe, 1981, or $H_2PO_4^-$, Nandi *et al.* 1987), or even hyperbolic (e.g. Rb^+, Carey & Berry, 1978). 'Critical temperatures' for root ion uptake have been suggested on the basis of 'breaks' in Arrhenius plots of the temperature response (Holmern *et al.* 1974; Carey & Berry, 1978; Clarkson & Warner, 1979; Bravo-F & Uribe, 1981). However, it should be noted that Arrhenius plot discontinuities may be eliminated at high external ion concentrations (Holmern *et al.* 1974; Bravo-F & Uribe, 1981). An example of an Arrhenius transformation is shown in Fig. 4 where the experimental points appear to fall into two distinct groups which can be fitted by 2 straight-line segments. While the validity of such a plot can be upheld by statistical criteria (such as the maximum likelihood analysis) interpretation, in terms of activation energies and membrane phase transitions may not be straightforward. The values for break points may resemble those found in the plots of K^+-stimulated Mg-ATPase activity versus temperature, and these may further correlate with changes in lipid ordering. Both maize and barley roots show discontinuities in the uptake of Rb^+ at about 10°C (Carey & Berry, 1978); it is possible to ascribe these to bulk membrane phase changes in the case of maize or to clustering of phospholipid head groups in barley (Caldwell & Whitman, 1987).

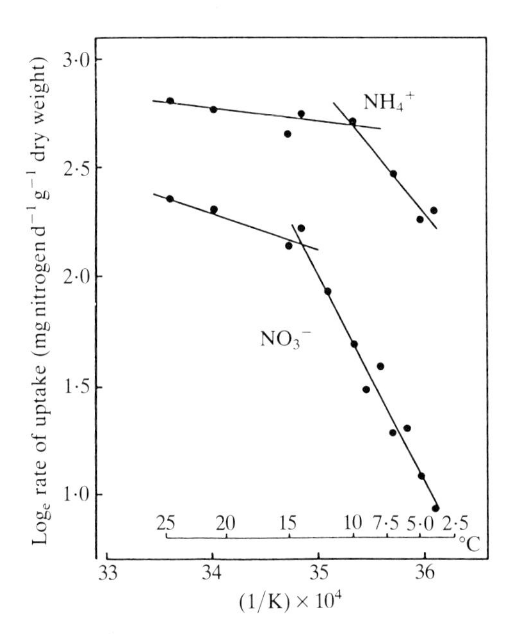

Fig. 4. Arrhenius plots of temperature response of NO_3^- and NH_4^+ uptake by roots of *Lolium multiflorum* showing changes in apparent activation energy. (Clarkson & Warner, 1979.)

Long-term changes of ion uptake in response to low temperatures

Temperature compensation; regulation of nutrient uptake at different temperatures

In organisms as diverse as the alga *Chara* (Raven & Smith, 1978; Sanders, 1981) and the grass *Lolium perenne* (Clarkson *et al.* 1986) tissue concentrations of nutrients are maintained at the more or less constant levels required for growth or homeostasis in individuals grown at widely differing temperatures. Table 3 shows data for the simple situation of unicellular algae in a non-growing state (Raven & Smith, 1978), the concentrations of K^+, Na^+ and Cl^- do not differ significantly for cells maintained in the range of 5 to 25 °C.

The response to temperature is more complex in growing, multicellular plants with roots and shoots; the total demand for nutrients may be reduced at low temperatures since growth is usually slowed down. Table 4 shows that the relative

Table 3. *Nutrient composition of cells of* Chara corallina *grown at three temperatures. Nutrient concentrations (mol m^{-3}) with SEs in parentheses. (Data from Raven & Smith (1978).)*

	Growth temperature		
Ion	25 °C	15 °C	5 °C
Cl^-	137 (15)	14) (14)	130 (4)
K^+	70 (13)	70 (4)	69 (5)
Na^+	94 (9)	90 (5)	82 (10)

Table 4. *Daily relative accumulation rates of dry matter (RGR), nitrogen (RAR_N) and potassium (RAR_K) of cooled plants relative to plants grown in warm conditions*

Species	Treatment	RGR (DM)	RAR_N	RAR_K
Wheat cv. Mardler; Cooper (1986)	DG[1]	90 %	100 %	93 %
	CG[2]	57 %	43 %	46 %
Rye cv. Rheidol; White (1987)	DG[1]	76 %	75 %	75 %
	CG[2]	40 %	39 %	49 %
Maize double cross hybrid 'campo'; Stamp (1984)	CG[3]	34 %	36 %	41 %

[1] RAR of plants grown with roots at 8 °C and shoots at 20 °C relative to RAR of plants grown with roots and shoots at 20 °C.
[2] RAR of plants grown with roots and shoots at 8 °C relative to RAR of plants grown with roots and shoots at 20 °C.
[3] RAR of plants grown with roots and shoots at 22/18 °C (day/night) relative to RAR of plants grown at 15/11 °C.

accumulation rate of nutrients is affected to the same extent as the relative growth rate. The growth rates of wheat, rye and maize have very different responses to temperature but, despite this, nutrient accumulation is in balance with growth in all cases. The percentage of dry matter often increases in plants grown at low temperature; such changes in dry weight/fresh weight ratio explain the small discrepancies between growth and nutrient accumulation rates seen in some treatments.

It is apparent from these observations that, although ion influxes in many tissues are temperature-dependent, internal controls can compensate for any effect of temperature. This was established many years ago in *Nitella* by Hoagland *et al.* 1926) but has been rather lost sight of in intervening years. They found that Br^- uptake had a large temperature sensitivity (Q_{10} = 3·5) in cells taken from ambient and cooled to 10 °C for 6 h. As time passed, however, the temperature coefficient became smaller; even in these relatively sluggishly growing organisms, therefore, there were signs of temperature acclimatization. Similarly, in *Chara*, Raven & Smith (1978) found that Cl^- influx was at least two-fold lower at 5 °C than at 15 °C in the first 24 h after the temperature was changed, but after acclimation for 96 h the temperature dependence was virtually eliminated. If *Chara* cells acclimated to 4 °C were warmed up to 20 °C the influx of Cl^- was 6–10-fold greater than in cells kept at 20 °C (Sanders, 1981). This enhanced Cl^- influx decayed over the subsequent 8–13 h at 20 °C. This type of behaviour is very similar to that seen for K^+ uptake by the roots of barley (Clarkson *et al.* 1974) and rye (Clarkson, 1976; White *et al.* 1987) although it must be emphasized that the *Chara* internodal cells were not growing and their 'demand' for nutrients would have been of a rather different character to that of barley plants.

Studies specifically concerned with either the availability of ATP or the membrane potential in cells growing at different temperatures are hard to find.

Table 5. *Examples of kinetic parameters measured in plants of high and low nutrient status*

Species	Nutrient and status	Km (μM)	V_{max} (μmol g^{-1} FW)
Hordeum vulgare (Drew & Saker, 1984)	High K	133	1
	Low K	36	7
Hordeum vulgare (Lee, 1982)	High P	6·6	0·26
	Low P	4·9	0·48
Solanum tuberosum (Cogliatti & Clarkson, 1983)	High P	22	0·72
	Low P	2	1·84
Macroptilium atropurpureum (Clarkson *et al.* 1983)	High S	5	0·34
	Low S	8	1·95

Such published work as we are aware of makes it seem unlikely that disruption of the energy supply for primary active transport persists for long after a change in environmental temperature has occurred. In *Nitella* it has been shown that the variation in the rate of cytoplasmic streaming (a process dependent on ATP) at temperatures in the range 10–30 °C can be explained completely by changes in cytoplasmic viscosity (Tazawa, 1968). The leaves of *Brassica napus* have a higher concentration of ATP and a greater adenylate energy charge at 2 °C than when grown at 20 °C (Sobczyk & Kacperska-Palacz, cited in Kacperska-Palacz, 1978). In roots of barley an ATP-'expensive' process such as nitrate reduction can be appreciably more active in roots acclimated to 10 °C than in those grown at 20 °C (Deane-Drummond *et al.* 1980).

Raven & Smith (1978) concluded that the ion fluxes into *Chara* cells were kinetically controlled and that this probably involved changes in the number or the kinds of carriers present in the cell membranes. Conclusions of this kind are less easily reached when we consider higher plants where nutrient intake is restricted to the roots. Clearly there is coordination between shoot and root so that net nutrient uptake satisfies the growth of both parts of the plant. Plants are able to vary the relative size of the two parts; thus, if the root becomes relatively larger the required nutrient uptake moves into the plant across a larger surface so that a lower flux is required, and vice versa if the root becomes relatively smaller. During the course of plant development from a seedling to a mature individual there are some major shifts in the ratio of absorbing surface to plant weight, and it can be deduced, from analysis of plant nutrient content at different times, that the nutrient fluxes across the root surface can vary greatly (e.g. Wild & Breeze, 1981). It has become clear that the cells of roots have a great deal of scope for varying the rate at which the uptake of ions can take place, and regulation can be ion-specific so that a particular nutrient requirement can be met (see Glass, 1983; Clarkson, 1985). As a general guide to the magnitude of these responses one can cite variations in the capacity of the high affinity transport systems of plants which are just entering the first stage of nutrient deficiency but have not yet had their rate of growth or metabolism severely impaired (Table 5). From this we conclude that

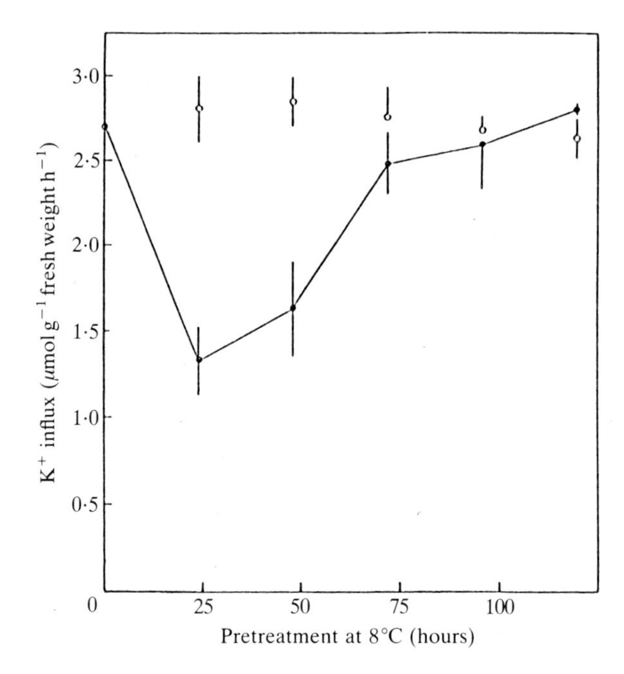

Fig. 5. Variation in (^{86}Rb)K^+ uptake in rye roots after transfer to a root temperature of 9°C from an initial temperature of 20°C. Assay was performed at 20°C. (White *et al.* 1987.)

plants would be able to adjust the kinetic controls on ion absorption if transfer to a lower temperature reduced nutrient content. Most evidence suggests that specific ion transporters respond in some way to the concentration of the ion, or some product of its assimilation in a regulatory pool. The question which concerns us here is whether, or in what circumstances, this ability to make adjustment to the activity of the transport system is brought into service when plants adapt to changing temperature, and to what extent are other, acclimatory processes specific to the response to low temperature involved.

After transfer of plants to low temperature, there is a transition period before a new balance between uptake and growth rates is achieved. Fig. 5 shows that when the roots of rye plants are transferred from 20° to 8° there is an initial reduction in K^+ uptake when assayed at 20°C, but that during a 2·5-day exposure at low root temperature the K^+ uptake is restored. We have indicated some of the biochemical, physiological and morphological changes that may take place during this transition period, but it is important to note that some of the changes that occur may not be acclimatory; they may represent direct effects of the changed temperature on growth rates. The existence of a transition phase does not necessarily imply, therefore, that acclimatory changes are taking place. Indeed, as will be discussed later, some morphological changes may actually increase the demand placed on the nutrient uptake system. Fig. 6 shows some results for growth and potassium uptake by *Lolium perenne* after transfer from a stable root

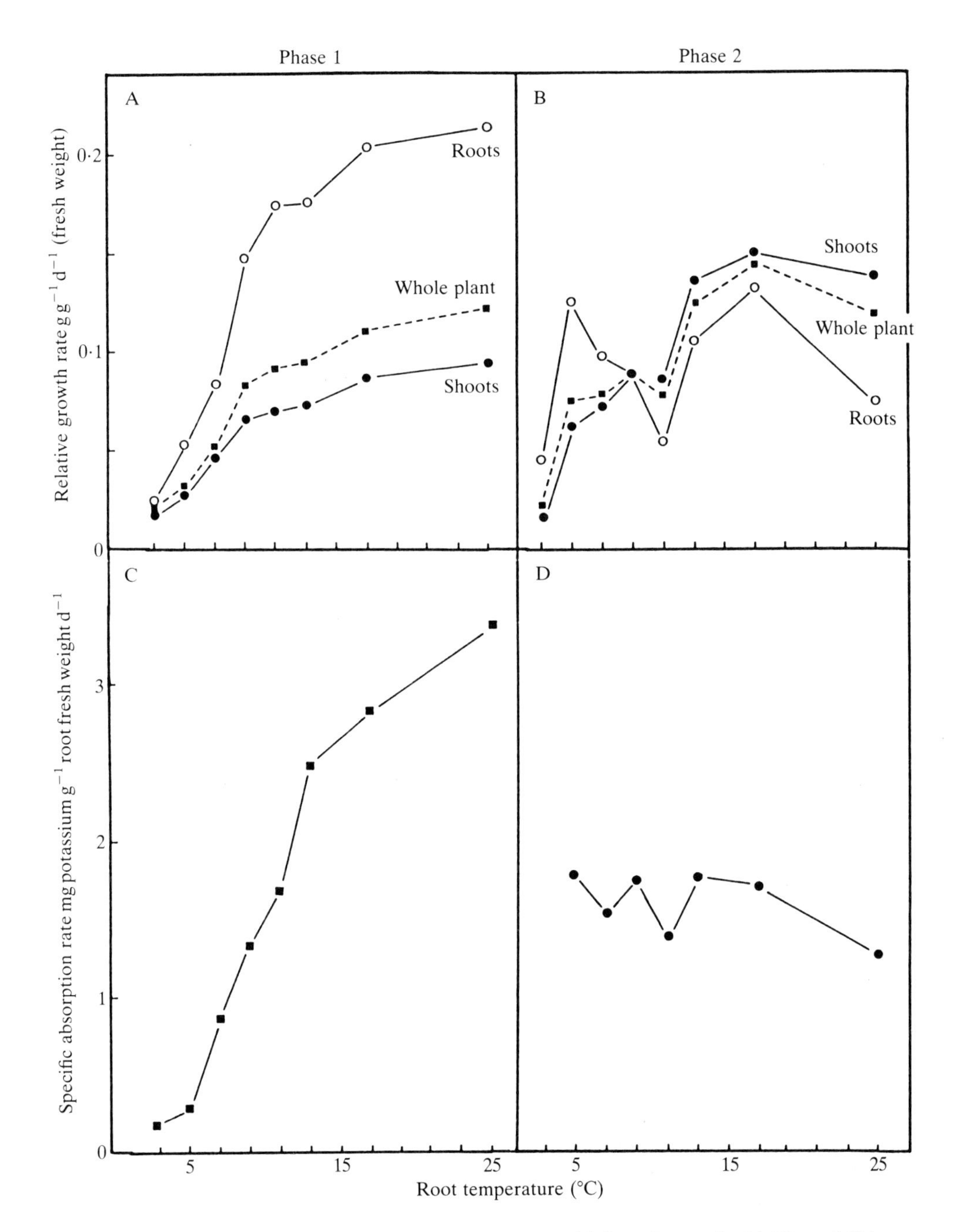

Fig. 6. Changes in the apparent temperature sensitivity of growth (A,B) and K^+ uptake (C,D) of *Lolium perenne* plants during adjustment to a temperature change. Pretreatment root temperature was 5 °C, subsequent root temperature is given on the *x*-axis; shoot temperatures for all treatments were 25 °C (day) and 15 °C (night). During phase 1 (A,C) plants were adjusting both growth rate and relative root size to the changed temperature. In phase 2 a relatively stable shoot/root ratio had been established. (Redrawn from Clarkson *et al.* 1986.)

temperature of 5°C to a range of root temperatures; shoots were maintained at temperatures of 25°C (day) and 15°C (night). During the first nine days after transfer (phase 1) there was a major change in the relative size of the root system; root growth rate was greater than shoot growth rate. At the end of this period new balances seem to have been struck between shoot and root size for each of the root temperature treatments. During phase 1, the uptake rate of potassium showed a marked dependence on temperature; the rate of uptake responded to temperature change to a greater extent than the growth of the whole plant. In the second phase, the uptake rate became much less sensitive to temperature.

This is one example of the many reports of the response of nutrient uptake to temperature and it illustrates the complexity of the system. The demand for nutrients per unit root is affected not only by growth rate, but also by the shoot/root ratio. The uptake rate is affected by changes in the number and activity of carriers per unit root and by the size of the absorbing surface. We will now examine in more detail some of the effects of temperature on growth and substrate levels and discuss some possible acclimatory mechanisms.

Changes in shoot/root ratio of plants exposed to low temperature

When a whole plant is transferred from a warmer environment to a cooler one there is usually an increase in the relative size of the root system (i.e. the S/R decreases). It is also found amongst wild plants that those growing in cooler environments tend to have larger root systems, though there is considerable variation in S/R amongst plants in all environments (Chapin, 1974; Körner & Larcher, 1988). Partitioning of dry matter in favour of root growth can be seen as an acclimatory or adaptive response; the increase in relative root size may compensate, in part, for the decrease in the rate of nutrient uptake per unit root seen in the short-term response.

The situation is more complex when the roots only are cooled relative to the shoots. Growth itself is very sensitive to temperature, and under these differential temperature conditions, the low temperature experienced by the root may lead to a slower root growth rate. The shoots, experiencing a higher temperature, continue growth at a higher rate and S/R may increase but will eventually reach a new steady state when shoot growth is being limited by some function related to root size. In fact, both this effect and the compensatory mechanism described above may operate, and both increases and decreases in S/R are reported in the literature (Fig. 7).

Some workers have found that if the root temperature is below the optimum for whole plant growth there is a greater partitioning of dry matter into root tissues (Fig. 7, lines a, c). This may be seen as an acclimatory response as in the situation where the whole plant is exposed to low temperatures; the demand for nutrients per unit root is decreased.

Other workers have found that the direct effects of temperature on root growth outweigh the compensatory mechanism of shoot–root partitioning (Fig. 7, lines b, d). In this situation the changes in S/R cannot be considered acclimatory, indeed

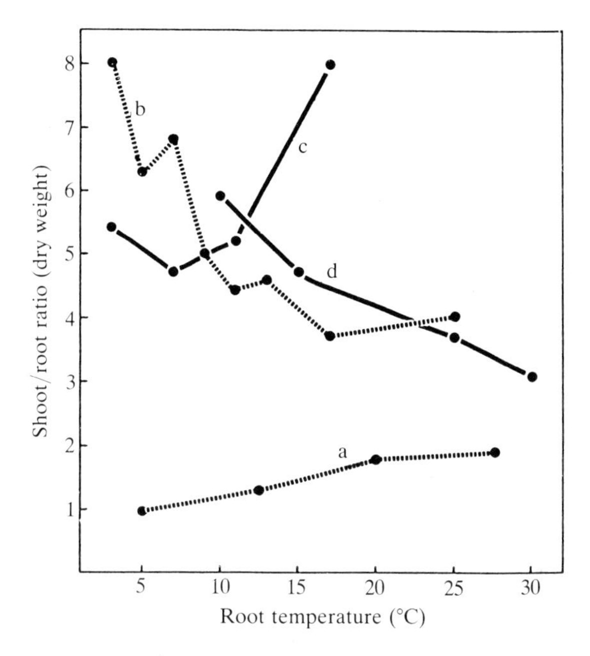

Fig. 7. Response of shoot/root (dry matter) ratio to shoot-root differential temperatures. Shoots were held at constant temperatures. Temperatures of roots are given on the *x*-axis. a, *Lolium perenne* (Davidson, 1969*a*); b, *Lolium perenne* cv. S23 (Clarkson *et al.* 1986); c, *Brassica napus* cv. bien venu (McDuff *et al.* 1987); d, *Brassica napus* cv. Emerald (Cumbus & Nye, 1982).

they add to the problem, requiring that uptake per unit root weight increases in order to satisfy plant demand. An extreme case was seen in maize roots cooled to 13–10°C while the shoots were held at 20°C (Clarkson & Gerloff, 1980). A very large S/R ratio resulted after 30 days of this treatment, creating a very high demand per unit root. This led to symptoms of severe nutrient deficiency in the plants since the ion transport systems were unable to meet this demand. The results of Stamp (1984; reported above, Table 4) show that, at only slightly warmer root temperatures, nutrient content can be maintained in *Zea mays* provided that an acceptable S/R is maintained. The importance of shoot/root ratio in determining uptake rates of plants within their range of tolerance is illustrated for rye in Fig. 8.

Acclimatory changes in carrier number or activity

There is often an increase in the rate of nutrient uptake per unit root surface area or weight following root cooling. It is unclear whether this results from a change in the number of carriers and/or their activity. The existence of a transition period of about 2·5 days in rye (Fig. 5) might suggest that carrier synthesis be involved and this view has been favoured on the basis that large changes in neither the Km (Siddiqi *et al.* 1984) nor the temperature sensitivity (Clarkson, 1976) of the uptake system occur during acclimation. However, acclimation via a change in

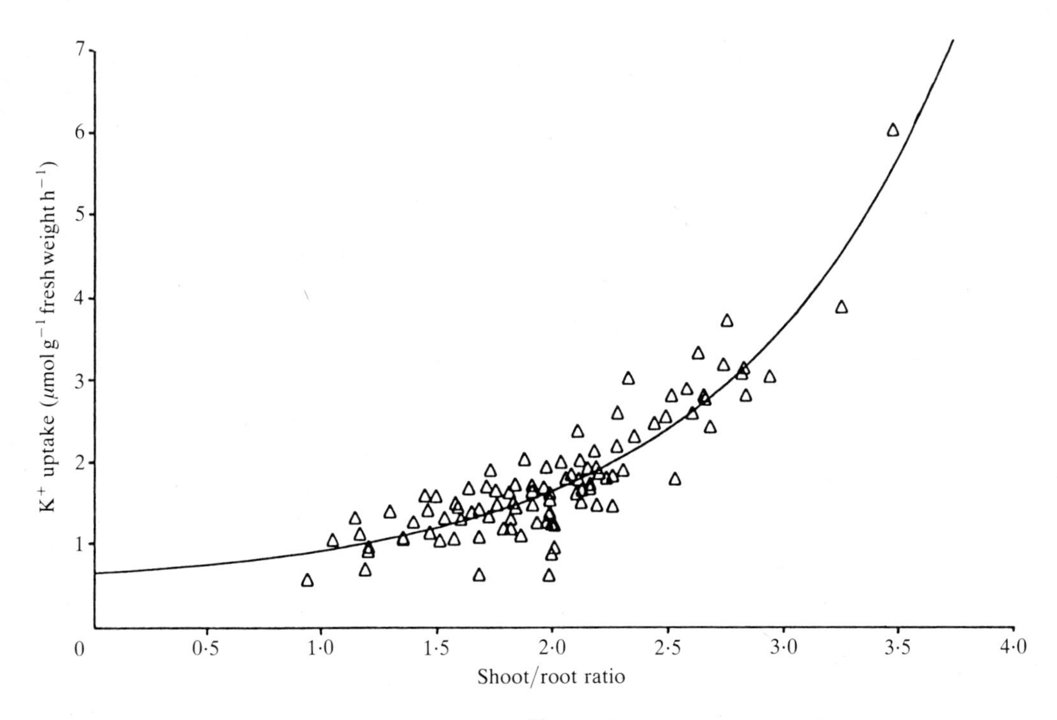

Fig. 8. Influence of shoot-root ratio on (^{86}Rb)K$^+$ uptake in rye plants grown and assayed at 20°C. (White *et al.* 1987.)

carrier activity remains an important possibility; changes in activity might be expected to be reflected by changes in either Km or V_{max} and regulation of carrier activity may be possible without influencing the temperature sensitivity of the carrier. In numerous publications Glass and his colleagues (see Glass, 1983) have proposed that the K$^+$-carrier system in roots of several species is allosterically regulated by the K$^+$ concentration in an internal pool. This regulation is accompanied by substantial shifts in the Km of the high affinity transport system (see also Table 5). Siddiqi *et al.* (1984) have proposed that the sensitivity of the K$^+$ carrier to the regulatory pool of K$^+$ in roots may decrease in cool conditions. White *et al.* (1987) have suggested that this would only be important at limiting external concentrations of K$^+$.

White *et al.* (1987) found that the K$^+$ uptake by rye roots was less sensitive to low root temperature in plants of high shoot/root ratio suggesting the presence of a temperature insensitive component of uptake. The contribution of this temperature-insensitive component increased with increasing shoot/root ratio and was greater in plants pretreated at low root temperatures. There are several reports of an increased hydraulic conductivity in cooled roots (Clarkson, 1976; Markhart *et al.* 1979; Stephens, 1981; White *et al.* 1987) and White *et al.* (1987) propose that this may be an acclimatory response to low temperatures; the high hydraulic conductivity, coupled with a high transpiration rate provided by a high relative leaf area, enhances ion uptake by mass flow and the sweeping away of ions in the xylem. Thus, the increased demand per unit root at high shoot/root ratios may be more easily met.

Increases in nutrient and carbohydrate contents of tissues exposed to low temperatures

It is a common observation that when roots are cooled there is an increased accumulation of carbohydrates in root tissues (e.g. Davidson, 1969*b*). This may result from a temporary imbalance in the supply of carbohydrates from the shoot and their utilization in the root. It is unlikely that this increase in the energy available for root activities explains the enhancement of ion uptake after low temperature pretreatment as the energy requirement for ion uptake (narrow sense costs; cf. Clarkson, 1985) is small compared to other root processes. Many reports (e.g. Lambers, 1979) indicate that there is normally an excess of respiratory substrates supplied to the root; if uptake is not limited by energy at higher temperatures an increase in the supply of energy cannot explain an increase in the uptake rate. Experimental observations show that the time course of the acclimatory increase of uptake rates following transfer of roots to low temperature is not paralleled by changes in carbohydrate content (Clarkson, 1976) and that while the increase in carbohydrate levels occurs nearer the root tips, the increase in ion-uptake rate occurs in the mid portion of the root (Bowen & Rovira, 1971; White, 1987).

There are also many reports of an increase in the mineral nutrient contents of plants exposed to low temperatures (for example, Körner & Larcher, 1988). We found increases of nitrogen and potassium concentrations on a fresh weight basis in young wheat plants exposed to low temperatures (Table 6), but not in rye. In this case, the increases occurred only in the tissues exposed directly to the low temperature. Thus, when the roots were cooled relative to the shoots, significant concentration increases were found in the roots but not in the shoots. Clarkson *et al.* (1986) also found higher N concentrations at lower root temperatures in *Lolium perenne*. In this case, the trend was observed in both shoots and roots. In both experiments, these differences were not apparent on a dry weight basis.

It has long been proposed that the accumulation of carbohydrates and other solutes has a role in osmotic adjustment during cold-hardening (see Levitt, 1980). Siddiqi *et al.* (1984) have suggested that the increased concentration of K^+ seen in

Table 6. *Nitrogen and potassium concentrations (μmols g^{-1} fresh weight) of wheat roots and shoots grown under three different temperature regimes. SEs are about 10%. (Data from Cooper, 1986.)*

	Temperature		Nitrogen		Potassium	
	Shoot	Root	Shoot	Root	Shoot	Root
WG	20	20	372	207	156	86
DG	20	8	431	369	170	155
CG	8	8	715	268	279	147

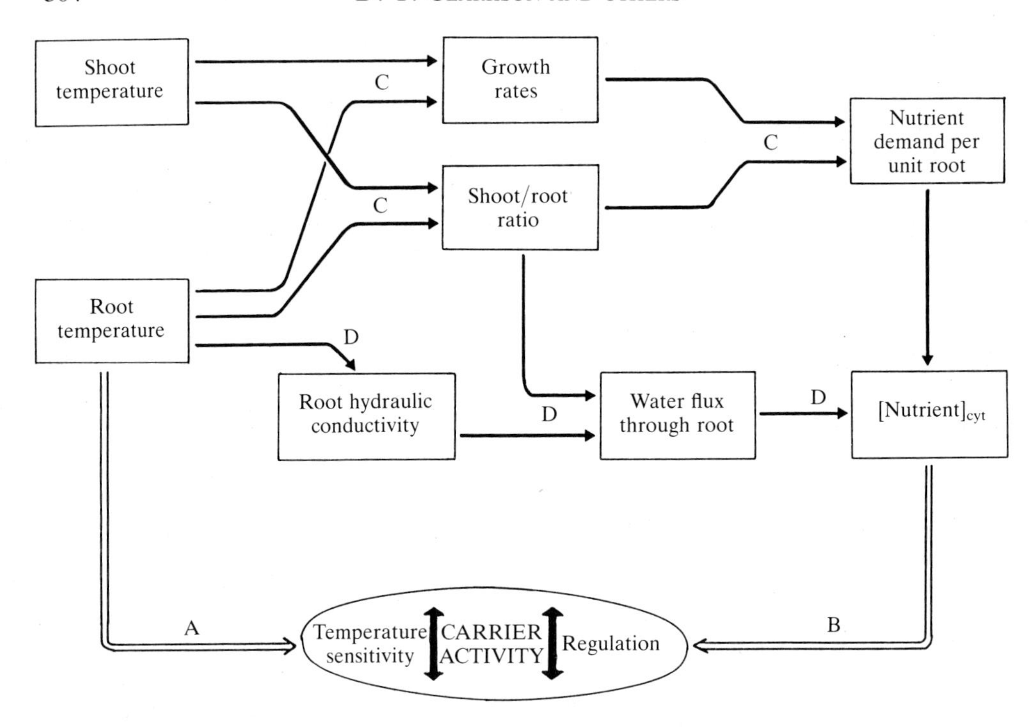

Fig. 9. Schematic representation of the effects of temperature on the activity of ion transport carriers. The direct, short-term effect of low root temperatures is to lower carrier activity (A; p. 284) but within the normal range of tolerance of the plant, regulatory mechanisms will operate (B; p. 297) so that carrier activity is increased to match the demand for nutrients. The demand itself may change as growth rates and shoot/root ratio are also influenced by temperature (C; p. 300). Increased root hydraulic conductivity and high shoot/root ratios may lead, via an increased water flux, to a lower ion concentration in an internal regulatory root pool (D; p. 302) and, therefore, to increased activity of the carrier acting under regulation by internal ion concentrations.

cooled roots may have such a role. However, there is doubt that even the much greater accumulation of carbohydrates is significant in this respect. Increases in inorganic nutrients and carbohydrates may merely reflect that growth processes are lowered to a greater extent by low temperature than either photoassimilatory or nutrient uptake processes (see also Farrar, 1988).

Conclusions

Much progress has been made in describing the effects of low temperature on artificial and isolated membrane systems but difficulties remain in relating these to changes in nutrient uptake by intact roots. Unicellular, non-growing algae provide useful model systems; temperature responses are much more complex in higher plants where growth rates and the relationship between shoot and root are also strongly influenced by temperature. It appears that, within the usual range of temperatures experienced by the growing plant, normal regulatory systems operate to match nutrient uptake with the growth-led demand. Fig. 9 suggests that the regulation of carrier activity by the internal concentration of nutrient in the root may be influenced by shoot/root ratio, growth rate and water flux. Characterization of carriers at a molecular level will allow progress in our understanding the mechanism of regulation at the cell level. Further studies of the long-distance transport of nutrients are also required in order to understand how demand for, and supply of, nutrients is integrated at the level of the whole plant.

References

AHLERS, J. (1981). Temperature effects on kinetic properties of plasma membrane ATPase from the yeast *Saccharomyces cerevisiae*. *Biochim. Biophys. Acta* **649**, 550–556.

ARMOND, P. A. & STAEHELIN, L. A. (1979). Lateral and vertical displacement of integral membrane proteins during phase transition in *Anacystis nidulans*. *Proc. Natl Acad. Sci.* **76**, 1901–1905.

ASHCROFT, R. G., COSTER, H. G. L., LAVER, D. R. & SMITH, J. R. (1983). The effect of cholesterol inclusion on the molecular organization of biomolecular lipid membranes. *Biochim. Biophys. Acta* **730**, 231–238.

BANGE, G. G. J. (1979). Loss of rubidium and potassium from barley roots on sudden chilling. *Pl. Physiol.* **64**, 581–584.

BEILBY, M. J. & COSTER, H. G. L. (1976). The action potential in *Chara corallina*: effect of temperature. *Aust. J. Pl. Physiol.* **3**, 275–289.

BENGA, G. & HOLMES, R. P. (1984). Interactions between components in biological membranes and their implications for membrane function. *Progr. Biophys. Mol. Biol.* **43**, 195–257.

BERRY, J. A. & RAISON, J. K. (1981). Responses of macrophytes to temperature. In *Encyclopedia of Plant Physiology*, NS **12A** (ed. O. L. Lange, P. S. Nobel, C. B. Osmond & H. Ziegler), pp. 277–338. Berlin: Springer-Verlag.

BIELESKI, R. L. & FERGUSON, I. B. (1983). Physiology and metabolism of phosphate and its compounds. In *Encyclopedia of Plant Physiology New Series Vol. 15A* (ed. A. Läuchli & R. L. Bieleski), pp. 422–449. Berlin: Springer-Verlag.

BLATT, F. J. (1974). Temperature dependence of the action potential in *Nitella flexilis*. *Biochim. Biophys. Acta* **339**, 382–389.

BOWEN, G. D. & ROVIRA, A. D. (1971). Relationship between root morphology and nutrient uptake. In *Recent Advances in Plant Nutrition*, **1** (ed. R. M. Samish), pp. 293–305. London: Gordon & Breach.

BRAVO-F, P. & URIBE, E. G. (1981). Temperature dependence on concentration kinetics of absorption of phosphate and potassium in corn roots. *Pl. Physiol.* **67**, 815–819.

CALDWELL, C. R. & HAUG, A. (1981). Temperature dependence of the barley root plasma membrane-bound Ca^{2+}- and Mg^{2+}-dependent ATPase. *Physiologia Pl.* **53**, 117–124.

CALDWELL, C. R. & WHITMAN, C. E. (1987). Temperature-induced protein conformational changes in barley root plasma-membrane enriched microsomes. I. Effect of temperature on membrane protein and lipid mobility. *Pl. Physiol.* **84**, 918–923.

CAREY, R. W. & BERRY, J. A. (1978). Effects of low temperature on respiration and uptake of rubidium ions by excised barley and corn roots. *Pl. Physiol.* **61**, 858–860.

CARRUTHERS, A. & MELCHIOR, D. L. (1986). How bilayer lipids affect membrane protein activity. *Trends Biochem. Sci.* **11**, 331–335.

CHAPIN, F. S. III (1974). Morphological and physiological mechanisms of temperature compensation in phosphate absorption along a latitudinal gradient. *Ecology* **55**, 1180–1198.

CLARKSON, D. T. (1976). The influence of temperature on the exudation of xylem sap from detached root systems of rye (*Secale cereale*) and barley (*Hordeum vulgare*). *Planta* **132**, 297–304.

CLARKSON, D. T. (1984). Ionic relations. In *Advanced Plant Physiology* (ed. M. B. Wilkins), pp. 319–353. Bath: Pitman.

CLARKSON, D. T. (1985). Factors affecting mineral nutrient acquisition by higher plants. *A. Rev. Pl. Physiol.* **36**, 77–115.

CLARKSON, D. T. & GERLOFF, G. C. (1980). Growth and nutrient absorption by roots of maize genotypes at low temperatures. In *Production and Utilization of the Maize Crop* (ed. E. S. Bunting), pp. 179–187. Chichester, UK: Packard Publishing.

CLARKSON, D. T., HALL, K. C. & ROBERTS, J. K. M. (1980). Phospholipid composition and fatty acid desaturation in the roots of rye during acclimatization to low temperature. *Planta* **149**, 464–471.

CLARKSON, D. T., HOPPER, M. J. & JONES, L. H. P. (1986). The effect of root temperature on the uptake of nitrogen and the relative size of the root system in *Lolium perenne*. I. Solutions containing both NH_4^+ and NO_3^-. *Pl. Cell Environ.* **9**, 535–543.

CLARKSON, D. T., SHONE, M. G. T. & WOOD, A. V. (1974). The effect of pretreatment temperature on the exudation of xylem sap by detached barley root systems. *Planta* **121**, 81–92.

CLARKSON, D. T., SMITH, F. W. & VANDEN BERG, P. J. (1983). Regulation of sulphate transport in a tropical legume, *Macroptilium atropurpureum*, cv. Siratro. *J. exp. Bot.* **34**, 1463–1483.

CLARKSON, D. T. & WARNER, A. (1979). Relationships between root temperature and the transport of ammonium and nitrate ions by Italian and perennial ryegrass (*Lolium multiflorum* and *Lolium perenne*). *Pl. Physiol.* **64**, 557–561.

COGLIATTI, O. H. & CLARKSON, D. T. (1983). Physiological changes in potato plants during the development of, and recovery from, phosphate stress. *Physiologia Pl.* **58**, 287–294.

COOPER, H. D. (1986). Uptake, assimilation and circulation of nitrogen compounds in cereals, with particular reference to the effects of environmental temperature. D.Phil. thesis, University of Oxford, 262pp.

CUMBUS, I. P. & NYE, P. H. (1982). Root zone temperature effects on growth and nitrate absorption in rape (*Brassica napus* cv. Emerald). *J. exp. Bot.* **33**, 1138–1146.

DAVIDSON, R. L. (1969*a*). Effect of root/leaf temperature differentials on root/shoot ratios in some pasture grasses and clover. *Ann. Bot.* **33**, 561–569.

DAVIDSON, R. L. (1969*b*). Effects of edaphic factors on the soluble carbohydrate content of *Lolium perenne* L. and *Trifolium repens* L. *Ann. Bot.* **33**, 579–589.

DEANE-DRUMMOND, C. E., CLARKSON, D. T. & JOHNSON, C. B. (1980). The effect of differential root and shoot temperature on the nitrate reductase activity, assayed *in vivo* and *in vitro*, in roots of *Hordeum vulgare* (barley). Relationship with diurnal changes in endogenous malate and sugar. *Planta* **148**, 455–461.

DOUGLAS, T. J. & WALKER, R. R. (1984). Phospholipids, free sterols and adenosine triphosphatase of plasma membrane-enriched preparations from roots of citrus genotypes differing in chloride exclusion ability. *Physiologia Pl.* **62**, 51–58.

DREW, M. C. & SAKER, L. R. (1984). Uptake and long-distance transport of phosphate, potassium and chloride in relation to internal ion concentrations in barley: evidence of non-allosteric regulation. *Planta* **160**, 500–507.

DUFOUR, J.-P. & GOFFEAU, A. (1980). Phospholipid reactivation of the purified plasma membrane ATPase of yeast. *J. biol. Chem.* **255**, 10 591–10 598.

DUFOUR, J.-P. & TSONG, T. W. (1981). Plasma membrane ATPase of yeast: activation and interaction with dimyristoylphosphatidylcholine vesicles. *J. biol. Chem.* **256**, 1801–1808.

EAST, J. M., JONES, O. T., SIMMONDS, A. C. & LEE, A. G. (1984). Membrane fluidity is not an important physiological regulator of the (Ca^{2+}-Mg^{2+})-dependent ATPase of sarcoplasmic reticulum. *J. biol. Chem.* **259**, 8070–8071.

FARRAR, J. F. (1988). Temperature and the partitioning and translocation of carbon. In *Plants and Temperature* (ed. S. P. Long & F. I. Woodward), *Symp. Soc. Exp. Biol.*, vol. 42, pp. 203–235. Cambridge: Company of Biologists.

GLASS, A. D. M. (1983). Regulation of ion transport. *A. Rev. Pl. Physiol.* **34**, 311–326.

GRAHAM, D. & PATTERSON, B. D. (1982). Responses of plants to low nonfreezing temperatures: proteins, metabolism and acclimation. *A. Rev. Pl. Physiol.* **33**, 347–372.

HOAGLAND, D. R., HIBBARD, P. L. & DAVIS, A. R. (1926). The influence of light, temperature and their conditions on the ability of *Nitella* cells to concentrate halogens in the cell sap. *J. gen. Physiol.* **10**, 121–146.

HOLMERN, K., VANGE, M. S. & NISSEN, P. (1974). Multiphasic uptake of sulphate by barley roots. II. Effects of washing, divalent cations, inhibitors and temperature. *Physiologia Pl.* **31**, 302–310.

HOPE, A. B. & ASCHBERGER, P. A. (1970). Effects of temperature on membrane permeability to ions. *Aust. J. biol. Sci.* **23**, 1047–1060.

HORVÁTH, I., VIGH, L., WOLTJES, J., FARKAS, T., VAN HASSELT, P. & KUIPER, P. J. C. (1987). Combined electron-spin-resonance, X-ray-diffraction studies on phospholipid vesicles obtained from cold-hardened wheats II. The role of free sterols. *Planta* **170**, 20–25.

IMBRIE, C. W. & MURPHY, T. M. (1984). Photo-inactivation of detergent-solubilized plasma membrane ATPase from *Rosa damascena*. *Pl. Physiol.* **74**, 617–621.

JENNINGS, P. H. & TATTAR, T. A. (1979). Effect of chilling on membrane potentials of maize and oat leaf cells. In *Low Temperature Stress in Crop Plants: The Role of the Membrane* (ed. J. M. Lyons, D. Graham & J. K. Raison), pp. 153–161. New York: Academic Press.

KACPERSKA-PALACZ, A. (1978). Herbaceous plant acclimation mechanisms. In *Plant Cold Hardiness and Freezing Stress* (ed. P. H. Li & A. Sakai), London: Academic Press.

KEENAN, T. W., LEONARD, R. T. & HODGES, T. K. (1973). Lipid composition of plasma membranes from *Avena sativa* roots. *Cytobios* **7**, 103–112.

KEIFER, D. W. & SPANSWICK, R. M. (1979). Correlation of adenosine triphosphate levels in *Chara corallina* with the activity of the electrogenic pump. *Pl. Physiol.* **64**, 165–168.

KINNEY, A. J. (1983). Effect of temperature on the turnover of membrane components of roots. D.Phil. Thesis, University of Oxford.

KINNEY, A. J., CLARKSON, D. T. & LOUGHMAN, B. C. (1987*a*). The regulation of phosphatidylcholine biosynthesis in rye (*Secale cereale*) roots. Stimulation of the nucleotide pathway by low temperature. *Biochem. J.* **242**, 755–759.

KINNEY, A. J., CLARKSON, D. T. & LOUGHMAN, B. C. (1987*b*). Phospholipid metabolism and plasma membrane morphology of warm and cool rye roots. *Pl. Physiol. Biochem.* **25**, 769–774.

KÖRNER, CH. & LARCHER, W. (1988). Plant life in cold climates. In *Plants and Temperature* (ed. S. P. Long & F. I. Woodward) *Symp. Soc. Exp. Biol.*, vol. 42, pp. 25–57. Cambridge: Company of Biologists.

KUIPER, P. J. C. (1972). Temperature response of adenosine triphosphatase of bean roots as related to growth temperature and to lipid requirement of the adenosine triphosphatase. *Physiologia Pl.* **26**, 200–205.

KUIPER, P. J. C. (1985). Environmental changes and lipid metabolism of higher plants. *Physiologia Pl.* **64**, 118–122.

LAMBERS, H. (1979). Energy metabolism in higher plants in different environments. Ph.D. thesis, University of Groningen, NL.

LEE, A. G. (1983). Lipid phase transitions and mixtures. In *Membrane Fluidity in Biology* (ed. R. C. Aloia), pp. 43–88. New York, London: Academic Press.

LEE, R. B. (1982). Selectivity and kinetics of ion uptake by barley plants following nutrient deficiency. *Ann. Bot.* **50**, 429–449.

LE GRIMELLEC, C. & LEBLANC, G. (1980). Temperature-dependent relationship between K^+ influx, Mg^{2+}-ATPase activity, transmembrane potential and membrane lipid composition in Mycoplasma. *Biochim. Biophys. Acta* **599**, 639–651.

LEVITT, J. (1980). *Responses of Plants to Environmental Stresses.* Vol. 1. Chilling, freezing and high temperature stresses. New York: Academic Press. 497pp.

LUCAS, W. J. (1984). How are plasmalemma transport processes of *Chara* regulated? In *Membrane Transport in Plants* (ed. W. J. Cram, K. Janacek, R. Rybova & K. Sigler), pp. 459–465. Prague: Academia.

LYONS, J. M. (1973). Chilling injury in plants. *A. Rev. Plant Physiol.* **24**, 445–466.

LYNCH, D. V. & STEPONKUS, P. L. (1987). Plasma membrane lipid alterations associated with cold acclimation of winter rye seedlings (*Secale cereale* L. cv. Puma). *Pl. Physiol.* **83**, 761–767.

MACDUFF, J. H., HOPPER, M. J., WILD, A. & TRIM, F. E. (1987). Comparison of the effects of root temperature on nitrate and ammonium nutrition of oil seed rape (*Brassica napus* L.) in flowing solution culture. I. Growth and uptake of nitrogen. *J. exp. Bot.* **38**, 1104–1120.

MARKHART, A. H. III, FISCUS, E. L., NAYLOR, A. W. & KRAMER, P. J. (1979). Effect of temperature on water and ion transport in soybean and broccoli systems. *Pl. Physiol.* **64**, 83–87.

MCKERSIE, B. D. & THOMPSON, J. E. (1979). Influence of plant sterols on the phase properties of phospholipid bilayers. *Pl. Physiol.* **63**, 802–805.

MELCHIOR, D. L. & STEIN, J. M. (1976). Thermotropic transitions in biomembranes. *A. Rev. Biophys. Bioeng.* **5**, 205–238.

MOELLER, C. H., MUDD, J. B. & THOMSON, W. W. (1981). Lipid phase separations and intramembranous particle movements in the yeast tonoplast. *Biochim. Biophys. Acta* **643**, 376–386.

MURATA, T. & TATSUMI, Y. (1979). Ion leakage in chilled plant tissues. In *Low Temperature Stress in Crop Plants: The Role of the Membrane* (ed. J. M. Lyons, D. Graham & J. K. Raison), pp. 141–151. New York: Academic Press.

NANDI, S. K., PANT, R. C. & NISSEN, P. (1987). Multiphasic uptake of phosphate by corn roots. *Pl. Cell Environ.* **10**, 463–474.

PLATT-ALOIA, K. A. & THOMSON, W. W. (1987). Freeze-fracture evidence for lateral phase separation in the plasmalemma of chilling-injured avocado fruit. *Protoplasma* **136**, 71–80.

POOLE, R. J. (1974). Ion transport and electrogenic pumps in storage tissue cells. *Can. J. Bot.* **52**, 1023–1028.

QUINN, P. J. (1981). The fluidity of cell membranes and its regulation. *Progr. Biophys. Mol. Biol.* **38**, 1–104.

QUINN, P. J. (1988). Effects of temperature on cell membranes. In *Plants and Temperature* (ed. S. P. Long & F. I. Woodward), *Symp. Soc. Exp. Biol.*, vol. 42, pp. 237–258. Cambridge: Company of Biologists.

QUINN, P. J. & CHAPMAN, D. (1980). The dynamics of membrane structure. *CRC Critical Rev. Biochem.* **8**, 1–117.

QUINN, P. J. & WILLIAMS, W. P. (1978). Plant lipids and their role in membrane function. *Progr. Biophys. Mol. Biol.* **34**, 109–173.

RAISON, J. K. & WRIGHT, L. C. (1983). Thermal phase transitions in the polar lipids of plant membranes – their induction by disaturated phospholipids and their possible relationship to chilling injury. *Biochim. Biophys. Acta* **731**, 69–78.

RAVEN, J. A. & SMITH, F. A. (1978). Effect of temperature on ion content, ion fluxes and energy metabolism in *Chara corallina*. *Pl. Cell Environ.* **1**, 231–238.

REA, P. A. & POOLE, R. J. (1986). Chromatographic resolution of H^+-translocating pyrophosphate from H^+-translocating ATPase of higher plant tonoplast. *Pl. Physiol.* **81**, 126–129.

ROBARDS, A. W. & CLARKSON, D. T. (1984). Effects of chilling temperatures on root cell membranes as viewed by freeze-fracture electron microscopy. *Protoplasma* **122**, 75–85.

ROTTEM, S., CIRILLO, V. P., DE KRUYFF, B., SHINITZKY, M. & RAZIN, S. (1973). Cholesterol in Mycoplasma membranes – correlation of enzymic and transport activities with physical state of lipids in *Mycoplasm mycoides* var. Capri adapted to grow with low cholesterol concentrations. *Biochim. Biophys. Acta* **323**, 509–519.

SANDERS, D. (1981). Physiological control of chloride transport in *Chara corallina*. I. Effects of low temperature, cell turgor pressure and anions. *Pl. Physiol.* **67**, 1113–1118.

SERRANO, R. (1985). *Plasma membrane ATPase of plants and fungi.* Boca Raton: CRC Press.

SIDDIQI, M. Y., MEMON, A. R. & GLASS, A. D. M. (1984). Regulation of K^+ influx in barley. Effects of low temperature. *Pl. Physiol.* **74**, 730–734.

SILVIUS, J. R. & MCELHANEY, R. N. (1980). Membrane lipid physical state and modulation by the Na^+, Mg^{2+}-ATPase activity in *Acholeplasma laidlawii* B. *Proc. Natl. Acad. Sci.* **77**, 1255–1259.

SINENSKY, M. (1974). Homeoviscous adaptation – a homeostatic process that regulates the viscosity of membrane lipids in *Escherichia coli*. *Proc. Natl. Acad. Sci.* **71**, 522–525.

SINENSKY, M., PINKERTON, F., SUTHERLAND, E. & SIMON, F. R. (1979). Rate limitation of $(Na^+ + K^+)$-stimulated adenosinetriphosphatase by membrane acyl chain ordering. *Proc. Natl. Acad. Sci.* **76**, 4893–4897.

SINGER, S. J. & NICOLSON, G. L. (1972). The fluid mosaic model of the structure of cell membranes. *Science* **175**, 720–731.

SLAYMAN, C. L., LONG, W. S. & LU, C. Y.-H. (1973). The relationship between ATP and an eletrogenic pump in the plasmamembrane of *Neurospora crassa*. *J. membrane Biol.* **14**, 305–338.

STAMP, P. (1984). Chilling tolerances in young plants demonstrated on the example of maize

(*Zea mays* L.). *Advances in Anatomy and Crop Science* No. 7 (Supplement 7 to Journal of Agronomy & Crop Science). Berlin: Paul Parey, pp. 83.

STEPHENS, J. S. (1981). Effects of temperature on hydraulic conductivity of the roots of *Zea mays*. Ph.D. thesis, University of Reading.

STUBBS, C. D. (1983). Membrane fluidity: structure and dynamics of membrane lipids. *Essays Biochem.* **19**, 1–39.

TÁNCZOS, O. G., VIGH, L., OLÁH, Z., HORVÁTH, I. & JÓO, F. (1984). Effect of catalyst-mediated hydrogenation on the fatty acid composition, membrane fluidity, and temperature dependence of ATPase activity in the microsomal fraction of wheat roots. *Abstracts 4th Congress of the Federation of European Societies of Plant Physiologists*, p. 584. Strasbourg: France.

TAZAWA, M. (1968). Motive force of the cytoplasmic streaming in *Nitella. Protoplasma* **65**, 207–222.

THILO, L., TRAÜBLE, H. & OVERATH, P. (1977). Mechanistic interpretation of the influence of lipid phase transitions on transport functions. *Biochemistry* **16**, 1287–1290.

THOMPSON, G. A. JR (1980). Regulation of membrane fluidity during temperature acclimation by *Tetrahymena pyriformis*. In *Membrane Fluidity: Biophysical Techniques and Cellular Regulation* (ed. M. Kates & A. Kuksis), pp. 381–397. New Jersey: Humana Press.

UEMURA, M. & YOSHIDA, S. (1984). Involvement of plasma membrane alterations in cold acclimation of winter rye seedlings (*Secale cereale* L. cv Puma). *Pl. Physiol.* **75**, 818–826.

VIGH, L., HORVÁTH, I., WOLTJES, J., FARKAS, T., VAN HASSELT, P. & KUIPER, P. J. C. (1987). Combined electron-spin-resonance, X-ray-diffraction studies on phospholipid vesicles obtained from cold-hardened wheats. I. An attempt to correlate electron-spin-resonance characteristics with frost resistance. *Planta* **170**, 117–122.

WHEELER, K. P., WALKER, J. A. & BARKER, D. M. (1975). Lipid requirement of the membrane sodium-plus-potassium ion-dependent adenosine triphosphatase system. *Biochem. J.* **146**, 713–722.

WHITE, P. J. (1987). The effects of temperature on ion transport and membrane properties in roots of rye (*Secale cereale* cv. Rheidal). Ph.D. thesis, University of Manchester.

WHITE, P. J., CLARKSON, D. T. & EARNSHAW, M. J. (1987). Acclimation of potassium influx in rye (*Secale cereale*) to low root temperatures. *Planta* **171**, 377–385.

WILD, A. & BREEZE, V. (1981). Nutrient uptake in relation to growth. In *Physiological Processes Limiting Plant Productivity* (ed. C. B. Johnson), pp. 331–344. London: Butterworths.

WILLEMOT, C. (1979). Chemical modification of lipids during frost hardening of herbaceous species. In *Low Temperature Stress in Crop Plants: The Role of the Membrane* (ed. J. M. Lyons, D. Graham & J. K. Raison), pp. 411–430. New York: Academic Press.

WRIGHT, L. C., MCMURCHIE, E. J., POMEROY, M. K. & RAISON, J. K. (1982). Thermal behaviour and lipid composition of cauliflower plasma membranes in relation to ATPase activity and chilling sensitivity. *Pl. Physiol.* **69**, 1356–1360.

YOSHIDA, S. & UEMURA, M. (1984). Protein and lipid compositions of isolated plasma membranes from orchard grass (*Dactylis glomerata* L.) and changes during cold acclimation. *Pl. Physiol.* **75**, 31–37.

YOSHIDA, S., KAWATA, T., UEMURA, M. & NIKI, T. (1986). Properties of plasma membrane isolated from chilling-sensitive etiolated seedlings of *Vigna radiata* L. *Pl. Physiol.* **80**, 152–160.

ZSOLDOS, F. & KARVALY, B. (1979). Cold shock injury and its relation to ion transport by roots. In *Low Temperature Stress in Crop Plants: The Role of the Membrane* (ed. J. M. Lyons, D. Graham & J. K. Raison), pp. 123–139. New York: Academic Press.

EFFECTS OF FREEZING ON PLANT MESOPHYLL CELLS

G. H. KRAUSE, S. GRAFFLAGE, S. RUMICH-BAYER and S. SOMERSALO

Botanisches Institut der Universität Düsseldorf, Universitätsstraße 1, D-4000 Düsseldorf 1, Federal Republic of Germany

Summary

Freezing and thawing of leaves of herbaceous plants leads to damage when the freezing temperature falls below a certain tolerance limit, which depends on the plant species and state of acclimation. Such damage is expressed as an irreversible inhibition of photosynthesis observed after thawing. In frost-damaged leaves the capacity of photosynthetic reactions of the thylakoid membranes is impaired. Particularly, the water-oxidation system, photosystems II and I are inhibited. However, it appears that CO_2 assimilation is more readily affected by freezing stress than the activity of the thylakoids. The inhibition of CO_2 fixation seen in initial stages of damage seems to be independent of thylakoid inactivation. This can be shown by chlorophyll fluorescence analysis made simultaneously with measurement of CO_2 assimilation. Fluorescence emission by leaves is strongly influenced by carbon assimilation activity, namely *via* the redox state of the photosystem II electron acceptor Q_A (Q_A-dependent quenching) and *via* energization of the thylakoid membranes depending on the transthylakoid proton gradient (energy-dependent quenching). Resolution of these components of fluorescence changes provides insight into alterations of the CO_2 fixing capacity of the chloroplasts and properties of the thylakoids.

The effects of freezing and thawing were studied in detail with isolated mesophyll protoplasts prepared from both non-hardened and cold-acclimated plants of *Valerianella locusta* L. Freezing damage was characterized by various parameters such as plasma membrane integrity, photosynthetic CO_2 assimilation, chlorophyll fluorescence emission and activities of thylakoids isolated from the protoplasts. All tests indicated a substantially increased frost tolerance of protoplasts obtained from cold-acclimated as compared to non-hardened leaves. CO_2 assimilation and related fluorescence changes were the most freezing-sensitive parameters in both types of protoplasts. Inactivation of CO_2 assimilation was correlated neither to the disintegration of the plasma membrane nor to inactivation of the thylakoids. Experimental data indicate that freeze-thaw treatment affected the light-regulated enzymes of the carbon reduction cycle, such as fructose-1,6-bisphosphatase, sedoheptulose-1,7-bisphosphatase and ribulose-1,5-bisphosphate carboxylase. Inhibition of light-activation of these enzymes may be based on altered properties of the chloroplast envelope.

General phenomena of cold acclimation and freezing damage

Plant cells may be injured and killed by exposure to sub-zero temperatures.

Many plants species, however, can acquire a frost resistance of varying degree by acclimation to low (usually above zero) temperatures (see Levitt, 1980). Cold-acclimated cells and tissues survive freezing stress, as long as the temperature stays above a certain limit, i.e. the 'frost-killing temperature'. According to definitions given by Levitt (1980), frost resistance is based either on tolerance to extracellular ice formation and concomitant severe cell dehydration, or on frost avoidance, particularly supercooling. The present article is concerned with photosynthetically active mesophyll tissue which, under natural conditions of cooling, is subjected to extracellular freezing (see Pearce & Willison, 1985*a*). In contrast, other plant tissues such as tree xylem parenchyma (see Wisniewski & Ashworth, 1985) or dormant bud primordia (see Ashworth, 1982, 1984) may exhibit deep supercooling.

To achieve protection against cell damage by extracellular freezing, two main strategies seem to be followed: (1) accumulation of 'cryoprotective' compounds in the cell compartments and (2) alterations of biomembrane composition and organization. Changes in metabolism leading to increased frost tolerance of leaves seems to be related to increased levels of abscisic acid (Lalk & Dörffling, 1985). Recent evidence supports the view that acquisition of frost tolerance is based on altered gene expression resulting in a changed pattern of protein synthesis (Guy & Haskell, 1987).

In numerous studies, increased concentrations of soluble carbohydrates have been found to be correlated with frost tolerance (see Levitt, 1980). Sucrose and raffinose accumulation in chloroplasts related to cold hardiness was described for cabbage (*Brassica oleracea* L.) leaves (Santarius & Milde, 1977). In spinach (*Spinacia oleracea* L.) leaves, levels of sucrose in the extraplastidic space and of sucrose, glucose, fructose and raffinose in the chloroplasts closely followed the degree of frost tolerance during a hardening and dehardening procedure (Krause, Klosson & Tröster, 1982). Similarly, an accumulation of free amino acids (proline, alanine, glutamine, glutamate, glycine and serine) in the chloroplasts of spinach leaves was observed in relation to frost hardening (Krause *et al.* 1984). Glycinebetaine has been identified as a further cryoprotective substance in spinach (Coughlan & Heber, 1982). Also polyamines (putrescine, spermidine) were suggested to play a role in cold hardening (Nadeau, Delaney & Chouinard, 1987; Kushad & Yelenosky, 1987).

Alterations of membrane lipids related to frost tolerance have been thoroughly studied. For instance, the total polar lipid and phospholipid content of winter wheat (*Triticum aestivum* L.) seems to increase during cold hardening (Horvath *et al.* 1980; Cloutier & Siminovitch, 1982). Similar results were obtained with needles of *Picea abies* L. (Senser, 1982) and *Pinus sylvestris* L. (Selstam & Öquist, 1985). Recently, a close negative correlation between frost tolerance of winter wheat and the ratio of total sterol to total phospholipid content was found (Horvath *et al.* 1987; see also Yoshida & Uemura, 1984). The authors postulated that a low sterol/phospholipid ratio in the plasma membrane would stabilize the lamellar bilayer lipid structure during freeze-dehydration. However, data available on changes in lipid contents and composition related to cold acclimation are contradictory. Lynch & Steponkus (1987) observed that in plasma-membrane-enriched fractions of cold-acclimated winter rye (*Secale cereale* L.) the relative proportions of both phospholipids and free sterols were increased, the latter at the expense of sterol derivatives; the ratio of free sterols to total phospholipids and the lipid/protein ratio were unchanged. Data published by Willemot (1980) on wheat and by Yoshida & Uemura (1984) on *Dactylis glomerata* showed little evidence for a role of sterol levels in frost hardening (see also Quinn, 1988). No major increase in lipid/protein or lipid/chlorophyll ratios in whole leaves or thylakoids of cold-grown winter rye was found in a study by Huner *et al.* (1987), nor was an increase in thylakoid fluidity observed. The study revealed a specific decrease in *trans*-Δ^3-hexadecenoic acid (*trans*-16:1) in phosphatidyldiacylglycerol (PG) that was found to influence the structure of the light-harvesting complex of photosystem (PS) II (Krupa *et al.* 1987). A report by Vigh *et al.* (1985) on the occurrence of unusual long-chain fatty acids in cold-hardy wheat could not be confirmed by Huner *et al.* (1987).

Injury of plant cells caused by extracellular freezing is considered primarily a damage of biomembranes. Effects of freezing on the plasma membrane have attained particular attention (see Levitt, 1980; Steponkus, 1984). According to Gordon-Kamm & Steponkus (1984*a*,*b*,*c*), freeze-dehydration of non-acclimated protoplasts at sub-zero temperatures above −5°C leads to deletion of single-bilayer vesicles from the plasma membrane, which causes expansion-induced lysis during thawing. At lower freezing temperatures (−10°C), the authors observed lamellar-to-hexagonal$_{II}$ phase transitions together with loss of osmotic responsiveness.

Singh, Iu & Johnson-Flanagan (1987) disputed this two-stage hypothesis on the basis of results showing a continuum of plasma membrane injury characterized by multilamellar vesicles that include material from the tonoplast. Similar conclusions were drawn earlier by Pearce & Willison (1985*b*) who did not detect lipids in the hexagonal$_{II}$ phase in lethally frozen leaf cells. These authors hypothesized that freezing damage of the plasma membrane is related to formation of areas free of 'intramembranous particles'. They suppose that those areas give rise to membrane folding, stacking, fusion, proliferation or related alterations, causing finally leakiness or rupture of the membrane. It should be noted that, furthermore, lethal freezing stress may cause alterations of the lipid composition (Borochov *et al.* 1987).

Besides the plasmalemma, other cell membranes are injured more or less simultaneously due to freezing stress. Electron micrographs of spinach leaves (Krause *et al.* 1984) showed after mild frost treatment (freezing and thawing) small groups of rather strongly damaged cells surrounded by areas of apparently unaltered tissue. The injured cells exhibited loss of turgor, damaged tonoplast (disappearance of the vacuole) and plasma membrane, swollen chloroplast stroma and intrathylakoid space and, in part, ruptured chloroplast envelopes. The response of mitochondria in spinach leaves to freezing stress was studied by

Thebud & Santarius (1981) who found that these organelles are damaged *in situ* in the same temperature range as thylakoid membranes. In contrast, Singh, de la Roche & Siminovitch (1977) observed a relatively high insensitivity of mitochondria in rye coleoptile cells.

It has long been known that photosynthetic activities of the thylakoids are readily affected by freezing stress *in vivo* and *in vitro* (see Heber, Tyankova & Santarius, 1973). Isolated thylakoids have been used in many studies as model systems to investigate the effects of freezing on biomembranes. However, as deduced from experiments with leaves, in the intact cell photosynthetic CO_2 assimilation seems to be more sensitive to freezing stress (when additional light stress is excluded) than the energy-conserving reactions on the thylakoids (Klosson & Krause, 1981*b*; Krause & Klosson, 1983; Krause *et al.* 1982, 1984; Strand & Öquist, 1985*a*,*b*; Bauer & Kofler, 1987). Experiments with isolated protoplasts of *Valerianella locusta* L. (Rumich-Bayer & Krause, 1986) showed that inhibition of CO_2 assimilation by frost-pretreatment was not related to plasma membrane and tonoplast disintegration. In the following, the effects of freezing on photosynthesis, in particular on CO_2 assimilation, will be discussed in more detail.

Effects of freezing on the photosynthetic apparatus

When isolated thylakoids are subjected to freezing and thawing in media containing inorganic salts (e.g. NaCl) as 'cryotoxic' agents, the primary manifestation of damage is an uncoupling of photophosphorylation that is based on dissociation of the coupling factor, CF_1 (see Heber *et al.* 1973; Heber *et al.* 1981; Garber & Steponkus, 1976; Volger, Heber & Berzborn, 1978). However, *in situ* this destabilization of the membranes by high levels of inorganic electrolytes, caused by the freeze-dehydration, appears to be a minor effect. Rather, freezing and thawing of leaves (Heber *et al.* 1973; Klosson & Krause, 1981*a*; Thebud & Santarius, 1981; Krause *et al.* 1982) caused primarily an inactivation of the electron transport system at several sites: water oxidation system, PSII, and PSI. Photophosphorylation was only slightly more sensitive, indicating that most of its inhibition was related to lowered electron transport capacity (see Fig. 1). Obviously, in the aqueous phase of the chloroplast stroma the solute composition is balanced, minimizing the dissociation of extrinsic membrane proteins by cryotoxic electrolytes. In fact, the freezing behaviour of thylakoids *in situ* could be simulated by freeze/thaw treatments of isolated thylakoids in complex media that resembled the solute composition of the chloroplast stroma (Grafflage & Krause, 1986; Santarius, 1986*a*,*b*,*c*, 1987). The protective effect of the sugars and amino acids accumulated in the stroma during hardening could also be demonstrated *in vitro* (Grafflage & Krause, 1986).

Freezing damage to thylakoids *in situ*, or *in vitro* in media simulating the chloroplast stroma, appears to be related to release of plastocyanin from the intrathylakoid lumen (Hincha *et al.* 1987*b*; Hincha, Heber & Schmitt, 1987*a*; see also Hincha, 1986). The authors used the term 'mechanical damage', assuming that during freeze-dehydration solutes may permeate into the intrathylakoid space and cause swelling and lesions, with subsequent resealing, during thawing. Hincha (1986) also considered membrane vesiculation (as postulated for plasma membranes, see above) as a cause of subsequent lesions. Such rupture would cause release of peripheral proteins, such as plastocyanin, from the inner thylakoid surface. Solute permeation could explain the swollen appearance of thylakoids and the chloroplast stroma, seen in lethally damaged leaf cells (Krause *et al.* 1984). However, as plastocyanin is slowly released from thylakoids even at 0°C (Hincha *et al.* 1987*b*), osmotic rupture may not be the only explanation of damage.

Even if the temperature stays above the limit of frost tolerance, the thylakoids may be damaged when cells in the frozen state are exposed to high light. In conifer needles, such photoinhibition appears to be a major factor of winter damage but usually can be repaired in the growing season (see Öquist, 1986; Öquist & Martin, 1986). As low above-zero temperatures frequently impose conditions of excess light due to low rates of photosynthetic carbon metabolism, cold acclimation is expected to include enforcement of protective systems against photoinhibition (see Baker, Long & Ort, 1988).

Frost-sensitivity of photosynthetic CO_2 assimilation, thylakoid activities and chlorophyll fluorescence yield

As shown by T_{50} values (minimum temperatures of frost treatment leading to 50% damage or inhibition) in isolated mesophyll protoplasts from *Valerianella locusta* L. (Fig. 1), photosynthetic CO_2-dependent O_2 evolution exhibited a considerably higher frost sensitivity than all other parameters measured, such as

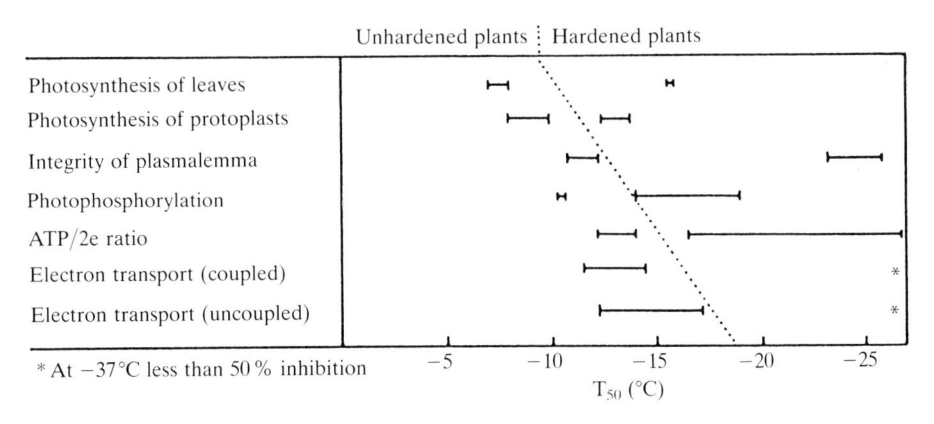

Fig. 1. T_{50} values of photosynthetic activities of leaves and isolated mesophyll protoplasts of *Valerianella locusta* L. T_{50} denotes the minimum temperature of frost treatment leading to 50% inhibition (or disintegration). Bars give the (double) standard deviation of T_{50} from the mean of 4–8 experiments (2 experiments in the case of photosynthesis of hardened leaves). Photosynthesis was measured as CO_2 fixation (leaves) or CO_2-dependent O_2 evolution (protoplasts). Plasma membrane integrity was determined by exclusion of Evans blue from the protoplasts. Thylakoid activities were determined after osmotic lysis of protoplasts subsequent to frost treatment. For details see Rumich-Bayer & Krause (1986).

integrity of the plasma membrane and tonoplast, and various activities of the thylakoids (see Rumich-Bayer & Krause, 1986). For comparison, T_{50} values of photosynthetic CO_2 fixation of the leaf material used for protoplast preparation are given in Fig. 1. In the protoplasts isolated from hardened plants, all parameters exhibited higher frost tolerance, as compared to non-acclimated material. In the hardy protoplasts, plasma membrane and electron transport systems showed much higher tolerance than expected from the freezing behaviour of intact leaves. This apparently indicates an effect of the cell wall on freezing injury (cf. Singh *et al.* 1987).

To determine thylakoid activities in the experiments of Fig. 1, the protoplasts had to be lysed. To detect and specify freezing damage in green plant cells, chlorophyll *a* fluorescence emission has been proved as a useful, *non-intrusive* tool (see Krause & Weis, 1984; Briantais *et al.* 1986). Various fluorescence parameters can be analysed for detection of injuries. For instance, damage to PSII including the water oxidation system (a major site of freezing injury) can be seen from the decrease in the ratio of maximum variable to total fluorescence, $F_{v(m)}/F_m$, which may serve as a measure of potential photochemical efficiency of PSII. In Fig. 2, data of initial (F_o) and variable fluorescence and of $F_{v(m)}/F_m$ ratios of frost-pretreated protoplasts are given. Damage to PSII is indicated by the data from non-hardy protoplasts. For hardy protoplasts, the high insensitivity to freezing stress of electron transport is confirmed by this non-destructive measurement. Fig. 3, depicting CO_2 fixation and fluorescence data from hardened frost-treated

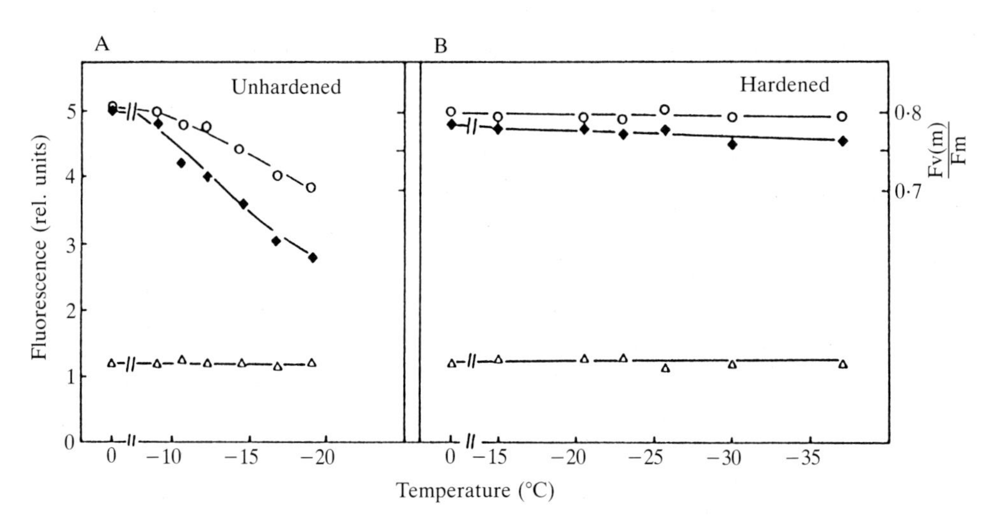

Fig. 2. Effects of freezing on chlorophyll fluorescence induction of mesophyll protoplasts isolated from unhardened (A) and cold-hardened (B) *Valerianella locusta* L. Fluorescence induction at 20°C was registered in the emission band at 686 nm upon illumination with broad-band blue light ($0{\cdot}8\,W\,m^{-2}$) in the presence of 25 μM 3-(3′,4′-dichlorophenyl)-1,1-dimethylurea (DCMU). Initial fluorescence, F_o (△), maximum variable fluorescence, $F_{v(m)}$ (◆) and $F_{v(m)}/F_m$ ratio (○) (F_m is the maximum total fluorescence) are given as function of minimum temperature of frost treatments. For details on plant material and frost treatment see Rumich-Bayer & Krause (1986).

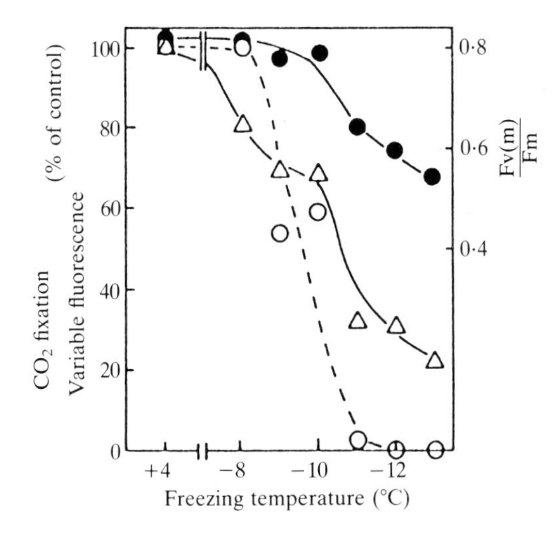

Fig. 3. Effects of freezing on photosynthetic CO_2 fixation (--○--) and fluorescence parameters $F_{v(m)}$ (– △ –) and $F_{v(m)}/F_m$ (– ● –) of cold-hardened spinach leaves. CO_2 fixation in 260 W m^{-2} white light and normal air was measured with an infrared analyser (data calculated from the sum of CO_2 uptake in the light and evolution in the dark; control rate, 40 μmol mg chlorophyll^{-1} h^{-1}). Fluorescence was recorded with a PAM 101 fluorometer (Walz) according to Schreiber, Schliwa & Bilger (1986) (*cf.* Krause & Laasch, 1987). Measurements were done at room temperature. The abscissa gives the minimum temperature of preceding frost treatments (control kept at +4°C). For the hardening procedure see Klosson & Krause (1981*a*).

spinach leaves, shows that inactivation of the electron transport system occurs, but the latter is significantly less sensitive than CO_2 assimilation.

The independent inhibition of CO_2 assimilation by freezing stress can be deduced from the response of fluorescence quenching (besides responses of light-dependent light scattering and absorbence changes). Recently developed techniques allow quantitative separation of the two main components of the decline in fluorescence seen during prolonged illumination: (1) photochemical quenching, q_Q, depending on the proportion of oxidized PSII electron acceptor Q_A and (2) energy-dependent quenching, q_E, which is related to the light-induced proton gradient across the thylakoid membrane.

As CO_2 is the final electron acceptor of the electron transport chain, an independent inhibition of carbon metabolism by freezing stress will tend to increase the proportion of reduced Q_A and thereby inhibit photochemical fluorescence quenching, q_Q. Fig. 4 shows for unhardened and hardened spinach leaves that under appropriate conditions, i.e. an intermediate light flux density below saturation, diminution of q_Q is closely correlated to inhibition of CO_2 fixation caused by preceding freezing/thawing treatment.

Energy-dependent fluorescence quenching, q_E, is supposed to reflect an increase in the rate-constant of thermal dissipation of absorbed light energy in PSII, which is induced by the transthylakoid proton gradient *via* a still unknown structural membrane alteration (see Krause, Briantais & Vernotte, 1983). Recently, q_E has been postulated to indicate a dynamic property of the thylakoid membrane that permits regulation of thermal energy dissipation (Krause & Behrend, 1986; Krause & Laasch, 1987; Weis & Berry, 1987). Figs 5 and 6 show that this membrane property is readily lost during freezing stress. Fig. 5 depicts the inhibition of the maximum q_E signal (observed at the beginning of illumination) in

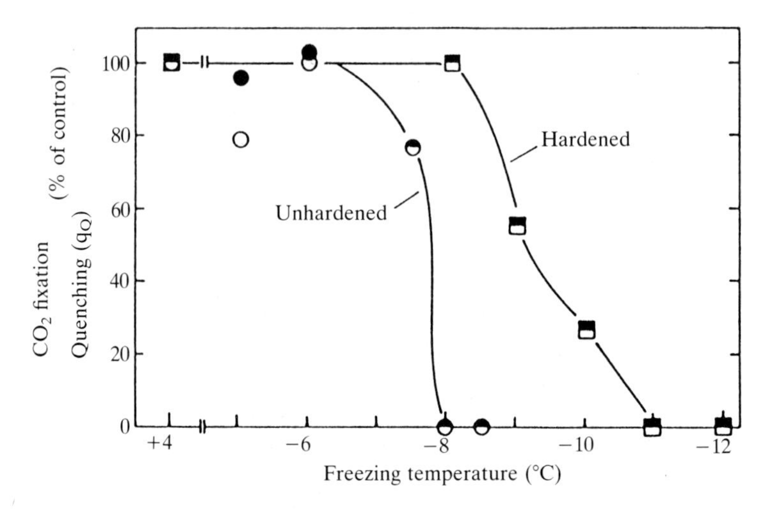

Fig. 4. Effects of freezing on CO_2 fixation (□ ○) and photochemical quenching of chlorophyll fluorescence (q_Q, ■ ●) in unhardened (○ ●) and cold-hardened (□ ■) spinach leaves. Data refer to the steady state in 200 W m^{-2} white light. For experimental details see legend to Fig. 3. Control values for unhardened (hardened) leaves were: rates of CO_2 fixation, 67 (53) μmol mg chlorophyll^{-1} h^{-1}; q_Q, 0·65 (0·72).

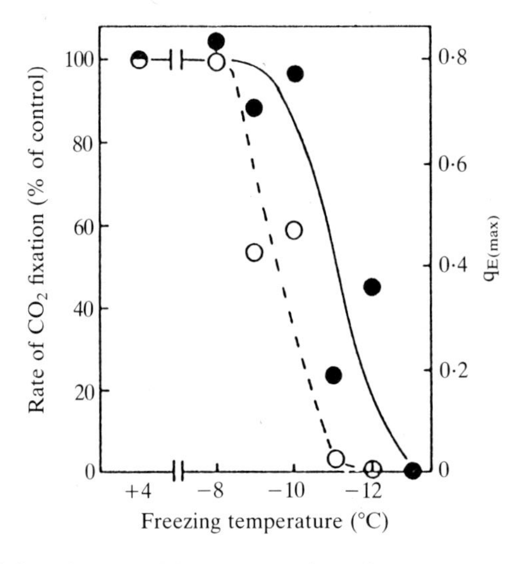

Fig. 5. Effects of freezing on CO_2 fixation (– – ○ – –) and maximum extent of energy-dependent quenching of chlorophyll fluorescence, $q_{E(max)}$ (– ● –), in cold-hardened spinach leaves. Data are from the same experiment as those of Fig. 3; $q_{E(max)}$ was determined (Schreiber *et al.* 1986) after 5-min illumination of dark-adapted leaves with 200 W m^{-2} while light.

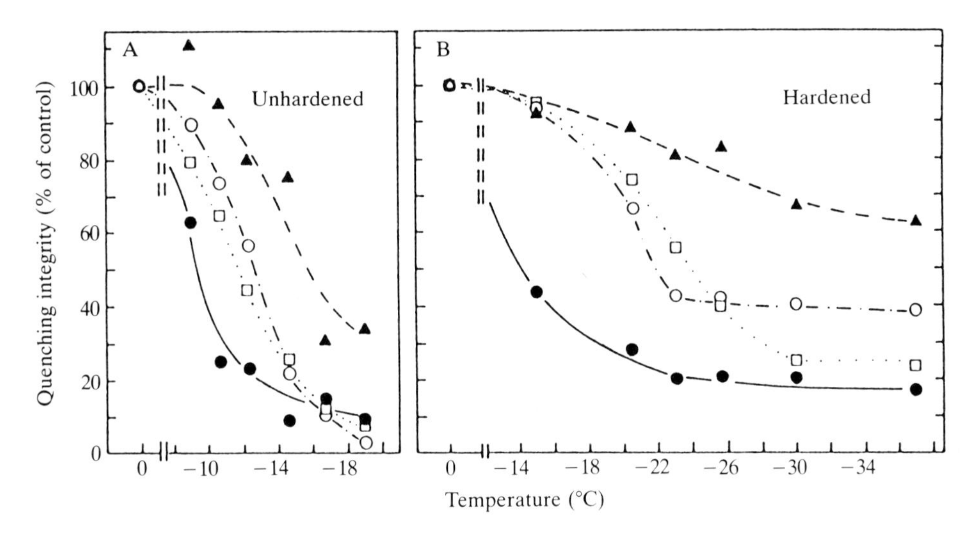

Fig. 6. Effects of freezing stress on chlorophyll fluorescence quenching (q_Q, –●– and q_E, –○·–), plasma membrane integrity (··□··) and trans-thylakoid proton gradient (-- ▲ --) in mesophyll protoplasts isolated from unhardened (A) and cold-hardened (B) *Valerianella locusta* L. The two components of quenching were determined in the steady state according to Krause, Vernotte & Briantais (1982) by addition of DCMU. Actinic light for fluorescence recording was 10 W m^{-2}; other details were as for Fig. 2. (*cf*. Fig. 1 for plasma membrane integrity). As a measure of the maximum size of the light-induced proton gradient, quenching of 9-aminoacridine fluorescence was determined after lysis of protoplasts (Rumich-Bayer & Krause, 1986). Controls from unhardened (hardened) material were: q_Q, 0·49 (0·32); q_E, 0·49 (0·72); integrity, 94 % (88 %); quenching of 9-aminoacridine fluorescence, 0·39 (0·68).

hardened spinach leaves. Energy-dependent fluorescence quenching (q_E) was less inhibited than CO_2 fixation but considerably more than the electron transport system, as seen by comparison with Fig. 3 ($F_{v(m)}/F_m$ ratio). This is confirmed with isolated frost-pretreated protoplasts from unhardened and hardened *Valerianella*, in which q_E and q_Q were determined in the steady state. Again, inhibition of q_Q, reflecting the decline in CO_2 fixation, was the most sensitive response to freezing. Because of the low light flux density applied, part of q_Q remained after full inhibition of CO_2 fixation; *cf*. Fig. 1. The inhibition of q_E was much stronger than that of the light-induced proton gradient, measured indirectly by the quenching of 9-aminoacridine fluorescence after osmotic rupture of the protoplasts. This shows that the property of the thylakoid membrane to form q_E in response to the ΔpH, rather than the build-up of the proton gradient was affected. Inhibition of q_E thus appears as the first indication of freezing injury of the thylakoids.

Inhibition of photosynthetic carbon metabolism in isolated protoplasts by frost treatment

Isolated mesophyll protoplasts in which CO_2 fixation was partially inhibited by preceding frost treatment exhibited characteristic changes in relative pool sizes of phosphorylated Calvin cycle intermediates (Rumich-Bayer, Giersch & Krause,

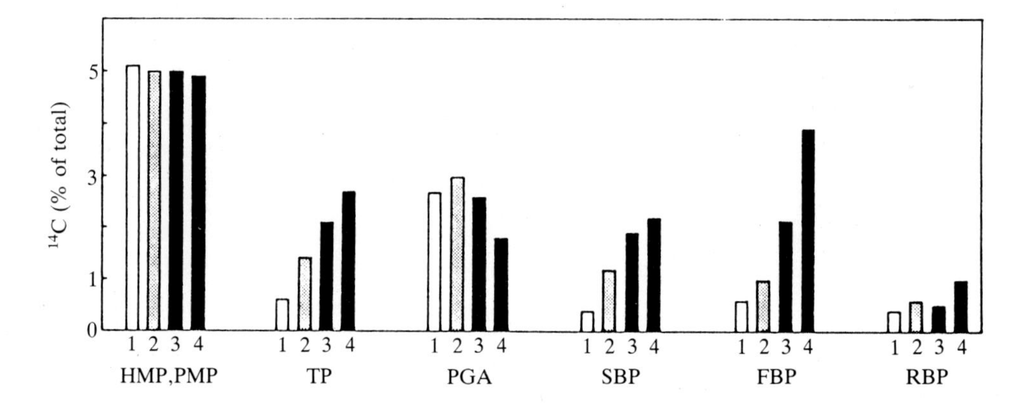

Fig. 7. Relative steady-state pool sizes of phosphorylated metabolites of the photosynthetic carbon reduction cycle in isolated protoplasts of unhardened leaves of *Valerianella locusta* L. Data are given in percent of total ^{14}C incorporation after 15 min of steady-state photosynthesis (250 W m^{-2} red light, 20°C) in the presence of (^{14}C)$NaHCO_3$. Bars denote controls preincubated at 0°C (1) and protoplasts pretreated at −7°C (2), −8·5°C (3) and −9·6°C (4). Abbreviations: HMP, hexosemonophosphates; PMP, pentosemonophosphates; TP, triosephosphates; PGA, 3-phosphoglycerate; SBP, sedoheptulose-1,7-bisphosphate; FBP, fructose-1,6-bisphosphate; RBP, ribulose-1,5-bisphosphate. (After Rumich-Bayer & Krause, 1986, altered.)

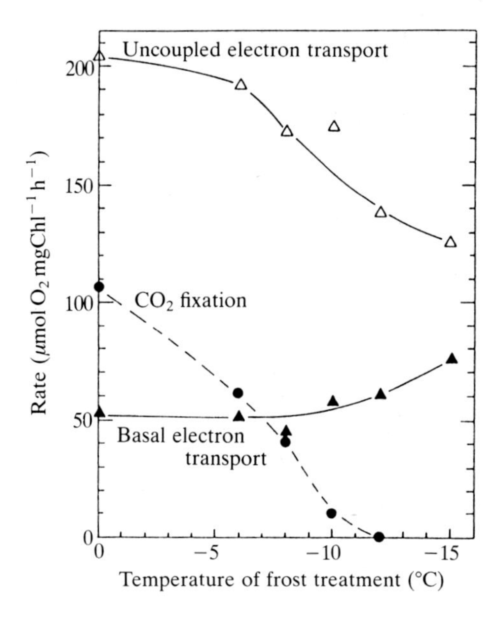

Fig. 8. Effects of freezing on CO_2 fixation and electron transport activities of mesophyll protoplasts from unhardened *Valerianella locusta* L. CO_2 fixation was measured as light-saturated CO_2-dependent O_2 evolution. After lysis of protoplasts, photosynthetic electron transport was measured as O_2 uptake mediated by methylviologen in the absence ('basal') and presence ('uncoupled') of 10 mM NH_4Cl (*cf.* Rumich-Bayer & Krause, 1986).

1987). Fig. 7 depicts relative levels of ^{14}C-labelled metabolites after 15-min steady-state photosynthesis. It can be seen that freezing stress caused an increase in proportions of triose phosphates, sedoheptulose-1,7-bisphosphate, fructose-1,6-bisphosphate and (slightly) of ribulose-1,5-bisphosphate (RuBP). This can be interpreted as an inactivation of fructose and sedoheptulose bisphosphatases and possibly of RuBP carboxylase in the chloroplasts. As these are light-regulated enzymes, an effect of freezing stress on the light-activation of the Calvin cycle can be assumed. In fact, when CO_2 fixation was strongly inhibited by preceding freezing and thawing, the activation of light-regulated enzymes by illumination of the protoplasts was impaired. This is shown by enzyme assays carried out subsequent to frost treatments. Fig. 8 depicts an experiment in which CO_2 assimilation was strongly inhibited in a temperature range of pretreatment from −6 to −10°C. These temperatures only slightly affected the capacity of electron transport and coupling to photophosphorylation, denoted by low 'basal' electron transport. Fig. 9 demonstrates that in this temperature range little effect was seen on the activity of subsequently extracted and activated RuBP carboxylase.

On the other hand, light activation, i.e. the difference between activity in light or dark-preincubated protoplasts, was strongly diminished. Fig. 10 shows that in the average of five experiments, light activation of RuBP carboxylase was

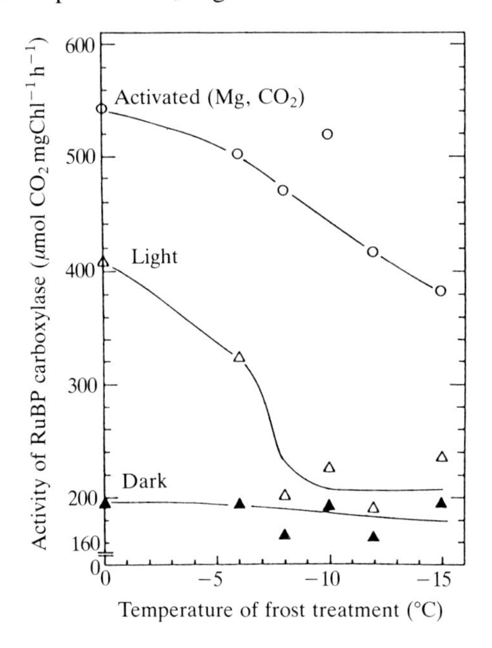

Fig. 9. Effects of freezing on RuBP carboxylase in protoplasts isolated from unhardened leaves of *Valerianella locusta* L. Subsequent to frost treatment (minimum temperatures given on the abscissa), the protoplasts were incubated for 7 min at 20°C in the dark (▲) or 250 W m^{-2} red light (△) and then lysed to determine the activity of RuBP carboxylase in the presence of 10 mM RuBP (enzymatic determination of the 3-phosphoglycerate formed). Alternatively, the protoplasts were lysed without preincubation and the RuBP carboxylase was activated with CO_2 and Mg^{2+} (○) according to Lorimer, Badger & Andrews (1976).

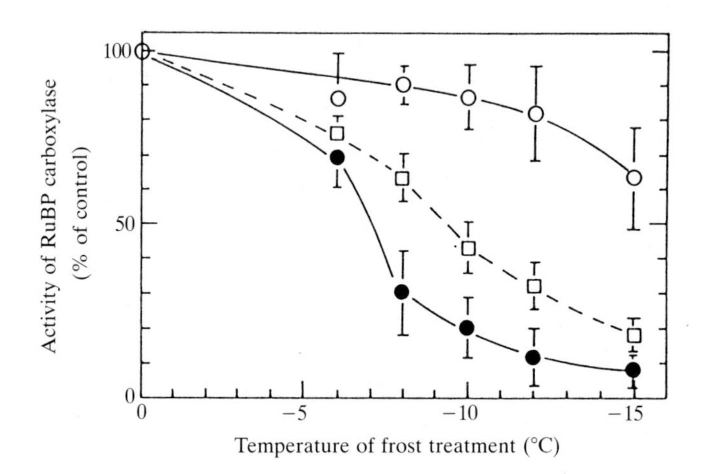

Fig. 10. Effects of frost treatment on light-activation of RuBP carboxylase in isolated protoplasts. Maximum enzyme activity after activation with CO_2 and Mg^{2+} (○), light-activation of the enzyme (●) (difference between activities in light and dark-preincubated protoplasts), and plasma membrane integrity (□) are presented as a function of the minimum temperature of the preceding frost treatment. Means of 5 experiments and standard deviations are given. For experimental details see legend to Fig. 9. Controls: enzyme activated with CO_2 and Mg^{2+}, 439; 'light', 333; 'dark', 186; 'light activation', 147 μmol CO_2 mg chlorophyll^{-1} h^{-1}; integrity, 87 %.

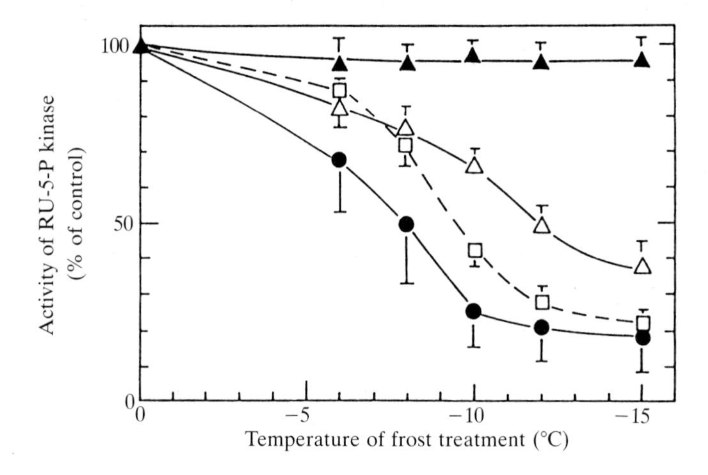

Fig. 11. Effects of frost treatment on light activation of ribulose-5-phosphate kinase in isolated protoplasts. The experiments were performed as those of Figs 9 and 10. Enzyme activity was determined according to Fischer & Latzko (1979). Means of 5 experiments and standard deviations are given. Curves show the enzyme activities of dark-preincubated (▲), and light-preincubated (△) protoplasts, 'light activation' (●) (difference between light and dark-preincubation), and plasma membrane integrity (□) as function of minimum freezing temperature. Controls: 'light', 634; 'dark', 257; 'light activation', 377 μmol mg chlorophyll^{-1} h^{-1}; integrity, 94 %.

significantly more affected than the integrity of plasmalemma and tonoplast. Similar results were obtained when activities of ribulose-5-phosphate kinase (Fig. 11) and fructose-1,6-bisphosphatase (not shown) were tested. The above data led to the conclusion that the inhibition of CO_2 fixation by freezing/thawing results from effects on the light-activation system of Calvin cycle enzymes.

At present, the cause of impairment of light activation is unknown. It may be speculated that freezing stress primarily affects properties of the chloroplast envelope that would impede the light-induced alkalization of the chloroplast stroma. Changes in pH are known as an important factor for enzyme activation (see Buchanan, 1980; Gardemann, Schimkat & Heldt, 1986) and stroma alkalization was found to be essential for light-activation of the Calvin cycle (Enser & Heber, 1980; Heber & Heldt, 1981). Inhibition of alkalization would be in agreement with the decrease in the 3-phosphoglycerate/triosephosphate ratio observed in frost-damaged protoplasts (Rumich-Bayer *et al.* 1987; *cf.* Fig. 7).

Conclusions

Freezing damage to plant cells appears as a complex process and at present cannot be traced back to a single primary site and mechanism. In green mesophyll cells, besides the plasma membrane and tonoplast, the photosynthetic and respiratory systems are readily affected by freezing stress. Inhibition of photosynthetic CO_2 assimilation is seen as the most sensitive response. This effect can be followed indirectly by analysis of photochemical chlorophyll fluorescence quenching, q_Q, and is primarily independent of thylakoid inactivation. The inhibition seems to be based on diminished light-activation of Calvin cycle enzymes, which again might be caused by adverse effects of freezing on a biomembrane, the chloroplast envelope. In the thylakoids, the first indication of freezing damage is a loss of the structural flexibility responsible for energy-dependent fluorescence quenching, q_E. More severe freezing stress leads to damage of PSII including the water-oxidation system and of PSI, while little primary uncoupling of photophosphorylation is observed.

Although present data on freezing injury of the various cell membranes are contradictory, the major cause of damage seems to lie in the severe dehydration and concomitant increase in solute concentrations occurring during extracellular ice formation. This supposedly alters membrane organization and leads to deletion of membrane material followed by lesions or leakiness during thawing. Additionally, in the case of smaller compartments such as the chloroplast stroma and intrathylakoid space, membrane leakiness during dehydration might allow influx of solutes causing subsequent swelling or rupture.

The study was supported by the Deutsche Forschungsgemeinschaft. S. Somersalo thanks the Academy of Finland for a research and travel grant.

References

ASHWORTH, E. N. (1982). Properties of peach flower buds which facilitate supercooling. *Pl. Physiol.* **70**, 1475–1479.

ASHWORTH, E. N. (1984). Xylem development in *Prunus* flower buds and the relationship to deep supercooling. *Pl. Physiol.* **74**, 862–865.

BAKER, N. R., LONG, S. P. & ORT, D. R. (1988). Photosynthesis and temperature, with particular reference to effects on quantum yield. In *Plants and Temperature* (ed. S. P. Long & F. I. Woodward) Symp. Soc. Exp. Biol., vol. 42, pp. 347–375. Cambridge: Company of Biologists.

BAUER, H. & KOFLER, R. (1987). Photosynthesis in frost-hardened and frost-stressed leaves of *Hedera helix* L. *Pl. Cell Env.* **10**, 339–346.

BOROCHOV, A., WALKER, M. A., KENDALL, E. J., PAULS, K. P. & MCKERSIE, B. D. (1987). Effect of a freeze-thaw cycle on properties of microsomal membranes from wheat. *Pl. Physiol.* **84**, 131–134.

BRIANTAIS, J.-M., VERNOTTE, C., KRAUSE, G. H. & WEIS, E. (1986). Chlorophyll *a* fluorescence of higher plants: Chloroplast and leaves. In *Light Emission by Plants and Bacteria* (ed. Govindjee, A. Amesz & D. C. Fork), pp. 539–583. New York: Academic Press.

BUCHANAN, B. B. (1980). The role of light in the regulation of enzymes. *A. Rev. Pl. Physiol.* **31**, 341–347.

CLOUTIER, Y. & SIMINOVITCH, D. (1982). Augmentation of protoplasm in drought- and cold-hardened winter wheat. *Can. J. Bot.* **60**, 674–680.

COUGHLAN, S. J. & HEBER, U. (1982). The role of glycinebetaine in the protection of spinach thylakoids against freezing stress. *Planta* **156**, 62–69.

ENSER, U. & HEBER, U. (1980). Metabolic regulation by pH gradients. Inhibition of photosynthesis by indirect proton transfer across the chloroplast envelope. *Biochim. Biophys. Acta* **592**, 577–591.

FISCHER, K. H. & LATZKO, E. (1979). Chloroplast ribulose-5-phosphate kinase: Light-mediated activation, and detection of both soluble and membrane-associated activity. *Biochem. Biophys. Res. Comm.* **89**, 300–306.

GARBER, M. P. & STEPONKUS, P. L. (1976). Alterations in chloroplast thylakoids during an in vitro freeze-thaw cycle. *Pl. Physiol.* **57**, 673–680.

GARDEMANN, A., SCHIMKAT, D. & HELDT, H. W. (1986). Control of CO_2 fixation. Regulation of stromal fructose-1,6-bisphosphatase in spinach chloroplast by pH and Mg^{2+} concentration. *Planta* **168**, 536–545.

GORDON-KAMM, W. J. & STEPONKUS, P. L. (1984*a*). The behaviour of the plasma membrane following osmotic contraction of isolated protoplasts: Implications in freezing injury. *Protoplasma* **123**, 83–94.

GORDON-KAMM, W. J. & STEPONKUS, P. L. (1984*b*). The influence of cold acclimation in the behaviour of the plasma membrane following osmotic contraction of isolated protoplasts. *Protoplasma* **123**, 161–173.

GORDON-KAMM, W. J. & STEPONKUS, P. L. (1984*c*) Lamellar-to-hexagonal$_{II}$ phase transitions in the plasma membrane of isolated protoplasts after freeze-induced dehydration. *Proc. Nat. Acad. Sci. U.S.A.* **81**, 6373–6377.

GRAFFLAGE, S. & KRAUSE, G. H. (1986). Simulation of in situ freezing damage of the photosynthetic apparatus by freezing in vitro of thylakoids suspended in complex media. *Planta* **168**, 67–76.

HEBER, U., TYANKOVA, L. & SANTARIUS, K. A. (1973). Effects of freezing in biological membranes in vivo and in vitro. *Biochim. Biophys. Acta* **291**, 23–37.

HEBER, U. & HELDT, H. W. (1981). The chloroplast envelope: Structure, function and role in leaf metabolism. *A. Rev. Pl. Physiol.* **32**, 139–168.

HEBER, U., SCHMITT, J. M., KRAUSE, G. H., KLOSSON, R. J. & SANTARIUS, K. A. (1981). Freezing damage to thylakoid membranes in vitro and in vivo. In *Effects of Low Temperatures on Biological Membranes* (ed. G. J. Morris & A. Clarke), pp. 262–283.

HINCHA, D. K. (1986). Sucrose influx and mechanical damage by osmotic stress to thylakoid membranes during an in vitro freeze-thaw cycle. *Biochim. Biophys. Acta* **861**, 152–158.

HINCHA, D. K., HEBER, U. & SCHMITT, J. (1987*a*). Stress resistance of thylakoids: Mechanical freezing damage and effects of frost hardening. In *Progress in Photosynthesis Research*, vol. 4 (ed. J. Biggins), pp. 107–110. Dordrecht: Martinus Nijhoff Publ.

Hincha, D. K., Höfner, R., Schwab, K. B., Heber, U. & Schmitt, J. M. (1987*b*). Membrane rupture is the common cause of damage to chloroplast membranes in leaves injured by freezing or excessive wilting. *Pl. Physiol.* **83**, 251–253.

Horvath, I., Vigh, L., Belea, A. & Farkas, T. (1980). Hardiness dependent accumulation of phospholipids in leaves of wheat cultivars. *Physiologia Pl.* **49**, 117–120.

Horvath, I., Vigh, L., Woltjes, J., Farkas, T., van Hasselt, P. & Kuiper, P. J. C. (1987). Combined electron-spin-resonance, X-ray-diffraction studies on phospholipid vesicles obtained from cold-hardened wheats. II. The role of free sterols. *Planta* **170**, 20–25.

Huner, N. P. A., Krol, M., Williams, J. P., Maissan, E., Low, P. S., Roberts, D. & Thompson, J. E. (1987). Low temperature development induces a specific decrease in *trans*-Δ^3-hexadecenoic acid content which influences LHCII organization. *Pl. Physiol.* **84**, 12–18.

Klosson, R. J. & Krause, G. H. (1981*a*). Freezing injury in cold-acclimated and unhardened spinach leaves. I. Photosynthetic reactions of thylakoids isolated from frost-damaged leaves. *Planta* **151**, 339–346.

Klosson, R. J. & Krause, G. H. (1981*b*). Freezing injury in cold-acclimated and unhardened spinach leaves. II. Effects of freezing on chlorophyll fluorescence and light scattering reactions. *Planta* **151**, 347–352.

Krause, G. H. & Behrend, U. (1986). ΔpH-dependent chlorophyll fluorescence quenching indicating a mechanism of protection against photoinhibition of chloroplasts. *FEBS Lett.* **200**, 298–302.

Krause, G. H., Briantais, J.-M. & Vernotte, C. (1983). Characterization of chlorophyll fluorescence quenching in chloroplasts by fluorescence spectroscopy at 77 K. I. ΔpH-dependent quenching. *Biochim. Biophys. Acta* **723**, 196–175.

Krause, G. H. & Klosson, R. J. (1983). Effects of freezing stress on photosynthetic reactions in cold acclimated and unhardened plant leaves. In *Effects of Stress on Photosynthesis* (ed. R. Marcelle, H. Clijsters & M. Poucke), pp. 245–256. The Hague: Martinus Nijhoff/Dr W. Junk Publ.

Krause, G. H., Klosson, R. J., Justenhoven, A. & Ahrer-Steller, V. (1984). Effects of low temperatures on the photosynthetic system in vivo. In *Advances in Photosynthesis Research*, vol. 4 (ed. C. Sybesma), pp. 349–358. The Hague: Martinus Nijhoff/Dr W. Junk Publ.

Krause, G. H., Klosson, R. J. & Tröster, U. (1982). On the mechanism of freezing injury and cold acclimation of spinach leaves. In *Plant Cold Hardiness and Freezing Stress* (ed. P. H. Li & A. Sakai), p. 55–75. New York: Academic Press.

Krause, G. H. & Laasch, H. (1987). Energy-dependent chlorophyll fluorescence quenching in chloroplasts correlated with quantum yield of photosynthesis. *Z. Naturforsch.* **42**C, 581–584.

Krause, G. H., Vernotte, C. & Briantais, J.-M. (1982). Photoinduced quenching of chlorophyll fluorescence in intact chloroplast and algae. Resolution into two components. *Biochim. Biophys. Acta* **679**, 116–124.

Krause, G. H. & Weis, E. (1984). Chlorophyll fluorescence as a tool in plant physiology. II. Interpretation of fluorescence signals. *Photosynth. Res.* **5**, 139–157.

Krupa, Z., Huner, N. P. A., Williams, J. P., Maissan, E. & James, D. R. (1987). Development at cold-hardening temperatures. The structure and composition of purified rye light harvesting complex II. *Pl. Physiol.* **84**, 19–24.

Kushad, M. M. & Yelenosky, G. (1987). Evaluation of polyamine and proline levels during low temperature acclimation of Citrus. *Pl. Physiol.* **84**, 692–695.

Lalk, I. & Dörffling, K. (1985). Hardening, abscisic acid, proline and freezing resistance in two winter wheat varieties. *Physiologia Pl.* **63**, 287–292.

Levitt, L. (1980). *Responses of Plants to Environmental Stresses*, vol. 1 (2nd edition). New York: Academic Press.

Lorimer, G. H., Badger, M. R. & Andrews, T. J. (1976). The activation of ribulose-1,5-bisphosphate carboxylase by carbon dioxide and magnesium ions. Equilibria, kinetics, and suggested mechanism, and physiological implications. *Biochemistry* **15**, 529–536.

Lynch, D. V. & Steponkus, P. L. (1987). Plasma membrane lipid alterations associated with cold acclimation of winter rye seedlings (*Secale cereale* L. cv. Puma). *Pl. Physiol.* **83**, 761–767.

Nadeau, P., Delaney, S. & Chouinard, L. (1987). Effects of cold hardening on the regulation

of polyamine levels in wheat (*Triticum aestivum* L.) and alfalfa (*Medicago sativa* L.). *Pl. Physiol.* **84**, 73–77.
ÖQUIST, G. (1986). Effects of winter stress on chlorophyll organization and function in Scots pine. *J. Pl. Physiol.* **122**, 169–179.
ÖQUIST, G. & MARTIN, B. (1986). Cold climates. In *Photosynthesis in Contrasting Environments* (ed. N. R. Baker & S. P. Long), pp. 237–293. Amsterdam: Elsevier.
PEARCE, R. S. & WILLISON, J. H. M. (1985*a*). Wheat tissue freeze-etched during exposure to extracellular freezing: distribution of ice. *Planta* **163**, 295–304.
PEARCE, R. S. & WILLISON, J. H. M. (1985*b*). A freeze-etch study of the effects of extracellular freezing on cellular membrane of wheat. *Planta* **163**, 304–316.
RUMICH-BAYER, S. & KRAUSE, G. H. (1986). Freezing damage and frost tolerance of the photosynthetic apparatus studied with isolated mesophyll protoplasts of *Valerianella locusta* L. *Photosynth. Res.* **8**, 161–174.
RUMICH-BAYER, S., GIERSCH, C. & KRAUSE, G. H. (1987). Inactivation of the photosynthetic carbon reduction cycle in isolated mesophyll protoplasts subjected to freezing stress. *Photosynth. Res.* **14**, 137–145.
SANTARIUS, K. A. & MILDE, H. (1977). Sugar compartmentation in frost-hardy and partially dehardened cabbage leaf cells. *Planta* **136**, 163–166.
SANTARIUS, K. A. (1986*a*). Freezing of isolated thylakoid membranes in complex media. I. The effect of potassium and sodium chloride, nitrate, and sulfate. *Cryobiology* **23**, 168–176.
SANTARIUS, K. A. (1986*b*). Freezing of isolated thylakoid membranes in complex media. II. Simulation of the conditions in the chloroplast stroma. *Cryo-Letters* **7**, 31–40.
SANTARIUS, K. A. (1986*c*). Freezing of isolated thylakoid membranes in complex media. III. Differences in the pattern of inactivation of photosynthetic reactions. *Planta* **168**, 281–286.
SANTARIUS, K. A. (1987). Freezing of isolated thylakoid membranes in complex media. IV. Stabilization of CF_1 by ATP and sulfate. *J. Pl. Physiol.* **126**, 409–420.
SCHREIBER, U., SCHLIWA, A. & BILGER, W. (1986). Continuous recording of photochemical and non-photochemical chlorophyll fluorescence quenching with a new type of modulation fluorometer. *Photosynth. Res.* **10**, 51–62.
SELSTAM, E. & ÖQUIST, G. (1985). Effects of frost hardening on the composition of galactolipids and phospholipids occurring during isolation of chloroplast thylakoids from needles of Scots pine. *Pl. Science* **42**, 41–48.
SENSER, M. (1982). Frost resistance in spruce (*Picea abies* L.): III. Seasonal changes in the phospho- and galactolipids of spruce needles. *Z. Pflanzenphysiol.* **105**, 229–239.
SINGH, J., DE LA ROCHE, A. I. & SIMINOVITCH, D. (1977). Relative insensitivity of mitochondria in hardened and nonhardened rye coleoptile cells to freezing *in situ*. *Pl. Physiol.* **60**, 713–715.
SINGH, J., IU, B. & JOHNSON-FLANAGAN, A. M. (1987). Membrane alterations in winter rye and *Brassica napus* cells during lethal freezing and plasmolysis. *Pl. Cell Env.* **10**, 163–168.
STEPONKUS, P. L. (1984). Role of the plasma membrane in freezing injury and cold acclimation. *A. Rev. Pl. Physiol.* **35**, 543–584.
STRAND, M. & ÖQUIST, G. (1985*a*). Inhibition of photosynthesis by freezing temperatures and high light levels in cold-acclimated seedlings of Scots pine (*Pinus sylvestris*). I. Effects on the light-limited and light-saturated rates of CO_2 assimilation. *Physiologia Pl.* **64**, 425–430.
STRAND, M. & ÖQUIST, G. (1985*b*). Inhibition of photosynthesis by freezing temperatures and high light levels in cold-acclimated seedlings of Scots pine (*Pinus sylvestris*). II. Effects on chlorophyll fluorescence at room temperature and 77K. *Physiologia Pl.* **65**, 117–123.
THEBUD, R. & SANTARIUS, K. A. (1981). Effects of freezing on spinach leaf mitochrondria and thylakoids *in situ* and *in vitro*. *Pl. Physiol.* **68**, 1156–1160.
VIGH, L., HORVATH, I., VAN HASSELT, P. R. & KUIPER, P. J. C. (1985). Effect of frost hardening on lipid and fatty acid composition of chloroplast thylakoid membranes in two wheat varieties of contrasting hardiness. *Pl. Physiol.* **79**, 756–759.
VOLGER, H., HEBER, U. & BERZBORN, R. J. (1978). Loss of function of biomembranes and solubilization of membrane proteins during freezing. *Biochim. Biophys. Acta* **511**, 455–469.
WEIS, E. & BERRY, J. A. (1987). Quantum efficiency of photosystem 2 in relation to 'energy' dependent quenching of chlorophyll fluorescence. *Biochim. Biophys. Acta* **894**, 198–209.
WILLEMOT, C. (1980). Sterols in hardening winter wheat. *Phytochemistry* **19**, 1071–1073.
WISNIEWSKI, M. E. & ASHWORTH, E. N. (1985). Changes in the ultrastructure of xylem

parenchyma cells of peach (*Prunus persica*) and red oak (*Quercus rubra*) in response to a freezing stress. *Am. J. Bot.* **72**, 1364–1376.
YOSHIDA, S. & UEMURA, M. (1984). Protein and lipid composition of isolated plasma membranes from orchard grass (*Dacytlis glomerata* L.) and changes during cold acclimation. *Pl. Physiol.* **75**, 31–37.

PLANTS AND HIGH TEMPERATURE STRESS

ENGELBERT WEIS[1] *and JOSEPH A. BERRY*[2]

[1]Botanisches Institut der Universität Düsseldorf, D-4000 Düsseldorf, Fed. Rep. Germany
[2]Carnegie Institution of Washington, Department of Plant Biology, Stanford, CA 94305, USA

Summary

The effect of high temperature on higher plants is primarily on photosynthetic functions. The heat tolerance limit of leaves of higher plants coincides with (and appears to be determined by) the thermal sensitivity of primary photochemical reactions occurring in the thylakoid membrane system. Tolerance limits vary between genotypes, but are also subject to acclimation. Long-term acclimations can be superimposed upon fast adaptive adjustment of the thermal stability, occurring in the time range of a few hours. Light causes an increase in tolerance to heat, and this stabilization is related to the light-induced proton gradient.

In addition to irreversible effects, high temperature may also cause large, reversible effects on the rate of photosynthesis. We report here some studies of photosynthetic gas exchange and chlorophyll fluorescence, designed to examine the energetic balance between photosynthetic carbon metabolism and light reactions during steady state photosynthesis with leaves of cotton plants at different temperatures. At temperatures exceeding the optimum for assimilation, but well below the tolerance limit, the feedback control of light reactions by carbon metabolism declines, as additional dissipative processes become important. Energy dissipated by photorespiration can exceed that consumed by CO_2 assimilation, and a reversible, temperature-induced non-photochemical 'quenching' process, related to 'spillover' of excitation energy to photosystem 1, decreases the efficiency of photosystem 2 with increasing temperature. However, despite the overall decline in the 'potential quantum efficiency', our analysis indicates that CO_2 assimilation may be limited, in part, at high temperature by an imbalance in the regulation of the carbon metabolism, which is reflected in a 'down-regulation' of the ribulose-1,5-bisphosphate carboxylase/oxygenase.

Introduction

Effects of high temperature on plants can be reversible or irreversible. The range of thermal tolerance of physiological functions varies with genotype and is also subject to acclimation processes. Most higher plants are stable within a range of 10°C to 35°C, but significantly higher tolerance limits can be found among plants adapted to warm climates. Heat tolerance limits are determined by the 'thermal stability' of structural entities used by physiological processes. Exposure of plants at temperatures that exceed the limit of tolerance results in loss of

This is CIW-DPB publication no. 1004.

physiological function and lethal damage. Usually, however, tolerance limits reflect the thermal environment, and peak temperatures experienced by a plant are within the range where structures are stable. Reversible effects of temperature on rate and efficiency of physiological functions are, thus, main factors determining the long-term physiological adaptation and productivity of a plant in a warm environment. Since, in a plant, reactions with different intrinsic responses to temperature may be connected to each other, temperature can have an influence on the position of 'limiting' steps within a reaction sequence. Therefore, changes in temperature affect regulation and the distribution of 'control' within a complex physiological system.

In this paper, we discuss the influence of high temperature on the complex leaf system of higher plants. We will review mechanisms, which may possibly be involved in irreversible heat damage of leaves. In plants, structural entities used in primary photosynthetic reactions appear particularly susceptible to thermal stress and may determine the overall tolerance limit of higher plants. The overall response of assimilation to temperature is complex and varies between genotypes. But in a large number of C_3 plants, the temperature optimum for assimilation is far below the thermal tolerance limit, i.e. even within the limits, where structures are stable, photosynthetic productivity at high temperatures is rather low. Using cotton plants, which are native to warm climates, we examined the rate and energy requirement of assimilation (by means of gas exchange and chlorophyll fluorescence). At temperatures exceeding the optimum for assimilation, photorespiration and dissipative processes at the pigment level decrease the quantum efficiency of CO_2 assimilation. Low rates of assimilation may also be caused by an imbalance in the regulation of the Calvin cycle at high temperature, as indicated by 'down-regulation' of the Rubisco. We will discuss the significance of stomatal conductance and heat exchange of leaves in limiting photosynthesis under heat condition.

Irreversible changes and thermal tolerance limit

In general, an increase in temperature tends to shift cellular structures to a state of higher 'disorganization'. Structural alterations induced by changes in temperature can be completely reversible and may even reflect a flexible adjustment to the thermal environment, as will be discussed below. Beyond a critical temperature limit, however, irreversible disturbance of cellular structures and functions takes place. This is thought to mark the 'thermal tolerance limit' of a plant. Specific structures of the thylakoid membrane involved in the primary energy conversion are known to be particularly sensitive to heat. Heat effects on the light-dependent reactions rather than thermal denaturation of enzymes of metabolism or alterations of compartmentation of the cell are primarily responsible for irreversible damage of leaves (Santarius, 1975; Krause & Santarius, 1975; Berry *et al.* 1975; Thebud & Santarius, 1982). Mitochondria are also found to be more stable than thylakoids (Thebud & Santarius, 1982) and cell respiration is known to increase with temperature beyond the level where photosynthesis declines (see review by Berry & Björkman, 1980).

Photophosphorylation is one of the thylakoid-bound processes which is relatively sensitive to heat stress (e.g. Emmett & Walker, 1973; Mukohata *et al.* 1973; Santarius, 1973, 1975; Stidham *et al.* 1982), but the exact mechanism of heat inactivation still needs to be elucidated. It has been concluded that a decline in the electrical potential across the thylakoid membrane, rather than the ability to create a proton gradient, is involved in 'thermal uncoupling'. A reversible depression of the electrochromic pigment absorbance shift at 515 nm has been observed upon very moderate levels of thermal stress (Weis, 1981*a*,*c*, 1982*a*) indicating a heat-induced decline in electrical field gradients at the thylakoid membrane. However, it is completely reversible and is apparently not involved in 'thermal uncoupling' of photophosphorylation. (The electrochromic change at 515 nm is, therefore, a very sensitive test for 'mild' thermal stress at sublethal levels.) The role of the ATP-synthesizing enzyme itself and of the complex mechanisms which regulate its activation in thermal damage to photophosphorylation is not clear and needs further examination.

While PS1 is stimulated, rather than inhibited by heat (Santarius, 1975; Pearcy *et al.* 1977; Stidham *et al.* 1982), a particular thermal sensitivity of the electron transport through PS2 has long been recognized. A series of detailed mechanistic analyses gives rise to the conclusion that thermal inactivation of PS2 activity can be related to three different effects:

1. Heat-inactivation of the oxygen evolving system (Katoh & San Pietro, 1967; Santarius, 1975). Damage to this mechanism interrupts electron donation to PS2 reaction centres as reflected in a decline in the variable part of chlorophyll fluorescence (Krause & Santarius, 1975), and substitute electron donors tend to relieve this effect (Pearcy *et al.* 1977; Weis, 1982*a*). In intact plants, heat inactivation of the oxygen evolving system is not necessarily irreversible, but can recover slowly following heat stress (e.g. Bauer & Senser, 1979; Weis *et al.* 1986).

2. Heat-induced disturbance of the lateral distribution of different pigment complexes in the thylakoid membranes (Armond *et al.* 1980; Gounaris *et al.* 1984). In the native state, thylakoid membranes are stacked and the related formation of grana serves to keep PS2 and PS1 well separated, PS2 being concentrated within the stacked region. This structure is suggested to depend on a delicate balance between Van Der Waal attractive interactions and repulsive ionic forces at the membrane surface (Barber, 1982). Controlled and moderate changes in that structure are assumed to be involved in the regulation of energy transfer between the different pigment structures (for a review see Fork & Satoh, 1986). Temperature interferes with that regulation, as will be discussed below.

3. A third mechanism, clearly irreversible and usually occurring above a very distinct temperature, is reflected in a sharp rise in the basal (F_o) fluorescence (Berry *et al.* 1975; Krause & Santarius, 1975; Schreiber & Armond, 1978; Armond *et al.* 1978). It has been interpreted as a blockage of the photochemical reaction centre of PS2, probably due to a functional detachment of the centre from the core

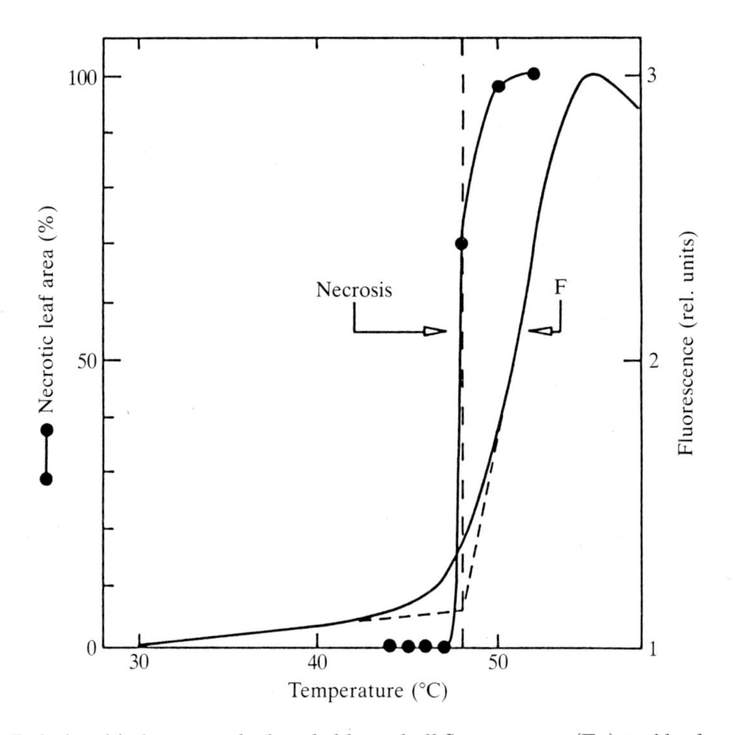

Fig. 1. Relationship between the basal chlorophyll fluorescence (Fo) and leaf necrosis, as induced by heat stress (redrawn from Bilger *et al.* 1984). The experiment demonstrates that the critical temperature limit, beyond which fluorescence starts to increase during the treatment (indicating primary damage at the PS2 pigment system), coincides with the overall tolerance limit of the leaf, indicated by necrosis of leaf tissue, developing in the time course of days after the heat stress. For experimental details see Bilger *et al.* 1984.

antenna (Armond *et al.* 1978, 1980; Schreiber & Armond, 1978). Clustering of separated chlorophyll a/b antennae and PS2 core particles has been observed upon elevated levels of thermal stress (Gounaris *et al.* 1984). In leaves, a correspondence between the break in the fluorescence versus temperature curve and an irreversible decline in photosynthetic capacity has been shown to hold for several species (Schreiber & Berry, 1977; Seemann *et al.* 1984, 1986). In addition, Bilger, Schreiber & Lange (1984), comparing a large number of species, showed that the rise in basal fluorescence, monitored on attached leaves immediately after imposing a thermal stress, corresponds with the threshold for necrotic damage of the total leaf tissue (developing several days after the stress; see Fig. 1). This overall breakdown in the integrity of leaves needs further examination. It appears, however, that the temperature effect is primarily on the light reactions of photosynthesis, and that other cellular processes decline as a *consequence* of this primary heat effect. Damage of PS2 causes an overall decline in the supply of reducing energy within the leaf, but may also affect enzyme reactions which are regulated by the 'light reactions'. For example, the light-dependent thioredoxin system, which controls key enzymes of leaf metabolism (for a review see Cseke &

Buchanan, 1986), is apparently inhibited at high temperature (Björkman *et al.* 1978). In addition, light absorbed by antennae connected to damaged reaction centres may cause photooxidative side reactions.

In the light of these observations, it appears likely that the thermal lability of a key reaction of primary photochemical energy conversion at the pigment level, is one of the main factors determining the overall heat tolerance limit of leaves. Therefore, the rise in basal fluorescence related to the primary damage is useful as a fast and reliable test for that tolerance limit (e.g., Schreiber & Berry, 1977; Smillie & Nott, 1979; Yordanov & Weis, 1984; Bilger *et al.* 1984; Seemann *et al.* 1984, 1986).

Adaptation and changes in thermal stability

There is a large variability in heat tolerance limits among different plants. Usually, plants native or grown in contrasting habitats have tolerance limits which reflect adaptation to the respective thermal environment. Adaptation can be the result of natural selection of genotypes (genotypic acclimation). Each genotype can also adjust its tolerance limit to seasonal changes in the thermal environment (phenotypic adaptation), but the ability to undergo this kind of acclimation varies between different plants. For example, desert evergreen species, which experience large seasonal changes in temperature show large changes in their thermal stability. Shifts in the upper tolerated temperature by 10°C and even more have been reported for such plants, while shifts of only a few degrees Centigrade are characteristic for genotypes native to more moderate climates (see Berry & Björkman, 1980; Berry & Downton, 1982).

Adaptation to contrasting climates and growth temperatures is correlated with changes in fatty acid composition and in the fluidity of the acyl lipids of the chloroplast membranes (Raison *et al.* 1980). In plants adapted to warmer environments, the proportion of unsaturated fatty acids tends to decrease (e.g. Pearcy, 1978; Süss & Yordanov, 1986) and the phase separation temperature of phospholipids tends to be higher (Pike *et al.* 1979; Pike & Berry, 1980; for a review see Berry & Downton, 1982). However, changes in the composition of the surrounding medium may also influence the thermal stability of thylakoid membranes. For example, substituting D_2O for H_2O or increasing sugar concentrations of the suspending medium increases the apparent thermal stability of thylakoids (Santarius, 1973; Armond & Hess, 1979; Seemann *et al.* 1986). Increased thermal tolerance of native plants is often correlated with the osmotic potential of their tissue (Hellmuth, 1971; Seemann *et al.* 1986).

Raison *et al.* (1980) discussed hydrophilic and hydrophobic interactions within the pigment–protein complexes and how they could be affected by temperature. The nature of these interactions, however, and the problem of whether changes in lipid fluidity *per se* determine the thermal stability of the pigment–protein complex, is still unclear. Gounaris *et al.* (1983, 1984) point out that the distribution between the bilayer and non-bilayer state of lipids may play a role for membrane

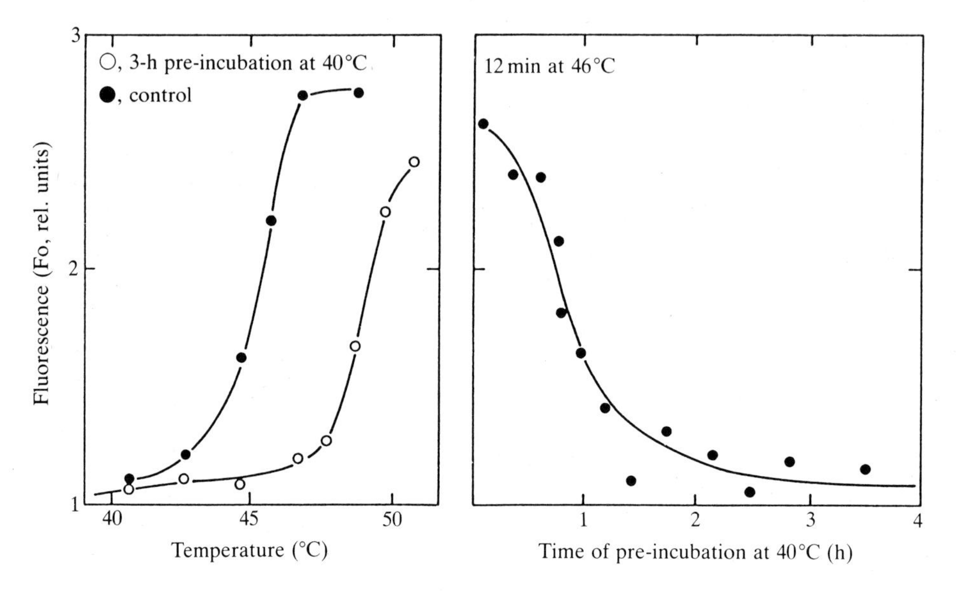

Fig. 2. Short-term adaptation of bean leaves (*Phaseolus vulgarus*) to high temperature. Plants grown at 25°C day temperature were incubated by 3 h in a 20°C (control) or 40°C chamber. Leaves were treated for 12 min at the indicated temperatures and then measured for the basal chlorophyll fluorescence (Fo). The increase in Fo indicates irreversible heat damage at PS2. The experiment shown on the left indicates that 3-h pre-adaptation at 40°C causes an increase in the tolerance limit by about 4°C. The experiment shown on the right indicates the time course of the transition from the control state to the high-temperature of leaves, by comparing the Fo level after 12 min at 46°C of leaves pretreated for various times at 40°C.

stability. They could demonstrate that upon severe heat stress non-bilayer lipid clusters are formed in the thylakoid membrane. The role of polar interactions at the surface in membrane stability has been studied with isolated membranes, and it has been suggested that at constant temperature ionization of mutually repulsive groups at the thylakoid membrane may have the same destabilizing effect as an increase in temperature (Weis, 1982*b*). This is consistent with an hypothesis, developed by Träuble & Eibl (1974), that an increase in the electrostatic forces at the membrane surface decreases its 'transition temperature', and *vice versa*.

Long-term acclimation is superimposed by another type of thermal adaptation, occurring in the time range of a few hours (Santarius & Müller, 1979). It may serve to adjust leaves to peak temperatures during a day. The tolerance limit can increase by several degrees Centigrade with a half-time of less than 1 h (Fig. 2), but the ability of this type of adaptation varies between genotypes. The molecular mechanisms leading to fast adaptive changes in tolerance are not yet known. A significant change in the composition of membrane lipids does not appear to be involved (Santarius & Müller, 1979). On the other hand, the ability to adapt rapidly declined and eventually disappeared in bean leaves, when the capacity for protein synthesis was very low (Yordanov & Weis, 1984). Protein phosphorylation and accumulation of 'heat shock proteins' in chloroplasts have been associated with changes in heat tolerance (Vierling *et al.* 1984; Süss & Yordanov, 1986; Ougham & Howarth, 1988), and may also be related to the rapid adaptive responses noted here.

When the energy absorbed by the pigment system exceeds that required by carbon metabolism (e.g., when Rubisco is limiting) excess protons are taken up into the thylakoid space. As a consequence, a 'high energy state' of the thylakoid membranes is established, indicated by a large decline in the chlorophyll fluorescence (e.g. Briantais *et al.* 1979; Kobayashi *et al.* 1982). This 'high energy quenching' of fluorescence (qE) is thought to reflect a deactivation mechanism which protects leaves against 'light stress' (Krause & Behrend, 1986; Weis *et al.* 1987; Weis & Berry, 1987; Krause *et al.* 1988*a*). Interestingly, this may also be the state where leaves are relatively protected against thermal stress. Light considerably increases the thermal stability of thylakoid membranes in leaves (Schreiber & Berry, 1977), and this stabilization is related to the uptake of protons into the thylakoid space (Weis, 1982*a*).

Steady state photosynthesis in cotton leaves

It is well known that most C_3 plants, even those native to a warm climate, show rather low photosynthetic capacity at high temperature. We examined rate and energy requirement of steady-state assimilation in cotton leaves at different temperatures to analyse factors limiting assimilation at high temperature. All changes between 10°C and 40°C (leaf temperature) were completely reversible. Light-saturated net CO_2 uptake, measured under slightly suboptimal CO_2 concentration (intercellular CO_2 was kept constant at 200 ppm), steeply declined at temperatures below 20°C, but showed a very broad 'temperature optimum' with only small changes over a wide range of temperatures (Fig. 3, lower part). In normal air, assimilation declined considerably above 30°C, while in air with 2% oxygen, almost no change in the rate was observed between 25 and 40°C. The apparent inhibition of CO_2 uptake by high oxygen (compared to the uptake at 2% oxygen) was small at low temperatures, but increased with temperature. This is a consequence of the competition between the carboxylation and oxygenation reactions catalysed by Rubisco, which controls the ratio between carbon fixation and photorespiratory carbon oxidation. With increasing temperature, the balance of the branched carbon metabolism is shifted towards photorespiration (see Berry & Björkman, 1980).

Applying a recently developed approach (Weis *et al.* 1987; Weis & Berry, 1987), we analysed the total photochemical activity in leaves from chlorophyll fluorescence. Coefficients for fluorescence quenching, q, ($0 < q < 1$) have been determined as described recently (Schreiber *et al.* 1986; Dietz *et al.* 1985). The variable part of fluorescence indicates excitation energy (singlet excited states) in PS2. Photochemical activity of centres in the open form, i.e. when electron acceptors become re-oxidized, is reflected in 'photochemical' quenching of fluorescence, qQ. Excitation energy can also be dissipated by non-photochemical

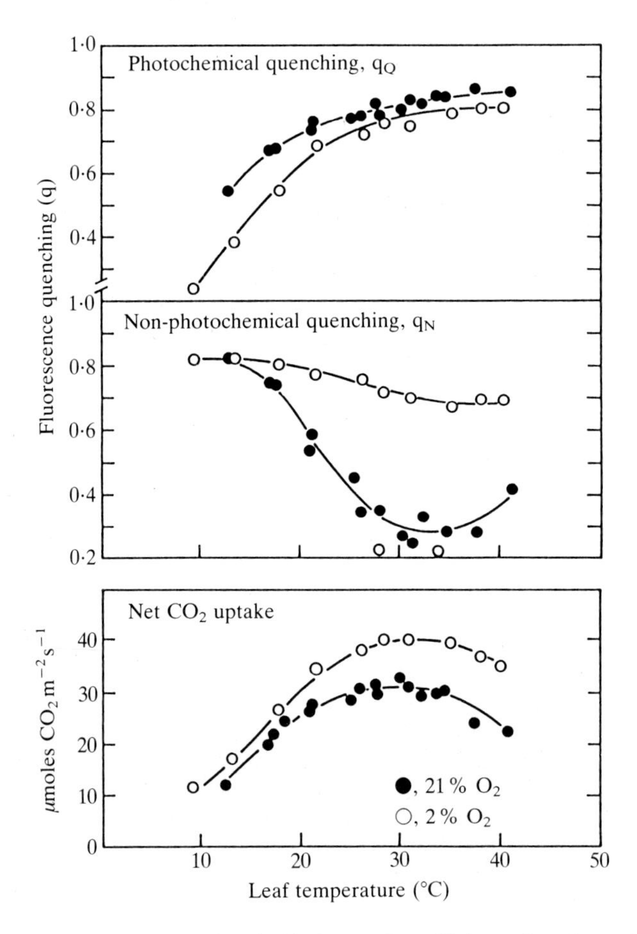

Fig. 3. Temperature response of assimilation and coefficients for photochemical and non-photochemical quenching of chlorophyll fluorescence in cotton leaves, measured in normal air and 2 % oxygen. CO_2 concentration was varied to keep the intercellular CO_2 concentration, Ci, at 200 ppm constant. White actinic light (1200 μE $m^{-2}\,s^{-1}$ PAR) was provided for photosynthesis. The coefficient qN represents total non-photochemical quenching, including light-induced qE quenching and qT.

deactivation of excited states at the pigment level causing 'non-photochemical quenching' of fluorescence, qN (see also Krause *et al.* 1988*b*). In 2 % oxygen, a high level of non-photochemical energy dissipation at the pigment level over the whole range of temperature is indicated by high values for qN, while in normal air qN declines at the temperature optimum and slightly increases again around 40 °C. qN consists mainly of 'high-energy quenching', qE, but it also includes effects on the transfer of excitation energy from PS2 to PS1.

qE quenching is not only a mechanism to protect leaves against light stress (see above). It is part of a feedback control that adjusts light reactions to the energetic balance of a leaf (Weis *et al.* 1987; Weis & Berry, 1987; Krause & Laasch, 1987; Krause *et al.* 1987). Apparently, this quenching is caused by the uptake of protons into the inner thylakoid space producing the so-called 'high energy state' and

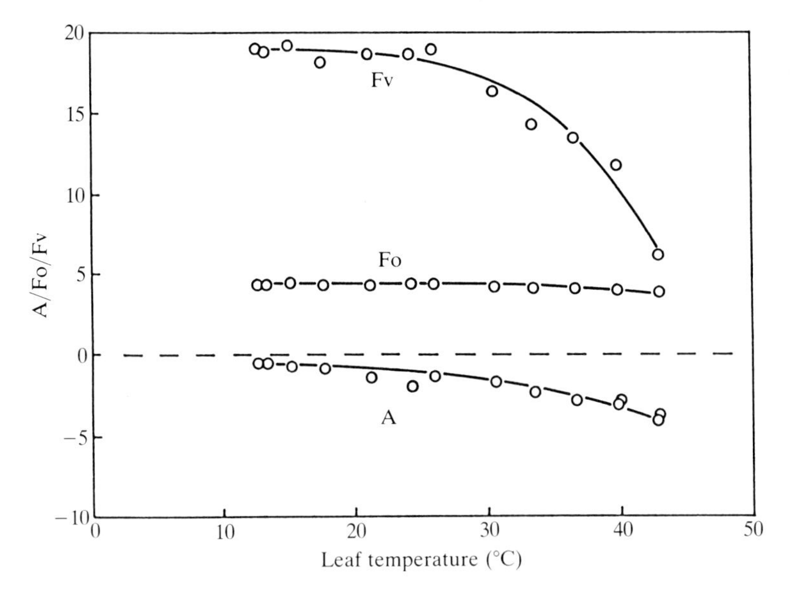

Fig. 4. Fluorescence and dark respiration of a cotton leaf in the dark as a function of the leaf temperature. The attached leaf was mounted in a leaf chamber in normal air and kept at each temperature for at least 15 min. CO_2 release (A, μmoles $CO_2\,m^{-2}\,s^{-1}$, negative values) represents dark respiration. Fluorescence, produced by a weak measuring beam, has been taken as basal fluorescence (Fo; when all PS2 centres are open). Short saturating light pulses have been given to obtain the maximal variable fluorescence (Fv; all PS2 centres closed). The ratio between Fv at a given temperature and Fv at 20°C has been taken as the quenching coefficient 'qT'. When the leaves were illuminated with a far-red background light, almost no decline in Fv was seen (not shown).

occurs when the absorbed energy exceeds that required by photosynthetic or photorespiratory metabolism. In this state, part of the 'excess' energy absorbed by the pigment system is dissipated by 'thermal' deactivation of excited states at PS2. A quantitative relationship between qE quenching and photochemical yield of PS2 could be shown for leaves (Weis *et al.* 1987; Weis & Berry, 1987) and isolated chloroplasts (Krause *et al.* 1988*a*).

A second type of non-photochemical fluorescence quenching considered in this analysis is not dependent on the energetic balance of the leaf, but is directly caused by exposure of plants to high temperature, even in the dark (Fig. 4). It can be reversed by far-red illumination and is thought to reflect a mild-heat-induced increase in the transfer of excitation energy from PS2 to PS1 (spillover), at the expense of PS2 excitation, as described for spinach leaves (Weis, 1984*a*,*b*, 1985). A reversible heat-induced conversion of PS2 into the less efficient β-form, as observed in leaves (Weis, 1984*b*) and isolated membranes (Sundby *et al.* 1986; Andersson *et al.* 1987) may be part of this temperature-induced 'state transition', caused by moderate thermal randomization of PS1 and PS2 within the membrane. This state may favour a PS1-dependent, proton-pumping cyclic reaction, while the quantum efficiency of the whole chain electron transport is thought to decline (Weis, 1985). A possible role in protecting PS2 against photoinhibition has been discussed (Andersson *et al.* 1987), but the physiological significance has not been tested. In this study, we will call this temperature-dependent quenching in fluorescence 'qT'.

If the flux of absorbed light energy exceeds the capacity of these different pathways for deactivation, then a high reduction level will be established in the electron transport system (redox feedback) and singlet excited states will accumulate in the antenna system. Deactivation of singlet states in the pigment system is reflected in emission of fluorescence (radiative deactivation). This state, indicated by low values of the quenching coefficient, qQ, may reflect 'light stress', as long lived excited states may cause harmful photoxidative side reactions at the pigment system, leading to photoinhibition.

Since the different fluorescence quenching processes have a well defined relationship to photochemistry, the total photosynthetic electron transport in a leaf can be calculated directly from fluorescence data on the basis of equations derived recently (Weis & Berry, 1987). Based on studies with isolated chloroplasts (not shown), we assume here, that the relationship between qT and photochemistry is similar to that of qE. The analysis for cotton (Fig. 5) shows that, in normal air and at high temperatures, where net uptake of CO_2 is inhibited (relative to the uptake in low oxygen) because of photorespiration, the total flux of electrons, Je, is considerably higher than at 2 % oxygen. This may be explained partially by a change in the balance between ATP and NADPH: since 3·5 ATP are required to regenerate ribulose-1,5-bisphosphate from the photorespiratory cycle (compared to 3 ATP in the reductive carbon cycle), an increase in temperature (and photorespiration) is expected to decrease the steady state level of ATP, compared to that of NADPH (see Berry & Downton, 1982). This may keep the proton gradient small, as actually indicated by low qE at high temperature in normal air, compared to that in 2 % oxygen. In the absence of photorespiration (low O_2), a high proton gradient may feed back and limit photochemistry. Other changes in the balance of regulation may also be involved.

From the above analysis, we may calculate the fraction of the photosynthetically active light energy absorbed by a leaf, that is de-excited by the different photochemical and non-photochemical pathways for deactivation. This is shown for a leaf in normal air (Fig. 6) and in 2 % oxygen (Fig. 7). Photochemical energy is consumed by whole chain electron transport linked to assimilation and photorespiration (and other oxidative processes). At temperatures below the optimum the total fraction of absorbed energy consumed by photochemical inactivation is rather small (because the maximal rate of stromal enzymes is low), while a large fraction of energy is dissipated in the pigment system, mainly by radiationless de-excitation at PS2 centres, caused by qE quenching (area b). At very low temperature, 'radiative' de-excitation, (indicated by decline in qQ; area a) becomes significant, especially in 2 % oxygen (Fig. 7). It reflects a high reduction level of the quinone pool and 'closure' of PS2 centres. As a conse-

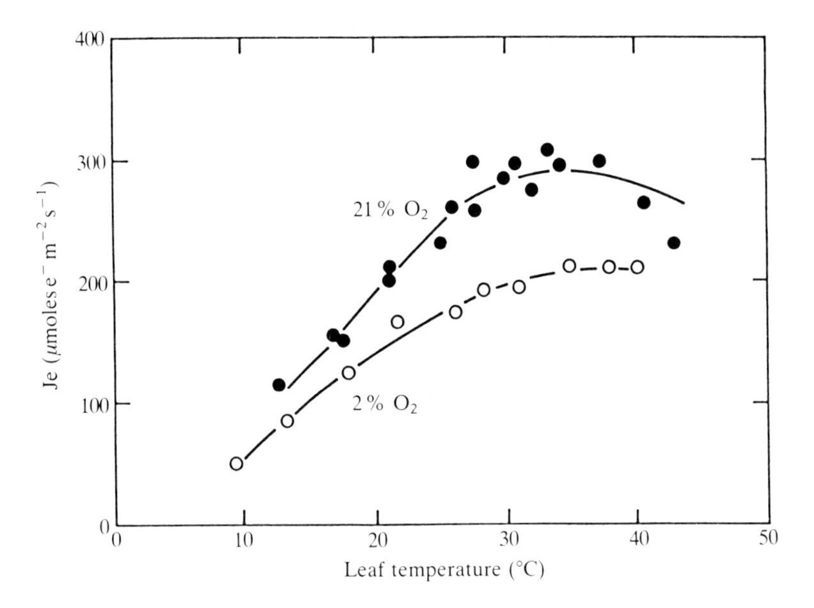

Fig. 5. Temperature response of total electron transport, Je, in cotton leaves, measured in normal air and 2 % oxygen. Je was calculated from fluorescence, using on the equation: $Je/I = qQ[qN\phi_e + \phi_o(1-qN)]$, where I is incident light; ϕ_e and ϕ_o are extrapolated yields of total electron transport in a leaf (independently determined from light response curves), when qN is 1 or 0, respectively. For details of the approach see Weis & Berry, 1987. Quenching coefficients have been taken from the experiments shown in Fig. 4. The rate of electron transport calculated in this way was highly correlated ($r^2 = 0{\cdot}92$) with the rate calculated by the analysis of net CO_2 exchange (data not shown).

quence, long-lived singlet excited states may accumulate within the pigment system. It apparently only occurs when the amount of absorbed energy exceeds several times that consumed by metabolism. It may then exceed the capacity of the qE-mechanism to dissipate 'excess' excitation energy and extended exposure to such 'light stress' may cause photooxidative damage. The analysis points to a high susceptibility of the chill-sensitive species, used here, to light stress at temperatures below the optimum for CO_2 assimilation (see also Baker, Long & Ort, 1988).

At moderately high temperatures, the fraction of energy dissipated by qE quenching becomes smaller with increasing temperature, especially in normal air, and that dissipated by 'radiative' deactivation becomes insignificant. In contrast, photorespiration (area d) increases with temperature (see experiment done in normal air, Fig. 6), while energy consumed by assimilation stays constant or even declines. In addition, 'quenching' of energy at the pigment level, related to qT (area c, Figs 6 and 7) becomes significant, especially at temperatures exceeding the optimum for assimilation. While qE quenching is a regulated process, depending on an 'internal' feedback between metabolism and light reactions (see above), both photorespiration and qT-quenching may represent dissipative processes induced directly by an increase in leaf temperature rather by feedback control.

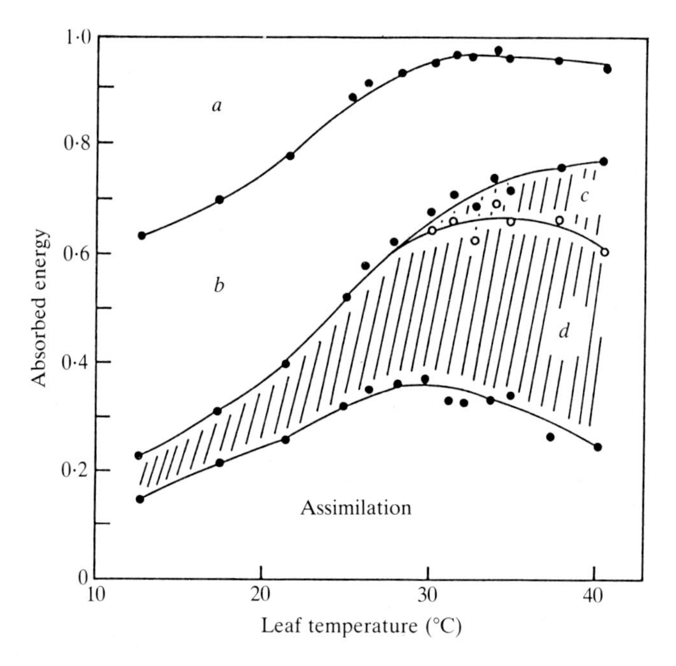

Fig. 6. Distribution of the absorbed energy between different pathways for deactivation in a cotton leaf as a function of the leaf temperature. Data from an experiment in normal air, 21 % O_2, 200 ppm Ci. The fraction of total absorbed energy, dissipated by radiative deactivation at closed PS2 centres (centres with reduced Q_A), is shown by the area (*a*); that by the light-dependent qE-quenching or by the temperature-dependent qT-quenching is shown by areas (*b*) and (*c*), respectively. The 'rest' is related to energy consumed by whole-chain electron transport (total photochemistry). This is divided into the fraction consumed by CO_2 assimilation and that consumed by non-assimilatory processes, mainly photorespiration. The curves separating the areas were calculated from coefficients for fluorescence quenching, according to $T_{\phi_a} = 1-qQ$; $\phi_b = \phi_a - 0{\cdot}7qE$; $\phi_c = \phi_b - 0{\cdot}7qT$, and $\phi_d = \phi_c(1-\gamma \cdot [O_2]/[CO_2])$, where γ is the specificity factor of Rubisco at the temperature: $[CO_2]$ and $[O_2]$ are the estimates of the concentrations in the chloroplasts; ϕ_a, ϕ_b, ϕ_c and ϕ_d are the four curves shown, top to bottom.

The fraction of total energy dissipated by these two processes (shaded areas, Fig. 6), markedly increases at high temperatures (more than 50 % at 40°C).

The analysis indicates that dissipative processes in metabolism and at the pigment level, caused by high temperature, become important controllers of assimilation under moderately high temperature stress, as they decrease the quantum efficiency of CO_2 assimilation. On the other hand, the loss in efficiency may partially be balanced by a gain in resistance against light stress, often occurring under conditions when temperature is high. This is very different to the situation at low temperature, where the 'efficiency' of assimilation is high. Its rate clearly is limited by the capacity of enzymes, and a large fraction of excess energy is dissipated by 'radiative' de-excitation. These factors may contribute to the increased sensitivity of plants to photoinhibition at low temperature (Powles *et al.* 1983; Baker *et al.* 1988).

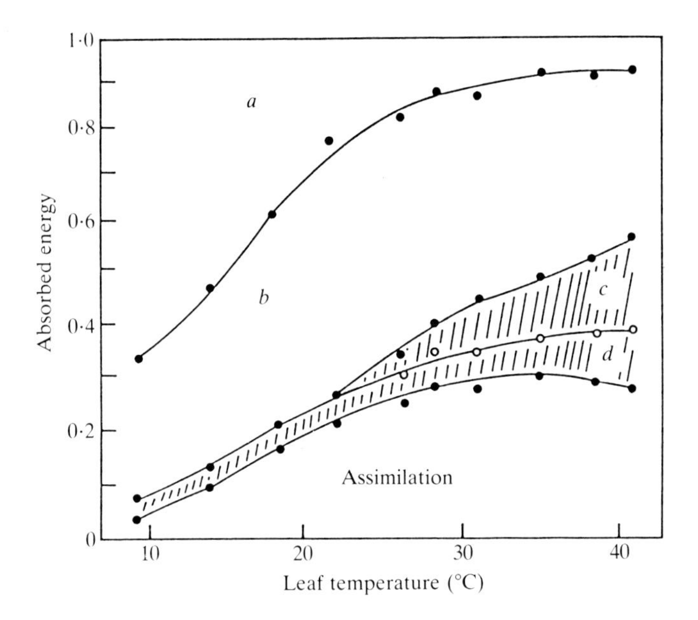

Fig. 7. Distribution of the absorbed energy between different pathways for deactivation in a cotton leaf as a function of the leaf temperature. Data from an experiment done in 2 % oxygen. Conditions and calculations as for Fig. 6.

The persistence of qE quenching even at superoptimal temperatures (Figs 6 and 7) indicates, however, that some additional factor, not related to the decline in 'efficiency', may limit photosynthetic rate at high temperatures. The catalytic activity of each single enzymatic reaction of photosynthetic metabolism should increase with temperature. However, little change in electron transport occurs over a wide range of high temperatures (Fig. 5), and the capacity of light reactions is still restricted to some degree by feedback from carbon metabolism. This is particularly significant in the experiment done in 2 % oxygen (Figs 3 and 7). Other evidence implicates the control of Rubisco activation in this response. After rapidly transferring spinach leaves from moderately high to lower temperatures, photosynthesis was markedly depressed, due to a blockage in the Calvin cycle, but recovered again rapidly (Weis, 1981*a*). Work with isolated chloroplasts (Weis, 1981*a*,*b*) and with whole leaves (Kobza & Edwards, 1987) showed that this reversible heat depression was related to a temperature-induced 'down regulation' of the Rubisco. In cotton leaves, we found at 38°C about 60 % activation of Rubisco, compared to its activation at 25 °C (100 ppm intercellular CO_2 concentration, 1200 $\mu E\ m^{-2}\ s^{-1}$; data not shown). Fig. 8 shows a temperature response of the apparent activation state of the Rubisco. This is taken as the ratio of the actual rate at each temperature to the expected Rubisco activity (based upon *in vitro* temperature dependence of Km and Vmax) calculated using the measured intercellular CO_2 and O_2 and assuming full activation at 25 °C. This ratio is constant between 20 °C and 30 °C and then declines with increasing temperature.

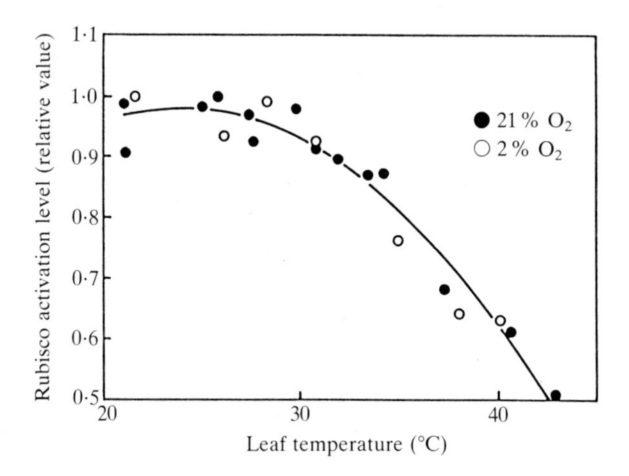

Fig. 8. Temperature response of the apparent activation state of Rubisco in a cotton leaf. The measured intercellular CO_2 concentration, and the in vitro temperature dependence of K_m and V_{max} has been used to calculate a theoretical temperature dependence of the carboxylation reaction in the leaf. The ratio between the actually measured rate of CO_2 fixation to that calculated from the kinetic expressions for the Rubisco activity has been taken as a relative measure for the apparent activation state in the leaf (assuming full activation at 25 °C).

The decline in activation counteracts the intrinsic response of the catalytic reaction to temperature.

The mechanism responsible for the inactivation is not known. There is indication that the Rubisco is regulated by an activase system (Portis *et al.* 1987), and it is possible that this activase is affected. Inactivation of this enzyme, however, is not necessarily a primary temperature effect. Labate & Leegood (1988) reported a close relationship between the temperature response of assimilation and the phosphate status of the mesophyll. Stitt (1987) discussed the possibility that photosynthesis at high temperature could be limited by depletion of metabolite pools of the Calvin cycle caused by too rapid export of triose-P from the stroma. 'Down regulation' of the Rubisco may then be a secondary effect, caused by an imbalance in regulation of the Calvin cycle or export reactions (Kobza & Edwards, 1987). The factors involved in this imbalance, however, may contribute to limitation of photosynthetic productivity at high temperature and need further examination.

At high temperatures, heat exchange of the leaf with its environment becomes important in controlling photosynthetic productivity. Any increase in leaf temperature at temperatures above the optimum can lead to a decrease in photosynthetic rate (see above). The leaf temperature is determined by the balance between the input of heat to the leaf (primarily by absorption of sunlight) and the dissipation of this heat (primarily by transpiration of water, emission of longwave radiation and sensible exchange with the air). Stomata regulate the rate of loss of water vapour and consequently have an influence on leaf temperature. For example, closure of stomata at high temperature would conserve water, but it would also increase leaf temperature leading to inhibition of photosynthetic rate or possible irreversible damage to photosynthetic membranes. An approach to analyse the control of assimilation by this complex set of physical and physiological factors has been presented (Woodrow *et al.* 1987).

Conclusions

The irreversible thermal disruption of thylakoid pigment structures will be a determinant of the upper thermal tolerance limit of plants. Little is known, however, about the factors which control the photosynthetic productivity of plants at high temperatures, approaching the tolerance limit. Temperature-dependent disturbances in the balance between different pathways of carbon metabolism appear to be important in controlling carbon assimilation under moderate thermal stress. Moreover, a reversible temperature-induced rearrangement of pigment–protein complexes within the thylakoid membrane appears to interfere with the efficiency and regulation of membrane-bound photochemical reactions. An integration of reversible temperature effects on thylakoid structure and regulation of photosynthesis into a general concept for adaptation of plants to high temperature is now required.

References

ANDERSSON, B., SUNDBY, C., LARSSON, U. K., MÄENPÄÄ, P. & MELIS, A. (1987). Dynamic aspects on the organisation of the thylakoid membrane. In *Progress in photosynthesis Research*, vol. 2 (ed. J. Biggins), pp. 669–676. Dordrecht, The Netherlands: Martinus Nijhoff Publishers.

ARMOND, P. A. & HESS, J. L. (1979). Enhancement of high temperature stability of protein-protein interactions by deuterium oxide. *Yearbk. Carnegie Instn Yb.* **78**, 168–171.

ARMOND, P. A., SCHREIBER, U. & BJÖRKMAN, O. (1978). Photosynthetic acclimation to temperature in the desert shrub, *Larrea divaricata* II. Light harvesting efficiency and electron transport. *Pl. Physiol.* **61**, 411–415.

ARMOND, P. A., BJÖRKMAN, O. & STAEHELIN, L. A. (1980). Dissociation of supramolecular complexes in chloroplast membranes. A manifestation of heat damage to the photosynthetic apparatus. *Biochim. Biophys. Acta* **601**, 433–442.

BAKER, N. R., LONG, S. P. & ORT, D. R. (1988). Photosynthesis and temperature, with particular reference to effects on quantum yield. In *Plants and Temperature* (ed. S. P. Long & F. I. Woodward). Symp. Soc. Exp. Biol., vol. 42, pp. 347–375. Cambridge: Company of Biologists.

BARBER, J. (1982). Influence of surface charges on thylakoid structure and function. *A. Rev. Pl. Physiol.* **33**, 261–295.

BAUER, H. & SENSER, M. (1979). Photosynthesis of ivy leaves (Hedera helix L.) after heat stress. II. Activity of ribulose bisphosphate carboxylase, Hill reaction and chloroplast ultrastructure. *Z. Pflanzenphysiol.* **91**, 359–369.

BERRY, J. A., FORK, D. C. & GARRISON, S. (1975). Mechanistic studies of thermal damage to leaves. *Carnegie Instn Yb.* **74**, 751–759.

BERRY, J. A. & BJÖRKMAN, O. (1980). Photosynthetic response and adaptation to temperature in higher plants. *A. Rev. Pl. Physiol.* **31**, 491–543.

BERRY, J. A. & DOWNTON, W. J. S. (1982). Environmental regulation of photosynthesis. In *Photosynthesis: Development, Carbon Metabolism and Plant Productivity*, vol. 2 (ed. Govindjee). New York: Academic Press.

BILGER, H.-W., SCHREIBER, U. & LANGE, O. (1984). Determination of leaf heat resistance: comparative investigation of chlorophyll fluorescence changes and tissue necrosis methods. *Oecologia* **63**, 156–262.

BJÖRKMAN, O., BADGER, M. R. & ARMOND, P. A. (1978). Thermal acclimation of photosynthesis: Effect of growth temperature on photosynthetic characteristics and components of the photosynthetic apparatus of *Nerium oleander*. *Carnegie Instn Yb.* **77**, 262–276.

BRIANTAIS, J.-M., VERNOTTE, C., PICAUD, M. & KRAUSE, G. H. (1979). A quantitative study of the slow decline of chlorophyll a fluorescence in isolated chloroplasts. *Biochim. Biophys. Acta* **538**, 128–138.

CSEKE, C. & BUCHANAN, B. B. (1986). Regulation and utilisation of photosynthate in leaves. *Biochim. Biophys. Acta* **853**, 43–63.

DIETZ, K.-J., SCHREIBER, U. & HEBER, U. (1985). The relationship between the redox state of Q_A and photosynthesis in leaves at various carbon-dioxide, oxygen and light regimes. *Planta* **166**, 219–226.

EMMETT, J. M. & WALKER, D. A. (1973). Thermal uncoupling in chloroplasts. Inhibition of photophosphorylation without depression of light-induced pH change. *Arch. Biochem. Biophys.* **157**, 106–113.

FORK, D. C. & SATOH, K. (1986). The control by state transitions of the distribution of excitation energy in photosynthesis. *A. Rev. Pl. Physiol.* **37**, 335–361.

GOUNARIS, K., BRAIN, A. P. R., QUINN, P. J. & WILLIAMS, W. P. (1983). Structural and functional changes associated with heat-induced phase-separation of non-bilayer lipids in chloroplast thylakoid membranes. *FEBS lett.* **153**, 47–52.

GOUNARIS, K., BRAIN, A. R. R., QUINN, P. J. & WILLIAMS, W. P. (1984). Structural reorganisation of chloroplast thylakoid membranes in response to heat stress. *Biochim. Biophys. Acta* **766**, 198–208.

HELLMUTH, E. O. (1971). Ecophysiological studies on plants in arid and semiarid regions in Western Australia. V. Heat resistance limits of photosynthetic organs of different seasons, their relation to water deficits and cell sap properties and the regeneration ability. *J. Ecol.* **59**, 365–374.

KATOH, S. & SAN PIETRO, A. (1967). Ascorbate-supported NADP photoreduction by heated euglena chloroplasts. *Arch. Biochem. Biophys.* **122**, 144–152.

KOBZA, J. & EDWARDS, G. E. (1987). Influences of leaf temperature on photosynthetic carbon metabolism in wheat. *Pl. Physiol.* **83**, 60–74.

KOBAYASHI, Y., KÖSTER, S. & HEBER, U. (1982). Light scattering, chlorophyll fluorescence and state of adenylate system in illuminated spinach leaves. *Biochim. Biophys. Acta* **682**, 44–52.

KRAUSE, G. H. & BEHREND, U. (1986). pH-dependent chlorophyll fluorescence quenching indicating a mechanism of protection against photoinhibition of chloroplasts. *FEBS lett.* **200**, 298–302.

KRAUSE, G. H. & LAASCH, H. (1987). Energy-dependent chlorophyll fluorescence quenching in chloroplasts correlated with quantum yield of photosynthesis. *Z. Naturforsch.* **42c**, 581–584.

KRAUSE, G. H. & SANTARIUS, K. A. (1975). Relative thermostability of the chloroplast envelope. *Planta* **127**, 285–299.

KRAUSE, G. H., LAASCH, H. & WEIS, E. (1988*a*). Regulation of thermal dissipation of absorbed light energy in chloroplasts indicated by energy-dependent fluorescence quenching. *Pl. Physiol. Biochem.* **26**, 000–000.

KRAUSE, G. H., GRAFFLAGE, S., RUMICH-BAYER, S. & SOMERSALO, S. (1988*b*). Effects of freezing on plant mesophyll cells. In *Plants and Temperature* (ed. S. P. Long & F. I. Woodward). Symp. Soc. Exp. Biol., vol. 42, pp. 311–327. Cambridge: Company of Biologists.

LABATE, C. A. & LEEGOOD, R. C. (1988). Limitation of photosynthesis by changes in temperature I. Factors affecting the response of carbon dioxide assimilation to temperature in barley leaves. *Planta*. **173**, 519–527.

MUKOHATA, Y., YAGI, T., HIGASHIDA, M., SHINOZAKI, K. & MATSUNO, A. (1973). Biophysical studies on subcellular particles – IV. Photosynthetic activities in isolated spinach chloroplasts after transient warming. *Pl. Cell Physiol.* **14**, 111–118.

OUGHAM, H. J. & HOWARTH, C. J. (1988). Temperature shock proteins in plants. In *Plants and*

Temperature (ed. S. P. Long & F. I. Woodward). Symp. Soc. Exp. Biol., vol 42, pp. 259–280. Cambridge: Company of Biologists.

PEARCY, R. W. (1978). Effects of growth temperature on the fatty acid composition of the leaf lipid in *Atriplex lentiformis* (Torr.) Wats. *Pl. Physiol.* **61**, 484–486.

PEARCY, R. W., BERRY, J. A. & FORK, D. C. (1977). Effects of growth temperature on the thermal stability of the photosynthetic apparatus of *Atriplex lentiformis* (Torr.) Wats. *Pl. Physiol.* **59**, 873–878.

PIKE, C. S. & BERRY, J. A. (1980). Membrane phospholipid phase separation in plants adapted to or acclimated to different thermal regimes. *Pl. Physiol.* **66**, 238–241.

PIKE, C. S., BERRY, J. A. & RAISON, J. K. (1979). Fluorescence polarization studies of membrane phospholipid phase separations in warm and cool climate plants. In *Low Temperature Stress in Crop Plants* (ed. J. M. Lyons, D. Graham & J. K. Raison), pp. 305–318. New York: Academic Press.

PORTIS, A. R. (JR), SALVUCCI, M. E., OGREN, W. L. & WERNEKE, J. (1987). Rubisco activase: A new enzyme in the regulation of photosynthesis. In *Progress in Photosynthesis Research*, vol. 3 (ed. J. Biggins), pp. 371–378. Dordrecht, The Netherlands: Martinus Nijhoff Publishers.

POWLES, S., BERRY, J. A. & BJÖRKMAN, O. (1983). Interactions between light and chilling temperatures on the inhibition of photosynthesis in chilling sensitive plants. *Pl. Cell Envir.* **6**, 117–123.

RAISON, J. K., BERRY, J. A., ARMOND, P. A. & PIKE, C. S. (1980). Membrane properties in relation to temperature stress. In *Adaptation of Plants to Water and High Temperature Stress* (ed. N. C. Turner & P. J. Kramer), pp. 221–233. New York: John Wiley.

SANTARIUS, K. A. (1973). The protective effect of sugars on chloroplast membranes during temperature and water stress and its relationship to frost, desiccation and heat resistance. *Planta* **113**, 105–114.

SANTARIUS, K. A. (1975). Sites of heat sensitivity in chloroplasts and differential inactivation of cyclic and noncyclic photophosphorylation by heating. *J. Therm. Biol.* **1**, 101–107.

SANTARIUS, K. A. & MÜLLER, M. (1979). Investigations on heat resistance of spinach leaves. *Planta* **146**, 529–538.

SCHREIBER, U. & ARMOND, P. (1978). Heat induced changes of chlorophyll fluorescence in isolated chloroplasts and related heat damage at the pigment level. *Biochim. Biophys. Acta* **502**, 138–151.

SCHREIBER, U. & BERRY, J. A. (1977). Heat-induced changes in chlorophyll fluorescence in intact leaves correlated with damage of the photosynthetic apparatus. *Planta* **136**, 233–238.

SCHREIBER, U., SCHLIWA, U. & BILGER, W. (1986). Continuous recording of photochemical and non-photochemical chlorophyll fluorescence quenching with a new type of modulation fluorometer. *Photosynth. Res.* **10**, 51–62.

SEEMANN, J. R., BERRY, J. A. & DOWNTON, W. J. R. (1984). Photosynthetic response and adaptation to high temperature in desert plants in comparison of gas exchange and fluorescence methods for plant studies of thermal tolerance. *Pl. Physiol.* **75**, 364–368.

SEEMANN, J. R., DOWNTON, W. J. S. & BERRY, J. A. (1986). Temperature and leaf osmotic potential as factors in the acclimation of photosynthesis to high temperature in desert plants. *Pl. Physiol.* **80**, 926–930.

SMILLIE, R. M. & NOTT, R. (1979). Heat injury in leaves of alpine, temperate and tropical plants. *Aust. J. Pl. Physiol.* **6**, 135–141.

STIDHAM, M. A., URIBE, E. G. & WILLIAMS, G.-J. (1982). Temperature dependence of photosynthesis in *Agropyron smithii* Rydb. II. Contribution from electron transport and photophosphorylation. *Pl. Physiol.* **69**, 929–934.

STITT, M. (1987). Limitation of photosynthesis by sucrose synthesis. In *Progress in Photosynthesis Research*, vol. 3 (ed. J. Biggins), pp. 685–692. Dordrecht, The Netherlands: Martinus Nijhoff Publishers.

SUNDBY, C., MELIS, A., MÄENPÄÄ, P. & ANDERSSON, B. (1986). Temperature-dependent changes in the antenna size of Photosystem II. Reversible conversion of Photosystem IIα and Photosystem IIβ. *Biochim. Biophys. Acta* **851**, 475–483.

SÜSS, K.-H. & YORDANOV, I. T. (1986). Biosynthetic cause of *in vivo* acquired thermotolerance of photosynthetic light reactions and metabolic responses of chloroplasts to heat stress. *Pl. Physiol.* **81**, 192–199.

THEBUD, R. & SANTARIUS, K. A. (1982). Effects of high temperature stress on various biomembranes of leaf cells *in situ* and *in vitro*. *Pl. Physiol.* **70**, 200–205.
TRÄUBE, H. & EIBL, H.-J. (1974). Electrostatic effects on lipid phase transitions: membrane structure and ionic environment. *Proc. Natn Acad. Sci. U.S.A.* **71**, 214–219.
VIERLING, E., MISKIND, M. L., SCHMIDT, G. W. & KEY, J. L. A. (1984). Specific heat shock protein in plants is transported into chloroplasts. *J. Cell Biol.* **99**, 452*a*.
WEIS, E. (1981*a*). Reversible heat-inactivation of the Calvin cycle: a possible mechanism of the temperature regulation of photosynthesis. *Planta* **151**, 33–39.
WEIS, E. (1981*b*). The temperature sensitivity of dark-inactivation and light-activation of the ribulose-1,5-bisphosphate carboxylase in spinach chloroplasts. *FEBS lett.* **129**, 197–200.
WEIS, E. (1981*c*). Reversible effects of high, sublethal temperatures on light-induced light scattering changes and electrochromic pigment absorption shift in spinach leaves. *Z. Pflanzenphysiol.* **101**, 169–178.
WEIS, E. (1982*a*). Influence of light on the heat sensitivity of the photosynthetic apparatus in isolated spinach chloroplasts. *Pl. Physiol.* **70**, 1530–1534.
WEIS, E. (1982*b*). The influence of metal cations and pH on the heat sensitivity of photosynthetic oxygen evolution and chlorophyll fluorescence in spinach chloroplasts. *Planta* **154**, 41–47.
WEIS, E. (1984*a*). Short-term acclimation of spinach to high temperatures: effect of chlorophyll fluorescence at 293 and 77 Kelvin in intact leaves. *Pl. Physiol.* **74**, 402–407.
WEIS, E. (1984*b*). Temperature-induced changes in the distribution of excitation energy between photosystem I and photosystem II in spinach leaves. In *Advances in Photosynthesis Research*, vol. III (ed. C. Sybesma), pp. 291–294. The Hague: Martinus Nijhoff/Dr W. Junk Publishers.
WEIS, E. (1985). Light- and temperature-induced changes in the distribution of excitation energy between Photosystem I and Photosystem II in spinach leaves. *Biochim. Biophys. Acta.* **807**, 118–126.
WEIS, E. & BERRY, J. A. (1987). Quantum efficiency of photosystem II in relation to 'energy'-dependent quenching of chlorophyll fluorescence. *Biochim. Biophys. Acta* **894**, 198–208.
WEIS, E., WAMPER, D. & SANTARIUS, K. A. (1986). Heat sensitivity and thermal adaptation of photosynthesis in liverwort thalli. *Oecologia* **69**, 134–139.
WEIS, E., BALL, J. T. & BERRY, J. A. (1987). Photosynthetic control of electron transport in leaves of phaseolis vulgaris: Evidence for regulation of photosystem 2 by the proton gradient. In *Progress in Photosynthesis Research*, vol. 2 (ed. J. Biggins), pp. 553–556. Dordrecht, Netherlands: Martinus Nijhoff Publishers.
WOODROW, I. E., BALL, J. T. & BERRY, J. A. (1987). A general expression for the control of the rate of photosynthetic CO_2 fixation by stomata, the boundary layer and radiation exchange. In *Progress in Photosynthesis Research*, vol. 4 (ed. J. Biggins), pp. 225–228. Dordrecht, Netherlands: Martinus Nijhoff Publishers.
YORDANOV, I. T. & WEIS, E. (1984). The influence of leaf-aging on the heat-sensitivity and heat-hardening of the photosynthetic apparatus in *Phaseolis vulgaris*. *Z. Pflanzenphysiol.* **113**, 383–393.

PHOTOSYNTHESIS AND TEMPERATURE, WITH PARTICULAR REFERENCE TO EFFECTS ON QUANTUM YIELD

NEIL R. BAKER, STEPHEN P. LONG and DONALD R. ORT[1]
Department of Biology, University of Essex, Colchester CO4 3SQ, UK

Summary

Previous reviews of the effects of temperature on *in vivo* photosynthesis have mainly concerned the effects of temperature on light saturated rates. The quantum yield of photosynthesis (ϕ), as a measure of light limited photosynthesis, has generally been regarded as temperature insensitive. At temperatures close to the minima and maxima at which plants can sustain photosynthetic CO_2 assimilation, light may damage the photosynthetic apparatus, an effect termed photoinhibition. A constant feature of photoinhibition is a reduction in ϕ. In maize, chilling-dependent photoinhibition reduces both ϕ and the light saturated rate of CO_2 assimilation (A_{sat}) and of O_2 evolution. Analysis of recovery of CO_2 uptake in these leaves suggests that whilst A_{sat} recovers in a few hours, ϕ may not be fully restored for days. Examination of mature crop canopies shows that only a small proportion of the leaves are likely to become light saturated and then only for part of the day. The relative significance of temperature-induced changes in A_{sat} and ϕ have also been tested in canopy models of maize crop photosynthesis. These suggest that whilst changes in either parameter will have similar effects on total canopy photosynthesis on the sunniest days of the year, for an average summer's day changes in ϕ will be of far greater importance. Consideration is therefore given to the factors associated with thylakoid membranes that may determine temperature-induced decreases in ϕ. Chilling of maize leaves under high light levels reduces the quantum yield of PSII and whole chain electron transport in concert with a decrease in the capacity of isolated thylakoids to bind atrazine, which is indicative of a loss or damage to the Q_B protein. Besides such classical symptoms of photoinhibition of PSII, chilling also induces the accumulation of a 31 kDa polypeptide in the thylakoids of maize leaves. This polypeptide fractionates with the light-harvesting chlorophyll a/b protein complex (LHCII) and has been tentatively identified as an unprocessed precursor of CP29 since it binds chlorophyll and is immunologically related to CP29. Accumulation of the 31 kDa polypeptide is associated with a modification in the energetics of LHCII, which may result in a decrease in excitation energy from LHCII to PSII and contribute to a decrease in ϕ. Examination is also made of how stress-induced modifications of interactions between PSII complexes, functioning of the cyt b_6/f complex, the

[1] Permanent address: USDA/ARS, Department of Plant Biology, University of Illinois, Urbana, Illinois 61801, USA.

permeability of the thylakoid membrane to protons and the activity of the coupling factor may contribute to decreases in ϕ.

Introduction

The light response curve of photosynthesis

The temperature response of photosynthetic CO_2 reduction by higher plants has been reported for a wide range of species (reviewed: Berry & Bjorkman, 1980; Berry & Raison, 1981). However, most studies have been concerned with the rate of CO_2 uptake at light saturation. Since leaves in their natural environment will experience a continually changing light environment, it is important to consider the interaction of temperature responses with light. The typical hyperbolic dependence of CO_2 uptake on light intensity is observed over a wide range of temperatures (e.g. see Fig. 1). At any given temperature the light response may be defined by three basic parameters: 1) the upper asymptote and light saturated rate of CO_2 uptake, A_{sat}; 2) the maximum quantum yield and initial slope of the response of CO_2 uptake to light, ϕ; and 3) a measure of the convexity of the hyperbola, θ, which may range in value from 0 for a slow transition from the initial slope to A_{sat}, to 1 for a sharp transition (Thornley, 1976; Jarvis & Sandford, 1986). At high light levels, A_{sat} will dominate, whilst at low light levels, ϕ will dominate. The responses of these two parameters to temperature varies markedly (Fig. 2). A_{sat} typically describes a bell-shaped response to temperature, with an optimum, and upper and lower thresholds for photosynthesis (Fig. 2A). The optimum temperature and the absolute value of A_{sat} will vary markedly between species as

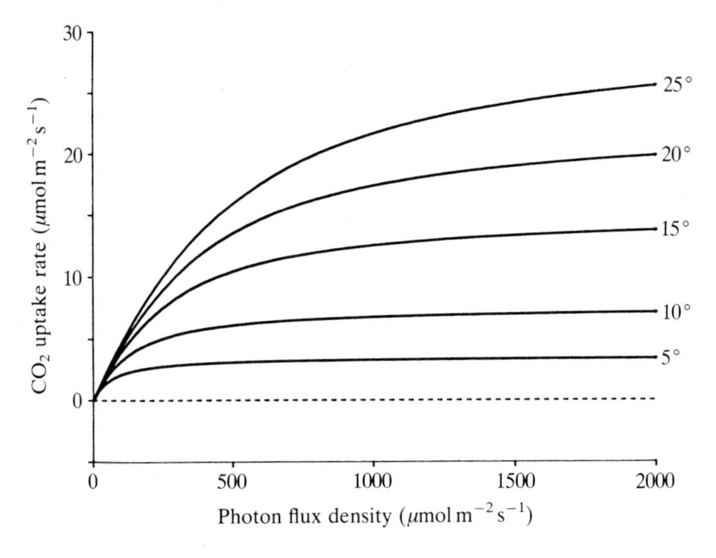

Fig. 1. The response of CO_2 uptake rate by young maize (*Zea mays*) leaves to photon flux density at five leaf temperatures. Curves are non-rectangular hyperbolae (Thornley, 1976) fitted to the data of Long *et al.* (1983) and corrected for dark respiration rates.

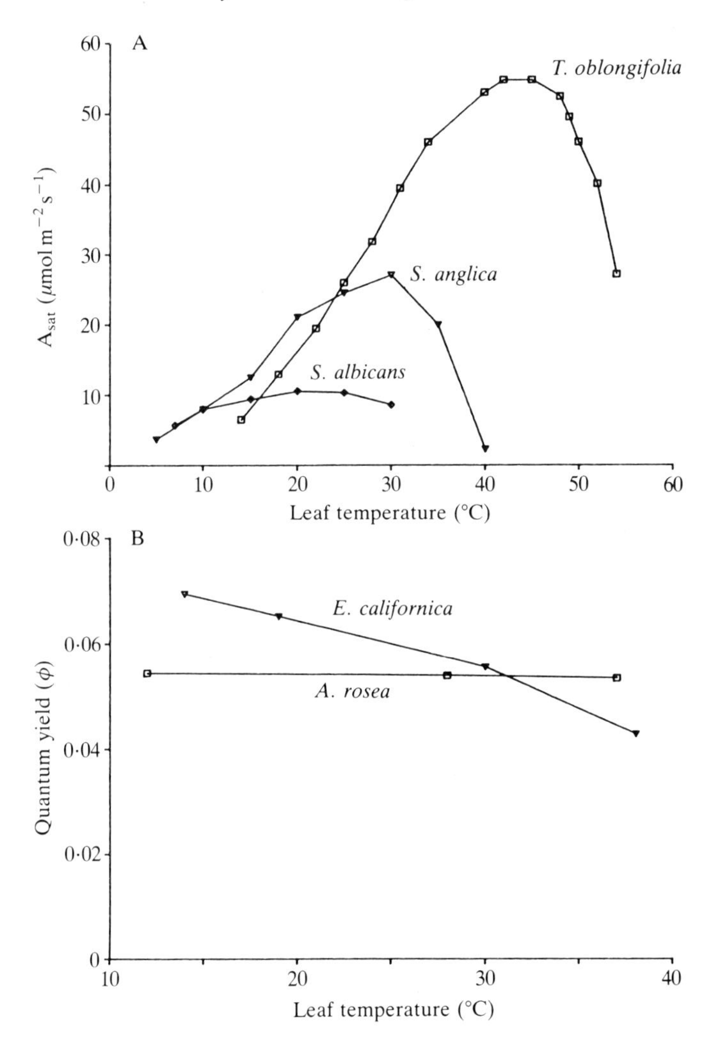

Fig. 2. A. The response of light-saturated rates of CO_2 uptake (A_{sat}) to leaf temperature in the desert C_4 dicot. *Tidestromia oblongifolia* (□), the cool temperate C_4 grass *Spartina anglica* (▽), and in the arctic-alpine C_3 grass *Sesleria albicans* (◇). Redrawn from the data Long *et al.* (1975) and Berry & Bjorkman (1980). B. The response of quantum yield (ϕ) for CO_2 uptake measured under light limiting conditions in the C_3 dicot. *Encelia californica* (▽) and the C_4 dicot. *Atriplex rosea* (□). Redrawn from the data of Ehleringer & Bjorkman (1977).

well as with climatic and edaphic growth environments (Fig. 2A,B). Quantum yield shows far less variation. In C_3 plants the quantum yield of CO_2 reduction decreases with increasing temperature because of the increase in the affinity of Rubisco for O_2 relative to CO_2 (Servaites & Ogren, 1978). In C_4 plants, and in C_3 plants when measured in an atmosphere of 1 % O_2, ϕ shows virtually no response to temperature above the ranges which will induce chilling injury and below that which will cause heat stress (Ehleringer & Bjorkman, 1977; Pearcy & Ehleringer,

1984). Thus, it is apparent that values of A_{sat} and ϕ reflect very different aspects of photosynthesis. A_{sat} is partially restricted by the diffusion of CO_2 through the boundary layer and stomata. Within the mesophyll, A_{sat} is limited by the amount of active carboxylase at low internal CO_2 concentrations (c_i) and by the rate of regeneration of CO_2 acceptor (J) at high values of c_i (Farquhar *et al.* 1980). Many plants appear to maintain a c_i during photosynthesis close to the point of transition between limitation by carboxylase and rate of regeneration of CO_2 acceptor, so that both processes will be co-limiting. J may be affected by a range of processes and although it is widely regarded to reflect the maximum coupled rate of chloroplast electron transport it could also be limited by the steps in photosynthetic carbon metabolism leading to regeneration of the acceptor. In theory, ϕ for CO_2 reduction reflects the maximum efficiency of utilization of captured photons in the reduction of CO_2. It will be influenced by any inefficiencies of excitation energy transfer within the chloroplast membrane as well as by any inefficiencies in the synthesis of ATP and NADPH. Also ϕ will be influenced by the requirements of ATP and NADPH within the pathway used for CO_2 reduction. As mentioned previously a rise in temperature increases the relative affinity of Rubisco for O_2, thereby increasing the proportion of carbon entering the photosynthetic carbon oxidation cycle and hence the demand for ATP and NADPH. As a result, ϕ of CO_2 reduction declines with temperature in C_3 plants. In C_4 plants, where oxygenase activity is inhibited, ϕ is independent of temperature, within the range of temperatures allowing normal growth. However, C_4 plants require more ATP for the assimilation of CO_2 than do C_3 plants. In consequence, when oxygenation is inhibited by low temperatures or under conditions of high atmospheric CO_2, C_3 plants show a higher ϕ than C_4 plants (Fig. 2). Since ϕ is the initial slope of the response of CO_2 uptake to absorbed light it cannot by definition be affected by either capacity for electron transport or carbon metabolism and thus is not influenced by the thermodynamic effects of temperature on these processes.

The emphasis on light-saturated rates of CO_2 uptake may be misleading when considering the responses of the canopies of plants and crops. Although A_{sat} may show a marked response to temperature, the response of a canopy of leaves may be considerably flatter. This is because in a crop canopy with a high leaf area index and a spherical arrangement of leaves, most of the leaves will be shaded from direct sunlight for most of the day. A_{sat} will only be of significance to the upper leaves of the canopy and then only in bright light. For the lower canopy all of the time and for the upper canopy on cloudy days, ϕ will dominate the rate of photosynthesis.

Photoinhibition and temperature stress

Exposure of plants to temperatures close to the upper or lower limits for photosynthetic CO_2 uptake does not only lower the instantaneous photosynthetic rate, but will often damage the capacity of leaves to assimilate CO_2 on return to optimal temperatures. The temperatures which will damage photosynthetic capacity will clearly vary from species to species, and depend on the environmental pre-history. Damage to photosynthetic capacity from chilling, i.e. exposure to temperatures in the range 0–12°C (Powles *et al.* 1983), has received considerable attention for the so-called chilling sensitive or warm weather crops, i.e. species of tropical or subtropical origin which have been introduced to temperate regions both as field crops, e.g. maize and Indian beans (Long *et al.* 1983; Powles *et al.* 1983) and protected crops, e.g. tomato and cucumber (Martin & Ort, 1985; Long *et al.* 1987). Chilling damage may also be of significance in evergreen crops of the Mediterranean, where leaves are subjected to chilling temperatures for much of the winter (Bongi & Long, 1987). Chilling-tolerant crops may show similar damage to photosynthetic capacity when exposed to temperatures just below 0°C; Scots pine (Strand & Oquist, 1985), winter rape and field beans (Farage & Long, unpublished data). The character of chilling and freezing damage is strongly influenced by interaction with light (see also Krause *et al.* 1988). Exposure of leaves to low temperature at night typically affects photosynthesis by lowering A_{sat} in the subsequent day; however, at chilling temperatures ϕ may be unaffected by the same treatment. In both tomato and olive it is possible to produce significant reductions in A_{sat} following chilling in darkness without any significant reduction of ϕ (Martin & Ort, 1982; Bongi & Long, 1987). Prolonged chilling in darkness (i.e. >24 h) will produce significant reductions in ϕ (Martin & Ort, 1982). Exposure of Scots pine needles at −7°C to high light for 3 h had effected no significant reduction in the subsequent A_{sat} relative to dark controls, whilst the same treatment for 1 h effected a 61 % decrease in ϕ relative to 25 % in needles frozen in darkness (Strand & Oquist, 1985). The presence of light during chilling greatly increases damage to A_{sat} and also depresses ϕ (Fig. 3; Table 1). The reduction in CO_2 uptake following this treatment is proportional to the light level during chilling, showing this damage to be a form of photoinhibition. The term photoinhibition is used in this article to indicate a light-dependent inhibition of photosynthetic CO_2 reduction. In chilling, this refers to the increased damage to photosynthesis when leaves are exposed to low temperatures in the presence of light, relative to controls chilled in darkness. Chilling-dependent photoinhibition has usually been demonstrated by short-term exposures (<24 h) to low temperatures and high light (*ca.* 1000 μmol m^{-2} s^{-1}). However, it has been shown that in 'chilling-tolerant' Scots pine prolonged exposure (6 weeks) to 4°C in low light (50 μmol m^{-2} s^{-1}) produced significant reductions in ϕ of *ca.* 40 % and changes in the photosynthetic apparatus characteristic of photoinhibition (Oquist & Strand, 1985).

Analysis of the response of CO_2 uptake rate (A) to the intercellular CO_2 concentration (c_i) suggests that chilling-dependent photoinhibition in maize and cucumber, and freezing-dependent photoinhibition in Scots pine, reduces both carboxylation efficiency (r_m) and A_{max} (Strand & Oquist, 1985; Long *et al.* 1987). Fig. 4 shows the recovery of r_m, A_{max} and ϕ in maize leaves, following chilling in high light. Carboxylation efficiency recovers within 1 h of return to non-chilling temperatures, indicating that reduction in r_m resulted from deactivation of the enzyme rather than a loss of protein. A_{max} recovered within 3 h of chilling,

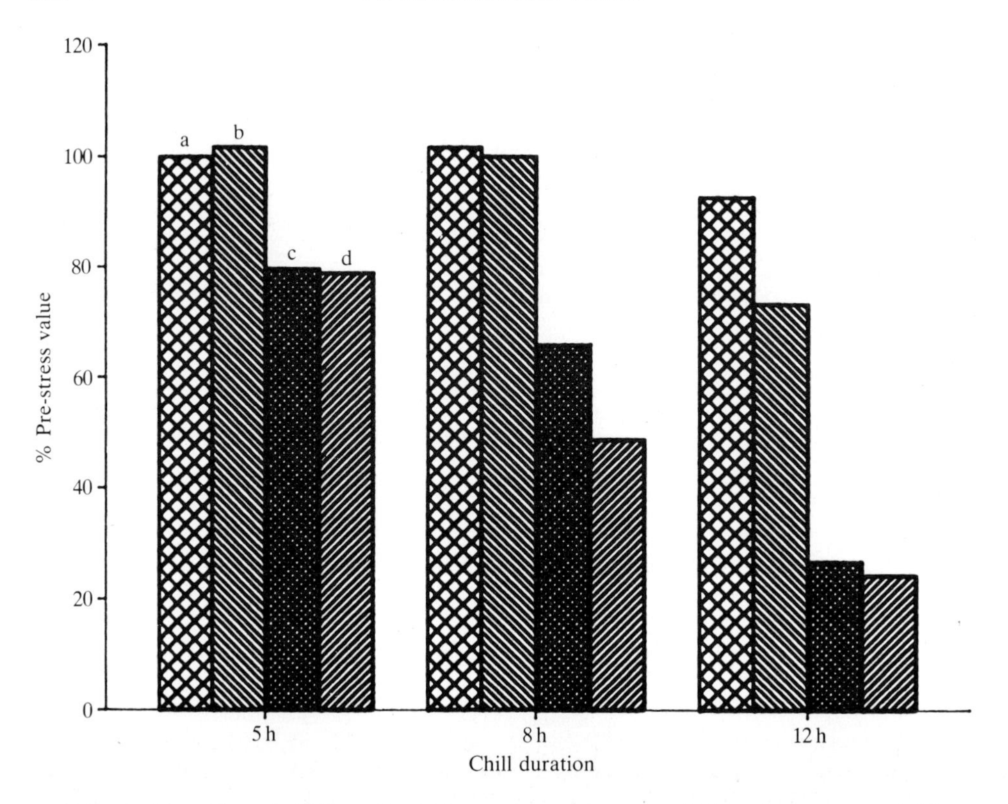

Fig. 3. The effect of chilling olive (*Olea europaea*) leaves at 5 °C on the quantum yield (ϕ) and light-saturated CO_2 uptake rate (A_{sat}), measured at 20 °C after chilling treatment. The first two columns show the changes in ϕ (a) and A_{sat} (b) of plants chilled in darkness. The second panel shows the changes in ϕ (c) and A_{sat} (d) when leaves were chilled in high light, i.e. a photon flux density of 1500 μmol m^{-2} s^{-1}. Data of Bongi & Long (1987).

possibly suggesting repair to damage to one or more components of the electron transport pathway. However, ϕ showed only a slow recovery and had not regained its pre-stress level even after 24 h. In olive, ϕ was similarly slow to recover following photoinhibition under chilling conditions. In contrast, recovery of chilling-dependent photoinhibition of ϕ in tomato was significantly more rapid than recovery of A_{sat} (Martin & Ort, 1985).

Most studies of chilling damage have concerned photosynthesis of individual leaves of plants under controlled environmental conditions. There is little information on the significance of this damage to crop canopies or to the individual plants in the field. In maize, light-dependent reductions in quantum yield were detected on cold bright mornings in the spring in southeast England (Fig. 5). These drops in quantum yield correlated with decreases in the efficiency of conversion of intercepted light by the crop into dry matter (Farage & Long, 1987). An assessment of the significance of chilling damage to photosynthesis by crop canopies may be achieved by the application of models to canopy photosynthesis.

Table 1. *Effects of chilling on the quantum yield of photosynthesis in different species*

Species	Chilling conditions: Temp. (°C)	Photon flux density $\mu mol\,m^{-2}\,s^{-1}$	duration (h)	% reduction ϕ
Cucumis sativa[1]	5	0	3	*n.s.*
Cucumis sativa[1]	5	1500	3	84
Gossypium hirsutum[2]	8	2000	3	42
Gossypium hirsutum[2]	8	70	3	*n.s.*
Lemna gibba[3]	3	1750	2	98
Lycopersicon esculentum[4]	4	1000	6	54
Lycopersicon esculentum[2]	7	2000	3	43
Nerium oleander[2]	6	2000	3	*n.s.*
Olea europaea[5]	5	1850	5	16
Olea europaea[5]	5	1850	12	73
Olea europaea[5]	5	90	12	*n.s.*
Phaseolus vulgaris[2]	5	2000	3	65
Phaseolus vulgaris[2]	5	70	3	*n.s.*
Pinus sylvestris[6]	−7	50	1	25
Pinus sylvestris[6]	−7	1300	1	61
Zea mays[7]	5	1500	6	56

The quantum yield of CO_2 assimilation (ϕ) is given as a percentage of control values. *n.s.* indicates no statistically significant change. [1] Long *et al.* (1987); [2] Powles *et al.* (1983); [3] Ogren *et al.* (1984); [4] Martin & Ort (1985); [5] Bongi & Long (1987); [6] Strand & Oquist (1985); [7] Long *et al.* (1983).

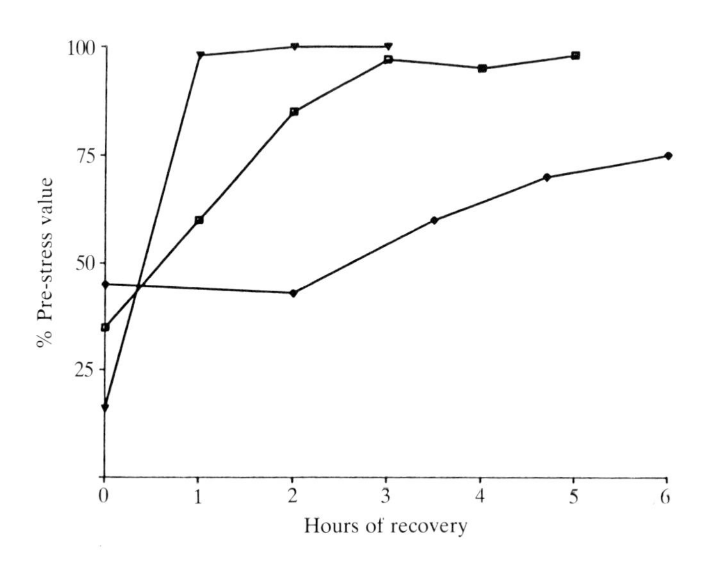

Fig. 4. Recovery of the light- and CO_2-saturated assimilation rate (□), carboxylation efficiency (▽) and quantum yield (◆) at 20 °C following chilling of young maize leaves for 6 h in high light, i.e. a photon flux density of 1500 $\mu mol\,m^{-2}\,s^{-1}$. Redrawn from Long *et al.* (1987).

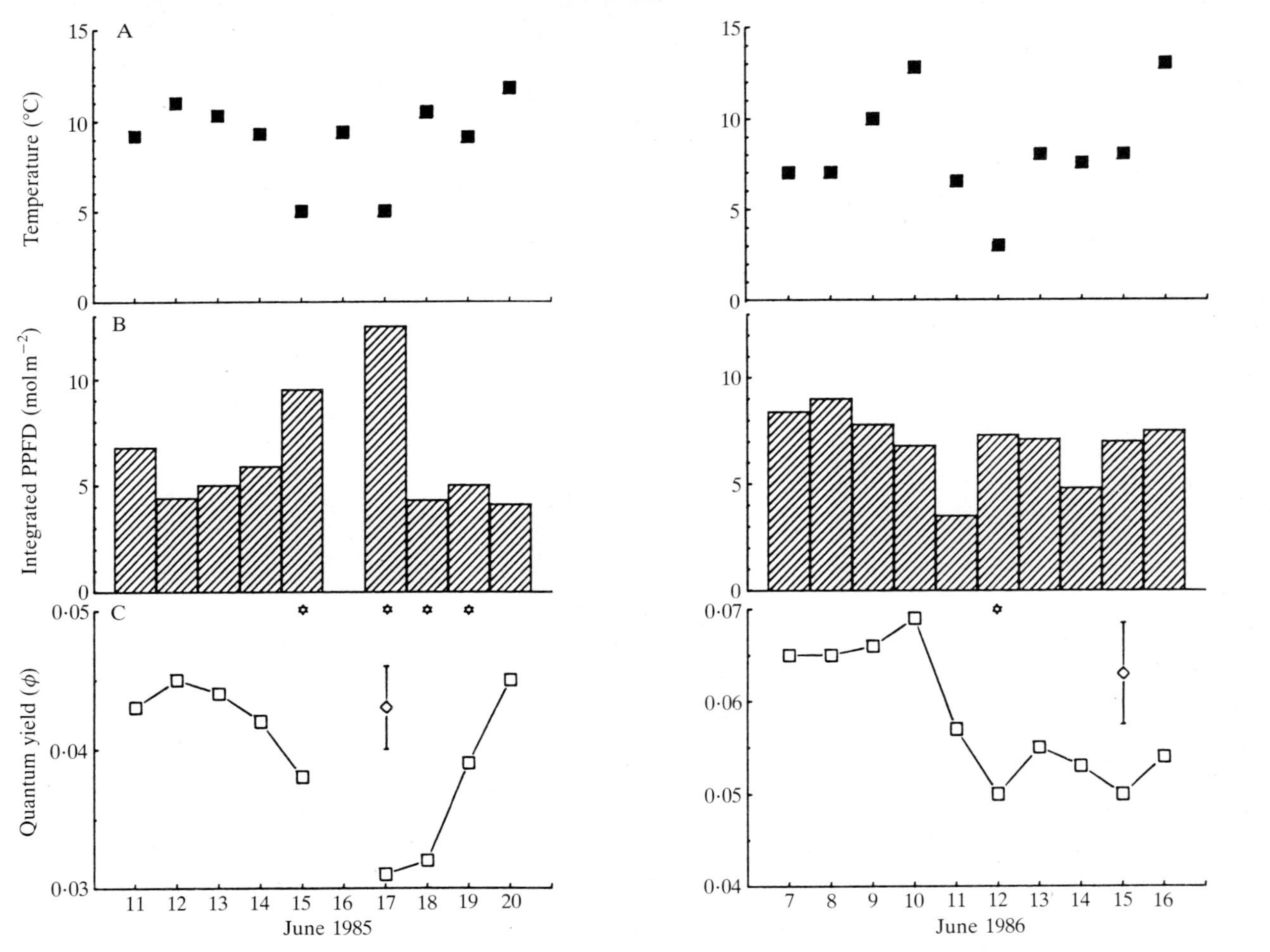

Fig. 5. A. Dawn canopy air temperature. B. Integrated photosynthetically-active photon flux density incident above the canopy during the 5 h following dawn. C. Quantum yield of CO_2 assimilation (ϕ) for *Zea mays* leaves in the field; the vertical bars indicate a A_{sat} ($P < 0{\cdot}05$). The asterisks show in 1985 significant differences in ϕ relative to 11 June and in 1986 when shaded and unshaded leaves showed a significant difference in ϕ.

Relating leaf effects to canopy performance

To assess the significance of changes in ϕ and A_{sat} at the leaf level to the crop level, a computer model for the calculation of maize canopy rates of CO_2 assimilation from rates for leaves was developed. Solar elevation was computed for 15-min intervals for 21 June at latitude 52.5° N, i.e. southern England, using the equations of Ross (1975). The incident radiation and ratios of diffuse: direct radiation were determined from solar elevation, the solar constant and atmospheric transmissivities equivalent to (i) a clear day and (ii) a day with 90 % cloud cover. The proportion of radiation in the photosynthetically active waveband was calculated from the tables given by Ross (1975) and mean energy of photons in this waveband for daylight was assumed to equal 2500 $kJ\,mol^{-1}$ (Kubin, 1971). The incident photon flux density, solar angle and proportion of diffuse light was then used in a model of light distributions in maize canopies provided by T. Arkebauer and J. M. Norman (University of Nebraska) to predict the proportions of the canopy which will be sunlit and shaded, and the mean photon flux densities of these two categories, given the leaf area and angle (Norman, 1980). The light levels of the sunlit and shaded leaf areas were then used to estimate the current leaf photosynthetic rates from the non-rectangular hyperbolic response of photosynthesis to light described by the quadratic equation of Thornley (1976). This equation is solved for its positive root.

$$A = \frac{\phi Q + A_{sat} - (\phi Q + A_{sat})^2 - 4\theta(\phi Q A_{sat})}{2\theta} \qquad \text{(Eqn 1)}$$

Where Q is the photon flux density incident on the leaf and ϕ, θ and A_{sat} are the quantum yield, convexity coefficient and light saturated rate of CO_2 uptake.

For the simulation, the parameters of the light response curve of unstressed leaves were assumed to be $\phi = 0{\cdot}055$, $\theta = 0{\cdot}5$ and $A_{sat} = 26\ \mu mol\,m^{-2}\,s^{-1}$, these values approximating to those for maize leaves grown under non-stress conditions (Long *et al.* 1983). When calculated for the sunlit and shaded foliage the two values of A were multiplied by the leaf areas of each category to obtain the total canopy photosynthetic rate. The upper curves of Figs 6A and 6B illustrate the computed canopy photosynthetic rates using these parameters.

To examine the relative significance of reduction in ϕ and A_{sat}, either or both were reduced by 50 % – this being similar to the reductions seen during chilling in high light for 6 h. The simulation was conducted both for a hypothetical bright June day in southern England and for a more realistic dull day. On a bright day, the photosynthesis of a canopy of LAI = 4 in which ϕ has been reduced by 50 % is significantly lower throughout the day. Although a 50 % reduction in A_{sat} results in a greater reduction at mid-day than does a 50 % reduction in ϕ, the converse is true in the 3 h following dawn and preceding dusk (Fig. 6A). However, on a dull day (total radiation 10 $MJ\,m^{-2}$) 50 % reduction in ϕ results in a very much lower canopy rate throughout the day (Fig. 6B). By integrating the area under the plots of canopy photosynthesis with time the daily integral of canopy photosynthesis is obtained (Fig. 7). Fig. 7 illustrates the interactions of LAI with the effects of

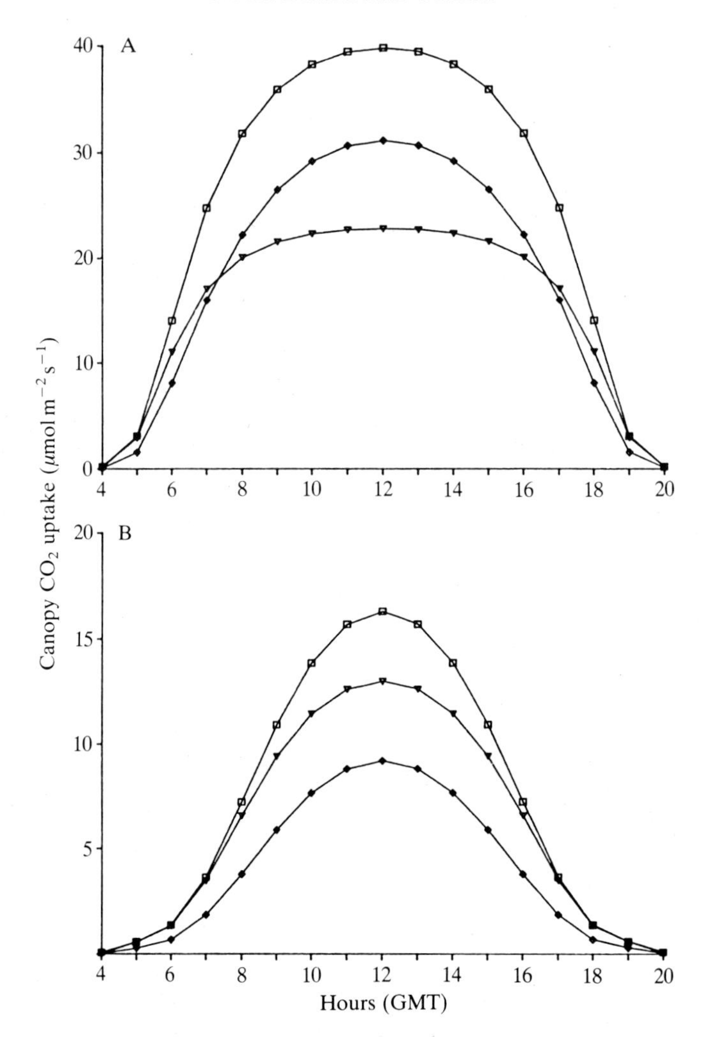

Fig. 6. A. Simulated hourly rates of gross CO_2 uptake per unit ground area for a maize canopy of leaf area index 2 and leaf inclination index 0, on 21 June at latitude 52° N assuming clear skies giving a total radiation receipt of 28 MJ. The upper curve (□) assumes that all leaves in the canopy show light-saturated rates of CO_2 uptake (A_{sat}) of 26 μmol m^{-2} s^{-1} and a quantum yield (ϕ) of 0·055. The two lower curves illustrate the effect of a 50% inhibition of A_{sat} (▽) and a 50% inhibition of ϕ (◇). B. As for A but simulated for a cloudy day and a total radiation receipt of 10 MJ.

reduction in ϕ and A_{max} with respect to daily integrals of canopy photosynthesis. In high light and with a low LAI a 50% reduction in A_{sat} results in a 40% reduction in canopy photosynthesis compared to a reduction of 25% in canopy photosynthesis resulting from a 50% reduction in ϕ. At higher LAIs the effects of ϕ and A_{sat} are similar for high light conditions (Fig. 7A). However, high light conditions reflect the most favourable conditions to the influence of A_{sat}, i.e. a clear sky on mid-summer's day. For the rest of the year and under duller

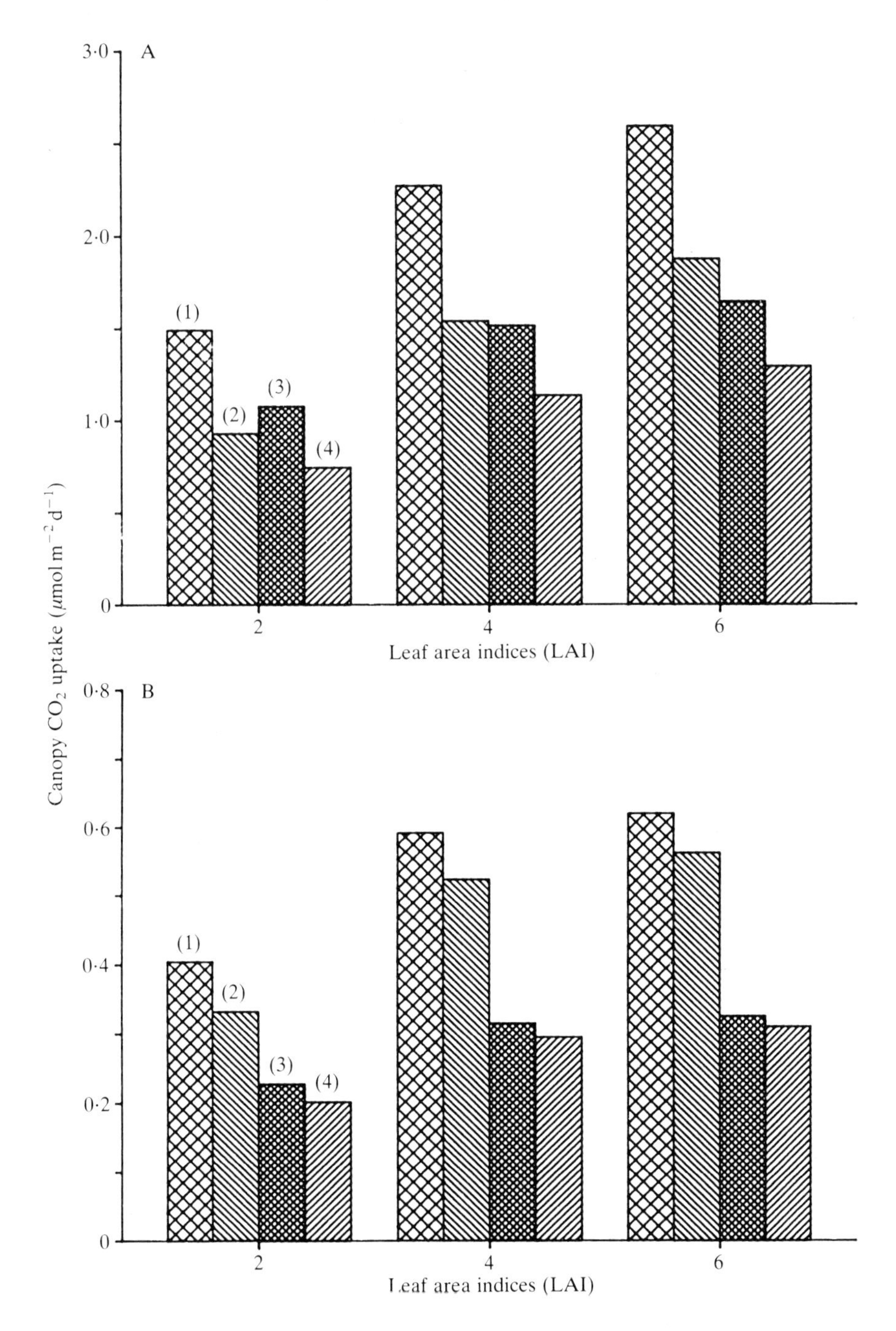

Fig. 7. A. Simulated daily integrals of gross CO_2 uptake per unit of ground area for maize canopies with a leaf inclination index 0 and leaf area indices (LAI) 2, 4 and 6. Parameters of leaf photosynthesis are as in Fig. 6. The four columns illustrate integrals, assuming: (1) no inhibition of either A_{sat} or ϕ at the leaf level, (2) 50 % inhibition of A_{sat}, (3) 50 % inhibition of ϕ, and (4) 50 % inhibition of both A_{sat} and ϕ. B. As for A but simulated for a cloudy day and a total radiation receipt of 10 MJ.

conditions the influence of ϕ relative to A_{sat} is increased. Indeed, when the simulation is conducted for more typical overcast sky conditions, it is clear that even a 50% reduction in A_{sat} has a minimal effect on canopy photosynthesis, whilst a 50% reduction in ϕ produces a much greater reduction in canopy photosynthesis (Fig. 7B). Indeed, when ϕ is reduced by 50%, simultaneous reduction in A_{sat} has remarkably little additional effect (Fig. 7B). These results show clearly that, except under conditions of bright sunshine and low LAI, reduction in ϕ following photoinhibition will have a far greater effect on canopy photosynthesis than reduction in A_{sat}.

Despite the emphasis which has been placed in previous reviews on the effects of temperature on A_{sat}, in the context of canopy photosynthesis effects of temperature on ϕ are likely to be of greater significance. This conclusion is drawn here from the example of maize canopy photosynthesis, but would be applicable to other crop canopies. Broad-leaved crops with more planophile canopies are likely to show dominance of ϕ at lower leaf area indices, since the proportion of shaded leaves will be increased. What mechanisms can explain the observed reductions in ϕ observed during chilling or freezing of leaves in the presence of light? The quantum yield (ϕ) is determined from the initial slope of the response of A to absorbed photon flux density. It is assumed, since A increases linearly with Q over this portion of the curve, that neither availability of acceptor for CO_2 fixation, active carboxylase nor CO_2 can be limiting. The value of ϕ will therefore depend on the quantity of absorbed light energy that is utilized in net CO_2 reduction. In C_3 species stomatal closure could reduce ϕ by restricting CO_2 diffusion, increasing the intercellular O_2:CO_2 ratio and hence the ratio of photorespiration: photosynthesis. However, stomatal limitation following chilling and freezing treatments is often unchanged or even diminished suggesting that this is unlikely to account for depressed ϕ (Martin & Ort, 1985; Strand & Oquist, 1985; Bongi & Long, 1987). In these studies stomatal limitations or conductance g_s were calculated as an average for the leaf or leaf portion under examination. Conclusions of a lack of stomatal effects assume that all parts of the leaf behave similarly. If patches of stomata close, ϕ could be reduced by CO_2 restriction in both C_3 and C_4 leaves, whilst g_s for the leaf as a whole may appear unaffected. This possibility of patchiness in stomatal response has been used to explain a reduction in ϕ of Xanthium leaves following abscisic acid treatment, which was not apparent when ϕ was determined at an elevated CO_2 concentration (see Farquhar, 1988). If patchiness in stomatal response does follow chilling treatment then a decrease in the convexity of the response of A to intercellular CO_2 concentration (c_i) would be expected, since variability in actual c_i across the leaf must increase. However, the curves of Long *et al.* (1987) for cucumber and for maize, and of Strand & Oquist (1985) for Scots pine show no evidence of a decrease in the convexity following exposure to high light at low temperatures. The reduction of ϕ must, therefore, be assumed to result from change in the efficiency of the thylakoid membrane in providing ATP and reducing power for CO_2 reduction. The remainder of this review will consider the possible mechanisms which could produce such changes.

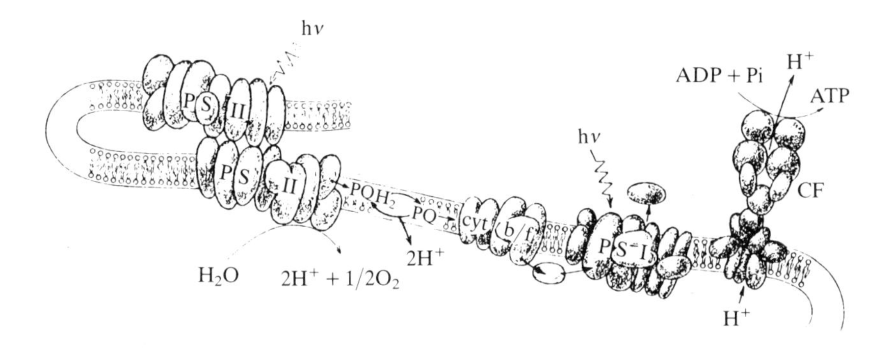

Fig. 8. A schematic sketch of the thylakoid membrane depicting the integral membrane polypeptide complexes involved in electron and proton transport. CF, coupling factor; cyt *b/f*, cytochrome b_6/f complex; PQ, plastoquinone; PQH_2, plastoquinol; PSI, photosystem I; PSII, photosystem II. The light-harvesting chlorophyll *a/b* protein complex (LHCII) is considered here to be part of PSII.

Thylakoid function and quantum yield

As commented upon briefly above, ϕ can be influenced by factors determining the conversion of absorbed light to chemical energy (ATP and NADPH), a process performed by the photosynthetic apparatus of the thylakoid membranes, which is diagrammatically represented in Fig. 8 (reviewed: Ort, 1986; Baker & Webber, 1987). Transduction of light to chemical energy by thylakoids can be considered to involve four distinct phases: (i) light absorption and transfer to the photochemical reaction centres of photosystems I and II (PSI and PSII); (ii) electron transfer, mediated by excited reaction centres, from electron donors to acceptors within the PSI and PSII polypeptide complexes; (iii) electron transfer between PSII and PSI and the associated transport of protons across the thylakoid membrane, and (iv) the conservation of a proton electrochemical potential differences across the thylakoid membrane and the transduction of this electrochemical potential via the membrane-bound ATPase enzyme complex (the coupling factor) into ATP. The potential involvement of each of these phases of light energy conversion in determining temperature-induced reductions in ϕ is examined below.

Light absorption and transfer

The light-harvesting apparatus of each photosystem consists of arrays of chlorophyll molecules and accessory pigments capable of transferring absorbed light energy to the photochemical reaction centre (reviewed: Baker & Webber, 1987). In higher plants the chlorophylls, and associated carotenoid accessory pigments, are intimately associated with proteins, which are specifically organized with a photosystem complex (see Thornber, 1986). In the thylakoids PSII complexes are often found to be associated with a light-harvesting chlorophyll a/b complex, LHCII, that has no photochemical activity and serves primarily as a secondary antenna for PSII. The interaction between LHCII and PSII appears to be physiologically important and regulated by phosphorylation of LHCII polypeptides (reviewed: Baker & Webber, 1987). It is believed that on phosphorylation of LHCII by a thylakoid protein kinase, the surface negative charge density of LHCII is increased causing a disconnection of LHCII from PSII, which in turn causes a decrease in the transfer of excitation energy from LHCII to PSII (Haworth *et al.* 1982; Horton, 1983; Bennett, 1984; Barber, 1985). This phenomenon is thought to be the basis of the State I–State II transition (Barber, 1982) and may regulate the distribution of excitation energy distribution between PSI and PSII (Allen *et al.* 1981; Haworth *et al.* 1982; Horton, 1983; Bennett, 1984; Barber, 1985). A change in the distribution of absorbed light energy at non-saturating light levels may be important in determining the rate of non-cyclic electron transport relative to cyclic, and thus the ratio of ATP: NADPH generated (Horton, 1985). On disconnection of phosphorylated LHCII from PSII, this portion of the LHCII population migrates from the PSII-rich appressed thylakoid membranes in the granal stacks to the non-appressed, PSI-rich stromal membranes, and in some cases can act as a secondary antenna for PSI, thus increasing the distribution of excitation energy to PSI (Baker & Webber, 1987). However, it is not necessary to invoke an increased interaction of LHCII with PSI to effect a change in the excitation energy distribution to PSI relative to PSII on phosphorylation of LHCII. Disconnection of phosphorylated LHCII from PSII alone will reduce the excitation rate of PSII relative to PSI (Baker & Webber, 1987). Dephosphorylation of LHCII by a membrane-bound phosphatase enables the previously phosphorylated LHCII to re-associate with PSII and restore a higher probability for PSII excitation (Haworth *et al.* 1982; Horton, 1983; Bennett, 1984; Barber, 1985).

Chilling of maize leaves has been shown to reduce the ability of thylakoids to respond to changes in the relative excitation of PSI and PSII (Fig. 9). Excitation of thylakoids from control leaves with 710 nm radiation, which preferentially excites PSI, produces a rapid decrease in PSII fluorescence followed by a slow increase, which is indicative of an increase in the excitation of PSII relative to PSI, i.e. a State II to State I transition. This phenomenon is inhibited by NaF (Fig. 9), an inhibitor of the phosphatase responsible for dephosphorylating LHCII. Thylakoids isolated from leaves chilled at 5°C for 5 h under a photon flux density of 1500 μmol m^{-2} s^{-1} show a considerable reduction in the ability to respond to changes in the relative excitation of PSI and PSII (Fig. 8). The chill-induced loss of this regulation of excitation energy distribution between the photosystems I and II would be expected to have consequences for the stoichiometry of ATP to NADPH production by the thylakoids and thus contribute to a decrease in ϕ.

Low-temperature-induced modifications to the light-harvesting apparatus may result from changes in the synthesis of chlorophyll–protein complexes and their targeting to the functional sites within the membrane. A light-dependent, chill-induced modification to LHCII in maize demonstrates an effect on chlorophyll protein synthesis. Thylakoids of maize leaves exposed to temperatures below 12°C in the light will accumulate a 31 kDa polypeptide, which is present in control leaves in only trace amounts (Fig. 10; Hayden *et al.* 1986; Covello *et al.* 1988). This

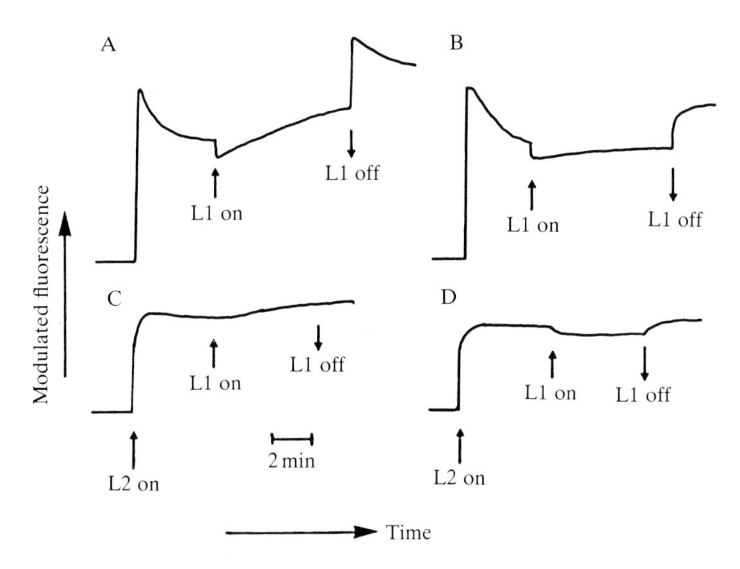

Fig. 9. Effect of 710 nm light (L1), which preferentially excites PSI, on the modulated fluorescence emission from PSII in thylakoids isolated from control (A) and maize leaves exposed to 5 °C at a photon flux density of 1500 μmol m^{-2} s^{-1} for 6 h (C). The effect of inhibiting the thylakoid phosphatase, responsible for dephosphorylation of phosphorylated LHCII, with NaF is shown for control, B, and stressed, D, thylakoids. ATP (200 μM) was present in the reaction media (from Percival *et al.* 1987).

31 kDa polypeptide isolates with LHCII from the thylakoids (Fig. 10; Hayden *et al.* 1986) and appears to be an unprocessed precursor polypeptide of the 29 kDa chlorophyll–protein complex, CP29 (Hayden *et al.* 1988). The polypeptide binds chlorophyll and is immunologically related to CP29; its quantity in the thylakoid changes in an antiparallel fashion with the amount of CP29 (Hayden *et al.* 1988). Although CP29 is considered to be a minor chlorophyll protein, comprising of only *ca.* 8 % of the total thylakoid chlorophyll (Thornber *et al.* 1986), it has been suggested that it plays an important role in mediating excitation energy transfer between LHCII and PSII (Staehelin, 1986). Chlorophyll fluorescence emission spectra at 77 K of LHCII, containing the 31 kDa polypeptide, isolated from chilled maize leaves exhibit a marked difference from LHCII isolated from control leaves (Fig. 11), indicative of a modification in the energetics of the complex, which may be associated with a decrease in energy transfer between LHCII and PSII (Hayden *et al.* 1986). On returning chilled maize leaves, containing the 31 kDa polypeptide, to normal growth temperature the polypeptide disappears within 2 h; however, a significant modification of the 77 K fluorescence emission spectrum of LHCII still remains suggesting that the modification produced by the insertion of the polypeptide into LHCII is not recoverable upon disappearance of the polypeptide. Synthesis of new LHCIIs to replace the modified LHCIIs may be necessary to effect a complete recovery of the light-harvesting apparatus (see Hayden *et al.* 1988). Such a requirement for synthesis of new LHCII may represent a significant drain on the products of photochemistry, diverting ATP away from CO_2

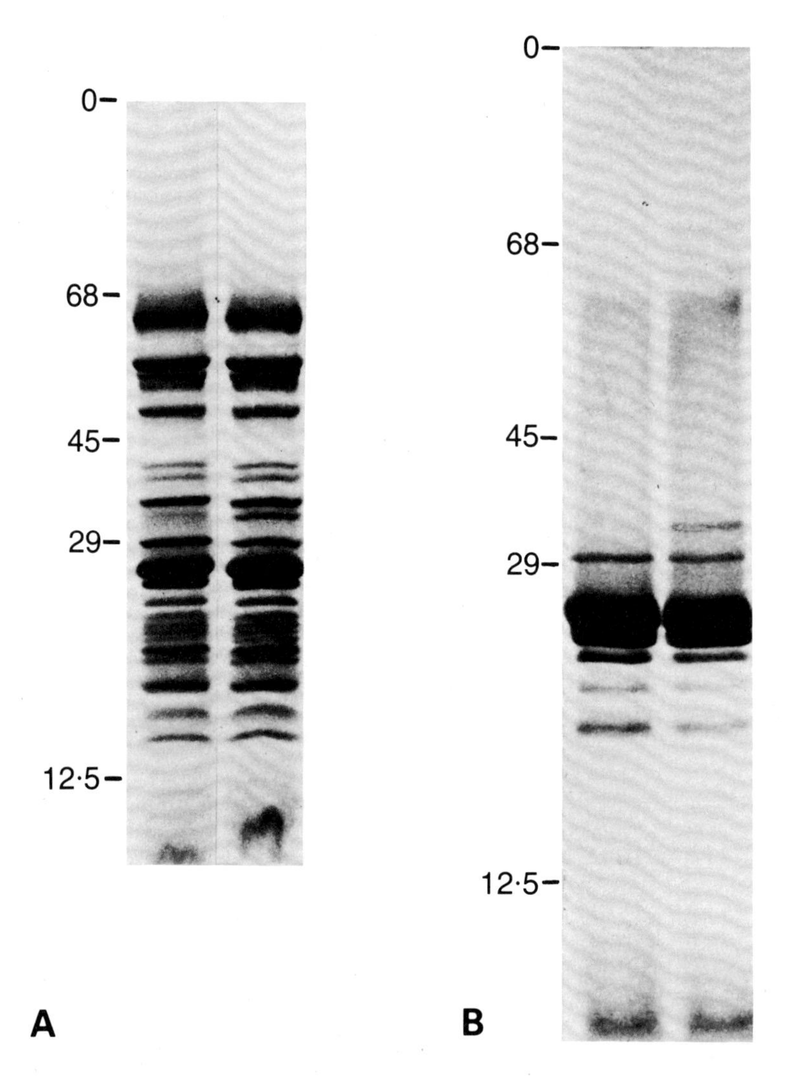

Fig. 10. Polypeptide profiles of thylakoids, A, and purified LHCII, B, isolated from control (left lane) and maize leaves chilled at 5°C under a photon flux density of 1500 μmol m^{-2} s^{-1} for 6 h (right lane). (From Hayden *et al.* 1986.)

metabolism and thus depressing ϕ. Although the exact relationship between such perturbation of the light-harvesting apparatus and ϕ has yet to be determined for maize exposed to low temperatures, it is evident from the preceding discussion that modification or damage to the light-harvesting apparatus can potentially contribute to a loss of quantum yield of carbon assimilation.

Primary electron transport

It is frequently observed that PSII is particularly vulnerable to damage when plants are exposed to environmental stresses. Electron transport from the oxidation of water to the reduction of the secondary quinone acceptor, Q_B, is

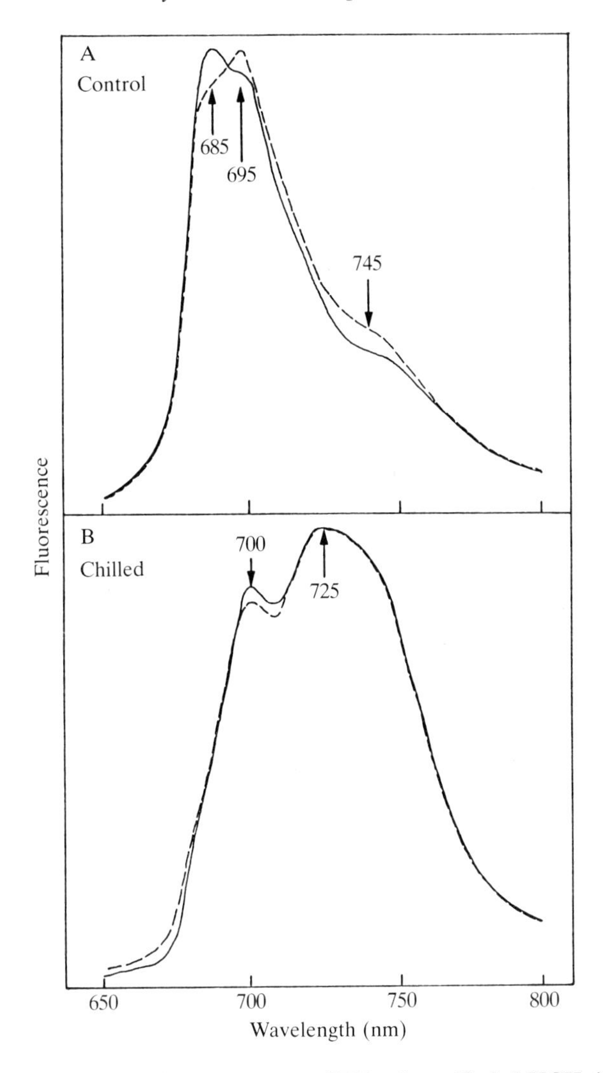

Fig. 11. Fluorescence emission spectra at 77 K of purified LHCII isolated from thylakoids of, A, control and, B, chilled maize leaves on excitation with 440 (——) and 470 (---) nm radiation. (From Hayden *et al.* 1986.)

carried out by the PSII polypeptide complex (Satoh & Butler, 1978; Diner & Wollman, 1980). Each PSII complex operates independently, with no electron sharing between complexes, but eventually acts to reduce a common plastoquinone acceptor pool. This occurs via plastoquinone bound to the Q_B protein of the PSII complex which on reduction is released and diffuses away from the complex to become part of the free plastoquinone pool within the membrane (reviewed: Ort, 1986). There is a strict functional association between PSII reaction centres and the water-oxidizing polypeptides associated with the PSII complex such that failure of one water-oxidizing system leads directly to the effective loss of photochemical activity of one PSII reaction centre. Such a loss of PSII reaction centre activity can be restored by supplying PSII with an artificial electron donor, e.g. catechol or diphenylcarbazide. Chilling tomato plants in the dark produces some loss of PSII electron transport from water to phenylenediamine (Fig. 12), which can be largely restored by the addition of catechol (Table 2). Chilling in the light produces a more rapid loss of PSII activity in tomato (Fig. 13), which may

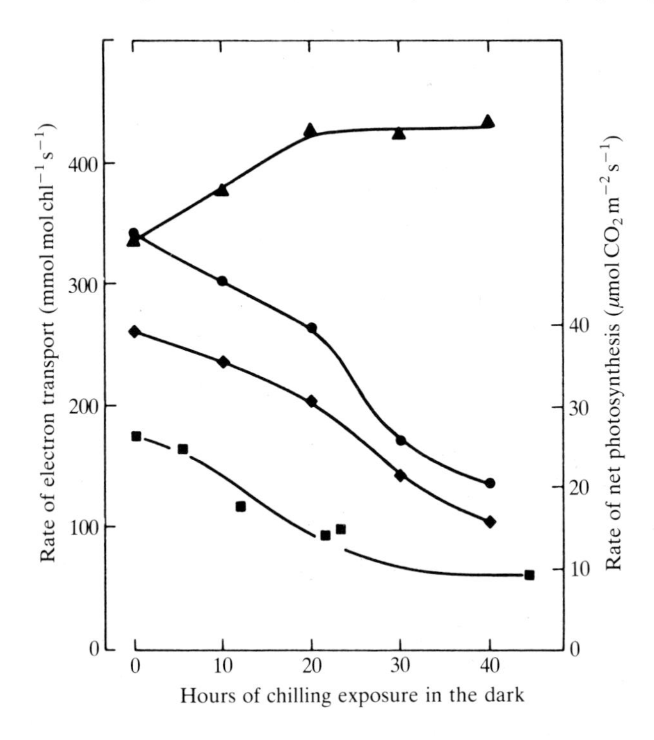

Fig. 12. Dependence of the rate of photosynthetic electron transport and of the rate of CO_2 assimilation on the chilling duration of intact tomato plants at 1°C in the dark. Electron transport rates were measured at saturating light intensity and CO_2 assimilation at a photon flux density of 1000 μmol m^{-2} s^{-1} in attached tomato leaves immediately after the end of the low temperature exposure. (◆) water to methyl viologen, (●) water to oxidized phenylenediamine, (▲) duroquinol to methyl viologen, (■) CO_2 assimilation (from Kee *et al.* 1986).

Table 2. *The rate of photosynthetic electron transport from water to methyl viologen measured in thylakoids in the presence and absence of catechol*

Treatment	Rate of electron transport (mmol electron mol chl^{-1} s^{-1}) −catechol	+catechol
Untreated	265 ± 20	275 ± 15
Dark chilled, 40 h	85 ± 5	180 ± 10

Thylakoids were isolated from untreated plants and from plants that had been chilled in the dark at 1°C for 40 h. 5-mM catechol and 0·5-mM ascorbate were added where the presence of catechol is indicated. (From Kee *et al.* 1986.)

involve a rather different type of damage to PSII complexes than that occurring during chilling in the dark. Indeed in maize, such light-dependent loss of PSII electron transport cannot be even partially restored by addition of diphenyl carbazide. This light- and chilling-induced loss of PSII activity occurred simultaneously with a loss of the ability of thylakoids to bind atrazine (Table 3), implying that at least part of the damage to PSII is associated with a loss of modification to the Q_B protein.

Damage to the Q_B protein is one of the best documented lesions associated with photoinhibition of PSII activity and can occur in many environmental stress situations when the photosynthetic apparatus cannot utilize the absorbed light energy through the normal metabolic and other dissipative pathways (Kyle *et al.* 1984; Bradbury & Baker, 1986). Loss of variable fluorescence is another diagnostic characteristic of photoinhibitory damage to PSII (Baker & Horton, 1987) and is particularly useful in detecting damage in intact leaf tissue. The large decrease in variable (F_v) relative to maximal (F_m) fluorescence after chilling maize leaves (Table 4) demonstrates that PSII photochemistry is damaged in situ. This suggests that decreases in PSII electron transport observed for isolated thylakoids are not attributable to stress-induced modifications to PSII that make PSII complexes more susceptible to damage during the isolation procedure.

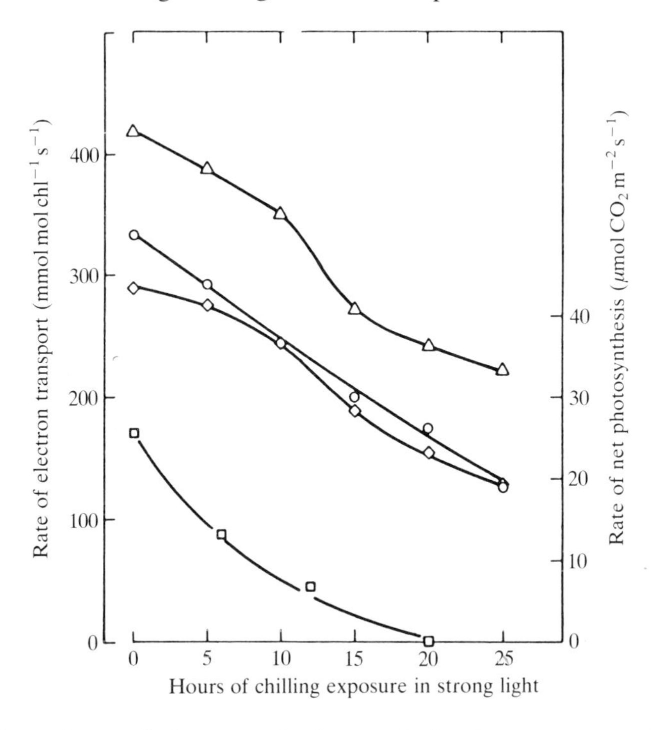

Fig. 13. Dependence of the rate of photosynthetic electron transport and CO_2 assimilation on the chilling duration of intact tomato plants at 4°C and a photon flux density of 1000 μmol m^{-2} s^{-1}. (◇) water to methyl viologen, (○) water to oxidized phenylenediamine, (△) duroquinol to methyl viologen, (□) CO_2 assimilation. (From Kee *et al.* 1986.)

Table 3. *Changes in thylakoid electron transport activities and atrazine binding capacity induced by exposing maize leaves to 5°C and a photon flux density of 1500 µmol m^{-2} s^{-1} for 6 h (from Percival* et al. *1987)*

Parameter	Control	Stress
Electron transport, water to methyl viologen		
(i) quantum yield (%)	100	52·0 ± 16·7
(ii) maximum rate (µmol O_2 mg^{-1} chl h^{-1})	153·9 ± 13·1	90·9 ± 19·8
Electron transport, water to silico molybdate		
(i) quantum yield (%)	100	64·4 ± 15·1
(ii) maximum rate (µmol SiMo mg^{-1} chl h^{-1})	62·4 ± 4·3	40·9 ± 10·5
Atrazine binding (nmol mg^{-1} chl)	1·61 ± 0·12	0·76 ± 0·08

Inhibitors of PSII electron transport, such as DCMU or atrazine, lower the quantum yield of CO_2 assimilation. Loss of PSII activity should have a substantially more severe effect on quantum yield than on the maximum rate of CO_2 assimilation because the rate-limiting step(s) in light- and CO_2-saturated photosynthesis is clearly not associated with PSII activity. Consequently, damage to a small fraction of PSII complexes would immediately be reflected in quantum yield measurements, whereas any effect on light- and CO_2-saturated rates of photosynthesis would be absent or substantially less than those on ϕ. A pertinent illustration of this point is given by a triazine-resistant biotype of *Senecio vulgaris*. The biochemical basis of this triazine resistance is a single base substitution in the chloroplast gene coding for the D_1 polypeptide of the PSII reaction centre (Hirschberg & McIntosh, 1983). The single amino-acid change resulting from the mutation produces a large decrease in the binding of the triazine molecules to the D_1 protein (Pfister & Arntzen, 1979; Mullet & Arntzen, 1981). However, this amino-acid change that confers resistance also dramatically alters the redox properties of the PSII reaction centre (Bowes *et al.* 1980), causing a markedly slower rate of electron transfer between the reaction centre quinone electron acceptors Q_A and Q_B (see Fig. 8). This slower rate of Q_B reduction is very likely the dominant factor in the 20-fold decrease in the equilibrium constant for electron transfer between Q_A and Q_B (Vermaas *et al.* 1984; Robinson, 1986). That is, for an

Table 4. *Effect of chilling maize leaves at 5°C under a photon flux density of 1500 µmol m^{-2} s^{-1} for 6 h on the quantum yield of PSII primary photochemistry as estimated from the ratio of variable to maximal fluorescence at 685 nm (F_v/F_m), measured from dark-adapted leaf tissue at 20°C infiltrated with 100 µM DCMU and at 77 K (from Baker* et al. *1983)*

Treatment	$(F_v/F_m)_{DCMU}$	$(F_v/F_m)_{77K}$
Control	0·79 ± 0·01	0·77 ± 0·02
Chilled	0·60 ± 0·04	0·63 ± 0·02

electron shared between Q_A and Q_B, the probability of the electron residing on Q_A at any given moment would be nearly 50 % greater in the resistant biotype. This would result in a large decrease in the probability of absorbed photons reaching an open PSII reaction centre and being utilized for photochemistry. The consequent increase in excitation energy loss from the PSII pigment matrices by non-photochemical dissipative processes would be associated with a proportional reduction in the quantum yield of PSII electron transport. Fig. 14 shows the manifestation of this PSII inefficiency on ϕ in susceptible and resistant biotypes of *Senecio vulgaris* (Ireland *et al.* 1987). A 20–25 % decrease in ϕ (see Fig. 14) is characteristic of triazine biotypes (e.g. Sims Holt *et al.* 1981; Ort *et al.* 1983; Ireland *et al.* 1988); however, the differences in photosynthetic performance between susceptible and resistant biotypes all but disappear at saturating light levels.

The magnitude by which a loss of PSII activity will reduce ϕ will be dependent upon the extent of excitation energy transfer between PSII complexes. The quantum efficiency of Q_A reduction would be predicted to be greater in a population of connected PSII complexes than in a PSII population in which intrasystem excitation energy transfer cannot occur, since excitation energy

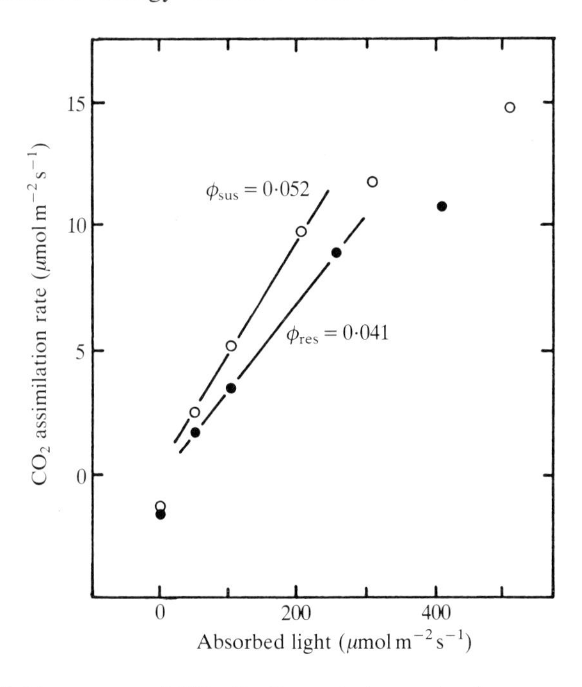

Fig. 14. The light response of CO_2 fixation rate in an atrazine-susceptible wild type and an atrazine-resistant mutant of *Senecio vulgaris*. Absorptivity of the leaf was measured with an integrating sphere allowing the flux absorbed at any incident photon flux density to be calculated. Points plotted are the mean of 3 replicates. The quantum efficiency of carbon assimilation (ϕ) is the slope of the linear portion of the relationship above the light compensation point (i.e. between the absorbed light fluxes of 50 and 150 μmol m^{-2} s^{-1}) and was obtained by a linear regression analysis. (From Ireland *et al.* 1988.)

reaching a closed reaction centre in a 'connected' PSII population has a probability for transfer to another, possibly open, PSII complex, whereas this is not the case in a 'disconnected' PSII population. This can be demonstrated by examining the quantum efficiency of Q_A reduction as a function of the proportion of closed PSII reaction centres (Fig. 15). Thus, it is evident that the magnitude of the reduction in ϕ due to damage to PSII will be dependent upon the degree of connectivity between PSII complexes; the effect of the damage will be minimized in membranes with a high degree of PSII connectivity. An additional complexity of inhibition of PSII complexes which may have a dramatic influence on ϕ is the possibility that PSII reaction centres, when exposed to excess light, can become strong, non-photochemical quenchers of excitation energy. Photochemically-inactive PSII reaction centres could thus still remain effective sinks for excitation energy and contribute to a decrease in the quantum yield of PSII electron transport. In this situation connectivity between PSII complexes could serve to accentuate the reduction in ϕ.

Inactivation of PSI reaction centres would also result in a decline in ϕ. A number of reports have suggested that PSI activity is considerably less affected than PSII after chilling both in the dark and the light (reviewed: Oquist & Martin, 1986); however, there is some evidence to suggest that PSI activity may be impaired to a similar degree to PSII when tomato leaves are chilled in the light (Fig. 13, Kee *et al.* 1986).

Intersystem electron transport and proton pumping

Electron transport between PSII and PSI is mediated by cytochrome b_6/f complexes (see Fig. 8). Plastoquinol is oxidized by the cytochrome b_6/f complex and electrons are transferred to plastocyanin, which in turn reduces a PSI reaction centre. The limiting step in this process is the oxidation of plastoquinol (reviewed: Ort, 1986; Baker & Webber, 1987). Since there is a free sharing of electrons

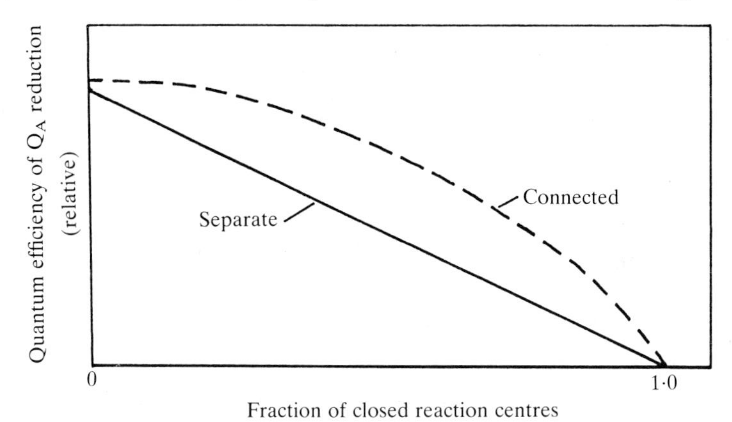

Fig. 15. Comparison of the quantum efficiency of Q_A reduction as a function of the fraction of closed PSII reaction centres for PSII complexes that can (---) and cannot (——) transfer excitation energy to neighbouring complexes. (Adapted from Clayton, 1980.)

between plastoquinol and cytochrome b_6/f complexes, i.e. in theory any plastoquinol molecule could be oxidized by any cytochrome b_6/f complex in the membrane, partial inactivation of the population of cytochrome b_6/f complexes would not be expected to modify the quantum yields of non-cyclic electron transport and CO_2 assimilation, although the light-saturated rate of CO_2 assimilation could be reduced.

Concomitant with electron transport through the cytochrome b_6/f complex is a transport of protons from the stroma to the thylakoid lumen. Until recently it had been thought that each electron removed from plastoquinol went directly to PSI via plastocyanin and one proton was released into the thylakoid vesicle (Ort, 1986). However, the experimental observation that proton to electron ratios above unity can be associated with cytochrome b_6/f turnover (e.g. Velthuys, 1978; Graan & Ort, 1983) prompted renewed interest in proposals for a more intricate mechanism of plastoquinol oxidation (Mitchell, 1975; Cramer *et al.* 1987). The operation of a Q cycle allows two protons to be translocated across the membrane for each electron transported through the cytochrome b_6/f complex (Ort, 1986) and appears to be dependent upon the magnitude of the electrochemical potential across the thylakoid. When ΔpH is low, the cytochrome b_6/f complex translocates close to two protons per electron whereas only one proton per electron is translocated under high energization conditions (Hangarter *et al.* 1987). The higher H^+/e ratio operating when the energization level is low allows for a higher ATP/2e ratio and thus a higher quantum yield of CO_2 assimilation than would otherwise be possible. If operation of the Q cycle is disturbed by environmental stresses, such a low temperature, then a decrease in ϕ would be unavoidable. To date this question has not been addressed experimentally; however, such a phenomenon is one of the possible explanations for low temperature-induced decreases in ϕ that cannot be attributed to the decreases in PSII electron transport.

Energy conservation and photophosphorylation

Photophosphorylation is among the most labile of chloroplast activities. From what is known of the mechanisms of photophosphorylation, its lability is not surprising (see Ort, 1986). The fragile and highly integrated nature of the process makes it a prime candidate for disruption by unfavourable environmental conditions such as chilling. If photophosphorylation is indeed impaired in chloroplasts isolated from stressed leaves, the possible cause of the inhibition should fall into one of three categories.

The first possibility is that the ability of the chloroplast to form an energized membrane may be diminished or absent. This condition is most likely to result from an electron transport activity impaired by the stress through changes considered in the two preceding sections. The second possibility is that the thylakoid is unable to maintain an energized state, so that electron transport is said to be uncoupled from ATP synthesis. In most cases uncoupling results from an increase in conductance of the membrane to protons. A third possibility in which photophosphorylation activity can be impaired pertains to the terminal reactions, i.e. the actual formation of the covalent bond in ATP by the membrane-bound ATPase enzyme complex. If the 'soluble' subunits (CF_1) of the enzyme complex become detached from the transmembrane proton channel (CF_0), the resulting inhibition of ATP synthesis is superficially similar to uncoupling. To function properly, the enzyme complex must bind substrates (ADP and P_i), release products (ATP) and conduct protons. Furthermore, these processes must occur in a rigid sequence that apparently involves various conformational rearrangements of the enzyme. In most cases, malfunctioning of the ATPase enzyme complex results in considerable inhibition of electron transport because the major process which dissipates the energized state of the membrane is impaired. For instance, when sunflower leaves are water-stressed the capacity for photophosphorylation and the activity of the coupling factor are decreased as a consequence (Ortiz-Lopez *et al.* 1987). Low temperature stress may also produce perturbations of the coupling factor's organization and activity. The effects of chilling in the dark on photophosphorylation in tomato leaves have been examined by measurement of the flash-induced electrochromic absorption band shift determined as an absorption change at 518–540 nm (Ort & Boyer, 1985). The membrane-depolarizing proton efflux through the coupling factor associated with phosphorylation results in an acceleration of the decay of the electric field across the membrane, which is indicated by an absorption change. Further acceleration of proton efflux by uncoupling agents results in even more rapid membrane depolarization, but proton efflux, slowed by inhibition of phosphorylation of ADP on the coupling factor causes relaxation of the field to be slowed (Witt, 1971). The time constants of the relaxation of the flash-induced absorbance change at 518–540 nm, for an attached cucumber leaf before and during 4 h chilling in the light (1000 μmol $m^{-2} s^{-1}$ at 5 °C) are given in Fig. 16. This sensitive monitor of energy conservation properties of in situ thylakoid membranes shows that no significant inhibition of the capacity for ATP synthesis accompanies the more than 50 % inhibition of ϕ (Fig. 16). These data indicate that altered energy conservation is unlikely to be an important factor contributing to the inhibition of ϕ induced in cucumber by chilling the light. There is no effect on ϕ by chilling in the dark.

Conclusion

It is increasingly apparent that chill-induced decreases in the photosynthetic productivity of crops may be attributable to a decline in ϕ. Clearly, potential exists in the future for the improvement of chilling-sensitive crop species by selecting and genetically manipulating varieties for tolerance of ϕ decline during low temperature stress. Unfortunately, it is unlikely that a common molecular basis exists for all chill-induced decreases in ϕ; differences exist between species and between chilling in the light and the dark for a single species.

Damage to thylakoid membrane components, such as Q_B, may make a major contribution to chill-induced decreases in ϕ, as discussed above, however, it

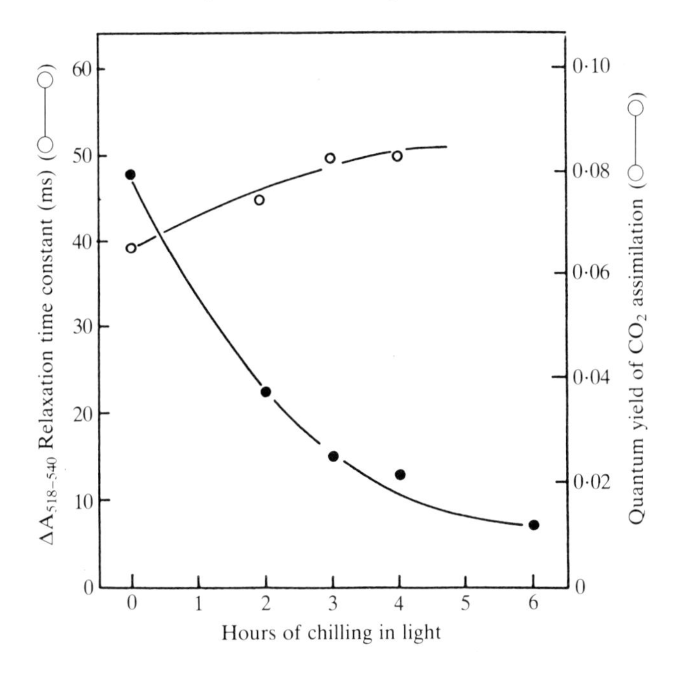

Fig. 16. Time constant of the relaxation of the flash-induced absorbance change at 518–540 nm (○) and photosynthetic quantum yield (μmol CO_2 evolved m^{-2} s^{-1} per μmol quanta absorbed m^{-2} s^{-1}) (●) of a cucumber leaf as a function of chilling (5 °C) in the light (1000 μmol m^{-2} s^{-1}). Plants were chilled for the time indicated then rewarmed in the dark for 30 min prior to measurement. The relaxation time constant was determined from a kinetic trace of the average of the response to 8 actinic flashes (given at 2 Hz) monitored at 518 nm minus 8 averaged flashes at 540 nm (also 2 Hz).

should be emphasized that ϕ could also be depressed for many other reasons. Specific organization of components within the thylakoid membranes is essential for the operation of subtle control mechanisms essential for the regulation of excitation energy transduction by the thylakoids, e.g. State 1–State 2 transition, and modifications to membrane composition and component distribution could have important consequences on ϕ. Complex interactions occur between the photosynthetic apparatus of the thylakoid membranes and the metabolic pathways associated with energy metabolism and may be important in the regulation of the rate of carbon assimilation under non-saturating light levels (reviewed: Horton, 1985). Perturbation of energy metabolism could interfere with regulatory processes such as LHCII phosphorylation and coupling factor activation and thus decrease ϕ. Also a decrease in the ratio of CO_2 reduction versus O_2 reduction would decrease ϕ by an increase in the oxygenase activity of Rubisco (Ehleringer & Bjorkman, 1976) or an increase in the rate of direct reduction of O_2 by PSI (or possibly even PSII – see Baker & Webber, 1987) would reduce ϕ. Decreases in ϕ can also occur due to damage to cellular systems which are not directly associated with photosynthetic metabolism. For example, on return to normal growth temperature, after chill-induced damage to cell components has occurred, an increased demand for ATP and reductants is likely to occur in order to repair the damage. Such energy requirements for repair, if of sufficient magnitude, could reduce ϕ, although it should be noted that little effect may be observed on the quantum yield of photosynthetic O_2 evolution. It is possible that the most important aspect of chill-induced depression of ϕ, with respect to crop photosynthetic productivity, will be the time taken to recovery from the stress on return to normal growth temperatures. Clearly the rate of repair of chill damage will be related to the energy requirements for repair, which in turn could contribute to the depression of ϕ during the recovery period.

Resolution of the factors determining the chill-induced decline and recovery of ϕ in crop species will not be an easy task and will require integration of information from field, whole leaf and subcellular studies. However, the potential gains to increased crop productivity of chilling-sensitive plants when grown in temperate regions and the possibility of improvement of chill-tolerance of subtropical and tropical crops to enable them to be grown productively in the field in temperate regions, rather than only in glasshouses, can be considered sufficient justification for addressing this complex physiological problem.

We would like to thank M. Bradbury, P. B. Beckwith, G. Bongi, J. Boyer, P. S. Covello, T. M. East, P. K. Farage, D. B. Hayden, C. R. Ireland, S. C. Kee, B. Martin, M. P. Percival and R. Wise for their part in obtaining the data presented in this article. A large proportion of the work discussed was supported by grants from the UK Agricultural and Food Research Council to NRB and SPL and the US Department of Agriculture to DRO.

References

ALLEN, J. F., BENNETT, J., STEINBACK, K. E. & ARNTZEN, C. J. (1981). Chloroplast protein phosphorylation couples plastoquinone redox state to distribution of excitation energy between photosystems. *Nature* **291**, 1–5.

BAKER, N. R. & WEBBER, A. N. (1987). Interactions between photosystems. *Adv. Bot. Res.* **13**, 1–66.

BAKER, N. R. & HORTON, P. (1987). Chlorophyll fluorescence quenching during photoinhibition. In *Photoinhibition* (ed. D. J. Kyle, C. J. Arntzen & C. B. Osmond). Amsterdam: Elsevier.

BARBER, J. (1982). Influence of surface charges on thylakoid structure and function. *A. Rev. Pl. Physiol.* **33**, 261–295.

BARBER, J. (1985). Thylakoid membrane structure and organization of electron transport components. In *Photosynthetic Mechanisms and the Environment* (ed. J. Barber & N. R. Baker), pp. 91–134. Amsterdam: Elsevier.

BENNETT, J. (1984). Chloroplast protein phosphorylation and the regulation of photosynthesis. *Physiol. Pl.* **60**, 583–590.

BERRY, J. A. & BJORKMAN, O. (1980). Photosynthetic response and adaptation to temperature in higher plants. *A. Rev. Pl. Physiol.* **31**, 491–543.

BERRY, J. A. & RAISON, J. K. (1981). Responses of macrophytes to temperature. In *Physiological Plant Ecology. I. Encyclopaedia of Plant Physiology.* New Series Vol. 12A (ed. O. L. Lange, P. S. Nobel, C. B. Osmond & H. Ziegler), pp. 277–338. Berlin: Springer-Verlag.

BONGI, G. & LONG, S. P. (1987). Light-dependent damage to photosynthesis in olive leaves during chilling and high temperature stress. *Pl. Cell Env.* **10**, 241–249.

Bowes, J., Crofts, A. R. & Arntzen, C. J. (1980). Redox properties on the reducing side of PSII in chloroplasts with altered herbicide binding properties. *Arch. Biochem. Biophys.* **200**, 303–308.

Bradbury, M. & Baker, N. R. (1986). The kinetics of photoinhibition of the photosynthetic apparatus in pea chloroplasts. *Pl. Cell Environ.* **9**, 289–297.

Clayton, R. K. (1980). *Photosynthesis: Physical Mechanisms and Chemical Patterns.* Cambridge: Cambridge University Press.

Covello, P. S., Hayden, D. B. & Baker, N. R. (1988). The roles of low temperature and light in accumulation of a 31 kDa polypeptide in the light-harvesting apparatus of maize. *Plant Cell Environ.* (in press).

Cramer, W. A., Widger, W. R., Black, M. T. & Girvin, M. E. (1987). Structure and function of photosynthetic cytochrome b-c_1 and b_6-f complexes. In *The Light Reactions* (ed. J. Barber), pp. 447–494. Amsterdam: Elsevier.

Diner, B. A. & Wollman, F. A. (1980). Isolation of highly active photosystem II particles from a mutant *Chlamydomonas reinhardtii. Eur. J. Biochem.* **110**, 521–526.

Ehleringer, J. & Bjorkman, O. (1976). Carbon dioxide and temperature dependence of the quantum yield for CO_2 uptake in C_3 and C_4 plants. *Carnegie Instn Yk.* **75**, 418–421.

Ehleringer, J. & Bjorkman, O. (1977). Quantum yields for CO_2 uptake in C_3 and C_4 plants: dependence on temperature, CO_2 and O_2 concentrations. *Pl. Physiol.* **59**, 86–90.

Farage, P. K. & Long, S. P. (1987). Damage to maize photosynthesis in the field during periods when chilling is combined with high photon fluxes. In *Progress in Photosynthesis Research*, Vol. IV (ed. J. Biggins), pp. 139–142. Dordrecht: Martinus Nijhoff.

Farquhar, G. D., von Caemmerer, S. & Berry, J. A. (1980). A biochemical model of photosynthetic CO_2 assimilation in leaves of C3 species. *Planta* **149**, 78–90.

Graan, T. & Ort, D. R. (1983). Initial events in the regulation of electron transfer in chloroplasts. The role of the membrane potential. *J. biol. Chem.* **258**, 2831–2836.

Hangarter, R. P., Jones, R. W., Ort, D. R. & Whitmarsh, J. (1987). Stoichiometrics and energetics of proton translocation coupled to electron transport in chloroplasts. *Biochim. Biophys. Acta* **890**, 106–115.

Haworth, P., Kyle, D. J., Horton, P. & Arntzen, C. J. (1982). Chloroplast membrane protein phosphorylation. *Photochem. Photobiol.* **36**, 743–748.

Hayden, D. B., Baker, N. R., Percival, M. P. & Beckwith, P. B. (1986). Modification of the photosystem II light-harvesting chlorophyll *a/b* protein complex in maize during chill-induced photoinhibition. *Biochim. Biophys. Acta* **851**, 86–92.

Hayden, D. B., Covello, P. S. & Baker, N. R. (1988). Characterization of a 31 kDa polypeptide that accumulates in a light-harvesting apparatus of maize leaves during chilling. *Photosynth. Res.* **15**, 257–280.

Hirschberg, J. & McIntosh, L. (1983). Molecular basis of herbicide resistance in *Amaranthus hybridus. Science* **22**, 1346–1349.

Horton, P. (1983). Control of chloroplast electron transport by phosphorylation of thylakoid proteins. *FEBS Lett.* **152**, 47–52.

Horton, P. (1985). Interactions between electron transfer and carbon assimilation. In *Photosynthetic Mechanisms and the Environment* (ed. J. Barber & N. R. Baker), pp. 135–187. Amsterdam: Elsevier.

Ireland, C. R., Telfer, A., Covello, P. S., Baker, N. R. & Barber, J. (1988). Studies on the limitations to photosynthesis in leaves of the atrazine-resistant mutant of *Senecio vulgaris* L. *Planta* **173**, 459–469.

Jarvis, P. G. & Sandford, A. P. (1986). Temperate forests. In *Photosynthesis in Contrasting Environments* (ed. N. R. Baker & S. P. Long), pp. 63–102. Amsterdam: Elsevier.

Kee, S. C., Martin, B. & Ort, D. R. (1986). The effects of chilling in the dark and in the light on photosynthesis of tomato: electron transfer reactions. *Photosyn. Res.* **8**, 41–51.

Krause, G. H., Grafflage, S., Rumich-Bayer, S. & Somersalo, S. (1988). Effects of freezing on plant mesophyll cells. In *Plants and Temperature*, vol. 42 (ed. S. P. Long & F. I. Woodward), pp. 311–327. *Symp. Soc. Exp. Biol.* Cambridge: Company of Biologists.

Kubin, S. (1971). Measurement of radiant energy. In *Photosynthetic Production: Manual of Methods* (ed. Z. Sestak, J. Catsky & P. G. Jarvis), pp. 702–763. The Hague: Dr W. Junk.

Kyle, D. J., Ohad, I. & Arntzen, C. J. (1984). Membrane protein damage and repair: selective loss of a quinone-protein function in chloroplast membranes. *Proc. natn. Acad. Sci. U.S.A.* **81**, 4070–4074.

LONG, S. P., EAST, T. M. & BAKER, N. R. (1983). Chilling damage to photosynthesis in young *Zea mays*. I. Effects of light and temperature variation photosynthetic CO_2 assimilation. *J. exp. Bot.* **34**, 177–188.

LONG, S. P., NUGAWELA, A., BONGI, G. & FARAGE, P. K. (1987). Chilling dependent photoinhibition of photosynthetic CO_2 uptake. In *Progress in Photosynthesis Research*, Vol. IV (ed. J. Biggins), pp. 131–138. Dordrecht: Martinus Nijhoff.

MARTIN, B. & ORT, D. R. (1982). Insensitivity of water-oxidation and photosystem II activity in tomato to chilling temperatures. *Pl. Physiol.* **70**, 689–694.

MARTIN, B. & ORT, D. R. (1985). The recovery of photosynthesis in tomato subsequent to chilling exposure. *Photosyn. Res.* **6**, 121–132.

MITCHELL, P. (1975). The proton motive Q-cycle: A general formulation. *FEBS Lett.* **59**, 137–139.

MULLET, J. E. & ARNTZEN, C. J. (1981). Identification of a 32–34 kDa polypeptide as a herbicide receptor in photosystem II. *Biochim. Biophys. Acta* **635**, 236–248.

NORMAN, J. M. (1980). Interfacing leaf and canopy light interception models. In *Predicting Photosynthesis for Ecosystem Models*, Vol. II (ed. J. D. Hesketh & J. W. Jones), pp. 49–67. Boca Raton: CRC Press.

OGREN, E., OQUIST, G. & HALLGREN, J. E. (1984). Photoinhibition of photosynthesis in *Lemna gibba* as induced by the interaction between light and temperature. I. Photosynthesis *in vivo*. *Physiol. Pl.* **62**, 181–186.

OQUIST, G. & MARTIN, B. (1986). Cold climates. In *Photosynthesis in Contrasting Environments* (ed. N. R. Baker & S. P. Long), pp. 237–293. Amsterdam: Elsevier.

OQUIST, G. & STRAND, M. (1985). Effects of frost hardening on photosynthetic quantum yield, chlorophyll organization, and energy distribution between the two photosystems in Scots pine. *Can. J. Bot.* **64**, 748–753.

ORT, D. R. (1986). Energy transduction in oxygenic photosynthesis: an overview of structure and mechanism. In *Encyclopedia of Plant Physiology, New Series*, vol. 19 (ed. L. A. Staehelin & C. J. Arntzen), pp. 143–196. Berlin: Springer-Verlag.

ORT, D. R., AHRENS, W. H., MARTIN, B. & STOLLER, E. W. (1983). Comparison of photosynthetic performance in triazine-resistant and susceptible biotypes of *Amaranthus hybridus*. *Pl. Physiol.* **72**, 925–930.

ORT, D. R. & BOYER, J. S. (1985). Plant productivity, photosynthesis and environmental stress. In *Changes in Eukaryotic Gene Expression in Response to Environmental Stress* (ed. B. G. Atkinson & D. B. Walden), pp. 279–313. New York: Academic Press.

ORTIZ-LOPEZ, A., ORT, D. R. & BOYER, J. S. (1987). *In situ* measurements of the inhibitory effects of low leaf water potentials on photophosphorylation. In *Progress in Photosynthesis Research*, Vol. IV (ed. J. Biggins), pp. 153–156. Dordrecht: Martinus Nijhoff.

PEARCY, R. W. & EHLERINGER, J. (1984). Comparative ecophysiology of C_3 and C_4 plants. *Pl. Cell Environ.* **7**, 1–13.

PERCIVAL, M. P., BRADBURY, M., HAYDEN, D. B. & BAKER, N. R. (1987). Modification of the photochemical apparatus in maize by photoinhibitory stress at low temperature. In *Progress in Photosynthesis Research*, Vol. IV (ed. J. Biggins), pp. 47–50. Dordrecht: Martinus Nijhoff.

PFISTER, K. & ARNTZEN, C. J. (1979). The mode of action of photosystem II – specific inhibitors in herbicide-resistant weed biotypes. *Z. Naturforsch.* **34**c, 996–1009.

POWLES, S. B., BERRY, J. A. & BJORKMAN, O. (1983). Interaction between light and chilling temperature on the inhibition of photosynthesis in chilling-sensitive plants. *Pl. Cell Environ.* **6**, 117–123.

ROBINSON, H. (1986). Non-invasive measurements of photosystem II reactions in the field using flash fluorescence. In *Advanced Agricultural Instrumentation, Design and Use* (ed. W. G. Gensler), pp. 92–106. The Netherlands: Martinus Nijhoff.

ROSS, J. (1975). Radiative transfer in plant communities. In *Vegetation and the Atmosphere*, Vol. 1 (ed. J. L. Monteith), pp. 13–56. London: Academic Press.

SATOH, K. & BUTLER, W. L. (1978). Low temperature spectral properties of subchloroplast fractions purified from spinach. *Pl. Physiol.* **61**, 373–379.

Servaites, J. C. & Ogren, W. L. (1978). Oxygen inhibition of photosynthesis and stimulation of photorespiration in soybean leaf cells. *Pl. Physiol.* **61**, 62–67.

Sims Holt, J., Stemler, A. J. & Radosevich, S. R. (1981). Differential light responses of photosynthesis by triazine-resistant and triazine-susceptible *Senecio vulgaris* biotypes. *Pl. Physiol.* **67**, 744–748.

Staehelin, L. A. (1986). Chloroplast structure and supramolecular organization of photosynthetic membranes. In *Encyclopaedia of Plant Physiology*, New Series, Vol. 19, *Photosynthesis III* (ed. L. A. Staehelin & C. J. Arntzen), pp. 1–84. Berlin: Springer-Verlag.

Strand, M. & Oquist, G. (1985). Inhibition of photosynthesis by freezing temperatures and high light levels in cold-acclimated seedlings of Scots pine (*Pinus sylvestris*). I. Effects on the light-limited and light-saturated rates of CO_2 assimilation. *Physiol. Pl.* **64**, 425–430.

Thornber, J. P. (1986). Biochemical characterization and structure of pigment-proteins of photosynthetic organisms. In *Encyclopedia of Plant Physiology*, New Series, Vol. 19 (ed. L. A. Staehelin & C. J. Arntzen), pp. 98–142. Berlin: Springer-Verlag.

Thornber, J. P., Peter, G. F., Nechustai, R., Chitnis, P. R., Hunter, F. A. & Tobin, E. M. (1986). Electrophoretic separation of chlorophyll-protein complexes and other apoproteins. In *Regulation of Chloroplast Differentiation* (ed. G. Akoyunoglou), pp. 249–258. New York: Alan R. Liss, Inc.

Thornley, J. M. M. (1976). *Mathematical Models in Plant Physiology*. London: Academic Press.

Velthuys, B. R. (1978). A third site of proton translocation in green plant photosynthetic electron transport. *Proc. natn. Acad. Sci. U.S.A.* **75**, 6031–6034.

Vermaas, W. F. J., Renger, G. & Dohnt, G. (1984). The reduction of the oxygen-evolving system in chloroplasts by thylakoid components. *Biochim. Biophys. Acta* **764**, 194–202.

Witt, H. T. (1971). Coupling of quanta, electrons, fields, ions and phosphorylation in the functional membrane of photosynthesis. Results by pulse spectroscopic methods. *Q. Rev. Biophys.* **4**, 365–477.

EFFECTS OF LOW TEMPERATURE ON THE RESPIRATORY METABOLISM OF CARBOHYDRATES BY PLANTS

T. ap REES[1], *M. M. BURRELL*[2], *T. G. ENTWISTLE*[1], *J. B. W. HAMMOND*[3], *D. KIRK*[1] *and N. J. KRUGER*[3]

[1] Botany School, University of Cambridge, Cambridge, CB2 3EA, UK
[2] Twyford Plant Laboratories, Cambridge Science Park, Cambridge, CB4 4WA, UK
[3] AFRC Institute of Arable Crops Research, Rothamsted Experimental Station, Harpenden, Herts, AL5 2JQ, UK

Summary

The effects of lowering the temperature from 25 °C to 2–8 °C on carbohydrate metabolism by plant cells are considered. Particular emphasis is placed on the mechanism of cold-induced sweetening in tubers of potato (*Solanum tuberosum*).

Temperatures between 0 and 10 °C were shown to cause a marked reduction in the rate of respiration of a wide range of plant tissues. At these temperatures the ability of suspension cultures of soybean (*Glycine max*), and callus cultures and tubers of potato to metabolize [^{14}C]glucose was appreciably diminished. The detailed distribution of ^{14}C showed that lowering the temperature decreased the proportion of the metabolized [^{14}C]glucose that entered the respiratory pathways and increased the proportion converted to sucrose. Pulse and chase experiments, in which [^{14}C]glucose was supplied to potato tubers at 2 and 25 °C, showed that lowering the temperature led to accumulation of label in hexose 6-phosphates, which were subsequently converted to sucrose. The patterns of $^{14}CO_2$ production from specifically labelled [^{14}C]glucose supplied to soybean suspension cultures and disks of potato tuber suggested that lowering the temperature reduced the activity of glycolysis more than that of the oxidative pentose phosphate pathway. It is argued that the above experiments demonstrate that lowering the temperature not only reduces the rate of carbohydrate metabolism but also alters the relative activities of the different pathways involved. A disproportionate reduction in glycolysis at the lower temperatures is suggested.

Mature tubers of many varieties of potato accumulate sucrose and hexose when stored between 2 and 10 °C. Starch is the source of carbon for this synthesis of sugar. We could not detect cytosolic fructose-1,6-bisphosphatase in potato tubers and suggest that carbon for sugar synthesis in the cold leaves the amyloplast, not as triose phosphate, but probably as a six-carbon compound.

Evidence is presented that phosphofructokinase (EC 2.7.1.11) plays a major role in regulating the entry of hexose 6-phosphates into glycolysis in potato tubers. Phosphofructokinase was purified from potato tubers and shown to consist of four forms. Three of these forms were shown to have higher Q_{10} values over the range 2–6 °C than over the range 12–16 °C and are regarded as being cold-labile. No such cold-lability was detected for the key enzymes involved in sucrose synthesis and the oxidative pentose phosphate pathway.

We suggest that in potatoes stored at 2–8 °C the cold-lability of phosphofructokinase leads to a greater reduction in glycolysis than in other pathways that consume hexose 6-phosphates. The increased availability of the latter is seen as leading to increased synthesis of sucrose. Support for this view is provided by the observation that a new breeding clone of potato that did not show cold-lability of phosphofructokinase did not accumulate significant amounts of sugar in the cold.

Introduction

In this article we intend to illustrate the complexities of the interactions between low temperature, 2–8 °C, and metabolism by concentrating on the metabolism of hexose 6-phosphates in the non-photosynthetic cells of higher plants. We pay particular attention to the hypothesis that lowering the temperature from 25 °C to 2–8 °C reduces the rate of glycolysis to a greater extent than the rates of other reactions that consume hexose 6-phosphates. We consider such a differential effect of temperature as a major cause of cold-induced sweetening, the accumulation of sucrose and hexose at 2–8 °C (ap Rees *et al.* 1981).

The most obvious effect of low temperature on the metabolism of hexose 6-phosphates by plant cells is that, as with enzyme-catalysed reactions in general, a small drop in the absolute temperature causes a very marked decrease in the rate of metabolism. This may be illustrated by considering the effect of cold on the rate of respiration, one of the major processes responsible for the metabolism of hexose phosphates. Table 1 shows that, although lowering the temperature greatly reduced respiration in all the instances quoted, the extent of this decrease varied with the tissue, and the range over which the temperature was reduced.

The question on which we wish to concentrate is whether lowering the temperature affects each pathway of hexose 6-phosphate metabolism equally or whether the effect is differential, with certain reactions being more affected than others. In the former situation, lowering the temperature would lead to a lower rate of metabolism in which the relative rates of the different pathways remained the same. In the latter situation, the reduced rate of metabolism would be accompanied by changes in the relative rates of the different pathways with a consequent re-direction of metabolism.

Effects of low temperature on metabolism of [^{14}C]glucose

The extent to which lowering the temperature affects hexose phosphate metabolism proportionately or differentially was studied by determining the distribution of label after supplying cells with [^{14}C]glucose at different temperatures. Data from two such experiments, carried out aseptically, for callus cultures of potato (*Solanum tuberosum*) and suspension cultures of soybean (*Glycine max*) are summarized in Table 2. To compare labelling patterns at different tempera-

Table 1. *Effects of temperature on respiration rate of plant tissues*

Plant	Tissue	Rate of gas exchange, as percentage of that at 10°C, at:						Reference
		0	3	5	20	25	30°C	
Cassiope tetragona	Leaf	30			286		625	Wager (1941)
Empetrum nigrum		24			263		555	
Ranunculus pygmaeus		53			344		588	
Saxifraga cernua	Leaf disks			67		114		McNulty & Cummings (1987)
Oxyria digyna	Whole plant				188		313	Morney & Billings (1961)
Atriplex polycarpa		31				335		Chatterton, McKell & Strain (1970)
Vicia faba					185			Breeze & Elston (1978)
Zea mays			56			323		Briggs, Kidd & West (1920)
Glycine max	Suspension cell cultures					417		ap Rees & Kirk (unpublished)

Table 2. *Effects of temperature on metabolism of [^{14}C]glucose by callus cultures of potato and suspension cultures of soybean*

Cell fraction	Percentage of metabolized ^{14}C recovered per fraction			
	Potato callus		Soybean cells	
	25°C	2°C	25°C	7°C
CO_2	22	4	34	16
Water-insoluble material	12	5	30	16
Protein			17	5
Water-soluble material				
Basic components	11	7	12	15
Acidic components	30	65	4	10
Hexose phosphates	12	45	1	4
Sucrose	20	13	15	41
Fructose	5	5	5	2
^{14}C metabolized				
as d.p.m. $\times 10^{-6}$	0·36	0·23	1·85	1·26
as % ^{14}C supplied	16·5	2·1	76	52

Potato cells were incubated for 3 h in [U-^{14}C]glucose; details and data from Pollock & ap Rees (1975*b*). Soybean cells were grown as described by Macdonald & ap Rees (1983), harvested and re-suspended (6 mg protein/2·5 ml) in the above growth medium in which sucrose was replaced with 83 mM sorbitol and 0·5 mM [6-^{14}C]glucose (0·48 Ci mol^{-1}). After incubation for 19 h at 25 or 7°C the cells were killed and analysed as described by Pollock & ap Rees (1975*b*).

tures, the ^{14}C recovered in each fraction of a sample of cells is expressed, not as a percentage of the ^{14}C supplied, but as a percentage of the ^{14}C metabolized by that particular sample. ^{14}C metabolized is defined as the sum of the ^{14}C recovered in compounds other than [^{14}C]glucose at the end of the incubation. For both types of cells reduction in temperature significantly reduced the amount of the supplied [^{14}C]glucose that was metabolized. This is entirely expected: what is less expected is that lowering the temperature significantly altered the manner in which the [^{14}C]glucose was metabolized by the cell cultures. The most striking difference is that the proportion of metabolized [^{14}C]glucose converted to CO_2 was greatly reduced at the lower temperatures. Incorporation into insoluble material, protein in particular, was also reduced. In the shorter (3 h) incubation used with the potato callus, the reduced labelling of CO_2 and insoluble material was offset by a sharp rise in the labelling of the acidic components of the water-soluble substances. Most of the latter label was in hexose phosphates. In the longer (19 h) incubation used with soybean cells there was also an increase in the labelling of the acidic fraction but the major increase was found in sucrose.

For the soybean samples in Table 2 we may estimate the proportion of the metabolized [^{14}C]glucose that entered the respiratory pathways. We do this by summing the ^{14}C recovered in the following fractions: CO_2, protein, basic components (mostly amino acids), acidic components minus that present as hexose phosphates. At 25°C 66 % of the metabolized label entered the respiratory pathways: at 7°C the value was 42 %. These results strongly suggest that lowering the temperature affected metabolism differentially by reducing the entry of hexose phosphate into the respiratory pathways to a greater extent than other aspects of hexose phosphate metabolism. The data for the potato callus indicate that the same phenomenon occurred in these cells (Table 2).

In the soybean cells the fall in the proportion of metabolized [^{14}C]glucose that entered the respiratory pathways, 24 % (66−42), is comparable to the rise in the proportion converted the sucrose, 26 % (41−15). Thus lowering the temperature appears to divert hexose phosphates away from respiration and into sucrose. This hypothesis was tested in detail with mature tubers of potato (cv. Record). A 50-μl micro-capillary pipette was used to make a hole in a mature tuber. The hole was then filled with [U-^{14}C]glucose. After 2 h (the pulse) the [^{14}C]glucose was replaced with glucose and the incubation continued for a further 18 h (the chase). Incubation were carried out at 2°C and 25°C and the detailed distribution of label was determined at the end of the pulse and the chase (Table 3).

The analyses reported in Table 3 were conducted without significant loss of label and allow the following conclusions to be drawn. At the end of the pulse at 25°C 36 % of the metabolized ^{14}C had entered the respiratory pathways (^{14}C in CO_2, organic acids, amino acids and protein) and 10 % was in hexose phosphates. At 2°C only 21 % had entered the respiratory pathways but 39 % was in hexose phosphates. This confirms the conclusion drawn from the experiments described in Table 2. During the chase at 2°C there was a marked drop in the labelling of hexose phosphates. Very little of the label that moved out of the hexose

Table 3. *Effects of temperature on metabolism of [^{14}C]glucose by callus cultures of potato and suspension cultures of soybean*

	Percentage of metabolized ^{14}C recovered per fraction			
	25 °C		2 °C	
Cell fraction	Pulse	Chase	Pulse	Chase
CO_2	1	11	0·2	1
Water-insoluble material	28	38	32	8
Protein	6	12	8	1
Starch	13	24	13	3
Non-starch polymeric glucose	4	2	0·2	1
Polymeric galactose	5	2	0	0·1
Water-soluble material	70	60	65	96
Sucrose	28	35	11	57
Hexose	3	5	3	11
Hexose phosphates	10	4	39	11
Organic acids	5	8	3	4
Amino acids	24	8	10	13

Data from Dixon & ap Rees (1980*b*).

phosphates during the chase at 2°C entered the respiratory pathways: the percentage of metabolized ^{14}C recovered in respiratory products did not increase during the chase. The only compounds to show an increase in labelling, comparable to the drop in that of the hexose phsophates, were the free sugars, particularly sucrose.

Collectively, our experiments provide compelling evidence that, at least in the plant species we studied, lowering the temperature affects the metabolism of hexose phosphates differentially by restricting their entry into respiration and promoting their conversion to sucrose. This view is strengthened by the observation that, in a wide variety of plants, transfer to 2–8°C leads to appreciable accumulation of sucrose or closely related compounds (ap Rees *et al.* 1981). Specific proof of such cold-induced sweetening has been obtained for the cultivar of potato (Record) that we used in the experiments reported in Table 3. Three weeks storage at 2°C leads to a fivefold increase in sucrose and an increase in reducing sugars, the latter almost certainly arising from the hydrolysis of sucrose by invertase (Pollock & ap Rees, 1975*a*).

Further examination of the data in Tables 2 and 3 strongly suggests that temperature affects not only the proportion, but also the fate, of hexose phosphate that enters the respiratory pathways. Respiratory substrate in plants is divided between provision of intermediates for biosyntheses, and complete oxidation to CO_2. The former includes incorporation of ^{14}C-labelled substrate into protein, amino acids and organic acids: the latter is represented by the labelling of CO_2. For soybean (Table 2) and potato tuber (Table 3) the ^{14}C recovered in protein, amino acids (basic components), organic acids (acidic components – hexose phosphates) may be summed to give the total ^{14}C that entered the respiratory

Table 4. *Effects of temperature on $^{14}CO_2$ production from specifically labelled [^{14}C]glucose supplied to suspension culture of soybean*

Position of label in [^{14}C]glucose supplied	$^{14}CO_2$ production as % ^{14}C metabolized	
	25 °C	2 °C
Carbon 1	43	25
Carbon 2	31	15
Carbons 3 and 4	52	27
Carbon 6	33	16

Cells were grown, harvested, incubated and analysed exactly as described in Table 2. Values are means of triplicates.

pathways. For example, for soybean at 25 °C (Table 2) this total is 66 % of the ^{14}C metabolized: the total being made up of the label in protein (17 %), amino acids (12 %), organic acids (3 %), and CO_2 (34 %). The percentage of label that entered the respiratory pathways and was subsequently converted to CO_2 is obtained by expressing the label in CO_2 (34 %) as a percentage of the total that entered the respiratory pathways (66 %) and in this instance is 52 %. For soybean at 7 °C only 28 % of the label that entered the respiratory pathways was converted to CO_2 (Table 2). Similar calculation for potato tubers at the end of the chase (Table 3) show that 28 and 5 % of the label that entered the respiratory pathways was converted to CO_2 at 25 and 2 °C, respectively. Thus, lowering the temperature appears to diminish the proportion of material entering the respiratory pathways that is completely oxidized to CO_2.

Hexose phosphates that enter the respiratory pathways do so via either glycolysis or the oxidative pentose phosphate pathway. The relative activities of two pathways vary according to the tissue and its environment. As yet it has not proved practicable to measure the fluxes through these two pathways accurately, but the patterns of $^{14}CO_2$ production from specifically labelled [^{14}C]glucose may be used as a rough guide (ap Rees, 1980*a*). The extent to which the yield from carbon 1 exceeds that from carbon 6 may reflect the activity of the oxidative pentose phosphate pathway. The yield from carbons 3 and 4 is likely to reflect the total amount of glucose metabolized by the two pathways together.

We have determined the effects of temperature on the patterns of $^{14}CO_2$ production from [^{14}C]glucose supplied to suspension cultures of soybean (Table 4) and to 'aged' disks (ap Rees & Beevers, 1960) of potato (Table 5). The patterns at 25 °C are those expected of non-photosynthesizing cells of plants in general. They reflect the activity of both pathways with a dominance of glycolysis (ap Rees, 1980*a*). In both soybean and potato, lowering the temperature increased the yield from carbon 1 relative to that from carbon 6. In both instances the difference between the yields from carbon 1 and carbon 6, when expressed as a percentage of the yield from carbons 3 and 4, is greater at the lower temperature. This is

Table 5. *Effects of temperature on $^{14}CO_2$ production from specifically labelled [^{14}C]glucose supplied to disks of potato (Record) tubers*

Temperature (°C)	Position of ^{14}C in [^{14}C]glucose	Percentage of added [^{14}C]glucose recovered as $^{14}CO_2$ in: 1 h	2 h	3 h	4 h	19·5 h
25	1	4·70	14·5	27·5	32·2	38·6
	2	0·81	3·2	7·0	10·8	24·6
	3·4	3·42	11·8	24·1	29·4	36·1
	6	0·88	2·91	5·0	7·4	18·3
2	1	0·144	0·400	0·728	1·140	6·49
	2	0·023	0·061	0·148	0·253	3·03
	3·4	0·050	0·158	0·321	0·515	5·19
	6	0·009	0·035	0·087	0·182	2·64

Disks (1 cm × 1 mm) of mature tubers were aged for 22 h at 25 °C. All samples were from the same batch of disks. Samples of 30 disks were suspended in 4·0 ml 50 mM KH_2PO_4, pH 5·2, that contained 0·5 mM [^{14}C]glucose (0·48 Ci mol^{-1} except that [3,4-^{14}C]glucose was 0·24 Ci mol^{-1}). Values are cumulative yields and are means of triplicates.

consistent with the relative activities of the two pathways being shifted away from glycolysis as the temperature falls.

The central feature of our experiments with labelled glucose is that they provide appreciable evidence that lowering the temperature from 25 °C to 2–8 °C affects not merely the rate of hexose phosphate metabolism, but also the direction of that metabolism. Specifically, our studies suggest that in at least two plant tissues glycolysis is particularly susceptible to temperatures between 2 and 8 °C. It thus seems likely that some key reaction or reactions of glycolysis are particularly sensitive to temperatures below 10 °C.

Effects of low temperature on enzymes of carbohydrate metabolism

To discover how low temperature causes the alterations in hexose phosphate metabolism, outlined in the previous section, we studied the effects of temperature on the activities of several of the key enzymes of carbohydrate metabolism in extracts of potato tubers (Pollock & ap Rees, 1975*a*). Table 6 shows the Q_{10} values of these enzymes measured over two ranges of temperature, 10–25 °C and 2–10 °C. Although we found that each enzyme showed the expected high Q_{10} the precise value, over a given range of temperature, varied with the enzyme. Such variation was appreciable: at 10–25 °C it ranged from 1·14 for pyruvate kinase to 2·48 for glyceraldehydephosphate dehydrogenase. More significantly, we also found that the Q_{10} of an enzyme can vary with the temperature and that the extent of this variation differs according to the enzyme. Such observations are not peculiar to potato enzymes. Hochachka & Somero (1984) discussed many other comparable examples and also showed that the response of an enzyme to temperature may depend upon species and the substrate concentration. Thus, it seems clear that, whilst all enzymes are very sensitive to relatively small changes in temperature, the

Table 6. *Temperature coefficients of enzymes of carbohydrate metabolism from tubers of potato (Record)*

Enzyme	Temperature coefficient		Fisher's *P* value
	Q_{10} (10–25°C)	Q_{10} (2–10°C)	10–25°C vs. 2–10°C
Sucrose synthase	1·86 ± 0·03	1·96 ± 0·04	N.S.
Sucrose phosphate synthase	1·92 ± 0·01	2·25 ± 0·24	N.S.
Glucose-6-phosphate dehydrogenase	1·97 ± 0·01	2·49 ± 0·23	N.S.
Phosphofructokinase	1·47 ± 0·05	3·33 ± 0·12	<0·001
Aldolase	2·35 ± 0·08	2·65 ± 0·52	N.S.
Glyceraldehydephosphate dehydrogenase	2·48 ± 0·06	4·96 ± 0·38	<0·002
Pyruvate kinase	1·45 ± 0·05	2·38 ± 0·12	<0·001

Values are means ± S.E. of measurements made with extracts of five different tubers. Data from Pollock & ap Rees (1975*a*).

extent of this sensitivity varies with the enzyme, the species, the temperature range, and the substrate concentration. This being so, it is to be expected that changing the temperature will often have differential effects on metabolism.

Closer inspection of Table 6 shows that three particular enzymes were found to be strikingly sensitive to low temperature. These are phosphofructokinase, EC 2.7.11 [PFK(ATP)]; glyceraldehydephosphate dehydrogenase, EC 1.2.1.12; and pyruvate kinase, EC 27.1.40. All three enzymes are likely to make major contributions to the control of glycolysis in general. Specific evidence that the conversion of fructose 6-phosphate (Fru-6-*P*) to fructose-1,6-bisphosphate (Fru-1,6-P_2), and the metabolism of phosphoenolpyruvate, are major points of control in potato tubers has been obtained (Dixon & ap Rees, 1980*a*). When glycolytic flux was varied the amounts of Fru-6-*P* and phosphoenolpyruvate in the tubers changed in the opposite direction to the flux.

The evidence that PFK(ATP) plays a key role in the regulation of glycolysis in potato tubers led to a closer examination of the effects of temperature on this enzyme (Dixon, Franks & ap Rees, 1981). The evidence for cold-lability was supported with studies with a purer preparation of the enzyme. Examination of the latter by differential scanning calorimetry provided evidence that the enzyme owes its cold-lability to denaturation of a multi-subunit enzyme complex. This partial dissociation was shown to be exothermic. As the corresponding entropy change is negative it seems very likely that hydrophobic interactions, which contribute to the stability of the native enzyme at ordinary temperatures, are weakened to the extent that the enzyme complex partially dissociates spontaneously at lower temperatures.

From Table 6 it can be seen that lowering the temperature reduced the activities of the regulatory enzymes of glycolysis to a greater extent than those of sucrose phosphate synthase and glucose-6-phosphate dehydrogenase. The latter two enzymes play major roles in the consumption of hexose 6-phosphates by sucrose synthesis and the oxidative pentose phosphate pathway, respectively (ap Rees,

1980*b*). This differential response to temperature shown by the enzymes that control the use of hexose 6-phosphates in potato tubers parallels very closely the differential effects of temperature on the manner in which the tubers metabolize [^{14}C]glucose.

Cold-induced sweetening of potatoes

Our observations on the effects of low temperature on the metabolism and enzymes of potato tubers led to the proposal that the cold-lability of PFK(ATP) caused a differential reduction in glycolysis at temperatures of 2–8 °C, and that this increased the availability of hexose phosphates for sucrose synthesis (Pollock & ap Rees, 1975*a*; ap Rees *et al.* 1981). We suggested that this is a major cause of cold-induced sweetening, which we see as a clear example of the differential effect of low temperature on plant metabolism. Since this hypothesis was put forward there have been a number of developments that complicate the original proposal.

Pyrophosphate: fructose 6-phosphate 1-phosphotransferase

Our view that the cold-lability of PFK(ATP) is a prime cause of cold-induced sweetening must be related to the subsequent discovery that plants contain another enzyme that is capable of converting Fru-6-*P* to Fru-1,6-P_2. This is pyrophosphate: fructose 6-phosphate 1-phosphotransferase, EC 2.7.1.90 [PFK(PP_i)]. This enzyme is confined to the cytosol, almost completely dependent upon activation by fructose-2,6-bisphosphate (Fru-2,6-P_2), and catalyses the following readily reversible reaction (ap Rees, 1985):

$$PP_1 + \text{Fru-6-}P \rightleftharpoons \text{Fru-1,6-}P_2 + P_1$$

PFK(PP_i) is present in a very wide range of plant tissues (Carnal & Black, 1983) and appreciable activities have been demonstrated in extracts of potato tubers (Van Schaftingen, Lederer, Bartrons & Hers, 1982; Morrell & ap Rees, 1986*b*). There is also substantial evidence that plants contain appreciable amounts of PP_i (Edwards *et al.* 1984) and that this is almost entirely confined to the cytosol (Weiner, Stitt & Heldt, 1987). Estimates of the mass-action ratio of PFK(PP_i) in plants (ap Rees, Green & Wilson, 1985; Weiner *et al.* 1987) suggest that the enzyme catalyses a near-equilibrium step *in vivo*. Thus PFK(PP_i) could act as a means whereby hexose 6-phosphate could enter glycolysis independently of PFK(ATP).

The view that the cold-lability of PFK(ATP) is a major cause of sweetening must take into account the possibility that PFK(PP_i) could bypass PFK(ATP) in potatoes. Two points are made. First is that, except possibly for one highly specialized tissue, pineapple leaves exhibiting Crassulacean Acid Metabolism (Dancer, 1987), there is no convincing evidence that PFK(PP_i) does act as a glycolytic enzyme in plants (ap Rees & Dancer, 1987; ap Rees, 1988). Second is that we have recently determined the effect of temperature on the activity of partially purified PFK(PP_i) from potato tuber (Table 7). The results show that

Table 7. *Effects of temperature on PFK(PP_i) partially purified from tubers of potato (Record)*

Temperature range (°C)	Q_{10}
5·1–6·9	4·12
2·3–5·1	4·79
2·3–6·9	4·51
12·1–25·1	1·75
14·8–19·9	1·79
17·3–25·1	1·60
12·1–17·3	1·99

Each value is the mean of triplicate assays at each temperature. Data obtained with partially purified PFK(PP_i); specific activity 12 μmol min^{-1} mg^{-1} protein at 25 °C. Unpublished data of N. J. Kruger.

PFK(PP_i) is cold-labile. Activity was very substantially reduced in the temperature range at which sweetening occurs. The values for Q_{10} are comparable to those of the cold-labile glycolytic enzymes shown in Table 6. Thus, even if PFK(PP_i) is an important glycolytic enzyme in potatoes, it would not provide a cold-resistant bypass to PFK(ATP). Should PFK(PP_i) be found to act glycolytically in potatoes, then our hypothesis would have to be extended to include the cold-lability of PFK(PP_i) as an additional cause of differential reduction of glycolysis in the cold.

Intracellular compartmentation

The careful analyses by Isherwood (1973) establish that the hexose phosphates converted to sucrose during cold-induced sweetening come from the breakdown of starch. This starch is made in plastids and sucrose synthesis occurs in the cytosol. Plastids from non-photosynthetic cells of plants, including amyloplasts, contain a significant proportion of many of the enzymes of glycolysis (ap Rees, 1987). Amyloplasts from soybean suspension cells (Wragg, 1987) and wheat endosperm (Entwistle, unpublished results) contain appreciable activities of each of the glycolytic enzymes required to convert hexose 6-phosphates to triose phosphate. In wheat endosperm it is likely that 33 % of the total cellular PFK(ATP) is in the amyloplasts and the rest is in the cytosol.

The fact that glycolysis, at least as far as triose phosphate, but not sucrose synthesis, can occur in the amyloplast affects the hypothesis that cold-lability of PFK(ATP) is a cause of sweetening. The crucial question is which product or products of starch breakdown move out of the amyloplast to support sucrose synthesis in the cytosol. The lack of any definitive evidence often leads to the assumption that transport of carbon across the amyloplast envelope occurs primarily as triose phosphate via the phosphate translocator (Boyer, 1985). If this is so in potato tubers, then it is difficult to see how cold-lability of PFK(ATP) could account for sweetening. If plastidic PFK(ATP) were cold-labile and carbon moved to the cytosol as triose phosphate, then, in the cold, hexose phosphates would accumulate in the amyloplast and thus not be available for sucrose synthesis in the cytosol. If the cytosolic PFK(ATP) were cold-labile then this would not greatly affect the availability of hexose 6-phosphates for sucrose synthesis. This is because the cytosolic hexose 6-phosphates used in sucrose synthesis would be synthesized from triose phosphate exported from the amyloplast, in which case a cytosolic fructose-1,6-bisphosphatase (FBPase) would probably be the key enzyme controlling the level of cytosolic hexose 6-phosphates.

Two types of plant tissues are known to be able to catalyse a net flux from triose phosphate to hexose 6-phosphate: photosynthetic cells and those involved in gluconeogenesis from storage fat or protein. Both are characterized by high activities of a highly regulated cytosolic FBPase. Thus, if the route from starch to sucrose during sweetening proceeds via triose phosphate, then potatoes should contain appreciable activity of a cytosolic FBPase. We investigated whether this was so.

The detection and measurement of FBPase in plant cells is complicated considerably by the presence in most plant cells of PFK(PP_i). Both the above enzymes can convert Fru-1,6-P_2 to Fru-6-P. In the presence of inorganic pyrophosphatase, which shows high activity in most plant extracts, both enzymes would appear to release P_1 from Fru-1,6-P_2. Finally, many plant extracts, and at least some commercial preparations of Fru-1,6-P_2, may contain enough P_1 to support some PFK(PP_i) activity.

We have investigated whether potato tubers contain cytosolic FBPase. We worked with unfractionated homogenates to guard against loss of activity during fractionation. We also assayed such homogenates after they had been desalted with Sephadex. We did this to see whether lack of activity was due to the presence in the extracts of low molecular weight substances that interfered with the assay, and because desalting would remove P_i and Fru-2,6-P_2 from the extract. Assays were conducted in the presence and absence of 20 μM Fru-2,6-P_2. If Fru-6-P formation was due to PFK(PP_i) then it would be stimulated by Fru-2,6-P_2. If activity was due to FBPase then Fru-2,6-P_2 should inhibit the cytosolic form and have relatively little effect on the plastidic form (Stitt *et al.* 1982). Each extract was assayed under two sets of conditions, optimized, respectively, for wheat leaf cytosolic (pH 7·1) and plastidic (pH 8·1) FBPase.

Table 8 summarizes our attempts to demonstrate FBPase activity in extracts of potato (Record) tuber. No activity could be detected for the cytosolic enzyme in unfractionated extracts of potato: very slight production of Fru-6-P was found in desalted extracts. This activity was increased by Fru-2,6-P_2, which also evoked very low activity in the unfractionated extracts. The results strongly suggest that such activity as we found was due to PFK(PP_1) not FBPase. The most favourable argument that could be made for a cytosolic FBPase is to accept the values for the desalted enzyme assayed at pH 7·1 in the absence of Fru-2,6-P_2 as representing FBPase. An average value of 9·3 nmol min^{-1} g^{-1} fresh weight is obtained. This is less than 10 % of the activity of the key enzyme of sucrose synthesis, sucrose phosphate synthase (Pollock & ap Rees, 1975*a*). Assuming a Q_{10} of 2, the rate for FBPase at 2°C would be 2 nmol min^{-1} g^{-1}. Estimates of the minimal rate of

Table 8. *Assay for fructose-1,6-bisphosphatase in extracts of potato (Record) tubers and pea leaves*

		Fru-6-*P* production (nmol min^{-1} g^{-1} fresh wt)			
		Untreated extracts		De-salted extracts	
Tissue	pH of assay	− Fru-2,6-P_2	+	− Fru-2,6-P_2	+
Potato tuber	7·1	0,0,0	0,8,10	14,13,1	22,20,14
	8·1	9,43,0	12,4,0	6,0·2,16	6,0·5,20
Pea leaf	7·1	227	31	155	69
	8·1	1186	773	888	581
Pea leaf + potato tuber	7·1	236		118	
	8·1	1203		1173	

Unpublished results of T. G. Entwistle. Samples, 0·4–1·0 g fresh wt, of potato tuber, leaves of 11-day-old seedlings of pea, and a mixture of equal weights of potato and pea leaf were homogenized in 100 mM Tris-HCl, 20 mM EDTA, 20 mM cysteine-HCl, 20 mM diethyldithiocarbamate, pH 7·6, except for the first two experiments with potato when 70 mM glycylglycine, pH 7·5 was used. All homogenates were made in the presence of 0·3% (w/v) bovine serum albumin and polyvinyl pyrrolidon at 250 mg per g fresh wt of tissue. Homogenates were centrifuged at 30 000 g for 30 min and assayed at once and after de-salting with Sephadex G-25 (coarse). Two assay mixtures were used (Stitt *et al.* 1982). that at pH 7·1 designed for the cytosolic enzyme and that at pH 8·1 for the plastidic enzyme. Rates for the mixed sample are expressed per g of pea leaf.

sweetening at 2°C (Pollock & ap Rees, 1975*a*) give a value of 4·1 nmol anhydro-hexose min^{-1} g^{-1} fresh weight. Thus even if the activity is FBPase there is not enough of it to mediate sweetening.

Table 8 shows that our assays for FBPase worked satisfactorily when applied to extracts of pea leaves, a tissue known to contain appreciable activity of PFK(PP_i) (Carnal & Black, 1983). We also assayed extracts made from mixtures of equal weights of pea leaf and potato tuber. The activities found were closely comparable to those observed when pea leaves alone were assayed. These experiments provide substantial evidence that our failure to detect significant FBPase in extracts of potato was not due to any inability of the assays to detect FBPase in the presence of PFK(PP_i) or the extract of potatoes.

Our results strongly suggest that potato tubers lack significant activities of cytosolic FBPase and that the carbon for sugar synthesis during cold-induced sweetening does not leave the amyloplast as triose phosphate. Thus it seems likely that export occurs as a six-carbon compound or, conceivably, as maltose.

Two additional pieces of evidence support the view that carbon can cross the amyloplast envelope as a six-carbon compound. First, the labelling of the glucosyl units of starch made by wheat grains supplied with [1-^{13}C]- and [6-^{13}C]glucose showed that there had been very little equilibration of label between carbons 1 and 6 during the conversion of the labelled glucose to starch (Keeling *et al.* 1988). If entry into the amyloplast had occurred as triose phosphate then the high activities of both cytosolic and amyloplast triosephosphate isomerase would be expected to

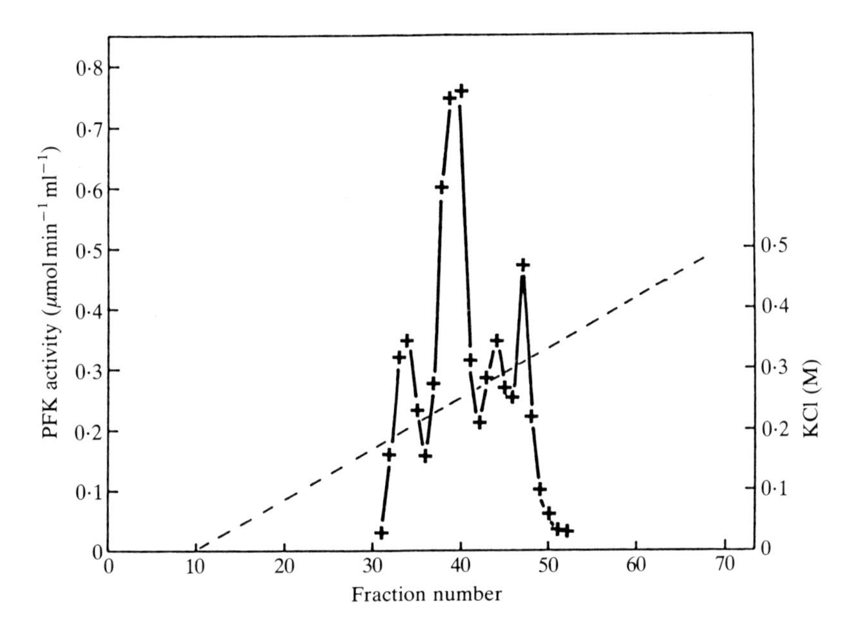

Fig. 1. Ion-exchange FPLC of potato tuber (Record) PFK(ATP). Purified PFK(ATP) was fractionated on a Mono Q column using a 0 to 0·48-M KCl linear gradient. PFK(ATP) activity was measured in each 0·5 ml fraction.

have produced extensive equilibration between carbons 1 and 6. In fact, the degree of such equilibration in the glucosyl units of starch was almost identical to that found in the glucosyl moiety of sucrose. This strongly suggests that both starch and sucrose were formed directly from the cytosolic pool of hexose 6-phosphates. The second piece of evidence that indicates that movement across the amyloplast envelope occurs as a six-carbon compound is the observation that intact amyloplasts from wheat endosperm incoporate glucose 1-phosphate into starch (Tyson, 1987).

If the carbon for cold-induced sweetening leaves the amyloplast as a six-carbon, or larger, compound then cold-lability of either, or both, cytosolic or plastidic PFK(ATP) could divert hexose phosphates from respiration to sucrose. If the cytosolic enzyme is cold-labile, then reduction in its activity could cause accumulation in the cytosol of hexose phosphates exported from the amyloplast. If the amyloplast PFK(ATP) were cold-labile, this could cause hexose phosphates to be diverted from the amyloplast to the cytosol.

Properties of potato tuber PFK(ATP)

The key role that we have ascribed to PFK(ATP) in cold-induced sweetening led us to examine the properties of this enzyme in more detail. The enzyme from potato (Record) tubers was purified to homogeneity and a specific activity of 200 μmol min^{-1} mg^{-1} protein. Ion-exchange fast protein liquid chromatography (FPLC) resolved the purified preparation into four forms (Fig. 1). Re-chromatography of these four forms, either individually or in combination, showed that

Table 9. *Effects of temperature on the four forms of PFK(ATP) purified from two varieties of potato tubers*

Variety	Temperature range (°C)	Q_{10}			
		PFK I	PFK II	PFK III	PFK IV
Record	12–16	3·00	2·46	2·61	2·68
	2–6	3·15	3·96	4·20	3·10
13737	12–16	2·93	2·67	2·69	2·22
	2–6	3·31	2·93	2·72	2·42

each eluted in its original position and that recovery was at least 85 %. A similar pattern of four peaks of PFK(ATP) has been demonstrated in partially purified extracts of tubers and of leaves of other cultivars of potato, and of the related wild species *Solanum commersonii*. The relative molecular masses of three of the forms of PFK(ATP) from tubers of cv. Record was determined by size-exclusion FPLC to be about 200 000, whilst that of the fourth form was greater than 800 000.

When the purified PFK(ATP), without separation into its four forms by FPLC, was analysed by SDS–polyacrylamide gel electrophoresis, four polypeptides were found. These are called PFK_a, PFK_b, PFK_c, and PFK_d and have M_r values of 46 300, 49 500, 50 000 and 53 000, respectively. When PFK_{a-d} were treated with cyanogen bromide and the products were separated by SDS–polyacrylamide gel electrophoresis, each polypeptide yielded a different set of cleavage products. This suggests that the four PFK polypeptides found in the enzyme preparations are not formed by proteolysis during purification. The polypeptides PFK_a and PFK_b were mainly confined to the enzyme forms I and II; PFK_c and PFK_d were found only in the enzyme forms III and IV. Thus, potato tubers contain four forms of PFK(ATP), which differ in their polypeptide composition. Our data suggest that the forms represented by peaks I to III are tetramers, and that IV is a high molecular weight aggregate.

The manner in which the four forms of PFK(ATP), shown in Fig. 1, respond to temperature is clearly crucial to our hypothesis of cold-induced sweetening. Table 9 shows how the four forms of PFK(ATP) from Record tubers, the cultivar used in all our studies of sweetening, are affected by temperature. All the forms of PFK(ATP) from Record, except PFK I, showed the cold-lability characteristic of the PFK(ATP) activity observed in partially purified preparations (Table 6). From Fig. 1, it is clear that PFK I accounts for only a small percentage of the total PFK(ATP) in Record tubers. These data confirm that the vast bulk of PFK(ATP) in tubers of this cultivar is cold-labile.

The evidence that PFK I from Record tubers is not cold-labile suggests that cold-lability is not a universal feature of PFK(ATP). This is confirmed by the behaviour of the enzyme from tubers of a new breeding clone obtained from the Scottish Crops Research Institute and designated 13737. Our results (Table 9) suggest that none of the four forms of PFK(ATP) from 13737 is anything like as cold-labile as PFK II–IV from Record. This variation in the properties of

Table 10. *Effects of cold on sugar content of potato tubers (Record and 13737)*

		Sugar content (μmol g^{-1} fresh wt)		
Variety	Storage conditions	Sucrose	Glucose	Fructose
Record	2 weeks at 15 °C	8·83 ± 0·15 (3)	6·16 ± 2·80 (3)	1·74 ± 0·49 (3)
	6 weeks at 4 °C	3·06 ± 0·78 (6)	26·36 ± 3·83 (6)	15·53 ± 3·00 (6)
	12 weeks at 4 °C	2·06 ± 0·52 (6)	19·21 ± 3·91 (6)	10·07 ± 1·03 (6)
13737	2 weeks at 15 °C	7·66 ± 0·98 (3)	3·44 ± 3·14 (3)	0·71 ± 0·48 (3)
	6 weeks at 4 °C	2·64 ± 0·27 (6)	5·23 ± 1·31 (6)	3·54 ± 0·94 (6)
	12 weeks at 4 °C	1·23 ± 0·30 (6)	4·68 ± 1·59 (6)	2·66 ± 0·72 (6)

Sugars were extracted and assayed as described by Morrell & ap Rees (1986*a*). Unpublished data of M. M. Burrell.

PFK(ATP) provides a further means of testing our hypothesis for cold-induced sweetening. If cold-lability of PFK(ATP) is a major cause of this sweetening then 13737 tubers should not sweeten appreciably at low temperatures. The data in Table 10 suggest that this is so.

Conclusion

We suggest that our data establish that lowering the temperature not only leads to a drastic reduction in the rate of plant metabolism but can also lead to substantial redirection of metabolism through the differential effect of temperature on the component enzymes of metabolism. The implications from these arguments are that plant metabolism can be affected in two major ways as the temperature is lowered to the range 0–10 °C. First there is a danger that the overall rate of metabolism will fall to the point at which it is no longer able to sustain life. In this respect it is pertinent to recall Dixon's (1951) definition of death as occurring as when 'oxidation cannot start again, owing to lack of co-enzyme, and the co-enzyme cannot be resynthesized owing to lack of energy from the oxidation'. The second major effect of lowering the temperature is that differential response of enzymes to temperature will alter the relative activities of different metabolic pathways. Such alterations could well be deleterious and contribute to chilling damage. However, it is conceivable that lowering the temperature could redirect metabolism in a way that provided a means of adaptation to low temperature, such as the synthesis of cryoprotectant.

We gladly acknowledge the previously published crucial contributions that Drs C. J. Pollock and W. L. Dixon have made to the development of the concepts described in this essay. We also acknowledge the generous support given by the Agricultural Genetics Company, and Twyford Plant Laboratories for much of the unpublished work that we have presented. We are particularly grateful to Dr G. R. Mackay, Scottish Crop Research Institute, Pentlandfield for his gift of tubers of the breeding clone 13737.

References

AP REES, T. (1980*a*). Assessment of the contributions of metabolic pathways to plant respiration. In *The Biochemistry of Plants*, vol. 2 (ed. D. D. Davies), pp. 1–29. New York: Academic Press.

AP REES, T. (1980*b*). Integration of pathways of synthesis and degradation of hexose phosphates. In *The Biochemistry of Plants*, vol. 3 (ed. J. Priess), pp. 1–42. New York: Academic Press.

AP REES, T. (1985). The organization of glycolysis and the oxidative pentose phosphate pathway in plants. In *Encyclopedia of Plant Physiology*, New series, vol. 18, *Higher Plant Cell Respiration* (ed. R. Douce & D. A. Day), pp. 391–417. Berlin: Springer.

AP REES, T. (1987). Compartmentation of plant metabolism. In *The Biochemistry of Plants*, vol. 12 (ed. D. D. Davies), pp. 87–115. New York: Academic Press.

AP REES, T. (1988). Hexose phosphate metabolism by nonphotosynthetic tissues of higher plants. In *The Biochemistry of Plants*, vol. 14 (ed. J. Priess). New York: Academic Press (in press).

AP REES, T. & BEEVERS, H. (1960). Pentose phosphate pathway as a major component of induced respiration of carrot and potato slices. *Pl. Physiol.* **35**, 839–849.

AP REES, T. & DANCER, J. (1987). Fructose-2,6-bisphosphate and plant respiration. In *Plant Mitochondria: Structural, Functional and Physiological Aspects* (ed. A. L. Moore & R. B. Beechey), pp. 341–350. New York: Plenum Press.

AP REES, T., DIXON, W. L., POLLOCK, C. J. & FRANKS, F. (1981). Low temperature sweetening of higher plants. In *Recent Advances in the Biochemistry of Fruits and Vegetables* (ed. J. Friend & M. J. C. Rhodes), pp. 41–61. London: Academic Press.

AP REES, T., GREEN, J. H. & WILSON, P. M. (1985). Pyrophosphate fructose 6-phosphate 1-phosphotransferase and glycolyis in non-photosynthetic tissues of higher plants. *Biochem. J.* **227**, 299–304.

BOYER, C. D. (1985). Synthesis and breakdown of starch. In *Biochemical Basis of Plant Breeding*, vol. 1 (ed. C. A. Neyra), pp. 133–153. Boca Raton: CRC Press.

BREEZE, V. & ELSTON, J. (1978). Some effects of temperature and substrate content upon respiration and the carbon balance of field beans (*Vicia faba* L.). *Ann. Bot.* **42**, 863–876.

BRIGGS, G. E., KIDD, F. & WEST, C. (1920). A quantitative analysis of plant growth. *Ann. appl. Biol.* **7**, 202–223.

CARNAL, N. W. & BLACK, C. C. (1983). Phosphofructokinase activities in photosynthetic organisms. *Pl. Physiol.* **71**, 150–155.

CHATTERTON, N. J., MCKELL, C. M. & STRAIN, B. R. (1970). Intra-specific differences in temperature-induced respiratory acclimation of desert salt bush. *Ecology* **51**, 545–549.

DANCER, J. E. (1987). *Role of pyrophosphate: fructose 6-phosphate 1-phosphotransferase in plants*, Ph.D. thesis, University of Cambridge.

DIXON, M. (1951). *Multi-enzyme Systems*. Cambridge: Cambridge University Press.

DIXON, W. L. & AP REES, T. (1980*a*). Identification of the regulatory steps in glycolysis in potato tubers. *Phytochemistry* **19**, 1297–1301.

DIXON, W. L. & AP REES, T. (1980*b*). Carbohydrate metabolism during cold-induced sweetening of potato tubers. *Phytochemistry* **19**, 1653–1656.

DIXON, W. L., FRANKS, F. & AP REES, T. (1981). Cold-lability of phosphofructokinase from potato tubers. *Phytochemistry* **20**, 969–972.

EDWARDS, J., AP REES, T., WILSON, P. M. & MORRELL, S. (1984). Measurement of the inorganic pyrophosphate in tissues of *Pisum sativum* L. *Planta* **162**, 188–191.

HOCHACHKA, P. W. & SOMERO, G. N. (1984). *Biochemical Adaptation*. Princeton: Princeton University Press.

ISHERWOOD, F. A. (1973). Starch–sugar interconversion in *Solanum tuberosum*. *Phytochemistry* **12**, 2579–2591.

KEELING, P. L., WOOD, J. R., TYSON, R. H. & BRIDGES, I. G. (1988). Starch biosynthesis in developing wheat grain. Evidence against the direct involvement of triose phosphates in the metabolic pathway. *Pl. Physiol.* **87**, 311–319.

MACDONALD, F. D. & AP REES, T. (1983). Enzymic properties of amyloplasts from suspension cultures of soybean. *Biochim. biophys. Acta* **755**, 81–89.

McNulty, A. K. & Cummins, W. R. (1987). The relationship between respiration and temperature in leaves of the arctic plant *Saxifraga cernua*. *P. Cell Environ.* **10**, 319–325.

Morney, H. A. & Billings, W. D. (1961). Comparative physiological ecology of arctic and alpine populations of *Oxyria digyna*. *Ecol. Monogr.* **31**, 1–29.

Morrell, S. & ap Rees, T. (1986*a*). Control of the hexose content of potato tubers. *Phytochemistry* **25**, 1073–1076.

Morrell, S. & ap Rees, T. (1986*b*). Sugar metabolism in developing tubers of *Solanum tuberosum*. *Phytochemistry* **25**, 1579–1585.

Pollock, C. J. & ap Rees, T. (1975*a*). Activities of enzymes of sugar metabolism in cold-stored tubers of *Solanum tuberosum*. *Phytochemistry* **14**, 613–617.

Pollock, C. J. & ap Rees, T. (1975*b*). Effect of cold on glucose metabolism by callus and tubers of *Solanum tuberosum*. *Phytochemistry* **14**, 1903–1906.

Stitt, M., Mieskes, G., Soling, H. D. & Heldt, H. W. (1982). On a possible role of fructose-2,6-bisphosphate in regulating photosynthetic metabolism in leaves. *FEBS Lett.* **145**, 217–222.

Tyson, R. H. (1987). *The conversion of sucrose to starch in the developing wheat grain*. Ph.D. thesis, University of Cambridge.

Van Schaftingen, E., Lederer, B., Bartrons, R. & Hers, H. G. (1982). A kinetic study of pyrophosphate: fructose-6-phosphate phosphotransferase from potato tubers. *Eur. J. Biochem.* **129**, 191–195.

Wager, H. G. (1941). On the respiration and carbon assimilation rates of some arctic plants as related to temperature. *New Phytol.* **40**, 1–19.

Weiner, H., Stitt, M. & Heldt, H. W. (1987). Subcellular compartmentation of pyrophosphate and alkaline pyrophosphatase in leaves. *Biochim. Biophys. Acta* **893**, 13–21.

Wragg, C. J. (1987). *The compartmentation of carbohydrate oxidation in non-photosynthetic cells of higher plants*. Ph.D. thesis, University of Cambridge.

MODELS RELATING SUBCELLULAR EFFECTS OF TEMPERATURE TO WHOLE PLANT RESPONSES

G. D. FARQUHAR

Plant Environmental Biology Group, Research School of Biological Sciences, Australian National University, Canberra City, ACT 2601, Australia

Summary

The analysis of the temperature dependence of the rate of photosynthetic carbon assimilation by leaves of C_3 plants has involved subdivision into a diffusional (mainly stomatal) component and a biochemical one. The latter has been further analysed in terms of Rubisco activity and of the capacity for regeneration of RubP. Rubisco activity has been modelled in terms of the number of active (carbamylated) enzyme sites and the kinetics of these sites. RubP regeneration capacity has been modelled in terms of photosynthetic electron transport and photophosphorylation capacity and the rate of supply of inorganic phosphate. The short-term (minutes to hours) responses of assimilation rate to changes in temperature of *Eucalyptus pauciflora* leaves, in the range 15 to 35°C, are examined using this framework.

It is argued that the above approach has some useful features, even if these are mainly pedagogical. Problems with the simplified framework are also discussed, including heterogeneity of the light intensity and CO_2 concentration within the leaf. The latter is discussed in terms of the resistances to diffusion within the leaf and observations of non-uniformity ('patches') of stomatal opening.

The effects of temperature, couched in the above terms, appear to be mainly those on the supply of carbon as a substrate for growth. Yet the model (or description of carbon assimilation) can accommodate certain effects of demand for carbon (sink limitation). These occur, notably, *via* the supply of inorganic phosphate, but potentially also *via* Rubisco activation and other mechanisms. A striking example of effects of reduced temperature on assimilation and growth being caused by changed demand for carbon, is given for the tropical crop, *Arachis hypogaea*. This example prompts an appraisal of the effects of temperature on components of growth other than photosynthesis, for example, of leaf production rate.

Introduction

More than 15000 genes constitute the genome of higher plants. This means complexity and diversity in the effects of temperature on plant performance. Models of plant responses are, therefore, at present and for the forseeable future, gross simplifications of reality. Nevertheless, they are useful for summarizing knowledge of artificially simplified systems. By this, I mean that they can be used for assembling facts of one level of biological organization and making predictions at the next. The artificial simplification is that any test of these predictions has to be done under conditions where the vast majority of possible interactions are eliminated. An experimenter uses model experimental systems in the same way.

From a mathematical viewpoint, modelling of plant responses is difficult. The system is heterogeneous but we usually ignore this and 'lump' distributed parameters into single ones. For example, when modelling leaf photosynthesis we usually ignore the fact that some of the carboxylating machinery is in palisade cells and some in mesophyll cells and yet we assign it to a single heading, but so too does the experimenter – usually.

Non-linearities in equations present problems for modellers. The techniques available to analyse linear systems are well-developed and powerful, which explains why man-made control systems are designed using linear components. However, it is so easy for a linear system to slip into non-linearity. A single multiplication or division of two variables will produce non-linearity; only addition and subtraction of variables, or their multiplication by constants, is allowed in the ideal linear world. An example comes from the modelling of the kinetics of ribulose-1,5-bisphosphate carboxylase-oxygenase (Rubisco) described later in this paper. The rate of binding of carbon dioxide to the enzyme-RubP complex is proportional to the product of its concentration and that of the complex. When the concentration of CO_2 is low, the amount of the complex is virtually constant and the system is linear. As the concentration increases the amount of complex starts to be affected – to become a variable – and non-linearity sets in. An early model of Rubisco kinetics (Farquhar, 1979), underpins much of the theory discussed in this chapter. In this model all of the individual binding steps have similar simple dependencies, but through interactions, complexity develops rapidly.

The example serves to introduce another common problem in modelling biological systems: how to define the boundary conditions, i.e. what is happening at the 'edges'. For the Rubisco system, do we assume that there is a fixed concentration of ribulose-1,5-bisphosphate (RubP) free in solution, or do we take into account the amount that is bound to the enzyme and chelated to magnesium and assume that there is a fixed *total* amount (von Caemmerer & Farquhar, 1985)? These would be practical considerations for experimenters if they worked at the high concentrations of Rubisco found *in vivo*. In mathematical terms the above question is asked as follows: do we take free or total concentration as the independent variable? Sometimes more realistic definitions of boundary conditions define *fluxes* rather than concentrations. Considering Rubisco again as an example, it might be appropriate, at least *in vivo*, to take the rate of supply of RubP as an independent variable. In biology defining the boundary conditions is usually very difficult.

Computational power is available today for digital, iterative solutions and this can help to overcome some of the difficulties listed above, including non-linearities (e.g. the solution of optimal stomatal behaviour by Cowan & Farquhar, 1977) and distributed (as opposed to 'lumped') parameters; for example, the solution by

Parkhurst (1987) for three-dimensional diffusion of CO_2 inside the leaf. Such computations can sometimes give numerical answers, but with no explanation or understanding. A mathematician's goal is to derive an analytical solution. As the name suggests, this is a solution which can be written precisely in terms which only involve the independent variables; equations used in this chapter are of this form. This goal is similar to that of the experimentalist with an analytical approach. However, the distinction becomes blurred when the terms used in the analytical solution are in themselves complicated, and tedious to calculate, because then the machine is needed to convert the analytical solution into numbers.

The complexity of a system depends in part on the number of variables, and in part on the number of dynamic variables that it contains. In plant biology, dynamic elements usually consist of 'storage elements', analogous to capacitors in an electrical circuit. Examples are tissues which change their water content in a hydraulic system or metabolite pools in a biochemical system. Systems with two dynamic elements usually give rise to quadratic equations, i.e. of the form, $a_u + a_1 \cdot x + a_2 \cdot x^2 = 0$, and have analytical solutions, as do cubic equations (which have an additional term in $a_3 \cdot x^3$ and often describe systems with three dynamic elements). There are, however, no general solutions for quartics and higher order equations. In fact, the analytical solution of the cubic equation is too complicated for intuitive understanding. All this means that there is often a difficult choice to make – whether to 'simplify' the system to the stage where only relatively simple analytical solutions are required, or whether to maintain complexity, forego mathematical understanding and resort to the machine, when there are more than two variables, or more than two dynamic elements.

If modelling plant responses to perturbations is difficult in general, it is particularly so if the perturbation is in temperature, because temperature affects so many processes. What previously were constants, now change. The difficulty increases as the time domain is extended, and more processes are affected. Sometimes, Nature may be kind to modellers working in long time domains if acclimation allows a similar coordination, at the new temperature, to that at the old; such kindness is rare.

In what follows, a framework is presented which may be useful for analysing the effects of temperature on photosynthetic carbon assimilation and accumulation. The former is emphasized because I have more experience of it, and because an understanding is developing of how to link the biochemical, protein-based views of carbon assimilation to the physiological ones. Genetic manipulation of crop responses, whether by traditional or modern techniques, is *via* alteration of proteins (thereby closing this circle of plant science), yet protein-based mathematical descriptions of processes controlling demand for, and accumulation of, carbon are at present poorly developed.

Photosynthetic carbon assimilation

A logical starting place is with the effects of temperature on the properties of the world's most abundant protein, Rubisco, which uniquely prepares CO_2 for chemical reduction.

Ribulose bisphosphate carboxylase-oxygenase (Rubisco)

The fully activated (carbamylated), RubP-saturated rate of carboxylation, W_c, is given by

$$W_c = \frac{V_{cmax}\,C}{C + K_c\,(1 + O/K_o)} \qquad \text{(Eqn 1)}$$

where V_{cmax} is the maximum rate, C is the partial pressure of CO_2 at the sites of carboxylation, O is that of O_2, and K_c and K_o are the corresponding Michaelis constants.

Björkman & Pearcy (1971) measured an apparent activation energy for V_{cmax} of 67 kJ mol^{-1} (mean) and Badger & Collatz (1977) measured 65 kJ mol^{-1}. More recently, Kirschbaum & Farquhar (1984) calculated a value of 53 kJ mol^{-1} from the data of Jordan & Ogren (1984) obtained above 15°C. These activation values may be related to values for the Q_{10} at a particular temperature and correspond to 2·5, 2·4 and 2·1, respectively, at 25°C (using the relationship Q_{10} (25°C) = $\exp(13{\cdot}6 \times 10^{-6}E)$, where E (J mol^{-1}) is the activation energy).

The temperature dependence is different when the partial pressure of CO_2 is not saturating, as indicated by the presence of the additional terms in the denominator of Eqn 1. Even in the unlikely event that V_{cmax} has an Arrhenius-like response, one would not expect such a response in W_c at subsaturating C.

The temperature response of Rubisco is often explained, in part, by the temperature-dependence of the solubilities of CO_2 and O_2. Such a description begs the question of what is the most appropriate measure of the activity of these gases. It is convenient to eliminate uncertainty about solubility (which is affected by the presence of different ions, as well as by temperature), by measuring partial pressures. Then the Michaelis 'constants' are partial pressures also. It may also be a more appropriate way, from a mechanistic point of view. The only compounds that appear to bind to the activated RubP-Rubisco site complex are gases. Perhaps the site in this condition is hydrophobic with only chemically bound water present. Haemoglobin, which reacts with O_2 and CO_2, appears to behave in this manner, and to be non-functional when the site is 'wet' (J. A. Berry, personal communication).

The rate of the Rubisco-catalysed reaction of RubP with O_2 V_o, is $2\Gamma_*/C$ times that with CO_2, V_c, where Γ_* is the CO_2 compensation point in the absence of non-photorespiratory ('dark') respiration, R_d (Laisk, 1977). Brooks & Farquhar (1985) found the following dependence of Γ_* on temperature:

$$\Gamma_* = 42{\cdot}7 + 1{\cdot}68(T - 25) + 0{\cdot}012(T - 25)^2 \qquad \text{(Eqn 2a)}$$

which, to a good approximation gives:

$$\Gamma_* = 1.7T \qquad \text{(Eqn 2b)}$$

where T is the leaf temperature in degrees Celsius. In theory and in practice Γ_* is linearly dependent on the concentration of O_2 and in Eqn 2 is measured in $\mu l\,l^{-1}$ at 21 % O_2.

The oxygenation of one mol RubP releases 0·5 mol CO_2 in the photorespiratory pathway. From Eqn 2 photorespiration increases *proportionally* faster than carboxylation with an increase of temperature at constant concentrations in the gas phase. Nevertheless, the increase as an *absolute* amount is predicted to be less, until comparatively high temperatures are reached.

Returning to the issue of the appropriate measure of activity of CO_2 and O_2, it is worth noting that, even at constant concentrations in the liquid, there is a temperature response (Brooks & Farquhar, 1985) – one expects enzymes to respond kinetically to temperature.

The rate of assimilation of CO_2 by a leaf, A, is given by:

$$\begin{aligned} A &= V_c - 0{\cdot}5V_o - R_d \qquad \text{(Eqn 3)} \\ &= V_c(1 - \Gamma_*/C) - R_d \end{aligned}$$

When the activity of Rubisco is limiting, V_c is replaced by W_c. Calculations based on Rubisco kinetics alone suggest then that A should increase with increases of temperature until eventually the gain in carboxylation is more than offset by the increase in photorespiration (Farquhar *et al.* 1980).

There are many complications in the modelling of Rubisco kinetics (von Caemmerer & Edmondson, 1986; von Caemmerer & Farquhar, 1985), caused by factors which decrease V_c below W_c. The most important ones occur when the capacity for regeneration of RubP is less than $W_c\ (1 + 2\Gamma_*/C)$ (the rate of RubP use when $V_c = W_c$). Even under these conditions, however, $V_o/V_c = 2\Gamma_*/C$.

RubP regeneration

Introduction

There are many other proteins besides Rubisco in the Calvin cycle. One would expect all of them to be limiting, to a greater or lesser extent, under certain conditions. Otherwise, I would argue, selection pressure would have acted to make fewer copies of the proteins involved. The limitation should be inversely related to the specific activity (rate per unit protein). On this basis, the limitations imposed by the amounts of proteins other than Rubisco and the complexes of the thylakoid should rarely be large. Nevertheless, we *expect* to see regulation, particularly by those partitioning the flux of reduced carbon between regeneration of RubP, formation of starch, and export.

When the capacity for regeneration under particular conditions (of light, temperature, etc.) is at a limiting level (let this equal X) then $(V_c + V_o)$ must be less than X, i.e. $(V_c\{1 + 2\Gamma_*/C\}) < X$. Thus V_c must be less than $X/\{1 + 2\Gamma_*/C\}$.

We denote this limit on the rate of carboxylation as W_r. Because V_c must also be less than W_c, the Rubisco-limited rate, we write

$$V_c \leqslant \min\{W_c, W_r\} \,, \qquad \text{(Eqn 4)}$$

where min {, } denotes 'minimum of'. If $V_c = W_r$ then

$$A = X(1 - \Gamma_*/C)/\{1 + 2\Gamma_*/C\} - R_d$$

$$= X(C - \Gamma_*)/(C + 2\Gamma_*) - R_d \qquad \text{(Eqn 5a)}$$

and to a good approximation in 21 % O_2

$$A = X(C - 1{\cdot}7T)/(C + 3{\cdot}4T) - R_d \,. \qquad \text{(Eqn 5b)}$$

Note that if X were insensitive to temperature, W_r and A would decrease as temperature and Γ_* increase. This is a mathematical way of saying that proportionally more of the regenerated RubP is being used for oxygenation and more carbon is being lost by decarboxylation of glycine.

The search for more temperature-sensitive steps now moves to two key processes, both of which involve membranes. The first is the thylakoid and the second is the chloroplast envelope.

Electron transport/photophosphorylation

Carbon in CO_2 is more oxidized than carbon in plant material. Four electrons must be extracted from water at photosystem (PS) II and donated to $NADP^+$ in order to reduce chemically the products of carboxylation of one molecule of RubP to the reduction level of RubP itself and of carbohydrate. Four electrons are also required following an oxygenation of RubP. So if the 'potential rate of electron transport', J, at a particular photon flux and temperature is limiting then $X = J/4$ and, from Eqns 5a and 5b,

$$A = J/4 \cdot (C - \Gamma_*)/(C + 2\Gamma_*) - R_d \qquad \text{(Eqn 6a)}$$

$$\simeq J/4 \cdot (C - 1{\cdot}7T)/(C + 3{\cdot}4T) - R_d. \qquad \text{(Eqn 6b)}$$

The phrase in inverted commas is rather cumbersome, but it is important to distinguish J from the (modelled) actual rate of electron transport, which may be limited by the availability of $NADP^+$ if Rubisco activity or other steps in carbon metabolism are relatively slow.

A simple case when the actual rate is probably close to the potential rate is at very low photon fluxes. Then one photon is absorbed by each of the photosystems and $J = I/2$, where I is the absorbed photon flux. For this condition, $X = I/8$ and $V_c = W_r = I/\{8(1 + 2\Gamma_*/C)\}$. An expression for rate of assimilation at low photon flux is then

$$A = I/8 \cdot (1 - \Gamma_*/C)/(1 + 2\Gamma_*/C) - R_d \qquad \text{(Eqn 7a)}$$

$$\simeq I/8 \cdot (C - 1{\cdot}7T)/(C + 3{\cdot}4T) - R_d. \qquad \text{(Eqn 7b)}$$

The coefficient of I in this equation is often called the quantum yield of carbon assimilation, provided the irradiance is above the low level at which R_d decreases with increase of irradiance (Kok effect). Its dependence on the partial pressure of CO_2 inside the chloroplast, C, can be easily calculated and is in good agreement with experimental measurements (Berry & Farquhar, 1978; Peisker, 1978; Kirschbaum & Farquhar, 1987).

The temperature dependence of Γ_*, Eqn 2b, can be used to predict the quantum yield of a leaf in an atmosphere containing 21 % O_2. To a good approximation it may be adapted from Eqn 7b as

$$Q_y = \frac{C - 1{\cdot}7T}{C + 3{\cdot}4T} \cdot \frac{1 - f}{8}, \qquad \text{(Eqn 8)}$$

where again T is the leaf temperature in degrees Celsius, and C (μmol mol^{-1}) is now the mole fraction of CO_2 in air containing 21 % O_2 in equilibrium with the dissolved concentration of CO_2 in the chloroplast stroma. The parameter f is a loss factor, as discussed below. Note that the measurements of Γ_*, approximated by Eqn 2b, were made in the range 15 to 30°C, so that extrapolation outside this range may not be valid. However, the prediction agrees well with the data of Ehleringer & Björkman (1977).

The absolute value of 8, which represents the minimum number of quanta required to fix a molecule of CO_2, and the loss factor, f, require comment. Some photons are absorbed ineffectively by photoinhibited or unconnected photosystems, or by other pigments in the leaf (c.f. discussion of modelling spectral effects on quantum yield by Farquhar & Kirschbaum, 1985). Some reductant is used for other purposes, such as making amino acids and lipids. From the elemental composition of plants, expressed in molar ratio to carbon, (CH_hO_x..), an overall equation of synthesis may be written:

$$CO_2 + h/2 \cdot H_2O \rightarrow yO_2 + CH_hO_x. \qquad \text{(Eqn 9)}$$

In the present case, ignoring nitrogen etc., the ratio of oxygen evolved to carbon dioxide assimilated, y, is given by McDermitt & Loomis (1981) as:

$$y = 1 + h/4 - x/2. \qquad \text{(Eqn 10)}$$

There are surprisingly few data published on elemental composition. Those for the aerial parts of *Arachis hypogaea* (Hubick *et al.* 1986) give $h = 1{\cdot}7$ and $x = 0{\cdot}63$, making $y = 1{\cdot}11$. On this basis, one expects 11 % more O_2 evolution than CO_2 uptake, and discrepancy should be even greater for lipid-rich tissue like the pods of peanuts (Hubick *et al.* 1986). Of course, for pods (and to some extent for leaves), this extra chemical reduction beyond the level of carbohydrate is not achieved directly by photosynthesis. Nevertheless, this and the other reasons outlined above may explain why 8 in Eqn 7 is an underestimate and needs to be replaced by 12 (according to typical measurements of CO_2 uptake) or 10 (according to typical measurements of O_2 evolution) making f equal to 0·3 and 0·2, respectively, in Eqn 8.

Eqns 6, 7 and 8 quanitfy how photochemistry, J, which is independent of temperature at low photon flux, nevertheless gives rise to a decreasing assimilation of carbon as temperature increases. The model is based on Rubisco kinetics and the conclusion is for C_3 species alone. The modelled quantum yield of C_4 leaves is virtually independent of temperature (Berry & Farquhar, 1978). However, as irradiance increases, enzymatic and diffusional limitation within the thylakoid become important. These processes speed up with increasing temperature and so, therefore, must J in the model. No detailed model exists for the light-saturated rate, J_{max}, and so it is empirically given a temperature dependence to match that of whole-chain electron transport, with the latter measured under conditions where the supply of acceptor ($NADP^+$, *in vivo*) is unlimiting. This temperature dependence can change, because of changes in thylakoid membrane viscosity and lipid composition, as the plants adapt to changes of temperature over a time-scale of days. (See discussion by Farquhar & von Caemmerer, 1982 and Quinn, 1988.) The equation printed in Farquhar *et al.* (1980) for the temperature-dependence of J_{max} (their Eqn 36 and erroneously reprinted in Farquhar & von Caemmerer (1982)) was not the one they actually used for simulations. The correct statement is:

$$J_{max} = 483 \exp[(T_k/298 - 1)E/RT_k]/(1 + \exp[(ST_k - H)/RT_k] \, ,$$

where T_k refers to the absolute temperature.

The equation for the temperature optimum which they derived was from the correct equation. The optimum was included because the electron transport data appear to show one. However, it may be that this is actually thermal damage rather than a truly reversible temperature response. Below this temperature, the data of Nolan & Smillie (1976), which were used to estimate the temperature dependence of J_{max}, may be approximated to fair accuracy by a linear dependence on the Celsius temperature:

$$J_{max} = zT, \qquad \text{(Eqn 11a)}$$

where z is chosen to give the appropriate rate at a standard temperature (e.g. 25 °C) or by using J_{mc}, the maximum rate at the critical temperature, T_c, at which damage becomes apparent:

$$J_{max} = J_{mc}T/T_c \, . \qquad \text{(Eqn 11b)}$$

A 'base temperature' other than zero may have to be introduced into Eqns 11 in other conditions.

Recapitulating briefly, the factor converting electron transport to net rate of CO_2 assimilation decreases with increasing temperature while, at low light, J itself is independent of temperature and, at saturating light, increases almost linearly with Celsius temperature. For intermediate irradiances J is found from a non-rectangular hyperbolic dependence on I (Farquhar & Wong, 1984).

$$\theta J^2 + (J_{max} + (1 - f)I/2)J + J_{max}\,(1 - f)I/2 = 0 \qquad \text{(Eqn 12)}$$

where θ is a convexity factor with a typical value of 0·7.

In all of the above, electron transport has been considered in terms of its role in providing NADPH (and reduced ferredoxin). ATP is also required for the photosynthetic carbon reduction and photorespiratory carbon oxidation cycles, the principles of limitation being analogous.

The view of temperature dependence of CO_2 assimilation rate presented thus far – in terms of effects on Rubisco activity, and electron transport capacity effects on RuBP regeneration capacity – has worked well to describe the short-term responses of A to T in leaves of *Eucalyptus pauciflora* (Kirschbaum & Farquhar, 1984). There are, however, complications.

Supply of inorganic phosphate

Carbon is exported from the chloroplast to the cytoplasm as triose phosphate, and is exchanged for orthophosphate to maintain the supply of phosphate (Walker & Robinson, 1978). If the latter process becomes limiting, assimilation rate, A, should become independent of C and O (Farquhar & von Caemmerer, 1982), and this has been observed in certain situations, and analysed by Sharkey (1985), Sharkey *et al.* (1986*a*,*b*) and Leegood & Furbank (1986).

A formal treatment in the present framework is simple (Kirschbaum & Farquhar, 1984). If the limitation to the rate of supply of inorganic phosphate from cytosolic reactions and starch formation is denoted S, then the rate of carboxylation must be less than W_p, given by

$$W_p = 3S/(1 - \Gamma_*/C) \ . \qquad \text{(Eqn 13)}$$

(Note that the factor 3 was mistakenly omitted in the reference above.) Eqn 4 is then extended to

$$V_c \leq \min \{W_c, W_r, W_p\} \ . \qquad \text{(Eqn 14)}$$

From the earlier discussion it would be expected that low O_2 partial pressure should greatly enhance electron transport – limited A when T is large, and very little when T is small (c.f. Eqn 6a with Γ_* large and small, respectively, representing high and low temperatures at normal oxygen levels, and with Γ_* being zero in the absence of oxygen). The small enhancement of A by low oxygen concentration is a common occurrence and easily understood. However, since the observations of Jolliffe & Tregunna (1968) there have been several reports of low oxygen concentrations, 2% typically, actually inhibiting photosynthesis at low temperatures in leaves from plants grown at higher temperatures. Phosphate is also involved here for it restores the stimulation by 2% oxygen at low temperatures (Leegood & Furbank, 1986). The inhibition appears to be a transient, recovering in less than an hour, in which case no steady-state solution could be given of the type used in the rest of this chapter. The distinction between steady state and transient is a little arbitrary, however, when one considers the longer term responses to change in temperature, and developmental changes including senescence.

Integrating the biochemistry and heterogeneity of light within the leaf

When a leaf always receives light from one side, the greatest rate of assimilation per unit leaf protein occurs when the amounts of photosynthetic machinery

(Rubisco, electron transport components, etc.) are distributed in proportion to the amount of light absorbed, i.e., there should be the greatest ratio of the amount of machinery to the amount of chlorophyll near the surface receiving the light. This conclusion is supported by the data of Terashima & Inoue (1985) and Terashima *et al.* (1986) on spinach and soybean, but not by that of Outlaw *et al.* (1976) on *Vicia faba*. Farquhar (1988) has shown that the above optimal solution enables the equations which apply at the level of a chloroplast to apply equally at the level of the whole leaf. The equation which relates local rate of assimilation to local activity of Rubisco, local capacity for electron transport, and local rate of absorption of quanta can be rewritten using total rate of assimilation, total activity of Rubisco, total capacity for electron transport, and total rate of absorption of quanta. Mathematically, the lumping of parameters works in this case, provided θ has the same value for all the chloroplasts.

For leaves, like *Eucalyptus*, which can receive light from either side, the conclusions are quite different. The concentration of machinery through the leaf should now be greatest near the leaf surfaces and diminish towards the centre, and the lumping of parameters introduces errors in modelling whole-plant photosynthesis (Kirschbaum, 1986).

Diffusion of carbon dioxide

So far in this chapter, the partial pressure of CO_2 at the sites of carboxylation, C, has been taken as a fixed, known parameter. However, C can have a strong dependence on temperature. If stomatal conductance is unchanging, then the draw-down or gradient of CO_2 across the epidermis is proportional to the assimilation rate. A large rate will therefore cause a low concentration inside the chloroplast, thereby providing passive negative feedback. However, the stomata themselves also respond, both to temperature and to changes in leaf-to-air vapour pressure difference. At a constant absolute humidity in the air, the vapour pressure difference increases rapidly with temperature, and this causes closure, while the effects of increased temperature *per se* are often the reverse. The former effect has been modelled in a crude mechanistic way (Farquhar, 1978), but not the latter. However, a different kind of modelling works reasonably well in both cases. That is the optimization of assimilation with respect to water loss, which yields conductance, and intercellular partial pressure of CO_2, p_i, as inferred parameters (Cowan & Farquhar, 1977; Hall & Schulze, 1980).

Intercellular diffusion

For leaves of some plants (e.g. olives, G. Bongi, personal communication), the gradient of CO_2 within the leaf may be important. Such effects have been modelled by Parkhurst (1986). We must also be aware of the possibility of significant gradients from the intercellular spaces to the sites of carboxylation. Measurements of carbon isotope discrimination may be useful for assessing non-stomatal resistances to diffusion (Evans *et al.* 1986).

Stomatal patches

The reduction in stomatal conductance which occurs with increasing temperature (at constant humidity) may potentially cause problems of interpretation of results obtained using gas-exchange systems. Farquhar *et al.* (1987) suggested that if the stomata closed non-uniformly one might be led to a mistaken conclusion that the mesophyll capacity for photosynthesis had decreased. The argument is as follows. If all the stomata on, say, the left side of a leaf were to close, and those on the right side remained open, conventional measurements of a whole leaf would show that the assimilation rate, A, and the conductance to diffusion of water vapour, g, had half their previous values. The intercellular partial pressure of CO_2, calculated (approximately) as

$$p_i = p_a - 1{\cdot}6AP/g \, , \qquad \text{(Eqn 14)}$$

(with P being absolute pressure), would appear to be unchanged at the value appropriate for the right hand side of the leaf, while the cells in the left side were in fact at the compensation point. With A halved and p_i apparently unchanged, one would naturally conclude that the photosynthetic capacity had been halved.

The argument above is put in an extreme form. However, if stomata close in patches, which are sufficiently large, then diffusion within the leaf will be too slow and a version of the above problem will occur (Farquhar, 1988). Terashima *et al.* (1988) have shown that this artefact explains the apparent inhibition of photosynthetic capacity by abscisic acid reported by Raschke (1982). Terashima *et al.* showed that the apparent inhibition of photosynthesis could be overcome by removing an epidermis or by giving sufficiently high concentrations of CO_2. The concentration required was very high, and this presumably closed the stomata even further, the result being that 10% CO_2 in an oxygen electrode was required. They also showed that application of abscisic acid to intact leaves of ordinary CO_2 concentrations led to starch accumulation in patches whereas it was accumulated uniformly in controls.

Growth

The foregoing sections show how temperature can have diverse effects on the supply of carbon for growth. Yet the supply of carbon is not always limiting. Bagnall *et al.* (1988) demonstrated this in *Arachis hypogaea*, a tropical crop that often grows slowly at temperatures below 25°C. Short-term responses of assimilation rate to lower temperatures were insufficient to explain this slow growth. Yet over several days assimilation rate declined, apparently to meet reduced sink demand. In plants which were exposed to low CO_2 concentrations during the same cool treatment, the assimilation rate was not reduced when measured at normal concentrations, suggesting that carbohydrate accumulation was responsible for the decline in assimilation. Cooling (to 19°C) most of the plant led, after three days, to a 70% decline in assimilation in the remaining leaves at 30°C. The converse treatment (30°C 'sink', 19°C 'source') resulted in a decline of only 17‰.

It is clear that, in this situation, it would be difficult to make a model of growth based on carbon assimilation and its control using parameters like V_{cmax} and J_{max} alone. One could use a model of growth that was independent of carbon supply and indeed such models exist. However, such a model would also lose generality when carbon supply is limiting. The two areas of plant physiology need to be brought together. Two obvious ways for sinks to exert control over the photosynthetic machinery are *via* the activation and inhibition of Rubisco and the supply of inorganic phosphate to the chloroplast. The steady activation (carbamylation) of Rubisco can be modelled (von Caemmerer & Farquhar, 1985), but it depends on the concentration of magnesium in the stroma, the control of which has not yet been modelled. This is an example of the difficulty of defining boundary conditions in biological systems. The effects and control of pH are equally important.

Even in the absence of sink effects on capacity, the link between modelling of phososynthesis and of growth at a mechanistic level is poorly developed. The controls of the levels of proteins per unit area of leaf, of leaf area development and of leaf appearance need to be tackled. This is not to deny the existence of models which can successfully predict aspects of growth and development. These have been reviewed by Curry & Eshel (1984).

Since the development of infra-red CO_2 analysers, the use of classical growth analysis seems to have gone out of vogue. A simple identity is useful for linking these areas, which have been unnecessarily separated. It relates relative growth rate, r (s^{-1}), to assimilation rate of the shoot, A ($mol\,CO_2\,m^{-2}\,s^{-1}$), such as would be measured in a gas-exchange chamber, by

$$r = Al(1 - \phi)/\rho \,, \quad \text{(Eqn 15)}$$

where l is the light period as a proportion of the day, ϕ is the proportion of net day-time shoot fixation of carbon lost by respiration, and ρ is the ratio of total plant carbon to photosynthetic area (Masle & Farquhar, 1988). This identity was used to analyse the increased water-use efficiency and decreased growth rate of wheat seedlings grown in soil of high strength, i.e. soils that are difficult to penetrate because of compaction or of low water content. These seedlings have smaller g and p_i, but greater assimilatory and Rubisco activity than the controls. They grow less rapidly because of an increase in the root/shoot ratio and partition factor, ρ. In some ways, this pattern resembles that of plants native to, and grown at, high altitude (see Körner & Larcher, 1988).

A commonality between the studies of seedlings grown at high soil strength and of the *A. hypogaea* grown in cool conditions was the reduction in leaf production rate under conditions where the supply of carbon did not appear to be limiting. There is a need for mechanistic models of the phyllochron, the time interval between appearance of successive leaves. More generally, there is a need for 'protein-based' modelling of demand for carbon, to link with those developed to describe its supply.

In conclusion, models are useful tools for relating short-term (hours) subcellular effects of temperature to effects at the whole leaf. More work needs to be done at other levels of organization and at other time scales. Nevertheless, it is worthwhile to quote the systems analysts/modellers, Gold & Raper (1984), who wrote that 'properly used, explicit models are valuable, and sometimes indispensable, aids to intuition, but almost never a valuable substitute'.

References

BADGER, M. R. & COLLATZ, G. J. (1977). Studies on the kinetic mechanism of ribulose-1,5-bisphosphate carboxylase and oxygenase reactions, with particular reference to the effect of temperature on kinetic parameters. *Carnegie Instn Yb.* **76**, 355–361.

BAGNALL, D. J., KING, R. W. & FARQUHAR, G. D. (1988). Temperature-dependent feedback inhibition of photosynthesis in peanut. *Planta* (in press).

BERRY, J. A. & FARQUHAR, G. D. (1978). The CO_2 concentrating function of C_4 photosynthesis. A bochemical model. In *Proc. 4th International Congress on Photosynthesis, Reading, England, 1977* (ed. D. Hall, J. Coombs & T. Goodwin), pp. 119–131. London: The Biochemical Society.

BJÖRKMAN, O. & PEARCY, R. W. (1971). Effects of growth temperature on the temperature dependence of photosynthesis *in vivo* and on CO_2 fixation by carboxydismutase *in vitro* in C_3 and C_4 species. *Carnegie Instn Yb.* **70**, 511–520.

BROOKS, A. & FARQUHAR, G. D. (1985). Effect of temperature on the CO_2/O_2 specificity of ribulose-1,5-bisphosphate carboxylase/oxygenase and the rate of respiration in the light. Estimates from gas-exchange measurements of spinach. *Planta* **165**, 397–406.

VON CAEMMERER, S. & EDMONSON, D. L. (1986). Relationships between steady-state gas exchange, *in vivo* ribulose bisphosphate carboxylase activity and some carbon reduction cycle intermediates in *Raphanus sativa. Aust. J. Pl. Physiol.* **13**, 669–688.

VON CAEMMERER, S. & FARQUHAR, G. D. (1985). Kinetics and activation of Rubisco and some preliminary modelling of RuP_2 pool sizes. In *Kinetics of photosynthetic carbon metabolism in C_3-plants, Proceedings of the 1983 Conference at Tallinn* (ed. J. Viil, G. Grishina & A. Laisk). Estonian Academy of Sciences.

COWAN, I. R. & FARQUHAR, G. D. (1977). Stomatal function in relation to leaf metabolism and environment. *Symp. Soc. exp. Biol.* **31**, 471–505.

CURRY, R. B. & ESHEL, A. (1984). Use of simulation as a tool in crop management strategies for stress avoidance. In *Crop reactions to water and temperature stresses in humid, temperate climates* (ed. C. D. Raper, Jr & P. J. Kramer), pp. 315–326.

EHLERINGER, J. & BJÖRKMAN, O. (1977). Quantum yields for CO_2 uptake in C_3 and C_4 plants. Dependence on temperature, CO_2 and O_2 concentrations. *Pl. Physiol.* **59**, 86–90.

EVANS, J. R., SHARKEY, T. D., BERRY, J. A. & FARQUHAR, G. D. (1986). Carbon isotope discrimination measured concurrently with gas exchange to investigate CO_2 diffusion in leaves of higher plants. *Aust. J. Pl. Physiol.* **13**, 281–292.

FARQUHAR, G. D. (1978). Feedforward responses of stomata to humidity. *Aust. J. Pl. Physiol.* **5**, 787–800.

FARQUHAR, G. D. (1979). Models describing the kinetics of ribulose biphosphate carboxylase-oxygenase. *Arch. Biochem. Biophys.* **193**, 456–468.

FARQUHAR, G. D. (1988). Models of integrated photosynthesis of cells and leaves. *Phil. Trans. Roy. Soc.* (Ser. B) (in press).

FARQUHAR, G. D., VON CAEMMERER, S. & BERRY, J. A. (1980). A biochemical model of photosynthetic CO_2 assimilation in leaves of C_3 species. *Planta* **149**, 78–90.

FARQUHAR, G. D. & VON CAEMMERER, S. (1982). Modelling of photosynthetic response to environmental conditions. In *Encyclopedia of Plant Physiology*, New Series Vol. 12B (ed. O. L. Lange, P. S. Nobel, C. B. Osmond & H. Ziegler), pp. 549–587. Heidelberg: Springer-Verlag.

FARQUHAR, G. D., HUBICK, K. T., TERASHIMA, I., CONDON, A. G. & RICHARDS, R. A. (1987). Genetic variation in the relationship between photosynthetic CO_2 assimilation rate and stomatal conductance to water loss. In *Progress in Photosynthesis Research* vol. IV (ed. J. Biggins), pp. 209–212.

Farquhar, G. D. & Kirschbaum, M. U. F. (1985). Environmental constraints on carbon assimilation. In *Regulation of sources and sinks in crop plants* (ed. B. F. Jeffcoat, A. F. Hawkins & A. D. Stead). British Plant Growth Regulator Group Monograph No. 12, pp. 87–97.

Farquhar, G. D. & Wong, S. C. (1984). An empirical model of stomatal conductance. *Aust. J. Pl. Physiol.* **11**, 191–210.

Gold, H. J. & Raper, C. D. Jr. (1984). Systems analysis and modeling in extrapolation of controlled environment studies to field conditions. In *Crop reactions to water and temperature stresses in humid, temperature climates* (ed. C. D. Raper, Jr & P. J. Kramer), pp. 315–326.

Hall, A. E. & Schulze, E.-D. (1980). Stomatal response to environment and a possible interrelation between stomatal effects on transpiration and CO_2 assimilation. *Pl. Cell Environ.* **3**, 467–474.

Hubick, K. T., Farquhar, G. D. & Shorter, R. T. (1986). Correlation between water-use efficiency and carbon isotope discrimination in diverse peanut (*Arachis*) germplasm. *Aust. J. Pl. Physiol.* **13**, 803–806.

Körner, Ch. & Larcher, W. (1988). Plant life in cold climates. In *Plants and Temperature*, vol. 42 (ed. S. P. Long & F. I. Woodward), pp. 25–57. Symp. Soc. Exp. Biol. Cambridge: Company of Biologists.

Jolliffe, P. A. & Tregunna, E. B. (1968). Effect of temperature CO_2 concentration, and light intensity on oxygen inhibition of photosynthesis in wheat leaves. *Pl. Physiol.* **43**, 902–906.

Jordan, D. B. & Ogren, W. L. (1984). The CO_2/O_2 specificity of ribulose-1,5-bisphosphate carboxylase/oxygenase. Dependence on ribulose bisphosphate concentration, pH and temperature. *Planta* **161**, 308–313.

Kirschbaum, M. U. F. (1986). The effects of light, temperature and water stress on photosynthesis in *Eucalyptus pauciflora*. Ph.D. thesis, Australian National University.

Kirschbaum, M. U. F. & Farquhar, G. D. (1984). Temperature dependence of whole leaf photosynthesis in *Eucalyptus pauciflora*. *Aust. J. Pl. Physiol.* **11**, 519–538.

Kirschbaum, M. U. F. & Farquhar, G. D. (1987). Investigation of the CO_2 dependence of quantum yield and respiration in *Eucalyptus pauciflora*. *Pl. Physiol.* **83**, 1032–1036.

Laisk, A. (1977). Modelling of the closed Calvin cycle. In *Biophysikalische Analyse pflanzlicher Systeme* (ed. K. Unger), pp. 175–182. Jena: Fischer.

Leegood, R. C. & Furbank, R. T. (1986). Stimulation of photosynthesis by 2% O_2 at low temperatures is restored by phosphate. *Planta* **168**, 84–93.

Masle, J. & Farquhar, G. D. (1988). Effects of soil strength on the relation of water use efficiency and growth to carbon isotope discrimination in wheat seedlings. *Pl. Physiol.* **86**, 32–38.

McDermitt, D. K. & Loomis, R. S. (1981). Elemental composition of biomass and its relation to energy content, growth efficiency and growth yield. *Ann. Bot.* **48**, 275–290.

Nolan, W. G. & Smillie, R. M. (1976). Multi-temperature effects on Hill reaction activity of barley chloroplasts.. *Biochim. Biophys. Acta* **440**, 461–475.

Outlaw, W. H., Schmuck, C. L. & Tolbert, N. E. (1976). Photosynthetic carbon metabolism in the palisade parenchyma and spongy parenchyma of *Vicia faba* L. *Pl. Physiol.* **58**, 186–189.

Parkhurst, D. F. (1987). Internal leaf structure: a three-dimensional perspective. In *On the economy of plant form and function* (ed. T. J. Givnish). New York: Cambridge University Press.

Peisker, M. (1978). A comment on the effects of carbon dioxide, oxygen and temperature on photosynthetic quantum yield in C_3 plants. *Acta Physiol. Pl.* **1**, 23–26.

Quinn, P. J. (1988). Effects of temperature on cell membranes. In *Plant and Temperature* (ed. S. P. Long & F. I. Woodward), pp. 237–258. Symp. Soc. Exp. Biol., vol. 42. Cambridge: Company of Biologists.

Raschke, K. (1982). In *Plant Growth Substances* (ed. P. F. Wareing), pp. 581–590. New York: Academic Press.

Sharkey, T. D. (1985). O_2-insensitive photosynthesis in C_3 plants. Its occurrence and a possible explanation. *Pl. Physiol.* **78**, 71–75.

Sharkey, T. D., Seemann, J. R. & Berry, J. A. (1986*a*). Regulation of ribulose-1,5-bisphosphate carboxylase in response to changing partial pressure of O_2 and light in *Phaseolus vulgaris*. *Pl. Physiol.* **81**, 788–791.

SHARKEY, T. D., STITT, M., HEINEKE, D., GERHARDT, R., RASCHKE, K. & HELDT, H. W. (1986). Limitation of photosynthesis by carbon metabolism. II. O_2 insensitive CO_2 assimilation results from triose phosphate utilization limitations. *Pl. Physiol.* **81**, 1123–1129.

TERASHIMA, I. & SAEKI, T. (1985). A new model for leaf photosynthesis incorporating the gradients of light environment and of photosynthetic properties of chloroplasts within a leaf. *Ann. Bot.* **56**, 489–499.

TERASHIMA, I., SAKAGUCHI, S. & HARA, N. (1986). Intra-leaf and intracellular gradients in chloroplast ultrastructure of dorsiventral leaves illuminated from the adaxial or abaxial side during their development. *Pl. Cell Physiol.* **27**, 1023–1031.

TERASHIMA, I., WONG, S. C., OSMOND, C. B. & FARQUHAR, G. D. (1988). Inhibition of leaf photosynthetic capacity by abscisic acid is an artefact caused by non-uniform stomatal closure. *Pl. Cell Physiol.* **29**, 385–394.

WALKER, D. A. & ROBINSON, S. P. (1978). Regulation of photosynthetic carbon assimilation. In *Photosynthetic carbon assimilation* (ed. H. W. Siegelman & G. Hind), pp. 43–59. New York, London: Plenum.

OVERVIEW: PROBLEMS AND PROPOSALS FOR THE UNDERSTANDING OF TEMPERATURE RESPONSES IN PLANTS

H. W. WOOLHOUSE

IPSR, John Innes Institute, Colney Lane, Norwich, NR4 7UH, UK

We might say that the symposium is ranged between fire and ice – those being roughly, though not exactly, the range of temperatures which plant life must endure.

In many previous SEB symposia it was a common practice to open with speakers who might deal with some of the more fundamental issues underlying the general theme of the symposium. Often it would be a mathematician, chemist, physicist or general theorist of one kind or another. In the present symposium only Nobel has recognized the need for such background in his treatment of energy budgets, but as I have read the excellent contributions to this symposium I still do just wonder. What might I have in mind? First, there is the evolutionary dimension, in relation to very different thermal regimes of past epochs. Periodic glaciations have continued into recent times of course, with successive glaciations over the past five million years. Pollen analysis has shown us that, in response to such changes, species migrate – they advance and retreat. Even where the cooling and warming changes are relatively slow, they are too rapid to allow the thermal adaptation of the species. Because temperature touches so many facets of metabolism, the time (in number of generations) taken for each of these to be altered and optimized relative to one another, at the level of their genetic determination, is enormous. Thus, it is worth noting that the numbers of plants that have evolved to grow in extremes of cold are relatively few. There are about 360 000 species of angiosperms on earth; in Britain 2000 species, in the tundra regions of the Northern Hemisphere fewer than 200 species and in the high latitudes of Antarctica fewer than 20 species.

In many habitats periodic cold is a problem, extremes of heat also occur and Nobel (1988) has shown very elegantly how physical principles can be applied to make some sense of the location and organization of growing points, transpiring surfaces and allied features, to cope with these extremes.

Woodward (1988) has entered the general field to which I have referred and he makes the very salient point that in considering life in extreme arctic habitats the element of unpredictability in the climate is an additional major factor: one sees this reflected in the often brutally violent fluctuations in the populations of arctic birds and mammals; snowy owls and caribou, arctic fox and polar bear provide very vivid examples. One option available to these species is a capacity for rapid migration which plants do not possess, and Woodward is at pains to point out the importance of a photoperiodic over-ride mechanism to avoid this trap in plants.

Körner & Larcher (1988) make two very salient points. They emphasize that dry matter allocation to different parts of plants under cold conditions does not show uniform trends, so that it is futile to make soft generalizations about strategies of adaptation. They also point emphatically to the mitotic cycle as the site of temperature action in the regulation of growth at this level. This of itself is not new. Monteith and his co-workers in the early nineteen sixties scratched the surface of this problem in their work on the growth of maize, but penetrating work on the subject is still lacking. From very different angles Pollock & Eagles (1988), and Francis & Barlow (1988) have also come to the problem of cell division at low temperatures in this symposium. I hope that soon we may see them pushing further along this track. It is worth reminding ourselves that the greatest progress in analysis of the mitotic cycle has been with yeast – where enetic dissection of the cycle has revealed an enormous complexity: in this field, temperature-sensitive mutants have played a major role. It is at this point that one feels that another fundamental paper, this time on the genetic analysis of temperature responses would have helped as a stronger pointer to future avenues of work. Pollock & Eagles (1988) made reference to the 'slender' mutant in barley and several people feel convinced that this represents an important model for the study of temperature effects: *now* note this is a single mendelian gene – we do not know what it is encoding, but there are powerful tools of molecular biology available for searching for needles in haystacks of this kind.

Before returning to matters genetical let us briefly touch upon an issue in the physiology of temperature responses of whole organisms which is at once remarkable, unexplained and of considerable economic significance. I refer to the fact that many gross physiological responses of plants are linear with temperature in the range of say 10–25°C even though underlying partial processes have non-linear responses to temperature. Examples in this volume include leaf elongation (Pollock & Eagles) CO_2 uptake (Baker *et al*; Weis & Berry), potassium uptake (Clarkson *et al.*) and seed germination (Roberts). It is this remarkable situation which allows Grace (1988) to make predications of 10% increases in dry matter production with a 1°C rise in mean temperature and to emphasize the empirical value of the concept of thermal time in predictions of productivity. In each of these and indeed other papers in the symposium one comes back to the authors' emphasis on how relatively little we know of underlying mechanisms.

I should like to conclude by taking two views of the prospects for understanding the thermal responses of plants, the pessimistic and the optimistic.

On the pessimistic side, there is the question of the thermal stability of proteins and the consequences for enzyme activity and the behaviour of structural elements such as the cytoskeleton, chromatin and various membrane constituents. We know at a coarse level that the various types of interaction which influence the conformational stability of proteins, such as covalent disulphide bonds, hydrophobic bonds, hydrogen bonds and electrostatic forces are differently affected by temperature. Disulphide bonds may help to preserve chain folding at elevated temperatures; hydrophobic bonds become weaker at lower temperatures so that enzymes with hydrophobically-bonded sub-units, for example, may fall apart and become inactive under cold conditions. The problem comes at the finer level of analysis; even when we know the complete amino-acid sequence of a protein and have its X-ray crystallographic structure at high resolution, we have no general theory of protein structure which allows us to predict the consequences for thermal stability or enzyme activity in relation to temperature, which would accrue from a single amino-acid substitution. The extent of the interactions within a protein molecule make it unlikely that the necessary theoretical framework will be readily achieved, so that the predictive engineering of proteins with altered thermal stabilities is a very remote prospect.

A more optimistic view of the possibilities for understanding temperature responses derives from the success of molecular genetic approaches, particularly with bacteria. We have heard in this symposium the exciting work on the formation of heat shock proteins in plants and of the continuing search for proteins specifically associated with tolerance of low temperatures (Ougham & Howarth, 1988). We do not yet know how the heat shock proteins work nor indeed how they are elicited, although studies with bacteria afford some interesting models. Thus, in *E. coli* there are mutants which lack the heat shock response; the gene involved encodes the sigma factor of an RNA polymerase, suggesting that the thermal signal may be targeted to the genes encoding the heat shock proteins by means of a change involving an RNA polymerase with altered promotor affinities. We have as yet no reports of a systematic search for higher plants lacking the heat shock response.

We know from the work of Geyl *et al.* (1977), also on *E. coli*, that there are mutations in genes encoding ribosomal proteins which confer low temperature sensitivity because the altered proteins are unable to assemble into intact ribosomes at low temperatures. Here the potential links with effects observable in flowering plants are strong; thus in barley Hoyer-Hansen & Cassandros (1982) have reported a cold-sensitive mutant, tigrina 034, in which the plants contain normal levels of 80S cytoplasmic and 70S chloroplastic ribosomes when grown at permissive temperatures but at lower temperatures chloroplast ribosomes are unable to assemble. It may be objected that this is an extreme or freak effect and so it is, but it is noteworthy that species differ markedly in their ability to form polysomes at low temperatures and one is led to wonder whether this may be due to smaller, but still significant, genetic changes leading to altered features of ribosome assembly which confer failure to form polysomes. Membrane properties are well known to change with temperature and the work of Raison *et al.* (1979) has shown how features of the lipid matrix may influence these properties.

I have referred to these studies in an attempt to encourage a more positive and rigorous use of molecular genetic approaches to the analysis of thermal adaptation and if this is to be done there arises the question of choice of material. Dr Pollock showed to the Symposium meeting photographs of *Dactylis glomerata* plants differing markedly in temperature response and commented, somewhat dejectedly, that the differences were polygenic – the traditional signal of an impasse in classical genetics. It is increasingly evident that what were previously referred to as polygenic traits may sometimes arise from subtle changes in promoter sequences between individual members of gene families. Such changes may be brought about by insertion and excision of transposons, some of which have temperature-dependent transposition thresholds. Through the operation of such elements, plants may have developed a means of modulating their mutation rates to increase the genetic load under conditions of thermal stress; in these circumstances the rate of adaptation may be tunable to the extent of the stress.

Finally, I should like to make it clear that in emphasizing the need for a greater commitment to molecular biology and genetics in the study of temperature responses in plants because they are, I feel, under-represented in this field of work in Britain and in this symposium, I am not wishing to suggest that such a shift of emphasis should be at the expense of physiological work. Farrar (1988) points very clearly to the potential interactions between plant organs when they are subjected to different temperatures; our ignorance of the details of signalling between organs and their temperature relations is very great and there are strong suggestions that sinks rather than sources may be a limitation to growth at low temperatures; but in truth we know very little of the mutual tuning of sink response to source capacity and *vice versa*. Some would argue that this element of ignorance is not for want of trying. Perhaps, again, we need new approaches? The 'slender' mutant of barley referred to earlier (Pollock & Eagles, 1988), in which a single mendelian gene mutation can bring about a large increase in the capacity of a major sink such as the growing leaf at low temperature, offers a convenient example of the type of approach I have in mind. A major effort on the isolation and characterization of this gene could yield a very valuable tool for the physiologist, since it would open up the possibility of introducing it via transgenic techniques into other species in which the sink size and temperature relations of the growing leaves are different. If we can press our analysis to this level we may even open the prospect of experimentally modifying temperature responses of plants by the introduction of specific genes from psychrophilic and thermophilic organisms. Whilst this may appear fanciful at the level of producing plants of improved economic performance in the foreseeable future it will almost certainly provide the physiologist and biochemist with an extremely powerful new tool for the systematic investigation of the underlying causes of the multiplicity of temperature responses and adaptations which are found in plants.

References

Baker, N. R., Long, S. P. & Ort, D. R. (1988). Photosynthesis and temperature, with particular reference to effects on quantum yield. In *Plants and Temperature* (ed. S. P. Long & F. I. Woodward). Symp. Soc. Exp. Biol., vol. 42, pp. 347–375. Cambridge: Company of Biologists.

Clarkson, D. T., Earnshaw, M. J., White, P. J. & Cooper, H. D. (1988). Temperature dependent factors influencing nutrient uptake: an analysis of responses at different levels of organization. In *Plants and Temperature* (ed. S. P. Long & F. I. Woodward). Symp. Soc. Exp. Biol., vol. 42, pp. 281–309. Cambridge: Company of Biologists.

Farrar, J. F. (1988). Temperature and the partitioning and translocation of carbon. In *Plants and Temperature* (ed. S. P. Long & F. I. Woodward). Symp. Soc. Exp. Biol., vol. 42, pp. 203–235. Cambridge: Company of Biologists.

Francis, D. & Barlow, P. W. (1988). Temperature and the cell cycle. In *Plants and Temperature* (ed. S. P. Long & F. I. Woodward). Symp. Soc. Exp. Biol., vol. 42, pp. 181–201. Cambridge: Company of Biologists.

Geyl, D., Bock, A. & Wittman, H. G. (1977). Cold sensitive growth of a mutant E. coli with an altered ribosomal protein S8: analysis of revertants. *Mol. gen. Genet.* **152**, 331–336.

Grace, J. (1988). Temperature as a determinant of plant productivity. In *Plants and Temperature* (ed. S. P. Long & F. I. Woodward). Symp. Soc. Exp. Biol., vol. 42, pp. 91–107. Cambridge: Company of Biologists.

Hoyer-Hansen, G. & Cassadoro, G. (1982). Unstable chloroplast ribosomes in the cold-sensitive barley mutant tigrina-O_{34}. Carlsberg Res. Comm. **47**, 103–108.

Körner, Ch. & Larcher, W. (1988). Plant life in cold climates. In *Plants and Temperature* (ed. S. P. Long & F. I. Woodward). Symp. Soc. Exp. Biol., vol. 42, pp. 25–57. Cambridge: Company of Biologists.

Nobel, P. S. (1988). Principles underlying the prediction of temperature in plants, with special reference to desert succulents. In *Plants and Temperature* (ed. S. P. Long & F. I. Woodward). Symp. Soc. Exp. Biol., vol. 42, pp. 1–23. Cambridge: Company of Biologists.

Ougham, H. J. & Howarth, C. J. (1988). Temperature shock proteins in plants. In *Plants and Temperature* (ed. S. P. Long & F. I. Woodward). Symp. Soc. Exp. Biol., vol. 42, pp. 259–280. Cambridge: Company of Biologists.

Pollock, C. J. & Eagles, P. W. (1988). Low temperature and the growth of plants. In *Plants and Temperature* (ed. S. P. Long & F. I. Woodward). Symp. Soc. Exp. Biol., vol. 42, pp. 157–179. Cambridge: Company of Biologists.

Raison, J. K., Chapman, E. A., Wright, L. C. & Jacobs, S. W. L. (1979). Membrane lipid transitions: their correlation with the climatic distribution of plants. In *Low Temperature Stress in Crops Plants* (ed. J. M. Lyons, D. Graham & J. K. Raison), pp. 177–186. New York: Academic Press.

Roberts, E. H. (1988). Temperature and seed germination. In *Plants and Temperature* (ed. S. P. Long & F. I. Woodward). Symp. Soc. Exp. Biol., vol. 42, pp. 109–132. Cambridge: Company of Biologists.

Weis, E. & Berry, J. A. (1988). Plants and high temperature stress. In *Plants and Temperature* (ed. S. P. Long & F. I. Woodward). Symp. Soc. Exp. Biol., vol. 42, pp. 329–346. Cambridge: Company of Biologists.

Woodward, F. I. (1988). Temperature and the distribution of plant species and vegetation. In *Plants and Temperature* (ed. S. P. Long & F. I. Woodward). Symp. Soc. Exp. Biol., vol. 42, pp. 59–75. Cambridge: Company of Biologists.

INDEX OF SUBJECTS